Springer Proceedings in Business and Economics

Springer Proceedings in Business and Economics brings the most current research presented at conferences and workshops to a global readership. The series features volumes (in electronic and print formats) of selected contributions from conferences in all areas of economics, business, management, and finance. In addition to an overall evaluation by the publisher of the topical interest, scientific quality, and timeliness of each volume, each contribution is refereed to standards comparable to those of leading journals, resulting in authoritative contributions to the respective fields. Springer's production and distribution infrastructure ensures rapid publication and wide circulation of the latest developments in the most compelling and promising areas of research today.

The editorial development of volumes may be managed using Springer Nature's innovative EquinOCS, a proven online conference proceedings submission, management and review system. This system is designed to ensure an efficient timeline for your publication, making Springer Proceedings in Business and Economics the premier series to publish your workshop or conference volume.

This book series is indexed in SCOPUS.

Francesco Schiavone • Nessrine Omrani
Heger Gabteni
Editors

Advanced Perspectives and Trends in Digital Transformation of Firms, Networks, and Society

2nd International Conference of the Digital Transformation Society, Naples, Italy, May 23-24, 2024

Springer

Editors
Francesco Schiavone
Parthenope University of Naples
Napoli, Napoli, Italy

Nessrine Omrani
Paris School of Business
Paris, Paris, France

Heger Gabteni
Paris School of Business
Paris, Paris, France

International Conference of the Digital Transformation Society dts, dts2024, DTS, 2,
Naples, Italy 23-05-2024 - 24-05-2024
ISSN 2198-7246 ISSN 2198-7254 (electronic)
Springer Proceedings in Business and Economics
ISBN 978-3-031-80691-9 ISBN 978-3-031-80692-6 (eBook)
https://doi.org/10.1007/978-3-031-80692-6

This Springer imprint is published by the registered company Springer Nature Switzerland AG
The registered company address is: Gewerbestrasse 11, 6330 Cham, Switzerland

Preface

In recent years, digital transformation has been one of the most disruptive forces in society, profoundly influencing industries and organizations. This transformation has been driven by the advancement of digital technologies such as artificial intelligence (AI), the metaverse, big data analytics, cloud computing, and the Internet of Things (IoTs). These technological advances are not only changing the way companies operate but also reshaping how they interact with consumers and create value in increasingly competitive markets.

The digital revolution is based on a data-driven approach, which has led to radical changes across many sectors, creating new growth and innovation opportunities. However, to fully realize these benefits, digital technologies must be strategically and coherently integrated into all aspects of a company. As previously mentioned, digital technologies are the primary enablers of transformation, but other factors also play a crucial role, including the evolution of organizational relationships, consumer expectations, digital competition, and data availability.

The effects of this transformation especially influence social dynamics and interpersonal relationships. The increasing convergence of the physical and digital worlds is transforming how we conduct business and interact with others, redefining the boundaries between personal and professional life, and creating new modes of connection and collaboration. Although there is not yet an adequate infrastructure to support large-scale immersive virtual worlds, research is exploring the potential transformations offered by the metaverse.

Moreover, digital transformation inevitably impacts the sustainability triangle of environmental, social, and economic aspects, which is currently a key concern. However, research that deeply examines the impact of digital technologies on environmental sustainability is still limited. Therefore, it is crucial to address this gap with more detailed studies on the implications of digital transformation for social responsibility, business performance, and environmental sustainability.

In light of the points raised, this book aims to explore the highlighted topics, inviting readers to reflect on how digital transformation is reshaping industries and society as a whole. Through a variety of theoretical and empirical contributions, this work seeks to illuminate both the opportunities that can be seized in the future and the challenges associated with integrating digital technologies into business models and organizational processes, with particular attention to the impacts on sustainability, ethics, and social relationships.

Napoli, Italy Francesco Schiavone
Paris, France Nessrine Omrani
Paris, France Heger Gabteni

Acknowledgments

The creation of this manual was made possible thanks to the valuable contribution and collaboration of all the authors, to whom we extend our heartfelt gratitude for their participation in the conference. The event involved as many as 140 international participants coming from 68 institutions around the world, with 110 papers presented.

Special thanks go to the publisher Springer, whose support was fundamental in bringing this volume to publication.

Our deepest thanks go to the local organizing committee, especially Anna Bastone, Federica Zeuli, and Annaluce Mandiello, as this project would not have been possible without their support. Special thanks also go to the keynote speakers Lou Shipley, Francesco Paolo Appio, Monica Masucci, and Vincenzo Corvello for their insightful contributions. We also extend our gratitude to the roundtable experts Valeria Sandei, Francesco Sodano, Alberto Tavani, Francesco Zaccariello, and Carmen Santoro for their valuable insights.

We would like to express our gratitude to the Department of Business and Quantitative Studies (DISAQ) at the University of Naples Parthenope, Paris School of Business Research Lab, Euram, and the Research Network on Innovation. Finally, our sincere thanks go to all the project partners: Know MedTech, Media Pro, Almawave, UCLM, Technovation, Technological Forecasting and Social Change, IEEE Transactions on Engineering Management, European Journal of Innovation Management, and IEEE Technology and Engineering Management Society.

We are pleased to confirm that the Digital Transformation Society conference will take place next year at the Paris School of Business on May 22–23, 2025!

Contents

Part I
Digital Transformation Across Industries

Chapter 1
Untangling the Life Cycle Impacts of Artificial Intelligence

Henrik Skaug Sætra and Kari-Anne Lyng

Abstract Artificial intelligence (AI) is increasingly being promoted as an instrumental part of the "twin transition," in which the digital and the green transition are seen as mutually reinforcing. However, there is a need for better methods and models for understanding the environmental impacts of AI systems. While positive impacts of AI for climate change mitigation and adaption are likely, these should only be pursued if they produce a net positive effect, and this requires a thorough understanding of both the climate-related impacts and impacts on, for example, resource depletion, water use, land use, and ecotoxicity. One method for unraveling these impacts is life cycle assessments (LCAs), and in this chapter, we identify and discuss methodological aspects that are relevant when performing LCA of AI systems. The study aims at contributing to taking the first steps toward methodological development and application of LCA to assess the environmental sustainability of AI systems.

Keywords Artifical intelligence · LCA · Life cycle assessment · Sustainability · Environment

1.1 Introduction

As artificial intelligence (AI) increasingly becomes a pivotal element in the twin transition toward a sustainable future, its environmental footprint has sparked a complex debate [1–4]. While digital technology can promote environmental

H. S. Sætra (✉)
Department of Informatics, University of Oslo, Oslo, Norway
e-mail: henrsae@ifi.uio.no

K.-A. Lyng
Department of Informatics, University of Oslo, Oslo, Norway

NORSUS Norwegian Institute for Sustainability Research, Fredrikstad, Norway

F. Schiavone et al. (eds.), *Advanced Perspectives and Trends in Digital Transformation of Firms, Networks, and Society*, Springer Proceedings in Business and Economics, https://doi.org/10.1007/978-3-031-80692-6_1

sustainability, this technology also comes with a range of environmental costs [5, 6]. AI is routinely championed for its potential to drive significant advancements in climate change mitigation and adaptation, but the broader environmental impacts of AI technologies, including resource depletion, water and land use, and ecotoxicity, remain underexplored. This narrative discrepancy underscores the urgency to develop a comprehensive understanding of AI's environmental impacts, ensuring that its deployment yields a net positive effect on environmental sustainability, while allowing for more robust and standardized sustainability evaluations and management of AI technologies by decision-makers in organizations and policymakers.

The aim of this chapter is to advocate for the application of life cycle assessments (LCAs) to AI systems to holistically evaluate and mitigate their environmental impacts. By establishing the foundations of a methodological framework for conducting LCAs on AI, we seek to contribute to the creation of a standardized approach that can measure and assess the environmental footprint of AI technologies, facilitating informed decision-making in the development and deployment of AI systems in line with sustainable development and the UN's Sustainable Development Goals (SDGs).

1.2 The State of the Art

Much attention is currently devoted to climate and AI [7] but also the environmental impacts of AI more broadly [8]. Many of the more visible efforts are headed by various public and policy-oriented organizations and partnerships, such as the OECD and the Global Partnership for AI (GPAI). While the reports produced by these entities reference some of the research on the environmental impact of AI, it is arguably often superficial and the analyses tend to rely on quite different methods and approaches.

As pointed out by the OECD [8], it is difficult to disentangle the impact of AI from digital technology more broadly, but it is nevertheless crucial to do so. Further exacerbating these challenges is how AI is central to many of the higher emitting uses of digital technology, such as video streaming through, for example, YouTube and Netflix. While AI might be an important part of these services, the entire impact should not be attributed to AI.

It is easy to find examples of "climate tech" initiatives highlighting use cases where AI is argued to have a potential positive impact. However, a growing literature is emerging on the cost side of this equation. With AI carbon costs now being universally acknowledged, other environmental costs of AI development and deployment are getting ever more attention. For example, water usage is now routinely mentioned, as the data centers running AI services require large amounts of water for cooling [9].

However, even if environmental issues are getting more attention, our experience indicates that there is still a gap regarding standardized and precise methods used for evaluating AI environmental impacts *and* public data allowing researchers and others to examine the actual impact of popular AI services. ChatGPT is a trending example, where a lack of data has led to widely diverging guesstimates regarding its greenhouse gas (GHG) emissions and other environmental impacts.

As we argue in this chapter, the LCA method is one possible approach that allows for more thorough and standardized evaluations, but little has been done in terms of LCAs on AI services and systems. There are some related examples, however, such as the theoretical work of Kaack et al. [10] who seek to identify sources of emissions, something that is useful as part of doing an LCA. Other relevant examples of early and partial work on LCAs or related approaches include, for example, work on cloud services and the Internet [11–13], and the carbon intensity of AI in *different* cloud instances [14].

AI's carbon footprint has thus far been most thoroughly covered by existing research, and while we'll argue that this gives an incomplete picture, it is certainly an important piece of the AI environmental cost puzzle. Other notable non-LCA work on GHG emissions from AI systems are, for example, found in Strubell et al. [15], who provided examples of estimated costs of training various large language models (LLMs), including versions of BERT and GPT-2. Costs related both to CO_2 equivalents (CO_2e) and monetary cloud compute costs were provided for some models, with, for example, *NAS* on P100x8 hardware, estimated to emit 626.155 lbs. CO_2e and costing between $942,973 and $3,201,722.

Another example of a study focused on the environmental costs of large language models (LLMs) is provided by Luccioni et al. [16], who focus on the life cycle of an AI model—without connecting this the formal LCA method. They take parts of the life cycle (focusing on "equipment manufacturing," "model training," and "model deployment") and omit early and final life cycle stages.

Considering the popularity of Microsoft's Azure cloud services, which is used by OpenAI for ChatGPT, it is also interesting to see that Microsoft researchers have pursued more insight and transparency regarding the carbon footprint of their services. In a recent project, AI2 and Microsoft (with others) used the *Software Carbon Intensity (SCI)* method to calculate the carbon footprint of training various AI models [17], and Microsoft says they have started providing their customers with metrics that allow them to better understand the energy use associated with their use of Microsoft's cloud services [18]. Such efforts could in theory be immensely valuable for increased transparency in the environmental impacts of AI, but as of yet it does not seem like these efforts have resulted in major users of Azure services, such as OpenAI, disclosing the environmental costs of their AI systems.

1.3 Life Cycle Assessments (LCAs) and AI

Much good work and many good intentions notwithstanding, we are still far away from having a universally accepted and standardized way to evaluate and communicate the environmental sustainability of AI. This is where LCA methods show

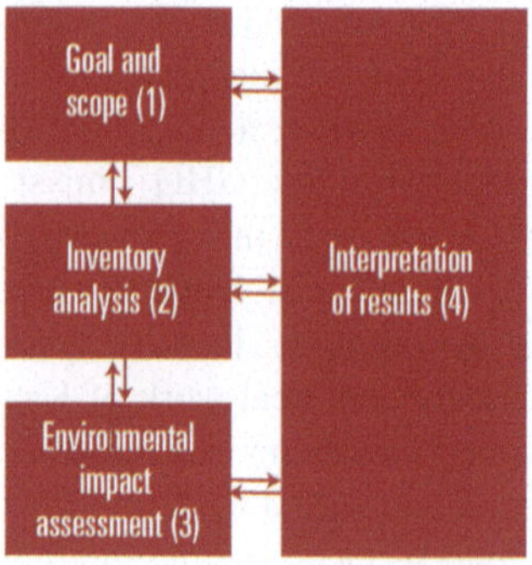

Fig. 1.1 Generic LCA framework

promise. Our approach involves a review of existing LCA methodologies and their applicability to AI systems, coupled with the proposal of a framework tailored to the unique characteristics of AI systems.

Key methodological considerations include defining the goal and scope of the LCA, selecting relevant impact categories, and establishing system boundaries. The proposed framework aims to capture the computational intensity and data-driven nature of AI development and deployment. This structured methodology is designed to address the methodological gaps identified in the existing analyses of the environmental sustainability of AI, paving the way for the development of more precise models that reflect the broader environmental impacts of these technologies.

The proposed LCA framework for AI, illustrated in Fig. 1.1, is based on the four steps of LCA projects described in ISO 14040 and 14044 and offers a novel approach to addressing these challenges. The framework provides a comprehensive method to evaluate the environmental footprint of AI systems throughout their life cycle.

Our investigation highlights a gap in the current understanding and application of LCAs to AI technologies, underlining the absence of standardized metrics and models specifically designed for AI. The complexity of AI applications complicates the environmental assessment, necessitating AI-specific analyses.

Due to non-existing guidelines and best practices related to LCA of AI systems, we assess and discuss the methodological aspects that must be resolved to perform robust LCA of AI systems that can serve as decision support. Our work is based on the general framework for LCA, such as the available ISO standards [19, 20] and the general guide for LCA [21], supplemented by relevant scientific literature related to digital/information technologies and LCA.

1.3.1 Phase 1: Goal and Scope!

The *Goal and Scope Definition (1)* phase of the LCA includes defining the system boundaries and the level of detail of the study, as well as defining clearly the intended use of the results from the LCA [19, 20]. The goal and scope will have impact on the methodological choices of the study, which in turn can impact the results. In the general guide for LCA, three types of LCAs are defined, based on the type of decision the study is intended to give input to: situation A (microlevel decision support), situation B (meso/macrolevel decision support), and situation C (accounting) [21].

As a framework for LCA of AI systems, we propose two main categories: product/service level and systems level (Fig. 1.2). While the product/service level is relevant for microlevel decision support (situation A), the systems level represents meso/macro decision support (situation B). An example of a product/service level is a study where the aim is to identify the potential environmental hotspots of an existing AI system, or for an AI system that is under development. These types of studies can be used by AI developers and operators to optimize the process or as input to design to reduce environmental impacts. These types of studies could also potentially be used by users of AI interested in being able to evaluate and choose systems based on environmental impacts. This will require clear guidelines or standardized methodology to ensure fair comparisons.

Examples of system-level assessments are studies that intend to give input to a decision or the debate about what AI should be used for and not, and whether the potential environmental benefits of applying AI are smaller or larger than the potential burdens. While studying the AI system itself may sometimes be sufficient at the product/service level, the systems level analyses require a comparison between preexisting solutions or alternatives covering similar functions.

The results from an LCA are presented relative to the functional unit, which is a quantified unit representing the function that the product or service provides to the user [19]. Defining the functional unit is a central part of the goal and scope phase of an LCA project. AI systems are, like other digital services, often by nature multifunctional [22], which means that they have different uses and can substitute many different services and activities. This substitution is sometimes referred to as indirect environmental effects [23], and examples include changes in transport patterns when digital services substitute conventional services or the reduction in paper use when use of digital devices substitutes more conventional communication systems. When AI is used in, for example, the transport, health, or education sectors, one can expect that it may result in reduction of products and services currently in use. This can be linked to the notion of scope 4 in GHG emissions accounting, at times referred to as "avoided emissions." In the GHG protocol, scope 1 refers to the direct emissions from own assets, scope 2 to the sourced energy emissions, while scope 3 is the up- and downstream emissions from the products and services sourced and the use of a company's products and services [24]. Scope 4, or the substitution or indirect environmental effects, refers to what the company's activities replace or come instead of, thus going beyond measuring the actual by comparing it to alternatives.

Fig. 1.2 AI LCA goal and scope

When performing system-level assessment, special attention must be put on the definition of the functional unit to enable a comparison between the AI system and the alternatives covering similar functions. In cases where it may be difficult to predict comparative alternatives, we recommend assessing a wide range of potential future scenarios.

In some cases, the function of the AI systems may not be clear, or it may not even have a clear useful function. In these cases, exploration of functional unit definitions may be necessary. For example, we propose testing functional units related to time use (or time wasted) or a functional unit related to sufficiency (see functional unit relative to time used (or time wasted) or a functional unit reflecting sufficiency LCA) [25].

1.3.2 Phase 2: Inventory Analysis of AI Systems

The *Inventory analysis (2)* phase involves an analysis of the input and output flows in each of the life cycle stages [19]. We suggest dividing the life cycle of AI systems into five main stages: raw material extraction, component production, model training, use, and end of life (Fig. 1.3). The first two steps involve quantifying or estimating the materials and energy required to produce the physical devices necessary to fulfill the functional unit. This involves first identifying all components required to develop and maintain the AI system, from servers in the data centers, digital devices utilized by the users as well as infrastructure to transfer data between units. As these components are often multifunctional [22], testing and development of allocation rules between the multiple services may be required to attribute the appropriate share of the environmental burden to the AI system. Among the key methodological aspects identified for future scenarios of information and communication sector are the number of units and the lifetime of key components [26]. In the inventory databases commonly used by LCA practitioners, there is generally a lack of data available for digital components, and the available datasets tend to have low data quality and become outdated quickly. This can be assumed to be caused by the rapid development when it comes to digital components, the large product complexity, and their global and untransparent value chains. The data availability and quality issues may be improved if the users of the digital infrastructure increase their demands

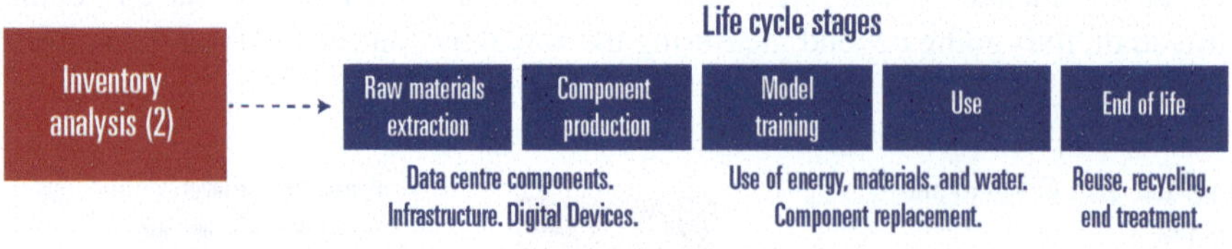

Fig. 1.3 AI LCA inventory analysis

related to transparency and environmental documentation. An important step toward improved transparency and data quality is to have detailed specifications of materials and production processes and energy use published by key producers for widely used products. An example could be Nvidia releasing detailed environmental product declarations (EPDs) for their new A100 and H100 graphics processing units (GPUs).

In the two next life cycle stages, model training and use, energy, water, and land use can be assumed to be important. This follows from data center requirements followed by the vastly increased compute required to train and run newer models, which means that data centers require more energy, more water is used for cooling, and more and larger data centers take up more land. When defining future scenarios, the growth in data traffic and energy efficiency improvements is likely to be key aspects [26].

The end-of-life phase requires the development of future scenarios related to collection and recycling of the materials and the components included in the two first life cycle stages.

1.3.3 Phase 3: Life Cycle Impact Assessment of AI Systems

In the third phase, environmental impact assessment, the inventory data for each life cycle stage is converted to potential environmental impacts. To avoid problem shifts, for example, by exacerbating one environmental problem while trying to solve another, we recommend including a broad range of environmental impact categories when assessing environmental impacts of AI systems. The scientific literature has highlighted a wide range of environmental issues, such as climate change [10], critical raw material consumption in digital components [27, 28], as well as land use [29], water footprint [9] related to data centers, and the energy intensity of cloud-based services [11]. An example range of impact categories is shown in Fig. 1.4.

As a minimum, we recommend including environmental impact categories that address these environmental problems (Fig. 1.5).

1.3.4 Phase 4: Interpretation of Results from LCA of AI Systems

The fourth phase of the LCA, interpretation of the results, must always be done in relation to the goal and scope (defined in phase 1). LCA of AI systems can potentially be used for different purposes, such as process optimization and design at product/service level, system improvement, and policy development (Fig. 1.6). For example, if the environmental hotspots of an AI service are identified to be related to the energy use during training and use stage, this indicates that impacts can be

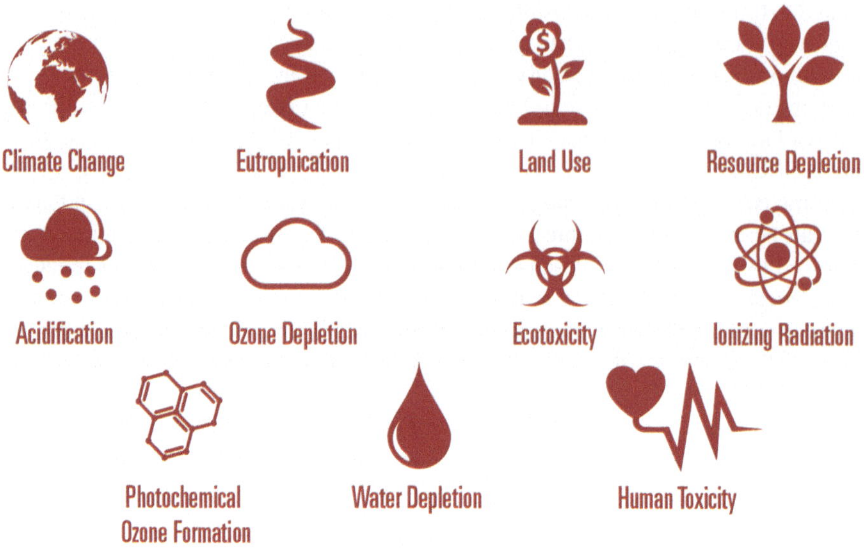

Fig. 1.4 Sample environmental impact categories

Fig. 1.5 AI LCA environmental impact assessment

Fig. 1.6 AI LCA interpretation and use of results

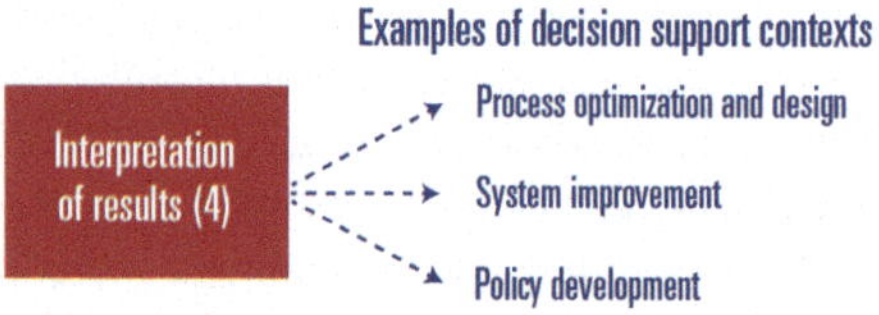

reduced by developing more resource-efficient algorithms or locating the data centers in areas where renewable energy can be sourced and in areas with less need for cooling. Previous studies have indicated that the choice of location for data centers may be of importance, as the geographic region of the data center plays a significant role in the carbon intensity [14].

System level LCAs of AI systems can potentially identify whether the environmental benefits of applying AI are larger than the potential burdens. This can serve as decision support for policymakers when trying to regulate the use of AI and developing environmental requirements for AI developers and providers. While a product/service level LCA requires high specificity when it comes to inventory data, the system level LCA can apply more generic data, but requires broader system boundaries.

1.4 Conclusion and Implications

Our assessment of the application of the general LCA framework on AI system has identified the following methodological aspects/challenges that needs to be resolved: defining an appropriate functional unit, definition of relevant future scenarios and alternative/reference scenarios to compare the AI systems with, improvement of data quality for the physical infrastructure in LCA inventory databases.

We propose a distinction between product/service level and systems level, as these two types of studies require different system boundaries and levels of detail and specificity.

The framework proposed in this chapter has significant implications for researchers, policymakers, and industry practitioners engaged. By establishing a methodological foundation for the application of LCAs to AI, we not only contribute to the scholarly understanding of AI's environmental impacts but also advocate for the integration of sustainability considerations in the development and deployment of AI technologies. The chapter underscores the need for collaborative efforts to advance LCA methodologies for AI, ensuring that the digital transition positively contributes to the environmental objectives of the SDGs and fosters a sustainable future.

References

1. Coeckelbergh, M., Sætra, H.S.: Climate change and the political pathways of AI: the technocracy-democracy dilemma in light of artificial intelligence and human agency. Technol. Soc. **75**, 102406 (2023)
2. Sætra, H.S.: AI in context and the sustainable development goals: factoring in the unsustainability of the sociotechnical system. Sustainability. **13**(4) (2021)
3. van Wynsberghe, A.: Sustainable AI: AI for sustainability and the sustainability of AI. AI Ethics, 1–6 (2021)
4. Brevini, B.: Artificial intelligence, artificial solutions: placing the climate emergency at the center of AI developments. In: Sætra, H.S. (ed.) Technology and Sustainable Development, pp. 23–33. Routledge, Milton Park (2023)
5. Brevini, B.: Is AI Good for the Planet? Polity (2021)
6. Sætra, H.S.: AI for the sustainable development goals. In: AI for Everything. CRC Press, Boca Raton (2022)
7. Clutton-Brock, P., et al.: Climate Change and AI. Recommendations for Government Action. In: GPAI, Climate Change AI. Centre for AI & Climate (2021)
8. OECD: Measuring the environmental impacts of artificial intelligence compute and applications: the AI footprint. In: OECD Digital Economy Papers (2022)
9. Mytton, D.: Data centre water consumption. NPJ Clean Water. **4**(1), 11 (2021)
10. Kaack, L.H., et al.: Aligning artificial intelligence with climate change mitigation. Nat. Clim. Chang. **12**(6), 518–527 (2022)
11. Ullrich, N., et al.: Estimating the resource intensity of the Internet: a meta-model to account for cloud-based services in LCA. Procedia CIRP. **105**, 80–85 (2022)
12. Marwah, M., et al.: Quantifying the sustainability impact of data center availability. ACM SIGMETRICS Perform. Eval. Rev. **37**(4), 64–68 (2010)

13. Itten, R., et al.: Digital transformation—Life cycle assessment of digital services, multifunctional devices and cloud computing. Int. J. Life Cycle Assess. **25**, 2093–2098 (2020)
14. Dodge, J., et al.: Measuring the carbon intensity of AI in cloud instances. In: Proceedings of the 2022 ACM Conference on Fairness, Accountability, and Transparency (2022)
15. Strubell, E., Ganesh, A., McCallum, A.: Energy and policy considerations for deep learning in NLP. arXiv preprint arXiv:1906.02243. (2019)
16. Luccioni, A.S., Viguier, S., Ligozat, A.-L.: Estimating the carbon footprint of BLOOM, a 176B parameter language model. arXiv preprint arXiv:2211.02001. (2022)
17. Buchanan, W., Dodge, J.: Measuring and mitigating AI carbon intensity. In: AI2. Medium (2022)
18. Buchanan, W.: Charting the path towards sustainable AI with Azure Machine Learning resource metrics. In: Green Tech Blog. Microsoft (2021)
19. ISO: ISO 14044:2006 Environmental Management - Life cycle assessment - Requirements and Guidelines. International Standardisation Organisation (2006)
20. ISO, ISO 14040:2006 Environmental Management — Life Cycle Assessment — Principles and Framework. 2006.
21. European Commission Joint Research Centre: ILCD Handbook. International Reference Life Cycle Data System. General guide for Life Cycle Assessment. Detailed Guidance. Joint Research Centre - Institute for Environment and Sustainability, Luxembourg (2010)
22. Itten, R., et al.: Digital transformation—Life cycle assessment of digital services, multifunctional devices and cloud computing. Int. J. Life Cycle Assess. **25**(10), 2093–2098 (2020)
23. Bieser, J.C.T., Hilty, L.M.: Assessing indirect environmental effects of information and communication technology (ICT): a systematic literature review. Sustain. For. **10**(8), 2662 (2018)
24. World Resources Institute. (2021). We Set the Standards to Measure and Manage Emissions. [cited 2021 August 8]; Available from: https://ghgprotocol.org
25. André, H.: "If less is more, how you keeping score?" Outlines of a life cycle assessment method to assess sufficiency. Front. Sustain. **5**, 1342223 (2024)
26. Furberg, A., Finnveden, G.: Towards the identification of key aspects for future scenarios of the information and communication technology sector's climate impact–Extended abstract. Proceedings http://ceur-ws.org ISSN. **1613**, 0073 (2023)
27. Xavier, L.H., Ottoni, M., Abreu, L.P.P.: A comprehensive review of urban mining and the value recovery from e-waste materials. Resour. Conserv. Recycl. **190**, 106840 (2023)
28. Charles, R.G., et al.: Towards increased recovery of critical raw materials from WEEE – Evaluation of CRMs at a component level and pre-processing methods for interface optimisation with recovery processes. Resour. Conserv. Recycl. **161**, 104923 (2020)
29. Cáceres, C.R., Sandberg, M., Sotoca, A.: Planning data center locations in Swedish municipalities. A comparative case study of Luleå and Stockholm. Cities. **150**, 105063 (2024)

Chapter 2
Overcoming Barriers to the Adoption of Anti-counterfeiting Technologies in the Italian Fashion SMEs: An Exploratory Analysis

Chiara Scrimieri, Antonio La Sala, Nicola Cucari, and Alberto Pastore

Abstract Counterfeiting is a pervasive problem, especially in countries like Italy, which boasts one of the most famous national brands in the world: "Made in Italy." This illegal activity not only jeopardizes the brand's reputation but also undermines the business results of companies, including small and medium-sized enterprises (SMEs). Although new digital technologies such as blockchain and the Internet of Things (IoT) offer opportunities in the fight against counterfeiting, their use is not widespread, especially in Italian SMEs. Our study investigates the barriers to the adoption of anti-counterfeiting technologies in Italian SMEs through an exploratory analysis. The preliminary results show the importance of traceability and trust within supply chains and shed light on the many challenges companies face when adopting anti-counterfeiting technologies. This study gives managers valuable insights into the complexities of deploying anti-counterfeiting technologies to protect brand integrity and business sustainability.

Keywords Digital technologies · Brand · Counterfeiting · "Made in Italy" · Case studies

2.1 Introduction

Companies need to develop successful strategies to gain a competitive edge in a constantly changing and digital competitive landscape [1, 2]. Branding is crucial in this context as it allows consumers to link specific value propositions with various

C. Scrimieri (✉) · A. La Sala · N. Cucari · A. Pastore
Faculty of Economics, Department of Management, Sapienza University of Rome, Rome, Italy
e-mail: chiara.scrimieri@uniroma1.it

F. Schiavone et al. (eds.), *Advanced Perspectives and Trends in Digital Transformation of Firms, Networks, and Society*, Springer Proceedings in Business and Economics, https://doi.org/10.1007/978-3-031-80692-6_2

companies. The brand is revealed to be a construct of significant complexity. "Distinctiveness" is the initial and essential aspect of a unique product, connecting supply and demand by embodying various qualitative and emotional characteristics that contribute to establishing a strong and recognizable brand identity [3]. The stronger and more recognizable a brand is, the higher the chance of imitation, and the "country branding" can also enhance this effect [4]. Accordingly, counterfeiting poses a major challenge for governments, businesses, and consumers [5, 6] and is increasing rapidly in terms of quantity, complexity, range of products, and worldwide impact according to the OECD/EUIPO report [7] and the Italian case is among the most relevant. Counterfeit products in Italy that violate registered trademarks contribute over 24 billion euros to global trade, accounting for 3.6% of total sales in the Italian manufacturing sectors, as reported in the MISE-UIBM-CENSIS report [8]. Counterfeiting jeopardizes the reputation of "Made in Italy," known globally as a megabrand for qualities such as sophistication, excellence, and uniqueness [9]. The expansion of the counterfeit market poses a threat to Italy's national competitiveness, particularly due to its reliance on small and medium-sized family enterprises that frequently lack the means to adequately address this issue [10, 11]. Research indicates that counterfeiting has a primarily adverse effect on businesses. The negative effects include decreased business performance, damage to brand reputation, and hindered market expansion opportunities [12]. Furthermore, the literature indicates that companies should approach the issue as a strategic choice rather than simply an external occurrence [13]. Furthermore, counterfeiting significantly undermines the positive effects of intellectual property rights. Counterfeit markets can significantly impact the revenue and brand reputation of individual firms [14]. Nevertheless, according to Pastore et al. [15], the current academic discussion on counterfeiting falls short in fully comprehending the main challenges it may pose. A glimpse of this could be found in digital technology adoption: while digital technologies are acknowledged as effective tools in the battle against counterfeiting, there exists a clear risk that counterfeiters may exploit them to gain knowledge and amplify their unlawful pursuits. The current academic debate offers a restricted comprehension of the complex nature of the counterfeiting challenge [15]. In this debate, digital technologies are recognized as beneficial for firms in combating counterfeiting, despite the possibility that counterfeiters may misuse them to acquire knowledge and facilitate counterfeiting. More recently, the bibliometric review by Marín-Palacios et al. [16] highlights the significant scientific interest in the development of new anti-counterfeiting technologies to reduce the number of counterfeits, make them more difficult to counterfeit, facilitate their identification, and prevent their entry into the supply chain.

A wide range of anti-counterfeiting technologies are identified and recommended in European and national reports [17, 18]. However, companies, especially small and medium-sized enterprises (SMEs), may not always be able to take advantage of these opportunities [19]. In addition, as noted by Kennedy [20], SMEs are often constrained by limitations in financial resources, human capital, and organizational capabilities, which can undermine their ability to establish brand protection programs. This prompts us to reflect on possible barriers to the implementation of these technologies, especially for Italian SMEs. So far, no study has investigated possible barriers that could prevent SMEs from using technologies such as blockchain or the Internet of Things (IoT) to combat counterfeiting. We, therefore, aim to fill this gap by addressing the following research question: *What are the barriers preventing SME brands from utilizing digital technologies to combat counterfeiting?*

To answer the research question, we studied two SMEs in the Italian fashion sector, one with a long-standing brand (Velasca) and the other with a newer brand (Inghirami Company), which serve as case studies to investigate the use of technology to fight counterfeiting. Using multiple cases allows us to gain insights from different perspectives and contexts. Given the vulnerability of Made in Italy companies to counterfeiting in the fashion and accessories industry, these case studies provide valuable insights into the challenges and opportunities associated with anti-counterfeiting measures. The findings show that SMEs are strongly interested in the adoption of anti-counterfeiting technologies that allow product traceability, but often encounter problems linked to a lack of know-how, economic support from institutions, and limited awareness of the problem.

The preliminary results have theoretical significance and can be utilized to develop specific research designs for each of the identified themes. They have managerial implications as they enable managers to focus on various business aspects that could impact counterfeiting.

The study is organized as follows: the literature review gives a summary of previous research areas and studies that focus on anti-counterfeiting business strategies, techniques, and possible digital technologies for counterfeiting mitigation. Section 2.3 focuses on the methodology employed in this study, while Sect. 2.4 sheds light on the main results. Significant implications, constraints, and prospects for future research close the study.

2.2 Literature Review

2.2.1 Counterfeiting: The Impact on Businesses and the Role of Digital Technologies

Counterfeiting and piracy are major threats to global competition and economic growth. Counterfeiting involves the unauthorized use of trademarks to sell goods, while piracy is the unauthorized reproduction and sale of copyrighted goods [21–23]. We focus specifically on counterfeiting as a form of knowledge theft. A widely used definition of counterfeiting and piracy is provided by the Agreement on Trade-Related Aspects of Intellectual Property Rights (TRIPs): "counterfeit trademark goods" include any goods bearing, without authorization, a trademark which cannot be distinguished in its essential aspects from the trademark registered for such goods, while "pirated copyright goods" include any goods bearing, without authorization, a trademark that cannot be distinguished in its essential aspects from the trademark registered for such goods. Staake et al. [23] expand the definition to include "trade in goods which, by virtue of their design, trademark, logo or company name, bear without authorization a reference to a trademark, a manufacturer or an organization which guarantees the quality or conformity of the goods, in such a way that the counterfeit goods could potentially be confused with goods which lawfully use that reference." In these definitions, the focus is on the physical figures of the product, while we want to emphasize the iconic dimension of the counterfeit product or brand and all the knowledge that makes the product that is stolen by the counterfeiter. Counterfeiting is a growing trend that has reached significant levels, according to the OECD/EUIPO report [7]. In 2019, the global trade in counterfeit products amounted to €412 billion, representing around 2.5% of world trade. According to the OECD/EUIPO report [7], this value increased significantly after the 2008 economic crisis. The phenomenon affects both established brands and SMEs, which may avoid entering the market due to concerns about counterfeiting [21]. Thus, companies that use trademarks and intellectual property face the threat of illicit trade, resulting in lost profits, reduced brand value and equity, damage to reputation, and increased costs for monitoring, legal action, and brand protection [13]. Cheap counterfeit products can reduce the perceived exclusivity of luxury goods [11, 24, 25] and consequently of goods from a particular country. Thus, counterfeiting has an overall negative impact on businesses. Berman [5] categorizes strategies that companies can employ to mitigate these effects as demand-side and supply side strategies aimed at deterring counterfeiting. One strategy involves implementing consumer and retailer education programs to enhance awareness regarding the purchase of counterfeit products. Another type of technology is authentication technologies that assist end users and retailers in identifying counterfeit products more efficiently. Supply side technologies like radio frequency identification (RFID) enable the tracking of individual products as they are transported between locations. A strategy that considers both supply and demand is price-based. Low prices can serve as a warning signal to consumers, prompting companies to adjust their prices accordingly. It is essential to have advertising campaigns that emphasize the financial, safety, and performance risks associated with purchasing counterfeit goods. These campaigns should ideally be backed by institutional organizations. Strategies to address counterfeiting from the supply side involve monitoring websites for fake products, legal actions to decrease counterfeiting, overseeing suppliers, and adopting traceable authentication technologies. The literature stresses the significance of overseeing external suppliers using HR strategies that prioritize safeguarding the company's core competencies, legal aspects like patents and contracts, and alliance processes that facilitate information exchange and partner access to information [26]. To reduce third-party resources created by outsourcers, choose service providers known for their integrity, perform surprise inspections of outsourcers, guarantee the retrieval of all confidential data after the outsourcing contract, and employ multiple outsourcers for the same product to prevent duplication. Companies may opt to conduct production internally utilizing a proprietary process or specialized equipment, a common practice among Italian small and SMEs.

Recent research highlights the significance of traceability using tracking technologies as a crucial business strategy. RFID technology has the potential to be utilized worldwide to prevent counterfeiting. RFID tags resemble regular tags, but they are computer chips enclosed in antennas. Each supplier or vendor adds a new product code to the chip as the product progresses through the supply chain. Sensors in the distribution center activate the tags using radio waves, which are then electronically scanned and tagged with a history of their location. Any tag displaying an inaccurate or incomplete list of product locations should be doubted, as genuine products are exclusively produced in specific locations and distributed through approved channels. In Italy, the Ministry of Made in Italy is supporting SMEs in the fashion industry to implement anti-counterfeiting technologies, contributing to the recognition of Made in Italy.

2.2.2 Counterfeiting and Digital Technologies in Fashion Industries: What Is the Obstacle to Adoption?

The digital transformation has significantly impacted the fashion industry, enhancing customer relations and operational efficiency [27]. Literature shows the profound influence of digital technology on various aspects of the fashion industry and highlights its potential to drive positive change and innovation (e.g., [28]).

Previous research on combating counterfeiting in fashion industries emphasizes the importance of implementing preventive measures [29, 30].

For example, [31] present novel nano-optics-based 3D multiplexing techniques for anti-counterfeiting, using both overt and covert technologies to enhance product security and deter counterfeiting. In addition, [32] advocate the use of anti-counterfeiting labels.

Recent literature highlights the effectiveness of information traceability management systems, particularly in branded apparel manufacturing, that leverage the unique characteristics of blockchain and/or the Internet of Things (IoT) (see, for example, [33]). Blockchain technology makes it possible to verify the authenticity and reliability of information related to the production of branded apparel, this makes it easier for end users to confirm the authenticity of the product, while allowing brands to ensure transparency and legitimacy. In this way, Blockchain technology has significantly improved the efficiency and ease of combating counterfeiting [34–36]. For example, Pun et al. [37] investigated how blockchain technology can be used by businesses and governments to combat counterfeiting by transforming misleading situations into accurate ones; the results suggest that government subsidies for blockchain technology would provide more benefits to consumers and society than government policy enforcement. While there is a growing body of literature examining the pros and cons of digital technologies in the anti-counterfeiting context, and governments are increasingly endorsing their use (e.g., Italy's National Plan of Territorial Anti-counterfeiting Actions and Europe's Digital Services Act), limited research has specifically examined the potential barriers businesses may face in adopting these technologies.

Identifying possible barriers and specific obstacles is critical for brands and governments fighting counterfeiting. Anticipated technological advancements, such as artificial intelligence, will have a significant impact on the fashion system. Understanding these dynamics is essential to developing effective strategies to protect brands, consumers, and markets from the threat of counterfeiting in an evolving fashion landscape.

2.3 Method

The study gathered empirical data using a multiple-case study approach [38, 39]. We chose to adopt this methodology due to the limited research on the perspectives of Made in Italy entrepreneurs and managers regarding the issue of counterfeiting. Selecting multiple (two) cases enables us to achieve a more widely applicable and strong perspective compared to just one case [40]. The study followed the qualitative research methodologies outlined in the literature by Yin [41, 42]. Data collection took place between January and February 2024 utilizing both primary and secondary data sources, including online public documents (such as company publications, presentations, Internet-based information, and newspapers) and semi-structured interviews. Semi-structured interviews were chosen as the primary data source because they represent a highly efficient way to gather rich empirical data [40]. The analysis was conducted in two steps: first, an analysis of within-case data, followed by the exploration of cross-case patterns [42].

We decided to interview the product development manager of Velasca srl and a member of the board of Inghirami Company. The two companies were selected based on the following criteria: (1) they operate in the Made in Italy fashion industry and emphasize “Made in Italy” as a strategic positioning lever; (2) they are currently undergoing digitalization of their business operations; (3) they are small and medium-sized enterprises (SMEs).

To ensure both reliability and validity, we developed an interview protocol based on existing literature and tailored to our research question [42]. The interviews were carried out via Google Meet and had durations of 90 and 60 min, respectively. The interviews were recorded, transcribed, and analyzed through thematic content analysis as outlined by King and Horrocks [43]. Three members of the research team independently conducted the coding process to assess the analysis’s reliability, which was then compared.

2.4 Findings

The study examined the issue of counterfeiting and the use of new technologies in two companies in the Made in Italy fashion sector. It highlighted the importance of traceability, not only to prevent and combat counterfeiting but also to improve quality production and pay attention to sustainability. The results also show the perceived benefits of new technologies in reducing counterfeiting along the supply chain.

The most recurring themes that emerged from the interviews are: (1) the importance of traceability; (2) trust along the supply chain; (3) economic and know-how barriers to adopting anti-counterfeiting technologies; which are reported in Table 2.1, along with coding and some examples of managers' responses to the open-ended questions.

Importance of Traceability

Counterfeiting has a pervasive impact in a range of sectors, particularly pharmaceuticals, technology, and fashion, and affects both large luxury brands and smaller companies. Our interviews show that the companies studied do not consider counterfeiting to be a primary threat but recognize its strong risk to the "Made in Italy" fashion industry. Both companies studied, consider product traceability a priority. However, the motivations for companies to make traceability of their products a priority are not the prevention of counterfeiting but others. The first is visibility for consumers: "a brand that offers transparency in its processes is seen better by consumers" while traceability is considered very important for communicating awareness of its sustainability. And only then does traceability become relevant to prevent counterfeiting. Velasca, for example, perceives counterfeiting as more relevant to large luxury brands than to smaller companies but attaches great importance to the

Table 2.1 Themes emerged from the interviews (Our elaboration)

Themes	Coding	Responses from interviewed
Importance of traceability	*Traceability*	*"a brand that offers transparency in its processes is seen better by consumers"; "traceability allows us to demonstrate our commitment to sustainability"; "counterfeiting is a problem we don't feel as much as the better-known brands. But we need to ensure the traceability of our products"*
Economic and know-how barriers of usage of anti-counterfeiting technologies	*Costs and know-how*	*"Government doesn't help us with* funds... *Even just to* do training courses... *Sometimes we don't even know what technologies exist"; "no one in the company* knows how *to use them"; "we are a small brand we do not have the ability to* bear the costs."
Trust along the supply chain	*Trust on suppliers*	"*We are concerned about counterfeiting* along the supply chain, *but..*" "*..We do not use technology... we* trust our suppliers*"; "we check our supplier with physic control, without use of technology"; "we are supplied by small Italian companies...They* don't use counterfeit material"

traceability of their products. Inghirami similarly stresses the importance of traceability for several factors, one of which is the fight against counterfeiting. The phenomenon of counterfeiting and the importance of traceability emerge more for the supply chain, where concerns become more important and the need to adopt anti-counterfeiting strategies is emphasized.

Trust along the Supply Chain
Concerns about counterfeiting are growing within the supply chain, particularly in the sourcing of raw materials and semi-finished products. While both companies in our study carefully monitor their supply chains to prevent encounters with counterfeit goods, they refrain from using digital tracking technologies, citing trust with suppliers as a critical factor. For example, Velasca does not feel the danger of counterfeiting because "All our stores are managed by Velasca directly, with our staff, because we want to have contact with the customer ourselves....".

The reliance on Italian SMEs within the supply chain presents a unique challenge, as it limits companies' control over the sourcing process. This "trust bias" hampers the adoption of anti-counterfeiting technologies and could require a re-evaluation of supply chain management practices.

Usage of Anti-counterfeiting Technologies
One of the brands studied, Velasca, uses RFID tags to authenticate its products, providing customers with a guarantee of authenticity. Although this solution is widely used by fashion companies, its potential to extend control to the entire supply chain remains untapped. For both companies interviewed, there is a strong interest in technologies such as blockchain, which promises to improve traceability and transparency. In particular, Inghirami, currently in its third generation of governance, is working to obtain traceability of the entire supply chain. The purposes are not only to prevent anti-counterfeiting but to guarantee the transparency of processes to have a competitive advantage that comes from an improved image. However, barriers such as lack of awareness and high costs are hindering its adoption. Among the difficulties encountered is the knowledge of tools that "are difficult to use" because "internally we do not have the skills to adopt these technologies." Furthermore, proximity to institutions is cited as a limiting factor in overcoming these challenges. In conclusion, our findings underline the multifaceted nature of the counterfeiting challenge for Made in Italy SMEs and highlight the need for further research to explore and address the barriers that hinder the adoption of digital anti-counterfeiting technologies. By understanding and mitigating these barriers, companies can effectively combat counterfeiting and increase their competitiveness in the global marketplace.

2.5 Discussion and Implication

Exploring the barriers to the adoption of digital anti-counterfeiting technologies in the context of a small Italian fashion brand provides critical insights into the strategic and operational dimensions of brand protection. The study highlights the complexity of the counterfeiting dilemma, particularly for SMEs promoting the "Made in Italy" brand. Our findings offer a multifaceted perspective on the intersection between technological innovation and brand integrity, shedding light on SMEs' awareness of counterfeiting and their strategic use of anti-counterfeiting technologies.

2.5.1 Theoretical Implications

This study expands the existing literature by shedding light on the barriers Italian SMEs face in combating counterfeiting and specifically using digital technologies to protect their brands. Previous studies have neglected SMEs in the study of protecting trademarks from counterfeiting. The exploratory study conducted highlights the importance of supply chain traceability and provides the basis for further studies on the adoption of anti-counterfeiting technologies in SMEs thanks Moreover, this research enriches the discourse on the role of trust in the supply chain by highlighting its impact on the adoption of anti-counterfeiting technology solutions. This is in line with existing theories on the dynamics of trust in supply chain management [44] and technology adoption in organizational contexts [45].

2.5.2 Managerial Implications

For practitioners, particularly those managing SMEs in the Made in Italy sector, this study offers actionable insights. First, it highlights the importance of having a deep understanding of the counterfeiting landscape and its potential impact on business. This means not only recognizing the risks associated with counterfeiting but also appreciating the strategic value of adopting digital technologies for brand protection. Managers are encouraged to critically evaluate their current stance on anti-counterfeiting technologies, considering not only the immediate costs but also the long-term impact on equity and trust. The study also suggests the need to strengthen trust-based relationships along the supply chain. This could include more rigorous due diligence processes, regular audits, and the use of technology to improve transparency and traceability. Accordingly, collaborative efforts between brands, technology providers, and institutions are needed to facilitate the adoption of anti-counterfeiting technologies.

Similar to luxury brands using technologies like blockchain to combat counterfeiting, small and medium-sized enterprises (SMEs) could emulate collaborative initiatives like the AURA Consortium. AURA is a consortium of luxury brands working together to improve product traceability and authenticity using blockchain solutions. SMEs could similarly form consortia to pool resources and adopt specialized technology solutions tailored to their needs, thereby strengthening their defenses against counterfeit products. By following AURA's example, SMEs can effectively address the challenges posed by counterfeiting and protect the integrity and reputation of their brands in the marketplace.

Finally, our study highlights several issues, which in turn may shed light on some possible barriers or obstacles to the use of digital technologies that could provide security against counterfeiting, both along the supply chain and on the finished product. SMEs could continue to grow without encountering phenomena that, as we have seen, can have negative effects on several fronts. Made-in-Italy brands, especially those belonging to the so-called four Made-in-Italy sectors [46], are subject to counterfeiting all over the world, and a combination of marketing strategies combined with the use of anti-counterfeiting technologies can ensure product authenticity and positive brand growth and performance.

2.6 Conclusion and Limitation

Our study calls for further research into the potential barriers preventing the adoption of digital technologies, such as blockchain and the Internet of Things (IoT), to combat counterfeiting. To this end, we propose an exploratory study aimed at investigating the barriers that prevent brands from adopting these technologies to combat counterfeiting. By studying two SMEs in the Italian fashion sector—Velasca, a long-established brand, and Inghirami Company, a newer brand—we aim to gain insights into the use of anti-counterfeiting technologies and identify economic and knowledge barriers that may hinder their implementation. Our results highlight the importance of understanding and addressing these barriers for SMEs not only for anti-counterfeiting purposes but also for traceability purposes, which emerged as the most relevant issue for the SMEs analyzed compared to mere counterfeiting. The adoption of technologies that allow traceability has important implications in terms of preventing counterfeiting and offering added value to businesses and consumers. Therefore, understanding how firms adopt (or do not adopt) digital technologies and the main reasons that sustain this choice offers wide, both theoretical and managerial, implications for companies. Exploring the barriers to the adoption of digital anti-counterfeiting technologies in a small Italian fashion brand goes in the direction of providing relevant observations on the strategic orientation and operational challenges faced by SMEs.

Nevertheless, while rich in contextual understanding and specificity, this research encounters inherent limitations that constrain its applicability and suggest avenues for further exploration. The primary limitation of this study stems from its methodological reliance on a single case study design. While this approach facilitates a deep dive into the nuanced realities of a specific organizational context, it also limits the breadth of applicability of the findings. This narrow focus may not fully capture the multiplicity of experiences, strategies, and barriers companies may experience. In addition, the qualitative nature of the study may overlook quantifiable aspects of technology adoption barriers and their impact on organizational performance. The subjective interpretation of data inherent in qualitative analysis further emphasizes the need for complementary research methods to triangulate findings and increase the robustness of the conclusions drawn.

References

1. Leão, P., da Silva, M.M.: Impacts of digital transformation on firms' competitive advantages: a systematic literature review. Strateg. Chang. **30**(5), 421–441 (2021)
2. Schiavone, F.: Strategic reactions to technology competition: a decision-making model. Manag. Decis. **49**(5), 801–809 (2011)
3. Fionda, A.M., Moore, C.M.: The anatomy of the luxury fashion brand. J. Brand Manag. **16**(5/6), 347–363 (2009)
4. Goldberg, M., Baumgartner, H.: Cross-country attraction as a motivation for product consumption. J. Bus. Res. **55**(11), 901–906 (2002)
5. Berman, B.: Strategies to detect and reduce counterfeiting activity. Bus. Horiz. **51**(3), 191–199 (2008)
6. Chaudhry, P.E., Cesareo, L.: Fake and pirated: do consumers care? J. Bus. Strateg. **38**(6), 11–19 (2017)
7. OECD/EUIPO: Illicit Trade: Global Trade in Fakes. A Worrying Threat. Organization for Economic Co-operation and Development, Paris (2021)
8. MISE-UIBM – CENSIS: La contraffazione in Italia. Studio sulle caratteristiche della domanda e stima del fatturato nel mercato interno e dei relativi impatti sull'economia del Paese (2021)
9. Marino, V., Mainolfi, G.: Country Brand Management Esperienze internazionali a confronto attraverso la ricerca qualitativa. Egea, Milano (2013)
10. Meraviglia, L.: Facing the crisis: "Craftsmanship", the competitive advantage of the Italian fashion system. In: Fashion in a Global World, pp. 171–192 (2011)
11. Meraviglia, L.: Counterfeiting, fashion and the civil society. J. Fashion Mark. Manag. **19**(3), 230–248 (2015)
12. Pittiglio, R.: Counterfeiting and firm survival: do international trade activities matter? Int. Bus. Rev. **32**(5), 102145 (2023)
13. Qian, Y.: Brand management and strategies against counterfeits. J. Econ. Manag. Strateg. **23**, 317–343 (2014)
14. Nia, A., Zaichkowsky, J.L.: Do counterfeit devalue the ownership of luxury brands? J. Prod. Brand Manag. **9**, 485–497 (2000)
15. Pastore, A., Cesaroni, F., Cesareo, L.: Counterfeiting and culture: consumer attitudes towards counterfeit products. J. Mark. Trends. **3**(1), 29–36 (2016)
16. Marín-Palacios, C., Botey-Fullat, M., Márquez, J.A.: A world of counterfeits: knowledge to decide. J. Consum. Behav., 1–20 (2024)
17. OECD/EUIPO. (2021). Global Trade in Fakes: A Worrying Threat. Illicit Trade. OECD Publishing, Paris. https://doi.org/10.1787/e39d8939-en
18. IBM-MISE: La Blockchain per la Tracciabilità del Made in Italy: Origine, Qualità, Sostenibilità (2019)

19. OECD/EUIPO: Risks of Illicit Trade in Counterfeits to Small and Medium-Sized Firms. Organization for Economic Co-operation and Development, Paris (2023)
20. Kennedy, J.: Proposed solutions to the brand protection challenges and counterfeiting risks faced by small and medium enterprises (SMEs). J. Appl. Secur. Res. **11**(4), 450–468 (2016)
21. Fink, C., Maskus, K., Qian, Y.: The economic effects of counterfeiting and piracy: a review and implications for developing countries. World Bank Res. Obs. **31**, 1–28 (2016)
22. Li, F., Yi, Z.: Counterfeiting and piracy in supply chain management: theoretical studies. J. Bus. Ind. Mark. **32**, 98–108 (2017)
23. Staake, T., Thiesse, F., Fleisch, E.: The emergence of counterfeit trade: a literature review. Eur. J. Mark. **43**, 320–349 (2009)
24. Cesareo L., Williams P., & Pastore A. (2017). Counterfeiting luxury goods. in Handbook of Research on Counterfeiting and Illicit Trade, 18(5), 193, Edvard Elgar Publishing.
25. D'Amato, I., Papadimitriou, T.: Legitimate vs illegitimate: the luxury supply chain and its doppelganger. Int. J. Retail Distrib. Manag. **41**, 986 (2013)
26. Norman, P.M.: Are your secrets safe? Knowledge protection in strategic alliances. Bus. Horiz. **44**(6), 51–60 (2001)
27. Bertola, P., Teunissen, J.: Fashion 4.0. Innovating fashion industry through digital transformation. Res. J. Text. Appar. **22**(4), 352–369 (2018)
28. Casciani, D., Chkanikova, O., Pal, R.: Exploring the nature of digital transformation in the fashion industry: opportunities for supply chains, business models, and sustainability-oriented innovations. Sustain. Sci. Practice Policy. **18**(1), 773–795 (2022)
29. Cho, S.H., Fang, X., Tayur, S.: Combating strategic counterfeiters in licit and illicit supply chains. Manuf. Serv. Oper. Manag. **17**(3), 273–289 (2015)
30. Lai, J., Zhang, Y., Li, Z., Wang, Y., & Li, M. (2022). Bioinspired Quasi-3D Multiplexed Anti-Counterfeit Imaging via Self-Assembled and Nanoimprinted Photonic Architectures. Advanced Materials. **34**(1), 2107243. https://doi.org/10.1002/adma.20210724
31. Yang, K., Chen, H., Hu, H., Sun, B., Zhao, H., Qie, Y., Luo, Z., Pan, Y., Chen, W., Lin, L., Guo, T., & Li, F. (2023). Dynamic Anti-Counterfeiting Labels with Enhanced Multi-Level Information Encryption. ACS Applied Materials & Interfaces. **15**(1), 2104–2111. https://doi.org/10.1021/acsami.2c17870
32. Gao, Y.: On the use of overt anti-counterfeiting technologies. Mark. Sci. **37**(3), 403–424 (2018)
33. Tyagi, A.K., Dananjayan, S., Agarwal, D., Thariq Ahmed, H.F.: Blockchain—Internet of Things applications: opportunities and challenges for Industry 4.0 and Society 5.0. Sensors. **23**, 947 (2023)
34. Choi, T.M.: Blockchain-technology-supported platforms for diamond authentication and certification in luxury supply chains. Transp. Res. E Logist. Transp. Rev. **128**, 17–29 (2019)
35. Choi, T.M., Ouyang, X.: Initial coin offerings for blockchain-based product provenance authentication platforms. Int. J. Prod. Econ. **233**, 107995 (2021)
36. Choi, T.M., Wen, X., Sun, X., Chung, S.H.: The mean-variance approach for global supply chain risk analysis with air logistics in the blockchain technology era. Transp. Res. E Logist. Transp. Rev. **127**, 178–191 (2019)
37. Pun, H., Swaminathan, J., Hou, P.: Blockchain adoption for combating deceptive counterfeits. Prod. Oper. Manag. **30**(4), 864–882 (2021)
38. Eisenhardt, K.M.: Building theories from case study research. Acad. Manag. Rev. **14**(4), 532–550 (1989)
39. Yin, R.K.: Discovering the future of the case study method in evaluation research. Eval. Pract. **15**(3), 283–290 (1994)
40. Eisenhardt, K.M., Graebner, M.E.: Theory building from cases: opportunities and challenges. Acad. Manag. J. **50**(1), 25–32 (2007)
41. Yin, R.K.: Case Study Research: Design and Methods. Sage, Thousand Oaks (2003)

42. Yin, R.K.: Case study Research and Applications: Design and Methods. Sage Publications, Washington DC (2017) (Author, F.: Contribution title. In: 9th International proceedings on proceedings, pp. 1–2. Publisher, Location (2010))
43. King, N., Horrocks, C.: Interviews in Qualitative Research, p. viii, 248. Sage Publications, Thousand Oaks (2010)
44. Handfield, R.B., Bechtel, C.: The role of trust and relationship structure in improving supply chain responsiveness. Ind. Mark. Manag. **31**(4), 367–382 (2002)
45. Rogers, E.M.: Diffusion of Innovations, 5th edn. Free Press, New York (2003)
46. Fortis, M.: Il Made in Italy. Il Mulino, Bologna (1998) (Collana: Farsi un'idea, Nr. 29)

Chapter 3
Assessing the Adoption of Gen-AI in the Italian Agri-Food Industry: An Empirical Analysis

Piloca Diletta, Luca Quaglieri, Francesco Mercuri, and Bernardino Quattrociocchi

Abstract The integration of Generative Artificial Intelligence (Gen-AI) into the agri-food sector marks a significant shift, enhancing the precision of production and distribution strategies, process optimization, and productivity. Gen-AI's capability to detect patterns enables agricultural stakeholders to make informed decisions, optimizing crop yields and minimizing waste. Simultaneously, it aids logistics operators in refining supply chain efficiency through improved process, inventory, and distribution management. Despite the growing adoption of Gen-AI in agri-food practices, the scholarly examination of its impact remains scarce, highlighting a critical need for further research to understand its transformative potential and implications within the sector's production paradigms. This study focuses on identifying the various factors influencing the adoption of Gen-AI technology in agri-food contexts, encompassing technological, environmental, and organizational dimensions. It aims to explore the drivers behind agri-food operators' intent to adopt Gen-AI and assess how organizational structure and leadership influence this adoption. By investigating these elements, the research seeks to provide a nuanced understanding of the integration of Gen-AI in the agri-food sector, offering insights into the challenges and opportunities that lie ahead for stakeholders navigating this technological evolution.

Keywords Gen-AI · Agri-food · Technologies · IA · Digital transformation

---

P. Diletta (✉) · F. Mercuri · B. Quattrociocchi
Faculty of Economics, Department of Management, Sapienza University of Rome, Rome, Italy
e-mail: diletta.piloca@uniroma1.it

L. Quaglieri
Faculty of Engineering, Department of Mechanical and Aerospace Engineering, Sapienza University of Rome, Rome, Italy

F. Schiavone et al. (eds.), *Advanced Perspectives and Trends in Digital Transformation of Firms, Networks, and Society*, Springer Proceedings in Business and Economics, https://doi.org/10.1007/978-3-031-80692-6_3

3.1 Introduction

In recent times, Artificial Intelligence (AI) has garnered increasing attention in various fields [1]. A significant portion of current research has recognized the importance of AI applications in enhancing supply chain efficiency and effectiveness [2], fostering innovation [3], promoting corporate sustainability [4], and improving operational performance [5, 6]. Among the research areas within the field of AI is Generative Artificial Intelligence, also known as Generative AI or Gen-AI, which utilizes machine learning and deep learning techniques to generate new data, including images, music, and text, that did not exist previously [7, 8]. Unlike discriminative AI, this type of AI is designed to develop an understanding of the presented data and create something new without making distinctions between inputs. Recent studies demonstrate the application of this technology in various sectors such as tourism, manufacturing, and education [9–11]. Although the effects of integrating digital technologies, including AI, in the agri-food sector are well-known and encompass concepts like "smart agriculture," "precision agriculture," "digital agriculture," "decision agriculture," and "agriculture 4.0" [12], there is still a limited and contextualized body of research investigating the effects of Generative Artificial Intelligence in this domain [13]. Consequently, there exists a discernible lacuna necessitating rigorous and contextually grounded research endeavors to unravel the transformative potential and nuanced ramifications of incorporating Generative AI within agricultural paradigms.

3.2 Objectives

The elucidation of Generative Artificial Intelligence's (GEN-AI) transformative potential within the agri-food sector constitutes a central aim of this study. To this end, our research endeavors to discern the multifaceted determinants that shape the widespread adoption of GEN-AI technologies in agri-food contexts, encompassing technological, environmental, and organizational dimensions. Drawing upon established frameworks within the fields of technology adoption and organizational theory, our investigation seeks to comprehensively evaluate the intricate interplay of these factors in influencing GEN-AI adoption dynamics within the agri-food domain [14]. In order to empirically ground our analysis and derive actionable insights, we employed a robust research methodology involving the execution of a large-scale survey targeting a diverse cohort of industry operators. This survey instrument was crafted to capture nuanced perspectives on GEN-AI utilization, enabling a nuanced examination of its current prevalence, as well as the attendant challenges and opportunities encountered by stakeholders across the agri-food value chain. By grounding our research in established theoretical paradigms, we aim to enrich our understanding of the complex dynamics shaping GEN-AI adoption within the agri-food sector, thereby contributing to both theoretical advancements and practical implications for industry stakeholders. Through this scholarly endeavor, we seek to shed light on the transformative potential of GEN-AI within the agri-food sector, offering insights that can inform strategic decision-making, policy formulation, and future research directions aimed at harnessing the full spectrum of benefits associated with GEN-AI technologies in agricultural and food production contexts.

3.3 Methods

Our research analyses the intention to adopt GEN-AI in the agri-food sector through an integrated model based on the Technology Acceptance Model (TAM) [15] and the Technology Organization-Environment model (TOE) [16] constructs. The integration of the models is perceived appropriately to interpret the social, environmental, and technological aspects of Artificial Intelligence adoption in business organizations [17]. The methodology employed in this study consisted of two distinct phases. During the first phase, data was collected through a survey distributed among administrative employees of agri-food companies. A total of 156 respondents participated in the survey. The questionnaire utilized a 5-point Likert scale, ranging from 1 ("completely disagree") to 5 ("completely agree"). The survey included an introduction to the "GEN-AI" technology and inquired about the participants' prior experiences with similar technologies. Several constructs were examined in the questionnaire. The first construct, Relative Advantage, consisted of three items and aimed to assess the degree to which management and employees perceived the new technology as superior and more effective than previous technologies [18]. The second construct, Compatibility, also consisting of three items, explored the ease of implementing the technology without encountering significant challenges in the field [19]. Top Management Support, the third construct, focused on the extent to which top management embraced, understood, and accepted the new technology [20]. The fourth construct, Organizational Availability, investigated the resources, both tangible and intangible, available within the companies to support technology adoption processes, programs, and procedures [21]. Customer Pressures, the fifth construct, considered the combined influence of customer needs, desires, and market trends [22]. The final three constructs were derived from the Technology Acceptance Model (TAM) proposed by Davis [15] and Venkatesh and Davis [23]. These constructs were Perceived Usefulness, Perceived Ease of Use [24], and Intention to adopt GEN-AI [17]. These constructs were assessed in accordance with the TAM model. During the second phase of the study, the collected data was analyzed using the AMOS software. Initially, confirmatory factor analysis (CFA) was conducted to validate the multi-item scales employed in the study [25]. The fitness of the CFA model was evaluated using various fit indices [26]. Subsequently, exploratory and confirmatory factor analyses were performed to assess the goodness-of-fit of the collected data with the proposed model [27]. The following image represents the hypothesis through which the model was developed (Fig. 3.1).

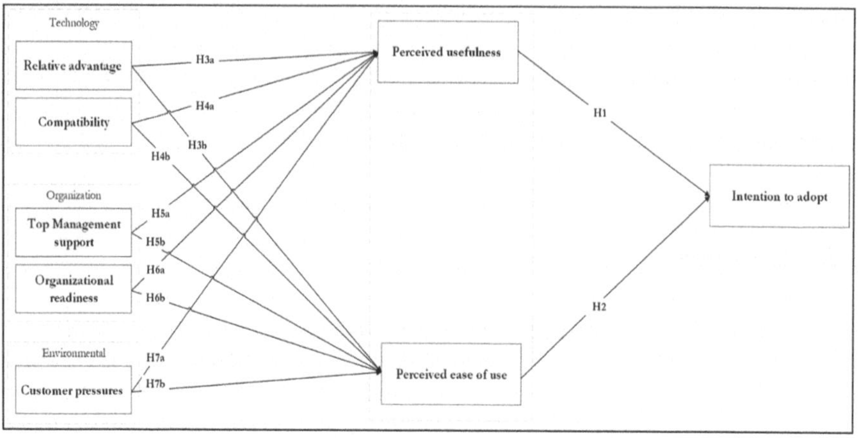

Fig. 3.1 The proposed model (Our elaboration)

3.4 Findings

Results show that the intention to adopt Gen-AI in the Agri-food sector is directly influenced by the constructs of TAM and indirectly influenced by the constructs of TOE, except for Top Management Support, which instead has a negative impact on the intention to adopt the new technology. These statements are supported by the fact that the data show strong statistical significance, thus underpinning the reliability and validity of the analytical framework utilized. Perceived Usefulness (PU) and Perceived Ease of Use (PEOU) from TAM, as well as Relative Advantage (RA), Compatibility (COM), Organizational Availability (OA), and Customer Pressure (CP) from TOE, play crucial roles in shaping the intention to integrate Gen-AI into the Agri-food Supply Chain. The unexpected negative influence of Top Management Support (TMS) on adoption intention raises intriguing questions for further investigation. Of particular note is the discernment that merely three out of the 12 hypotheses failed to garner empirical support, as they purportedly evinced a counterproductive influence on the Adoption Intention of Gen-AI within the Agri-food Supply Chain. The empirical findings gleaned from the analysis furnish invaluable insights into the validation of hypotheses constituting the research model, thereby contributing significantly to the scholarly discourse surrounding technology adoption within the Agri-food sector. The results of the research are summarized in Fig. 3.2.

HP.	RELATIONSHIP	DESCRIPTION	EVALUATION
H1	PU -> IA	Perception of usefulness positively influences adoption intention	Confirmed
H2	PEOU -> IA	Perception of ease of use positively influences adoption intention	Confirmed
H3a	RA - > PU	Relative advantage positively influences perceived usefulness	Confirmed
H3b	RA - > PEOU	Relative advantage positively influences perceived ease of use	Confirmed
H4a	COM -> PU	Compatibility positively influences the perception of utility	Confirmed
H4b	COM -> PEOU	Compatibility positively influences the perception of ease of use	Confirmed
H5a	TMS -> PU	Executive support positively influences perceived usefulness	Not Confirmed
H5b	TMS -> PEOU	Executive support positively influences perceived ease of use	Not Confirmed
H6a	OA -> PU	Organizational availability positively influences perceived usefulness	Confirmed
H6b	OA -> PEOU	Organizational availability positively influences perceived ease of use	Confirmed
H7a	CP -> PU	Costumer pressures positively influence perceptions of utility	Confirmed
H7b	CP -> PEOU	Costumer pressures positively influence perceived ease of use	Not Confirmed

Fig. 3.2 Testing of hypotheses (Our elaboration)

3.5 Implications

Generative AI presents significant possibilities for use in the agricultural domain, deserving more scrutiny. Moving from the assumptions of the model, our research presents both theoretical and empirical implications. From a theoretical point of view, this study contributes to the literature on consumer intention to adopt GEN-AI technology in agri-food sector, which is still poorly investigated. Precisely, the study investigated the adoption intention of GEN-AI applications through an integrated model based on TAM and TOE model, which was useful to analyzing how Perceived Usefulness and Ease of Use influence GEN-AI consumer intention to adopt. Results highlight that these two factors directly influence the GEN-AI intention to adopt within agri-food sector. From a managerial perspective, this research sheds light on the organizational dynamics and managerial challenges involved in fostering a culture of innovation and technological adoption within the agri-food sector. Despite the evolution of organizational models across various production sectors, our findings indicate the need for focused efforts to cultivate an entrepreneurial culture that supports the adoption of new technologies, including Generative AI. Specifically, our empirical analysis reveals that while agri-food companies exhibit organizational readiness, Top Management Support is not completely oriented toward the adoption and implementation of GEN-AI in the Agri-food Supply Chain, thus negatively and indirectly influencing the Adoption Intention of the technology. This emphasizes the importance of aligning organizational priorities and leadership support with the necessity of integrating innovative technologies to drive sustainable growth and competitiveness within the agri-food sector. Moreover, this study underscores the pivotal role played by perceived usefulness and ease of use in shaping GEN-AI adoption intentions within the agri-food domain. These two factors are revealed to exert direct and significant influences on the intention to adopt GEN-AI technologies, highlighting their critical importance in driving adoption behaviors among agri-food stakeholders.

References

1. Burger, B., Kanbach, D.K., Kraus, S., Breier, M., Corvello, V.: On the use of AI-based tools like ChatGPT to support management research. Eur. J. Innov. Manag. **26**(7), 233–241 (2023)
2. Bahoo, S., Cucculelli, M., Qamar, D.: Artificial intelligence and corporate innovation: a review and research agenda. Technol. Forecast. Soc. Chang. **188**, 122264 (2023)
3. Javaid, M., Haleem, A., Singh, R.P., Suman, R.: Artificial intelligence applications for industry 4.0: a literature-based study. J. Ind. Integr. Manag. **7**(01), 83–111 (2022)
4. Al-Qaysi, N., Granić, A., Al-Emran, M., Ramayah, T., Garces, E., Daim, T.U.: Social media adoption in education: a systematic review of disciplines, applications, and influential factors. Technol. Soc. **73**, 102249 (2023)
5. Mariani, M.M., Machado, I., Magrelli, V., Dwivedi, Y.K.: Artificial intelligence in innovation research: a systematic review, conceptual framework, and future research directions. Technovation. **122**, 102623 (2023)
6. Quattrociocchi, B., Boria, P., Mercuri, F., Sergiacomi, S.: La blockchain applicata alle imprese: un nuovo modello di business. Implicazioni tecniche, fiscali ed economiche. In: Collana di Economia Finanza, Diritto dell'impresa. Giappichelli Editore (2022)
7. Celik, A., Eltawil, A.M.: At the dawn of generative AI era: a tutorial-cum-survey on new frontiers in 6G wireless intelligence. IEEE Open J. Commun. Soc. (2024)
8. Goodfellow, I., Pouget-Abadie, J., Mirza, M., Xu, B., Warde-Farley, D., Ozair, S., et al.: Generative adversarial nets. Adv. Neural Inf. Proces. Syst. **27** (2014)
9. Doanh, D.C., Dufek, Z., Ejdys, J., Ginevičius, R., Korzynski, P., Mazurek, G., et al.: Generative AI in the manufacturing process: theoretical considerations. Eng. Manag. Prod. Serv. **15**(4), 76–89 (2023)
10. Dogru, T., Line, N., Mody, M., Hanks, L., Abbott, J.A., Acikgoz, F., et al.: Generative artificial intelligence in the hospitality and tourism industry: developing a framework for future research. J. Hosp. Tour. Res. **10963480231188666** (2023)
11. Ratten, V., Jones, P.: Generative artificial intelligence (ChatGPT): implications for management educators. Int. J. Manag. Educ. **21**(3), 100857 (2023)
12. Gebresenbet, G., Bosona, T., Patterson, D., Persson, H., Fischer, B., Mandaluniz, N., et al.: A concept for application of integrated digital technologies to enhance future smart agricultural systems. Smart Agric. Technol. **5**, 100255 (2023)
13. Dennehy, D., Griva, A., Pouloudi, N., Dwivedi, Y.K., Mäntymäki, M., Pappas, I.O.: Artificial intelligence (AI) and information systems: perspectives to responsible AI. Inf. Syst. Front. **25**(1), 1–7 (2023)
14. Sapkota, R.: Harnessing the power of AI based image generation model DALLE 2 in agricultural settings. arXiv preprint arXiv:2307.08789. (2023)
15. Davis, M.B.: Lags in vegetation response to greenhouse warming. Clim. Change. **15**(1–2), 75–82 (1989)
16. Tornatzky, L.G., Fleischer, M.: The Processes of Technological Innovation. Lexington Books, Lexington (1990)
17. Chatterjee, S., Rana, N.P., Dwivedi, Y.K., Baabdullah, A.M.: Understanding AI adoption in manufacturing and production firms using an integrated TAM-TOE model. Technol. Forecast. Soc. Chang. **170**, 120880 (2021)
18. Sharma, S., Singh, G., Islam, N., Dhir, A.: Why do SMES adopt artificial intelligence-based chatbots? IEEE Trans. Eng. Manag. (2022)
19. Bag, S., Rahman, M.S., Gupta, S., Wood, L.C.: Understanding and predicting the determinants of blockchain technology adoption and SMEs' performance. Int. J. Logist. Manag. **34**(6), 1781–1807 (2023)
20. Hashimy, L., Jain, G., Grifell-Tatjé, E.: Determinants of blockchain adoption as decentralized business model by Spanish firms–An innovation theory perspective. Ind. Manag. Data Syst. **123**(1), 204–228 (2023)

21. Iranmanesh, M., Lim, K.H., Foroughi, B., Hong, M.C., Ghobakhloo, M.: Determinants of intention to adopt big data and outsourcing among SMEs: organisational and technological factors as moderators. Manag. Decis. **61**(1), 201–222 (2023)
22. Abed, S.S.: Social commerce adoption using TOE framework: an empirical investigation of Saudi Arabian SMEs. Int. J. Inf. Manag. **53**, 102118 (2020)
23. Venkatesh, V., Davis, F.D.: A model of the antecedents of perceived ease of use: development and test. Decis. Sci. **27**(3), 451–481 (1996)
24. Ma, X., Huo, Y.: Are users willing to embrace ChatGPT? Exploring the factors on the acceptance of chatbots from the perspective of AIDUA framework. Technol. Soc. **75**, 102362 (2023)
25. Jain, V., Raj, T.: Modeling and analysis of FMS flexibility factors by TISM and fuzzy MICMAC. Int. J. Syst. Assur. Eng. Manag. **6**, 350–371 (2015)
26. Bagozzi, R.P.: Structural equation models are modelling tools with many ambiguities: comments acknowledging the need for caution and humility in their use. J. Consum. Psychol. **20**(2), 208–214 (2010)
27. Hair, M.L., Tripp, C.P.: Alkylchlorosilane reactions at the silica surface. Colloids Surf. A Physicochem. Eng. Asp. **105**(1), 95–103 (1995)

Chapter 4
Exploring Metaverse Innovations in Public Administration: A Focus on Human Resource Management

Filomena Buonocore, Davide de Gennaro, and Ludovica Del Barone

Abstract The metaverse, an expansive virtual shared space, is revolutionizing various sectors, including Public Administration (PA). This mini-review systematically explores the current initiatives and practices related to the metaverse in PA, with a particular focus on human resource management (HRM). By leveraging immersive technologies such as virtual reality (VR) and augmented reality (AR), public institutions can enhance employee engagement, streamline recruitment and onboarding processes, and provide more effective training and performance management. Despite the promising potential, the integration of these technologies also presents significant challenges, including legal, regulatory, and ethical issues.

A comprehensive literature search using scientific databases identified 24 key articles that discuss the implications of the metaverse for public sector HRM. The findings indicate that while the metaverse can transform HRM practices by offering more engaging and efficient processes, successful implementation requires addressing several hurdles. Key themes include the necessity for robust governance frameworks, the development of sustainable digital transformation practices, and the need for adaptive leadership styles in virtual environments.

This chapter contributes to the ongoing discourse on digital transformation in the public sector by highlighting both the opportunities and challenges of integrating metaverse technologies into HRM. It provides actionable insights for policymakers and practitioners and suggests directions for future research, including empirical studies to assess the long-term impacts and comparative analyses across different contexts. The study underscores the importance of continued innovation and ethical considerations to unlock the full potential of the metaverse in enhancing public service delivery and workforce management.

F. Buonocore (✉) · D. de Gennaro
Parthenope University of Naples, Naples, Italy

L. Del Barone
University of Salento, Lecce, Italy

F. Schiavone et al. (eds.), *Advanced Perspectives and Trends in Digital Transformation of Firms, Networks, and Society*, Springer Proceedings in Business and Economics, https://doi.org/10.1007/978-3-031-80692-6_4

Keywords Metaverse · Public administration · Human resource management · Virtual reality · Digital governance · Immersive technologies

4.1 Introduction

Imagine a world where the boundaries between the real and virtual are so blurred that you can conduct business meetings, hold training sessions, and even manage entire public departments without ever leaving your home. Welcome to the metaverse, a rapidly evolving digital frontier that promises to revolutionize the way we live and work. As this virtual realm expands, its potential to transform Public Administration (PA) is both fascinating and inevitable.

The metaverse, a collective virtual shared space created by the convergence of virtually enhanced physical reality and persistent virtual reality (VR) [1, 2], is emerging as a pivotal platform for innovation in various sectors, including PA [3]. This digital universe is not just a fantasy of gamers and tech enthusiasts; it is becoming a viable environment for public institutions to enhance their operations, engage with citizens, and streamline processes [4, 5]. PA in the metaverse can leverage immersive technologies such as VR and AR to create more interactive, efficient, and accessible public services [3, 5].

The advent of the metaverse brings forth unprecedented opportunities and challenges for PA [6]. As public institutions strive to keep pace with rapid technological advancements, integrating the metaverse into their operational frameworks presents a unique opportunity to innovate and improve service delivery [7]. The metaverse offers a dynamic platform where public services can be delivered in more engaging and efficient ways, overcoming traditional barriers of time and space [8]. For instance, virtual town halls, immersive training programs for public employees, and VR-based citizen services are just a few examples of how the metaverse can transform PA (e.g., [9]). This potential transformation extends to Human Resource Management (HRM) [10–12] within PA [3, 8], where virtual environments can be used for recruitment, onboarding, training, performance management, and employee engagement.

Despite the burgeoning interest and potential applications of the metaverse, there remains a significant gap in the academic and practical understanding of its implications for PA [13], particularly in the realm of HRM [3, 5]. Current literature has primarily focused on the technological advancements and theoretical frameworks of the metaverse (e.g., [14]), often overlooking the practical challenges and opportunities it presents for public sector HRM [3]. Questions about how virtual environments can be effectively integrated into public sector HR practices [15], how they can enhance employee engagement [16], training, and performance management [15], and what ethical and regulatory issues need to be addressed remain largely unexplored [17]. Additionally, there is limited empirical research on the effectiveness of metaverse-based HRM practices in the public sector, making it difficult to assess their impact and identify best practices [5].

The objective of this study is to bridge this gap by exploring the current state of initiatives related to the metaverse in PA, with a particular focus on HRM. This chapter aims to identify existing projects and practices, assess their effectiveness, and highlight areas where further research and development are needed. By providing a comprehensive overview of the intersection between the metaverse and PA-HRM, this study seeks to contribute to the ongoing discourse on digital transformation in the public sector and offer actionable insights for policymakers and practitioners looking to harness the potential of the metaverse. Through a mini-review (e.g., [18, 19]), we aim to shed light on the current landscape of metaverse initiatives in PA, identify the challenges and opportunities they present, and propose directions for future research and practice. This chapter will serve as a valuable resource for public sector leaders and HR professionals seeking to navigate the complexities of integrating the metaverse into their operations and unlock its full potential for improving public service delivery and workforce management.

This chapter is structured as follows: The next section "Methodology" outlines the steps taken to conduct this mini-review, including the research question framing, literature search strategy, and data analysis process. The Results section presents the key findings from the analyzed articles, highlighting the current state of metaverse initiatives in PA and their implications for HRM. This is followed by the Discussion, which interprets the findings in relation to existing literature, discusses theoretical and practical implications, and addresses the limitations of the study. Finally, the Avenues for Future Research section proposes directions for further investigation to advance our understanding and application of metaverse technologies in public sector HRM.

4.2 Methods

This mini-review aims to systematically explore the current landscape of metaverse initiatives in PA, with a specific focus on HRM. A mini-review is an appropriate methodology for this study as it allows for a concise yet comprehensive synthesis of existing literature on a focused topic [20, 21]. Unlike a full systematic review, which can be exhaustive and time-consuming, a mini-review provides a targeted overview of the most pertinent studies, making it ideal for emerging fields where literature is still developing, such as the application of the metaverse in PA [22].

This approach ensures a manageable scope while still providing valuable insights into current trends, challenges, and opportunities. We followed several steps for this review in line with the guidelines outlined by Hoon [23], ensuring a systematic and rigorous approach to the analysis of the literature.

The first step in this mini-review involved framing the research question to guide the review process. The primary research question was: "What are the current initiatives and practices related to the metaverse in PA, particularly concerning HRM?" This question was designed to capture both the breadth and depth of relevant studies, ensuring a comprehensive understanding of the subject matter. By defining a clear and focused research question, the review process was structured to address specific aspects of the metaverse's impact on HRM within PA, thereby providing a detailed exploration of this niche area.

Next, we located relevant research by conducting extensive searches using scientific search engines and databases. These included Scopus, Web of Science, EbscoHost Business Source Complete, PsycINFO, PsycARTICLES, and Google Scholar. The search strategy involved using a combination of keywords and phrases such as "metaverse," "Public Administration," "human resource management," "HRM," "virtual reality (VR)," "augmented reality (AR)," and "digital governance." This approach ensured the retrieval of a broad range of studies, encompassing various aspects of the metaverse and its application in PA. By employing multiple databases and search engines, we maximized the likelihood of capturing all relevant literature, thus enhancing the comprehensiveness of the review.

After gathering a substantial number of articles, the next step was to establish inclusion and exclusion criteria to filter the studies. Inclusion criteria were (1) articles published in peer-reviewed journals, (2) studies focusing on the metaverse in the context of PA, and (3) articles that discussed or investigated HRM practices within PA. Exclusion criteria included (1) articles not available in English, (2) studies focusing solely on the technical aspects of the metaverse without any application to PA, and (3) non-peer-reviewed sources such as opinion pieces or editorials. These criteria ensured that the review included only the most relevant and rigorous studies. By applying stringent inclusion and exclusion criteria, we ensured the quality and relevance of the selected articles, thereby maintaining the integrity of the review process.

Data extraction and coding were the subsequent steps. Each article that met the inclusion criteria was thoroughly read and analyzed. Key information was extracted, including the study's objectives, methodologies, findings, and conclusions related to the metaverse and HRM in PA. This data was then coded thematically to identify common patterns, themes, and discrepancies among the studies. Coding was performed manually to ensure a deep and nuanced understanding of the content. The thematic coding process involved categorizing the data into predefined themes and allowing new themes to emerge organically from the data, ensuring a comprehensive analysis that captured all relevant aspects of the studies.

The analysis phase involved examining different conceptualizations and making comparisons across the studies. This included comparing how different researchers and practitioners conceptualized the metaverse and its applications in HRM within PA. Attention was given to variations in methodologies, findings, and theoretical frameworks to identify areas of convergence and divergence. By systematically comparing these conceptualizations, the review was able to highlight commonalities and differences in how the metaverse is perceived and utilized in PA contexts, providing a richer understanding of the field.

Finally, synthesizing the findings involved integrating the insights from the various studies to provide a coherent narrative. This synthesis aimed to highlight the state of the art in metaverse initiatives in PA, identify gaps in the current knowledge, and propose directions for future research. Through this process, 24 articles were identified that specifically investigated or mentioned the metaverse in the context of PA. By analyzing these articles, the review aimed to uncover any existing HRM practices adopted within the metaverse framework and assess their effectiveness. The synthesis process involved summarizing the key findings of each study, identifying overarching themes, and critically evaluating the implications of these findings for PA and HRM.

4.3 Findings

This mini-review synthesizes the findings from 24 articles that investigate or mention the implications of the metaverse for PA, particularly in the context of HRM. The synthesis aims to identify existing initiatives, practices, and emerging themes related to the use of the metaverse in public sector HRM, providing insights into the current landscape and highlighting areas for further research.

The synthesis of these findings highlights the transformative potential of the metaverse for HRM in PA. By leveraging immersive technologies, public institutions can enhance employee engagement, streamline recruitment and onboarding processes, and provide more effective training and performance management. However, the successful integration of these technologies requires addressing legal, regulatory, and ethical challenges.

4.3.1 Integration of Metaverse Technologies in PA

The integration of metaverse technologies in PA is primarily driven by the potential to enhance service delivery, citizen engagement, and operational efficiency. Alahi et al. [24] discuss the role of IoT-enabled technologies and AI in creating smart city scenarios, emphasizing the importance of these technologies in the metaverse context. Similarly, Allam et al. [6] explore the metaverse as a virtual extension of smart cities, highlighting opportunities for environmental, economic, and social sustainability.

Anim [4] investigates the use of VR for citizen feedback on public space design, demonstrating how immersive technologies can facilitate better public engagement. This aligns with Kshetri et al. [5], who outline the prospects and challenges of the metaverse for advancing government functions, including HRM.

4.3.2 HRM Practices in the Metaverse

The application of metaverse technologies in HRM within PA includes various innovative practices such as virtual recruitment, onboarding, training, performance management, and employee engagement. Lnenicka et al. [3] provide a comprehensive overview of the requirements and suitability of the metaverse for delivering digital public services, including HR functions. Their findings suggest that the metaverse can offer more engaging and efficient HR practices.

Uzun [8] discusses metaverse governance, noting that virtual environments can streamline HR processes by offering immersive training programs and VR-based recruitment. This is further supported by the work of Popescu et al. [25], who highlight how AI can stimulate innovation in educational services and university programs related to PA.

4.3.3 Challenges and Opportunities

While the metaverse presents numerous opportunities for enhancing HRM in PA, several challenges need to be addressed. Mensah [26] identifies legal and regulatory challenges in implementing AI-augmented public health administration in sub-Saharan Africa, which can be extrapolated to the broader context of metaverse-based HRM. Bie et al. [27] examine the impact of government digital transformation on energy efficiency in Chinese cities, indirectly pointing out the need for sustainable practices in the digitalization of public services, including HRM.

Dia [28] emphasizes the importance of developing a governance framework that ensures the commercial success, inclusivity, and safety of the metaverse. This is crucial for HRM practices to be effective and equitable. The interdisciplinary approach proposed by Vigoda-Gadot [29] for intelligent governance also underscores the need for robust frameworks to manage the integration of human and machine interactions in PA.

4.3.4 Emerging Themes and Best Practices

Several emerging themes and best practices were identified from the analysis of the selected articles. The use of VR and AR for immersive training and development programs is a recurring theme. For example, So and Park [30] discuss the application of digital content technology in veterans' diplomacy, illustrating how virtual environments can enhance training and engagement.

Elsamani and Kajikawa [31] highlight the changing dynamics of the labor market and workforce due to teleworking adoption, which can be further transformed by metaverse technologies. This is echoed by Min [32], who discusses management styles under the Fourth Industrial Revolution, emphasizing the need for adaptive leadership in virtual environments.

4.4 Discussion

The primary aim of this mini-review was to investigate the current initiatives and practices related to the metaverse in PA, particularly focusing on HRM. Our findings indicate a significant transformative potential of the metaverse for HRM in PA. By leveraging immersive technologies such as VR and AR, public institutions can enhance employee engagement, streamline recruitment and onboarding processes, and provide more effective training and performance management. However, the integration of these technologies also presents several challenges, including legal, regulatory, and ethical issues.

The findings of this mini-review contribute to the existing body of knowledge by highlighting the practical applications of the metaverse in PA and HRM. Previous studies, such as those by Alahi et al. [24] and Allam et al. [6], have primarily focused on the technological advancements and theoretical frameworks of the metaverse. This review extends the understanding by providing empirical insights into how these technologies are currently being implemented in public sector HRM. The integration of IoT-enabled technologies and AI in smart city scenarios Alahi et al. [24] and the use of VR for citizen feedback [4] illustrate practical applications that were previously underexplored in academic literature.

Furthermore, the work of Lnenicka et al. [3] and Uzun [8] on the requirements and governance of the metaverse for public services provides a foundational framework for understanding how these technologies can be integrated into HRM practices. This review also aligns with the interdisciplinary approach proposed by Vigoda-Gadot [29], emphasizing the need for robust governance frameworks to manage the integration of human and machine interactions in PA.

From a practical standpoint, this review underscores the significant opportunities that the metaverse offers for PA-HRM. Public institutions can utilize VR and AR to create more engaging and efficient HR practices. For instance, virtual environments can facilitate immersive training programs and VR-based recruitment, as discussed by Popescu et al. [25]. These technologies can also enhance employee engagement and performance management by providing interactive and personalized experiences. However, the successful implementation of these technologies requires addressing several challenges. Legal and regulatory issues, as highlighted by Mensah [26], are critical considerations that need to be addressed to ensure the smooth adoption of metaverse technologies in HRM. Additionally, the need for sustainable practices in digital transformation, as pointed out by Bie et al. [27], is essential for the long-term success of these initiatives. Practical recommendations for public sector leaders and HR professionals include developing comprehensive governance frameworks, as suggested by Dia [28], to ensure the inclusivity, safety, and commercial success of the metaverse. Additionally, adaptive leadership styles, as emphasized by Min [32], are crucial for managing the dynamic and evolving virtual environments.

While this mini-review provides valuable insights into the integration of the metaverse in PA-HRM, it is not without limitations. The primary limitation is the scope of the review, which focused on a specific set of articles. This may have resulted in the exclusion of relevant studies that could provide additional perspectives. Additionally, the review relied on available literature, which may not fully capture the rapidly evolving nature of metaverse technologies and their applications.

4.4.1 Avenues for Future Research

The exploration of the metaverse in PA, particularly concerning HRM, is a burgeoning field with immense potential. However, several avenues for future research remain to be explored to fully harness this potential and address the existing gaps in the literature. Firstly, empirical studies are needed to assess the effectiveness of metaverse-based HRM practices. While theoretical frameworks and initial implementations have been discussed [3, 25], there is a lack of empirical evidence demonstrating the impact of these practices on employee performance, engagement, and overall organizational efficiency. Longitudinal studies that track these metrics over time would provide valuable insights into the long-term benefits and challenges of integrating metaverse technologies into HRM.

Secondly, there is a need for comparative studies that evaluate the effectiveness of metaverse-based HRM practices across different PA contexts. Different regions and administrative bodies may face unique challenges and opportunities when adopting these technologies. Comparative research can highlight best practices and identify contextual factors that influence the success of metaverse initiatives in HRM. For instance, studies comparing developed and developing countries could reveal critical insights into how resource availability and technological infrastructure impact the implementation of virtual environments in public sector HRM.

Moreover, the ethical and regulatory dimensions of metaverse integration in HRM warrant further investigation. As highlighted by Mensah [26], legal and regulatory challenges pose significant barriers to the adoption of AI-augmented public health administration in sub-Saharan Africa. Similar challenges are likely to arise with the use of metaverse technologies in HRM. Future research should focus on developing comprehensive regulatory frameworks that address issues such as data privacy, security, and equitable access. Ethical considerations, including the potential for digital exclusion and the implications of virtual surveillance, also need to be thoroughly examined to ensure that metaverse initiatives promote inclusivity and fairness.

Another critical area for future research is the development of robust governance frameworks for the metaverse. Dia [28] and Vigoda-Gadot [29] have emphasized the importance of governance in ensuring the commercial success, inclusivity, and safety of the metaverse. Future studies should explore governance models that can effectively manage the integration of virtual environments into PA-HRM. This includes investigating how existing governance structures can be adapted or new models can be created to oversee the ethical and efficient use of metaverse technologies.

Additionally, research should delve into the technological advancements and innovations needed to support metaverse-based HRM practices. While current studies have explored the potential applications of VR and AR [4, 8], there is a need for ongoing technological development to enhance the functionality and accessibility of these tools. Future research could focus on developing more immersive and user-friendly VR/AR platforms, as well as exploring the integration of other emerging technologies such as blockchain and IoT to further enhance HRM practices in the metaverse.

Furthermore, the impact of metaverse technologies on organizational culture and employee well-being is an important area for future research. The shift to virtual environments can significantly alter workplace dynamics and employee interactions. Studies should investigate how metaverse-based HRM practices affect organizational culture, employee morale, and mental health. Understanding these impacts is crucial for designing HRM strategies that not only leverage the benefits of the metaverse but also foster a positive and supportive work environment.

Finally, interdisciplinary research that combines insights from PA, technology, psychology, and ethics is essential for a holistic understanding of the metaverse's implications. Such research can provide a more comprehensive view of how metaverse technologies intersect with various aspects of PA and HRM. Collaborative efforts between academia, industry, and government can also drive innovation and ensure that research findings are translated into practical applications.

4.5 Conclusion and Limitations

This mini-review has explored the transformative potential of the metaverse in PA, with a specific focus on HRM. By synthesizing insights from 24 key articles, the review has highlighted innovative practices, such as virtual recruitment, onboarding, training, and performance management, while also identifying significant challenges, including legal, regulatory, and ethical issues. The findings underscore the necessity for continued empirical research, the development of robust governance frameworks, and the consideration of ethical dimensions to fully harness the benefits of the metaverse. This study contributes to the ongoing discourse on digital transformation in the public sector and provides valuable insights for policymakers and practitioners seeking to navigate the complexities of integrating the metaverse into their operations, ultimately aiming to enhance public service delivery and workforce management.

References

1. Dionisio, J.D.N., Iii, W.G.B., Gilbert, R.: 3D virtual worlds and the metaverse: current status and future possibilities. ACM Comput. Surv. **45**(3), 1–38 (2013)
2. Mystakidis, S.: Metaverse. Encyclopedia. **2**(1), 486–497 (2022)
3. Lnenicka, M., Rizun, N., Alexopoulos, C., Janssen, M.: Government in the metaverse: requirements and suitability for providing digital public services. Technol. Forecast. Soc. Chang. **203**, 123346 (2024)
4. Anim, P.: Exploring the Citizen Engagement Metaverse: Testing Virtual Reality for Citizen Feedback on Public Space Design Options. The University of Regina (Canada) (2022)
5. Kshetri, N., Dwivedi, Y.K., Janssen, M.: Metaverse for advancing government: prospects, challenges and a research agenda. Gov. Inf. Q. **41**(2), 101931 (2024)
6. Allam, Z., Sharifi, A., Bibri, S.E., Jones, D.S., Krogstie, J.: The metaverse as a virtual form of smart cities: opportunities and challenges for environmental, economic, and social sustainability in urban futures. Smart Cities. **5**(3), 771–801 (2022)
7. Distor, C., Dhaou, S.B., Nielsen, M.M.: Metaverse vs. metacurse: the role of governments and public sector use cases. In: Metaverse vs. Metacurse: the Role of Governments and Public Sector Use Cases. CEUR Workshop Proceedings (2023)
8. Uzun, M.M.: Metaverse governance. In: Metaverse: Technologies, Opportunities and Threats, pp. 231–244. Singapore, Springer Nature (2023)
9. del Campo, G., Saavedra, E., Piovano, L., Luque, F., Santamaria, A.: Virtual reality and internet of things based digital twin for smart city cross-domain interoperability. Appl. Sci. **14**(7), 2747 (2024)
10. Lyons, N.: Talent acquisition and management, immersive work environments, and machine vision algorithms in the virtual economy of the metaverse. Psychosociol. Issues Hum. Resour. Manag. **10**(1), 121–134 (2022)
11. Vochozka, M., Horak, J., Carey, B.: Virtual human resource management, workplace tracking and algorithmic monitoring systems, and cognitive and affective metrics in the immersive metaverse environment. Psychosociol. Issues Hum. Resour. Manag. **10**(2), 119–134 (2022)
12. Zvarikova, K., Cug, J., Hamilton, S.: Virtual human resource management in the metaverse: immersive work environments, data visualization tools and algorithms, and behavioral analytics. Psychosociol. Issues Hum. Resour. Manag. **10**(1), 7–20 (2022)
13. Chevalier, E., Sebastián, E.M.M.: Digitalisation and good administration principles. Eur. Rev. Digit. Adm. Law. **3**, 5–8 (2022)
14. Song, C., Shin, S.Y., Shin, K.S.: Exploring the key characteristics and theoretical framework for research on the metaverse. Appl. Sci. **13**(13), 7628 (2023)
15. Masenya, T.M.: Integrating talent and knowledge management practices in the new normal business environment: developing future leaders in public sector organizations. In: Navigating the New Normal of Business with Enhanced Human Resource Management Strategies, pp. 113–144. IGI Global (2022)
16. McNutt, K.: Public engagement in the Web 2.0 era: social collaborative technologies in a public sector context. Can. Public Adm. **57**(1), 49–70 (2014)
17. Berman, E.M., Bowman, J.S., West, J.P., Van Wart, M.R.: Human Resource Management in Public Service: Paradoxes, Processes, and Problems. CQ Press (2021)
18. Marditama, T., Yusliza, M.Y., Ghani, L.A., Saputra, J., Muhammad, Z., Bon, A.T.: Green human resource management and sustainable organization literature: a mini-review approach. In: 11th Annual International Conference on Industrial Engineering and Operations Management, pp. 3964–3979 (2021)
19. Zen, A.: Mini review: organization commitment, organization sustainability, resilience, community social behavior, transformational leadership. Formosa J. Sci. Technol. **2**(3), 875–890 (2023)
20. Paré, G., Trudel, M.C., Jaana, M., Kitsiou, S.: Synthesizing information systems knowledge: a typology of literature reviews. Inf. Manag. **52**(2), 183–199 (2015)

21. Pautasso, M.: Ten simple rules for writing a literature review. PLoS Comput. Biol. **9**(7), e1003149 (2013)
22. Grant, M.J., Booth, A.: A typology of reviews: an analysis of 14 review types and associated methodologies. Health Inf. Libr. J. **26**(2), 91–108 (2009)
23. Hoon, C.: Meta-synthesis of qualitative case studies: an approach to theory building. Organ. Res. Methods. **16**(4), 522–556 (2013)
24. Alahi, M.E.E., Sukkuea, A., Tina, F.W., Nag, A., Kurdthongmee, W., Suwannarat, K., Mukhopadhyay, S.C.: Integration of IoT-enabled technologies and artificial intelligence (AI) for smart city scenario: recent advancements and future trends. Sensors. **23**(11), 5206 (2023)
25. Popescu, R.I., Sabie, O.M., Trușcă, M.I.: The contribution of artificial intelligence to stimulating the innovation of educational services and university programs in public administration. Transylv. Rev. Adm. Sci. **19**(70), 85–108 (2023)
26. Mensah, G.B.: AI-augmented public health administration in sub-Saharan Africa: addressing challenges in Ghana's cyberlaws regimes for smooth and effective use. Int. J. Legal Sci. Innov. **5**(6), 26–54 (2023)
27. Bie, F., Zhou, L., Liu, S., Yang, T.: Government digital transformation, resource curse and green total factor energy efficiency in Chinese cities. Resourc. Policy. **92**, 105002 (2024)
28. Dia, H.: Developing a Governance Framework for a Commercially Successful, Inclusive, and Safe Metaverse. Doctoral dissertation, Walden University (2023)
29. Vigoda-Gadot, E., Mizrahi, S.: The digital governance puzzle: towards integrative theory of humans, machines, and organizations in public management. Technol. Soc. **77**, 102530 (2024)
30. So, B., Park, H.: Application of digital content technology for veterans diplomacy. J. Public Dipl. **3**(2), 35–52 (2023)
31. Elsamani, Y., Kajikawa, Y.: How teleworking adoption is changing the labor market and workforce dynamics? PLoS One. **19**(3), e0299051 (2024)
32. Min, X.U.: The management style under the fourth industrial revolution. J. Public Adm. Finance Law. **29** (2023)

Review Articles

Atianashie, M., Adaobi, C.C., Adinkrah-Appiah, K., Obeng-Ofori, D.: Digital technologies: a congregating pan-African public management on the fourth industrial revolution for employability and sustainable growth. J. Educ. Technol. **8**(1) (2024)

Balcerzak, A.P., Nica, E., Rogalska, E., Poliak, M., Klieštik, T., Sabie, O.M.: Blockchain technology and smart contracts in decentralized governance systems. Adm. Sci. **12**(3), 96 (2022)

Bibri, S.E., Huang, J., Krogstie, J.: Artificial intelligence of things for synergizing smarter eco-city brain, metabolism, and platform: pioneering data-driven environmental governance. Sustain. Cities Soc., 105516 (2024)

Khanina, O., Hladchenko, A., Lavrentii, D., Kuzmenko, H., Kruhlov, V., Kryzyna, O.: Mechanisms of state management of personnel development of the health care system. Econ. Aff. **69**, 187–195 (2024)

Park, S., Kudo, H.: Cultural institutions as knowledge-intensive public organisations (KIPOs) and their role: how digitalisation can change them providing sustainable and accessible public services. J. Public Budg. Account. Financ. Manag. (2024)

Qadir, A.M.A., Fatah, A.O.: Platformization and the metaverse: opportunities and challenges for urban sustainability and economic development. EAI Endorsed Trans. Energy Web. **10**(1) (2023)

Vărzaru, A.A., Bocean, C.G., Simion, D., Berceanu, D., Mangra, M.G.: Digital revolution, sustainability, and government revenues: a transversal analysis of how digital transformation and sustainable practices impact sustainable government revenues. Systems. **11**(11), 546 (2023)

Vigoda-Gadot, E.: Can governance be intelligent?: an interdisciplinary approach and evolutionary modelling for intelligent governance in the digital age. Elem. Public Nonprofit Adm. (2024)

Vrabie, C.: E-government 3.0: an AI model to use for enhanced local democracies. Sustain. For. **15**(12), 9572 (2023)

Yin, B., Shen, Y.: Compensatory beliefs in the internet gratification behavior: a study of game-based assessment. Front. Public Health. **11**, 997108 (2023)

Chapter 5
AI-Aided Creativity for Technological Innovation and Sustainability: Proof of Concept of a Digital Platform to Support Decisions in the Field of Daylighting Design

Federica Giuliani, Andrea Zingoni, Jurji Filieri, and Giuseppe Calabrò

Abstract Artificial intelligence (AI) is likely to be increasingly present in the future of building design practices. Indeed, its use offers significant advantages, including reducing project execution times and the replicability of design solutions. The primary application is through generative artificial intelligence (Gen AI). However, the indiscriminate use of Gen AI poses significant risks, including potential loss of control over the design process, lack of originality, and overconfidence in automation. As a result, a rethink is needed on how AI can be leveraged in architectural design contexts by imagining other applications to exploit its potential, which could open new and more human-centric scenarios. An alternative exploration pathway is utilising non-generative AI (Non-Gen AI) in architectural design. This chapter presents SOLARIA, an applied research project that proposes Non-Gen AI integrated into sustainable architectural design. SOLARIA mainly focuses on improving daylighting, even if it can be easily extended to other aspects of architectural design. The project output will be a digital platform, embedding a decision support system (DSS) that allows designers to make informed choices in the early design stage to increase daylighting and integrate sustainable design elements into their projects without delegating all the creative processes to the machine. A proof of concept for implementing the digital platform is provided here, highlighting its strengths, limitations, and expected results.

Keywords Artificial intelligence (AI) · Sustainable architecture · Generative AI · Non-generative AI · Decision support system (DSS) · Daylighting design

F. Giuliani (✉) · A. Zingoni · J. Filieri · G. Calabrò
Università degli Studi della Tuscia, Viterbo, Italy
e-mail: federica.giuliani@unitus.it

F. Schiavone et al. (eds.), *Advanced Perspectives and Trends in Digital Transformation of Firms, Networks, and Society*, Springer Proceedings in Business and Economics, https://doi.org/10.1007/978-3-031-80692-6_5

5.1 Introduction

Artificial intelligence (AI) in architecture offers excellent potential for designing sustainable and highly efficient buildings [1]. In this sense, generative AI or Gen AI is among the most popular trends [2]. Such a technology starts from specific constraints and predetermined objectives to optimise the architect's project's performance, energy efficiency, and functionality. Among the positive aspects are the optimisation of building standard verification parameters, the adaptability of solutions to the client's needs, the reduction of time in the design process, and the related improvement in user competitiveness, which are helpful and have progressively gained attention and success. In terms of sustainability, they can also move towards more eco-friendly solutions that consider the environmental impact of the building and its entire life cycle at an early stage of the design process. Despite this, the indiscriminate use of Gen AI in architecture entails risks that must be considered. First, there is the risk of a potential loss of control by designers over the entire design process. It can lead to a lack of originality in solutions and an undesirable uniformity of solutions due to the recurring compositional patterns that AI tends to propose [3]. Another risk is contextualisation, which involves complex "human factors" such as society, culture, and environment [4].

Moreover, there could be the risk of solutions that may be overly complex because AI response does not always follow a simplicity paradigm, and it is not easy to control, limit, or stop the designer at a certain point [5]. This could increase time and costs considerably. One must not underestimate the errors of algorithms, which, on the one hand, are not flawless yet and, on the other hand, are anyway subjected to human input faults. Finally, there is also a potential risk of an over-reliance on AI generative approaches, which could bring a dependency on this technology, leading to new generations of designers unable to control the design process and no longer able to provide a valid, autonomous, analytical, and cultural contribution [6].

Therefore, despite all the advantages AI can bring to architectural design, Gen AI presents many risks that must be addressed.

This chapter addresses the previously mentioned aspects and proposes a different way to apply AI in sustainable architectural design. This method includes humans in the loop and gives them a centric role in the creative process. It can be considered a Non-Gen AI methodology.

The SOLARIA project (an acronym in Italian that stands for *Sistema Operativo di supporto alle scelte progettuali per La progettazione ARchitettonica basata sull'Intelligenza Artificiale*, namely "an operative system to support project choices in architectural design, based on AI"), here presented, attempts to do this by developing a decision support system (DSS) to help (not overlap) human design choices, in favour of a sustainable approach in line with the Sustainable Development Goals (SDGs), through a customised AI algorithm that takes into account and tries to overcome the inherent risks present in this technology. SOLARIA aims to enhance environmental sustainability by using daylight in living spaces, although its approach can be extended to other fields.

5.1.1 Literature Review

In architecture, there have always been questions about the best way to design human environments effectively and innovatively. In this sense, there has always been a succession of design techniques and the introduction of new tools and methodologies to address increasingly complex challenges. This theme has reached a new frontier by exploring AI's role in architectural design. There are growing questions about how this technology can be used to optimise the creative process, improve building performance, and promote sustainability. This literature review analyses how architectural design techniques were interpreted before the advent of AI and how it is changing the paradigms in the AI era.

The architectural design before the AI era. The theories underlying architectural design have evolved since ancient times, with continual attempts to find new paradigms and valid interpretations of the science of designing, a complex endeavour encompassing methodological, cognitive, cultural, and many other aspects. Between the 1960s and 1970s, there were many significant theoretical developments in design methodologies in architecture, among which the most important were those of Morris Asimoz [7] and Christopher Jones [8]. Jones introduced two types of design that have succeeded in the past century.

(1) The era of "Design by Drawing", which spans from the Renaissance to the 1950s, now referred to as the "black-box" method, implies that the design process was not visible to anyone and led to a result without explicitly stating the cognitive path that led to this outcome.
(2) The era of "System Designing", which began in the twentieth century, where the design process emerged from the darkness of the "black box" and transformed into a conscious group effort. This phase represents a step towards greater complexity in the design process, as design based on drawing becomes a tool for solving a sub-problem within a more extensive system. More recently, this approach has been included in the "information-based design" category. Here, the design starts from having a certain amount of information that determines subsequent design choices. This type of design can be defined as a deterministic system because it seeks to reduce uncertainty, provide definite answers based on certain premises, and identify the parameters involved in the project [9].

However, there are two potential design risks [10]—the issue of information overload and distinguishing which information is necessary for the design process. Information overload has been recognised for decades under the term "information overload anxiety" [11]. To address this risk, simplified methods have emerged in the design field for the early stages of the design process, where having less but valuable information is conducive to making informed design choices. About daylighting, methods such as the LT Method [12, 13] and LightSolve [14, 15] have been developed to help the preliminary design consider lighting and energy aspects.

Another alternative to information-based design is evidence-based design, or EBD [16]. EBD is an approach that relies on using the best available evidence to make design decisions. It gained popularity through Ulrich's study (1984), which demonstrated the impact of views from windows on patient recovery [17]. Studies in this field have evolved, particularly in hospitals, showing how the physical environment can positively influence health outcomes [18]. EBD is closely related to performance-based building design (PBBD), which seeks to establish transparent statistical relationships between design decisions and satisfaction levels demonstrated by building systems [19, 20]. Other related approaches include evidence-based medicine or EBM [21].

The **architectural design in the AI era.** In recent years, significant progress in research and the development of generative models has radically transformed how designers conceive and realise their projects. A notable increase in papers reporting research on Gen AI in architectural design has been observed since 2020 [2]. Using the article classification criteria applied by Li et al., architectural design has been divided into six phases where AI and Gen AI can have a decisive intervention: (1) architectural preliminary 3D forms design, (2) architectural layout design, (3) architectural, structural system design, (4) detailed and optimisation design of architectural 3D forms, (5) architectural facade design, and (6) architectural imagery expression. Compared to these stages, most research is concentrated in the areas of (1) preliminary design of architectural 3D forms and (6) architectural imagery expression. In contrast, areas with limited investigation include architectural structural design (3), detailed optimisation of architectural forms (4), and architectural facade design (5), where Non-Gen AI models prevail due to their analytical support in aiding architects with data analysis, simulation, and visualisation.

Generative adversarial networks (GANs), variational autoencoders (VAEs), and diffusion models (DMs) are primarily involved in the rapid advancement of Gen AI models. A GAN generator creates images, and the discriminator evaluates their authenticity, while VAEs are probabilistic models that learn a latent representation of images [2].

Therefore, the GANs and VAEs underpin new methodological paradigms in architectural design, allowing designers to explore a broader range of creative solutions and optimise the efficiency of the design process.

However, the authors express concerns about the current state of Gen AI applications in architectural design. They emphasise that current Gen AI applications, especially those generating images used as architectural drawings, are distant from both architectural professional practice and the methodological process of architectural design. Current Gen AI models are limited in interpreting and adequately responding to these multidisciplinary complexities, generating responses based on learned patterns during training rather than a profound understanding of the design process [6].

Moreover, the widespread adoption of generative AI technologies has raised psychological and social concerns. Recent studies have revealed strong biases against AI-generated artworks, indicating a preference for works created by human artists. This phenomenon suggests an unease in accepting that machines can replicate creativity, challenging traditional conceptions of human uniqueness and artistic creativity [22, 23].

Not all researchers believe that the problem is only ethical and social, but that Gen AI could influence the design process, with architectural visualisation directly impacting it. Other ways of integrating AI into the design process could come from Non-Gen AI. Non-Gen AI does not involve the creative generation of new outputs such as images, text, and 3D models. Still, its capabilities suggest a broad spectrum of applications that could aid in the design process and management of architectural projects during documentation and construction phases [4, 5].

Others suggest that Gen AI could be integrated into the architectural design process in later stages rather than initiating it. They also propose some potential Gen AI applications that could add value to the architectural design process, emphasising the need to differentiate between the product of Gen AI models and the product of architect-controlled generative design [24, 25]. Chaillou argues that the creation of an appropriate pipeline is critical to the success of AI as an architectural tool, favouring "grey boxing" over "black boxing" approaches to allow users to intervene at different stages of the process [26].

In conclusion, there is no general consensus on what the role of AI should be in the architectural design process. There are concrete doubts about the correctness of using Gen AI in the process. At the same time, it is undeniable that Non-Gen AI can be a valid support in facing increasingly complex challenges and is also in line with the sustainability requirements for the future.

It is overcoming the potential "black-box" nature of Gen AI [cfr. 8, 26] and finding new applications for AI in the design process will be the challenge of the future [27].

5.2 The SOLARIA Project

The SOLARIA project aims to build an AI-based DSS to help designers make informed choices about integrating daylight and sustainable design elements into their projects.

The operational goal of SOLARIA is to harness AI's great potential and put it at the service of human creativity. The intent is to generate a system that can help provide helpful information early in the design process. The purpose is to create a design idea prompter that can trigger autonomous reflections in the designer and activate their search for original design solutions.

The application area of daylighting was chosen because, as many studies have shown, there is a significant gap in knowledge about this topic and a tendency to resort to design solutions that do not effectively control this valuable natural resource. The DSS should help answer the question, "What elements should a designer rely on to increase indoor daylighting levels?"

From a technological point of view, the goal is to satisfy two needs: understanding what to do concerning the complex design framework (cultural information) and choosing how to do it in the landscape of available and accessible options (technical information). Therefore, DSS provides a double level of information.

The SOLARIA DSS does not aim to provide "exact" answers but to broaden the cultural horizon of possibilities by giving information (not data) about the designer's real case, thus inspiring them in the creation process.

5.3 The DSS Methodology

The DSS aims to support the architect in the early stages of architectural design. Specifically, the model seeks to suggest design solutions (case studies) and technical solutions (devices) that the architect can evaluate for their project to increase natural light. The case studies are intended to be understood as WHAT could be used as a design strategy, while the devices are HOW these strategies could be implemented by applying specific equipment.

In application terms, this resulted in a DSS providing a double level of information in the form of two kinds of fact sheets:

1. The fact sheets on WHAT to do present case studies of existing architectures that show factors of relevance/effectiveness concerning daylighting for inspiration.
2. The fact sheets on HOW to do suggest technical solutions and architectural devices that can be used to achieve the above.

The designer will use the DSS to evaluate design alternatives based on previously implemented cases and assess how to apply them using technical information on existing devices. This approach is inspired by evidence-based design (EBD) and the case-based reasoning (CBR) theories [10, 20].

Such fact sheets are collected in two databases created ad hoc. Helpful data is taken from online archives, which are now being evaluated (Fig. 5.1).

An essential part of the model is how it extracts data from the platform. The architecture of this part mirrors how a designer thinks and seeks to provide helpful information. The data extraction is based on queries, and the cases provided answer the designer's design problem.

Queries are divided into three levels of problem approach: spatial issues, physical constraints, and quality analysis. Spatial issues concern actual facts of the environment where the project is located and are related to obstacles, glare, thermal solar gain, and other location characteristics that may influence the choices of strategies and devices. On the other hand, physical constraints refer to building characteristics that can be modified, such as the possibility of intervening on the roof (Table 5.1). If modifications are not possible, for example, some solutions like skylights cannot be used. Lastly, qualitative evaluations pertain to perceptual aspects of strategies and devices, which must be compared once the system has been applied to make informed choices. In other words, through qualitative assessments, a designer can determine why a particular device fits their case better. The designer responds to the questions based on their actual problem, and the system extracts only the cards that correspond to the designer's problem.

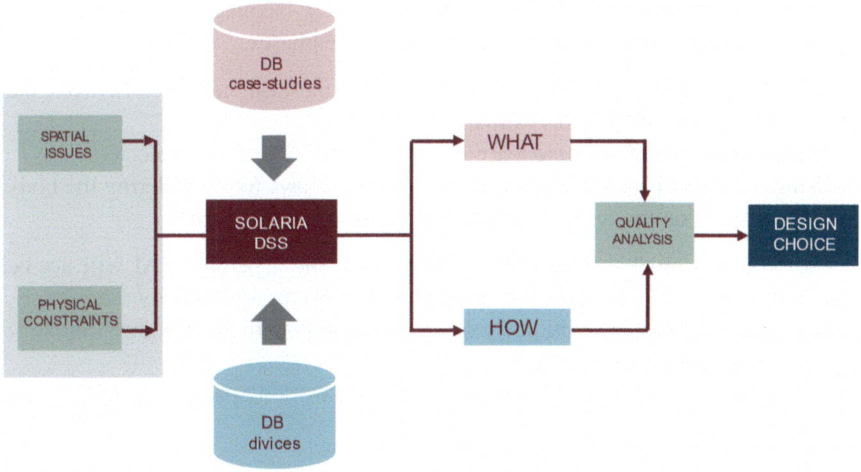

Fig. 5.1 SOLARIA DSS data extraction scheme from user/designer input

Table 5.1 List of queries on spatial problems and physical constraints contained in the DSS

Code	Parameter	Parameter
Spatial issue		
SI.01	Field of view	*Is there a visual obstruction?*
SI.02	Glare	*Is it possible to see a portion of the sky?*
SI.03	Solar gain	*Is there direct sunlight in the ambient?*
SI.04	Luminance	*Is there glare?*
SI.05	Illuminance	*Is there enough light in the room in general?*
SI.06	Fixed position	*Are there fixed positions in the room for residents* (e.g. *in the office)?*
Physical constraints		
PC.01	Orientation	*Room orientation*
PC.02	Ceiling	*Can the ceiling be modified?*
PC.03	Roof	*Can the roof be perforated?*
PC.04	Facade #1	*Can the façade be modified?*
PC.05	Facade #2	*Can elements be inserted into the façade?*
PC.06	Facade overhang	*May the fixtures be modified?*
PC.07	Opening size	*Can the size of the openings be modified?*
PC.08	Window frames	*Can the window frames be replaced?*
PC.09	Internal finishes	*Can the internal finishes be modified?*
PC.10	Shading systems control	*Is light flow control necessary?*

The model intends to use the SQL (structured query language) standard and will rely on artificial intelligence models applied in three phases.

3. Natural language processing (NLP) associates the designer's needs, from the questionnaire to specific keywords.
4. Using unsupervised machine learning algorithms to cluster the selected cases.
5. Using a content-based recommendation system allows for considering the body of knowledge on the topic in the final suggestion to the designer.

The proposed model wants to be non-deterministic. Therefore, AI will not be exploited to "provide" a specific solution decided purely algorithmically. Conversely, it will "suggest" possible solutions in similar cases to inspire the designer's creative process and validate their choices.

The construction of the DSS is divided into four working packages (WP):

- WP1: the creation of databases
- WP2: the construction of informative sheets
- WP3: the fine-tuning of the DSS architecture
- WP4: the validation of the model

Based on the initial results obtained, comparative analyses of data extraction methods will be conducted, and theoretical modifications to the model may be considered.

Finally, to validate the model, a phase of administering the tool to designers and students of design courses is planned, which, through a qualitative evaluation system, should make the tool more robust.

5.3.1 WP 1 Progress

The SOLARIA project is ongoing, and WP1 is being realised.

This initial investigation phase aims to build a repository of reliable case studies to train the machine.

Initially, it was decided to start with databases selected from the network that could provide examples of architectures (WHAT database). However, from an initial recognition, it was clear that it was not easy or even impossible to find case studies that contained meaningful information about the daylighting design of constructed buildings. Therefore, it was decided to start with websites of daylighting device productions, such as light pipes, windows, and solar mirrors (HOW database). This choice is motivated by the fact that starting from the devices, it was possible to indirectly implement the WHAT database backwards. Each device application is contained in a real case, bringing information about the buildings in which it is incorporated. Based on this re-modulation of the research principle, the double database of preliminary cases is currently being built.

Such first repository was considered reliable because all its case studies contain a dual source of information regarding HOW (built building) / WHAT (device applied in the actual example). This repository is called the Primary Repository (PR). It contains an initial sample of 700 cases from approximately 20 different commercial websites to train the machine.

The search for commercial websites with the required level of HOW / WHAT information cannot be automated because it is too complex and discretionary. Therefore, humans trained to recognise the appropriate characteristics needed in the case studies select sources.

Since the information in PR is not always sufficient to adequately inform a designer and make informed design choices, a second repository has been constructed. This is called the Secondary Repository (SR), which contains general but fundamental notions to increase the data gaps in the case studies included in PR.

This second repository comprises online architecture magazines, design studio websites, online archives of industry libraries, professional associations, and other dissemination sources.

The AI's task will be to select the case studies from PR, integrate them with the information in SR, and incorporate them into the various fields of an informational sheet template.

If a case study in PR does not find a corresponding data match in SP, it is discarded. Subsequently, the AI can conduct cyclical searches in the sources of PR and SP to add new cases to those initially built or discarded.

AI implementation in WP1. In the first phase, an NLP model will be exploited to connect the designer's needs, extracted through the questionnaire, to a first-course group of sheets presenting existing works related to such needs.

Thus, a 0/1 entries vector, as large as the number of questionnaire items, is created. Each entry corresponds to a specific item. An ONEDRIVE0 is assigned to each item answered with a "NO", whereas a one is assigned when the answer is "YES". Such a vector will be compared with a look-up table (LUT) in which each possible answer for each possible question is associated with one or more keywords related to the answer given to each question and with project requirements emerging for the bundle of answers. In this way, a group of meaningful keywords for the designer's choices is extracted.

Then, such keywords are given to an NLP model, which also receives the database as input and provides a score to each document based on the presence and the frequency of the selected keywords within. It is worth noting that the model searches for the keywords and their synonyms and semantically identical expressions. Thus, two steps are essentially performed by the NLP model, namely (i) increasing the number of keywords with other words/expressions that have the same meaning and (ii) finding them and calculating their frequency in each document. Pre-trained state-of-the-art NLP models and a new customised one will be considered to reach the optimum solution.

When each document has been classified with its score, the winning N will be given to the designer to be used by them as a source of inspiration. The value of N can be chosen by the designer, even if a range of recommended values will be provided, depending on the results of the scoring step.

5.3.2 Next WPs

(WP2) The construction of informative sheets: Once the databases have been completed and the machine learning process started, the information sheets will be constructed.

To create the WHAT cards, online pages will be outlined with fields the AI will fill in by linking data extracted from various reliable sources.

A specific system will be created to eliminate cards not covering the fields deemed essential for presenting the information sheet to the designer (project data, design technique involving natural light, association of the project with a specific device).

An estimated fact sheet/web page will be created in a repository for each valid case study. The machine will periodically review the repository to see if new cases have been added.

Similarly, HOW cards (devices) will be constructed; in this case, the construction criteria will differ according to the type of product presented.

(WP3) Fine-tuning the DSS architecture. Based on the initial results obtained, comparative analyses of data extraction methods will be conducted, and theoretical modifications to the model may be considered. Based on the questionnaire submitted to the user/designer, the algorithm for extracting the informative sheets will be studied in detail.

(WP4) The validation of the model. Finally, to validate the model, a phase of administering the tool to designers and students of design courses is planned, which, through a qualitative evaluation system, should make the tool more robust.

5.4 Discussion and Implication

From the initial results obtained with the SOLARIA project, several reflections can be made regarding the challenges of considering a different role for AI in the design process.

Many of them came from collecting reliable data and the complexity of extracting information from architectural case studies.

To provide designers with solid, informative sheets, it is essential first to understand what they need to know from the perspective of the design process. This implies a clear understanding of the design process, including which decisions are critical at each stage and how they should be provided. While this may seem like a straightforward process, all these steps require a high level of interdisciplinary and specialised knowledge and an elevated awareness of the cognitive processes involved in the creative process during the construction phase of the databases.

This implies that it is a process built by human intelligence and destined for human intelligence. The process is not easily automated because it would involve a high degree of approximation.

Since the SOLARIA method is potentially applicable to several design processes where cultural references are desired without directly influencing choices, once the model is completed, it can be replicated in other thematic areas (such as environmental energy) to assess whether there is a vital principle of replicability.

5.5 Conclusion

SOLARIA proposes an alternative approach to using AI in architecture to improve the sustainability of the built environment through a methodological approach that combines AI-based databases with design principles based on EBD and CBR.

The project provides designers with a DSS that facilitates informed choices regarding integrating daylight and energy-efficient design elements. This would lead to incorporating AI into sustainable development initiatives in a more human-centred way.

Rather than generating design solutions, the DSS invites case studies and application technologies. This forces the designer to research their design theme, making them more professionally informed and accurate in their choices.

Moreover, the proposed DSS is more comprehensive than its use in daylighting design. In the future, other areas of investigation, again with a view to sustainability, can be implemented by creating a specific set of databases and queries, such as improving solutions for the recovery of rainwater, biophilia, and active solar heating systems.

By guiding designers towards more informed design decisions, SOLARIA aims to promote sustainable architectural practices in line with the overall goals of the 2030 Agenda for Sustainable Development.

References

1. Newton, D.: Generative deep learning in architectural design. Technol.| Archit.+ Des. **3**(2), 176–189 (2019)
2. Li, C., Zhang, T., Du, X., Zhang, Y., Xie, H.: Generative AI for Architectural Design: A Literature Review. http://arxiv.org/abs/2404.01335 (2024)
3. Salem, A.A., Mansour, Y., Eldaly, H.: Generative vs. Non-Generative AI: analyzing the effects of AI on the architectural design process. Eng. Res. J. (Shoubra). **53**(2), 119–128 (2024)
4. Ayman, A., Mansour, Y., Eldaly, H.: Generative vs. Non-Generative AI: analyzing the effects of AI on the architectural design process. Eng. Res. J. (ERJ). **53**, 119–128 (2024) Retrieved from https://erjsh.journals.ekb.eg
5. Muehlbauer, M., Song, A., Burry, J.: Towards intelligent control in generative design. In: Future Trajectories of Computation in Design, Proceedings of CAAD Futures, pp. 630–647 (2017)
6. Simkute, A., Tankelevitch, L., Kewenig, V., Scott, A.E., Sellen, A., Rintel, S.: Ironies of Generative AI: Understanding and mitigating. arXiv preprint arXiv:2402.11364 (2024)
7. Asimov, M.: Principi di progettazione. Marsilio Editore, Padova (1968)

8. Jones, J.C.: Design Methods: Seeds of Human Future. John Wiley and Sons, New York and Chichester (1970)
9. Nesi, A.: Progettare con l'informazione. Percorsi e gestione delle informazioni tecniche per la promozione e il controllo dell'innovazione nei materiali e nel progetto di... Ediz. illustrata. Gangemi Editore (2009)
10. Morabito, G.: Scienza e arte per progettare l'innovazione in architettura: saggio su un processo progettuale alla Leonardo da Vinci. UTET libreria, Torino (2004)
11. Wurman, R.S.: Information Anxiety. Doubleday, New York (1989)
12. Baker, N., Steemers, K.: LT method 3.0 — A strategic energy-design tool for southern Europe. Energ. Buildings. **23**(3), 251–256 (1996). https://doi.org/10.1016/0378-7788(95)00950-7
13. Antonini, E., Zacchei, V.: A simplified tool to predict the shading effect of multi-functional surfaces and active coatings on the building energy behaviour. In: AES-ATEMA International Conference Series – Advances and Trends in Engineering Materials and Their Applications, pp. 57–68 (2013)
14. Andersen, M., Kleindienst, S., Yi, L., Lee, J., Bodart, M., Cutler, B.: An intuitive daylighting performance analysis and optimisation approach. *Build Res Inf.* ***36***(6), 593–607 (2008) Routledge
15. Kleindienst, S., Andersen, M.: Comprehensive annual daylight design through a goal-based approach. Build. Res. Inf. **40**(2), 154–173 (2012)
16. Stichler, J., Hamilton, D.K.: Evidence-based design: what is it? Health Environ. Res. Des. J. **1**(2), 3–4 (2008)
17. Ulrich, R.: View Through a Window May Influence Recovery from Surgery. Science, vol. 224, pp. 420–421, New York (1984). https://doi.org/10.1126/science.6143402
18. Cama, R., Herbert, C.L., Mare, G.C., Zimring, C.: Dublin Methodist Hospital: Applying evidence-based design in a race to revolutionise healthcare. Healthcare Design. https://healthcaredesignmagazine.com/trends/architecture/dublin-methodist-hospital-applying-evidence-based-design-race-revolutionize-healthcare/ (2007)
19. Morabito, G., Giuliani, F., Marrone, P., Zacchei, V.: Design training and education using an evolutionary process training experiences in technological design. Conf Proc. EAAE, Istanbul. 439–452, 17–19 June (2010)
20. Morabito, G., Marrone, P., Zacchei, V.: Designing Safer Buildings: towards Evidence-Based Decisions. CIB World Building Congress (2013)
21. Sackett, D.L., Rosenberg, W.M.C., Muir Gray, J.A., Haynes, R.B., Richardson, W.S.: Evidence-based medicine: what it is and what it isn't. Br. Med. J. **312**, 71–72 (1996)
22. Millet, K., Buehler, F., Du, G., Kokkoris, M.D.: Defending humankind: anthropocentric bias in the appreciation of AI art. Comput. Hum. Behav. **143**, 107707 (2023)
23. Ragot, M., Martin, N., Cojean, S.: AI-generated vs. human artworks. A perception bias towards Artificial Intelligence? In: Extended Abstracts of the 2020 CHI Conference on Human Factors in Computing Systems (CHIEA '20). Association for Computing Machinery. https://doi.org/10.1145/3334480.3382892
24. Bao, Y., Xiang, C.: Exploration of Conceptual Design Generation based on the Deep Learning Model-Discussing the Application of AI Generator to the Preliminary Architectural Design Process. Paper presented at the xArch symposium-Creativity in the age of digital reproduction, Xi'an Jiaotong-Liverpool University, Suzhou, China (2023)
25. Han, S., Jiang, Y., Huang, Y., Wang, M., Bai, Y., Spool-White, A.: Scan2drawing: for as-built model landscape architecture. J. Constr. Eng. Manag. **149**(5), 04023027 (2023)
26. Chaillous, S.: AI & Architecture—an Experimental Perspective. Harvard University, Cambridge (2019)
27. Main, J., Dillon, T.S., Shiu, S.C.K.: A tutorial on case based reasoning. In: Pal, S.K., Dillon, T.S., Yeung, D.S. (eds.) Soft Computing in Case Based Reasoning. Springer, London (2001). https://doi.org/10.1007/978-1-4471-0687-6_1

Chapter 6
AI Among the Elderly, Use and Adoption: A Domain-Based Literature Review

Gogiali Giulia, Nevi Giulia, Dezi Luca, and Gamma Federico

Abstract Artificial Intelligence (AI) has made great progress in several fields. However, its integration among the elderly remains an under-explored area in the existing literature. The increase in the world's elderly population is a huge challenge and the adaptability of AI makes it a promising tool to improve the mental and physical well-being of the elderly during the ageing process. This research aims to examine the prevalent applications of AI in elderly care across various domains and to investigate the factors that facilitate its adoption among older adults. Through a systematic literature review, the current landscape of AI use in the elderly population is analysed and summarized. In addition, a comparative analysis of the theories and concepts that constitute the theoretical frameworks in the literature is conducted to identify recurring factors influencing AI adoption. The study provides insights into contemporary trends in AI adoption among older adults to clarify the factors influencing it and relevant challenges. This systematic literature review serves as a basis for further exploration in this evolving field, suggesting potential future avenues of research to probe the expanding role of AI in the lives of older adults.

Keywords AI · Artificial Intelligence · Elderly adoption · Silver economy · Healthcare · Domain-based literature review

G. Giulia (✉) · D. Luca · G. Federico
Sapienza University of Rome, Rome, Italy
e-mail: giulia.gogiali@uniroma1.it

N. Giulia
Polytechnic University of Marche, Ancona, Italy

F. Schiavone et al. (eds.), *Advanced Perspectives and Trends in Digital Transformation of Firms, Networks, and Society*, Springer Proceedings in Business and Economics, https://doi.org/10.1007/978-3-031-80692-6_6

6.1 Introduction

The use of Artificial Intelligence (AI) has seen significant growth in recent years, proving effective in a wide range of tasks, including speech recognition, personalized recommendations, problem-solving, and data processing [1]. AI has found applications in different professional and business domains, from optimizing human resource management and customer relations to its use in medical practices and healthcare [2–5]. Despite the widespread adoption of AI in many areas of human activity, its usage among the elderly has been relatively unexplored. The increasing elderly population and people with disabilities pose a growing global challenge, exacerbated by rising life expectancy [6]. This lack of attention towards the use and motivating factors for the elderly to use AI represents a gap in current research on the subject [1] and calls for action to ensure social system stability and explore new business opportunities. The market for AI targeting the elderly population is steadily growing, currently valued at USD 719 million, and this figure is expected to increase due to the growing demand. The elderly population is projected to reach 2.1 billion in 2050 [7]. This demand presents both opportunities and challenges, including addressing health conditions and ensuring the elderly's ability to engage with increasingly advanced technology. The implications of aging are diverse, not only impacting economic growth through the rising demand for elderly-focused funding but also influencing the fiscal and social balance of nations [8]. Ensuring adequate well-being during the aging process is crucial to mitigate financial burdens and healthcare costs, emphasizing the need for elderly citizens to lead peaceful and comfortable lives. From promoting physical and mental health to facilitating an independent and active lifestyle, AI offers significant opportunities to support the elderly [9]. In this perspective, the term "Agetech" denotes the use of technologies and services aimed at promoting active aging, including AI [10]. However, challenges such as the digital divide and resistance to technology persist. Studies on the interaction between AI and the elderly are scarce and fragmentary [11]. On these bases and the need to develop new and better conceptualization on AI application for elderly, the study aims to address the following research questions (RQs):

RQ1: What are the main domains of the use of AI for elderly care?
RQ2: What are the factors influencing the adoption of these technologies by the elderly?

This study proposes a systematic review of literature to address its research questions [12] and using a domain-based approach [13], it provides insights into AI implementation in elderly care. While AI permeates various fields, its application in elderly care represents a crucial but under-explored frontier, which requires academic attention and strategic exploration. Through an in-depth analysis that includes the systematic categorisation of frameworks and measures applied in the adoption of AI, the study establishes a fundamental understanding for scholars, practitioners, and policy makers. The study highlights AI's potential to enhance both physical and mental well-being among the growing elderly population, alleviating strain on national systems. Acknowledging the heterogeneity of the elderly population and the importance of customized solutions, the study supports flexible and customized approaches in the development of AI systems, chatbots, and assistive robots. The paper starts with an overview of AI's use in elderly-related areas, followed by methodology, literature analysis, and coding to address RQ1. It conducts a comparative analysis of theoretical frameworks for RQ2. Results, implications, and limitations are discussed, concluding the paper and providing new opportunities for future research.

6.2 Literature Review

The term "AI" was first introduced in 1955 by John McCarthy, who proposed AI as a distinct research field, envisioning that 1 day machines would speak, form abstract concepts, and solve problems for which, until then, only humans had been able to find solutions [14]). AI, often compared or considered superior to human capabilities, shows flexibility and self-adaptation. While capable of handling different tasks, learning, making predictions and interacting with its environment, there is no comprehensive AI capable of performing all functions simultaneously [15]). AI solutions operate within defined domains, guided by human-created algorithms in a language understandable to AI [16]. Despite the concept of AI being a subject of research for decades, there is still no single universal definition. Some definitions have a mathematical and computational orientation, while others focus on intelligence and cognition [17]. According to Liu ([18], p. 1133), AI "constitutes a relevant component of modern computers". It can simulate human behaviour and thinking to deal more effectively with complex problems and simplify intricate situations. AI is characterized by strong integration, as it can assimilate various new technologies, including psychology, linguistics, and logic. According to Cusack and Quintos ([17], p. 4), AI "is the collection of mathematical and statistical methods operationalized in software allowing for outputs to be generated absent formal programming from inputted data". Furthermore, the term AI encompasses several related areas because AI is a broad term that includes subfields such as machine learning (ML), deep learning (DL), data mining, research, and optimization. Thanks to its learning and adaptation capabilities, AI can revolutionize the quality of life of the elderly, optimizing crucial aspects such as physical and mental health [19], promoting independence in daily life [20], and even facilitating social interactions [21].

Through a range of technologies such as virtual assistants, wearable devices, and remote monitoring, AI offers the possibility to create a personalized and proactive support environment [10]. AI capabilities have the potential to address the growing demand for healthcare by the elderly, bridging the gap created by inadequate healthcare resources. This, in turn, can significantly relieve pressure on the existing healthcare infrastructure, offering a more effective solution to the current challenges of the healthcare system. Moreover, information and communication technologies aid in maintaining independence, reducing isolation, and improving medication and transportation management, thereby enhancing quality of life and social inclusion for older individuals [22]. Nevi et al. [23] suggest further analysis of the applications of AI for elderly care, as although these technologies may be truly relevant for their care, the literature to date is really fragmented with a loss in terms of effectiveness in their adoption, development, and use. AI has the potential to enhance the quality of life and autonomy of the elderly across physical, mental, and social dimensions [19].

While academic literature often focuses on technological aspects, such as home automation and AI development models [21], increasing interest from management perspectives sees elderly populations as lucrative market segments with growing purchasing power and advanced needs. National and supranational organizations, like the United Nations, are also prioritizing elderly care, naming the decade 2021–2030 as the "Decade of Healthy Aging". However, despite the numerous technologies available, there's a lack of systematization regarding their domain of use and factors influencing their adoption, hindering social innovation efforts [22].

6.3 Methods

A systematic literature review is a robust methodology for comprehensively mapping existing knowledge in a specific academic area [12]. It is pre-defined to address research questions and provide insights for future studies. This study conducts a domain-based systematic literature review [24], which synthesizes studies within the same research domain to extend the literature [25] and allow to better focus on a specific sector despite other kind of review typologies. Authors systematically identify, evaluate, and synthesize relevant studies meeting pre-established inclusion criteria [26]. Two research objectives aligning with the proposed research questions were defined:

1. Identify and synthesize various theories and approaches related to the adoption of AI by the elderly in the scientific literature;
2. Highlights key factors influencing the adoption of AI by the elderly, based on theories and evidence from the analysed studies.

The search keywords and databases to be examined were identified during the review phase:

> "AI" OR "AI" AND ("ageing" OR "elderly" OR "Old" OR "senior" OR "Silver age" OR "Silver economy").

The keywords aim to capture the essence of the research, ensuring a precise selection of relevant articles that address the research questions comprehensively. The study focuses on AI and its interaction with the elderly. To ensure comprehensive coverage of relevant studies, several related terms such as "ageing", "elderly", "old", and "senior" were included. Additionally, the terms "Silver Age" and "Silver Economy" were incorporated to specifically capture literature related to the economic and social aspects of aging. Two online databases, Scopus and Web of Science, were utilized for data collection, which began with a search for English-language studies, limited to the "Business and Economics" field on Scopus and "Business or Management" and "topic" on Web of Science, with keyword searches limited to the title, abstract, or keywords. The initial search was conducted in July 2023 and 573 studies were founded, reciprocally 292 on Scopus and 281 on Web of Science (Fig. 6.1). An in-depth analysis of the abstracts and selection of the papers was conducted in August. Subsequently, the authors analysed the individual papers

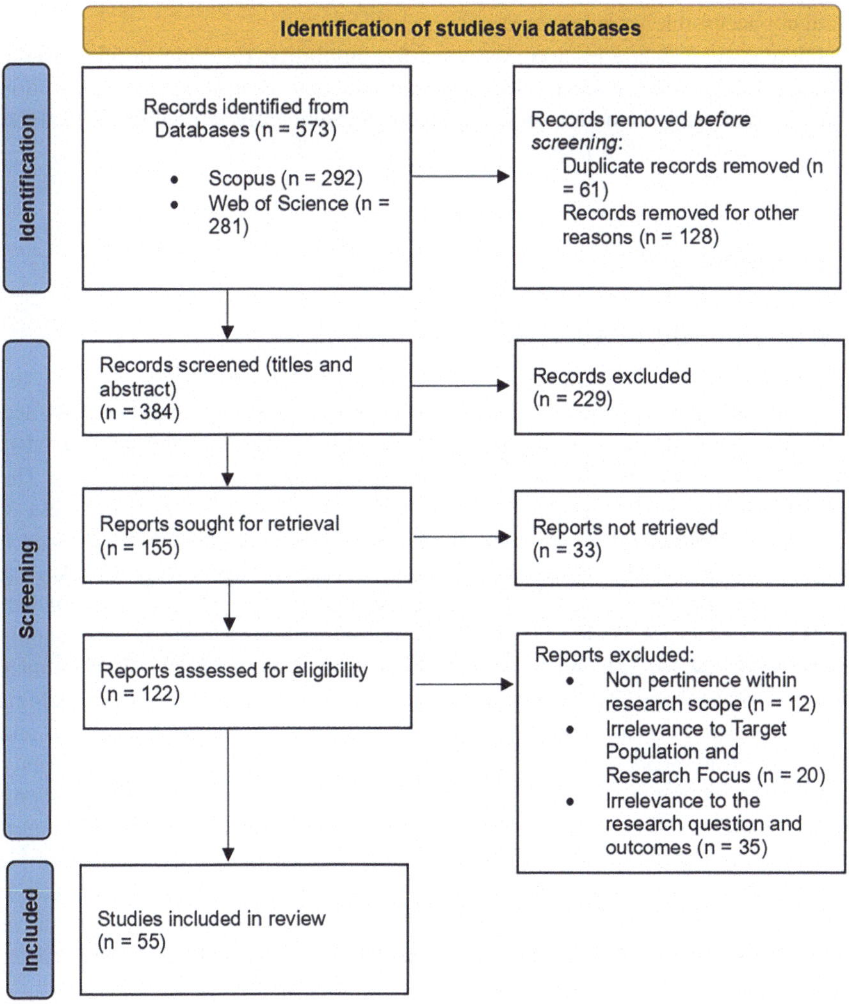

Fig. 6.1 PRISMA flow diagram

from September to October 2023 and, after a discussion on the different dimensions to considered, the research output was transcribed in November.

Exclusion criteria were applied, leading to the exclusion of studies related to books, book chapters, editorial material, conference papers, conference reviews, and studies withdrawn by the publisher, resulting in 384 papers related to proceeding papers, articles, and reviews.

Subsequent abstract and title screening phase reduced the pool to 155 articles for full-text review.

Articles addressing AI usage modes, benefits, barriers, adoption factors, research outcomes, and emerging trends relevant to the elderly were included. Articles that were not relevant to the research purpose, focused on a target other than the elderly, or addressed a topic that was unrelated were excluded. Additionally, 33 articles that were not accessible were removed.

Ultimately, 122 articles were assessed for eligibility, resulting in 55 analysed articles. Each article underwent descriptive analysis, examining the publication year, journal, and methodologies employed. Subsequently, a qualitative analysis compared the theoretical frameworks utilized in the studies.

6.4 Findings

6.4.1 Descriptive Analysis

The selected articles provide an overview of trends in the application of AI and related technologies across various sectors. Firstly, research in the field of AI is diversified in terms of reference industries, methodologies, and main topics. The studies cover a wide range of sectors, including Computer Science, Business & Economics, Medical Informatics, and Healthcare. The most frequently used research methods in the sample are qualitative (51% of selected papers, N = 28). Among these, 6 use interviews [15, 16, 27–30], 1 uses focus groups [31], while 10 conduct literature reviews [5, 17, 19, 32–38]. Notably, among these, 2 perform a systematic literature review [5, 32], and 1 conducts a hybrid systematic review that combines bibliometric analysis [37]. Eleven studies adopt additional qualitative methodological approaches such as exploratory research, the development of conceptual and theoretical frameworks, or case study analysis [1–3, 9, 10, 14, 18, 39–42]. Twenty two studies employ a quantitative methodology (40% of selected papers). Among these, 11 use questionnaires [6, 43–51], 4 rely on data collection through automation systems [52–55], and 5 of these studies conduct practical experiments [4, 56–59]. Two studies employ statistical methods to explore relationships between variables [8, 60]. Finally, 5 studies, or 9%, use mixed qualitative-quantitative approaches [20, 21, 61–63]. The main journals in which these articles appeared are listed in Table 6.1.

Table 6.1 Journal distribution of articles

Journal name	Number of articles
Technological Forecasting & Social Change	5
Journal of Business Research	3
Australasian Journal of Information Systems	2
Big Data and Cognitive Computing	2
Futures	2
International Journal of Engineering Business Management	2
International Journal of System Assurance Engineering and Management	2
Service Industries Journal	2
Technology in Society	2
ICEMIT 2018	1
Administrative Sciences	1
Australasian Accounting, Business and Finance Journal	1
BASIQ International Conference: New Trends in Sustainable Business and Consumption 2018	1
Clothing and Textiles Research Journal	1
Cornell Hospitality Quarterly	1
Human Resource Management	1
IEEE Engineering Management Review	1
Industrial Management and Data Systems	1
International Journal of Consumer Studies	1
International Journal of Recent Technology and Engineering	1
Internet Research	1
IOP Conference Series-Materials Science and Engineering	1
IZA Journal of Labor Policy	1
Journal of Distribution Science	1
Journal of Financial Services Marketing	1
Journal of Hospitality And Tourism Technology	1
Journal of Nursing Management	1
Journal of Organizational And End User Computing	1
Journal of Professions And Organization	1
Journal of Property Tax Assessment And Administration	1
Journal of Research In Interactive Marketing	1
Journal of Service Research	1
Management Research Review	1
Marketing and Management Of Innovations	1
Organizational Psychology Review	1
Procedia Computer Science	1
Proceedings on Engineering Sciences	1
Psychology & Marketing	1
Science and Engineering Ethics	1
The TQM Journal	1
Technology Innovation Management Review	1
Market-Tržište	1

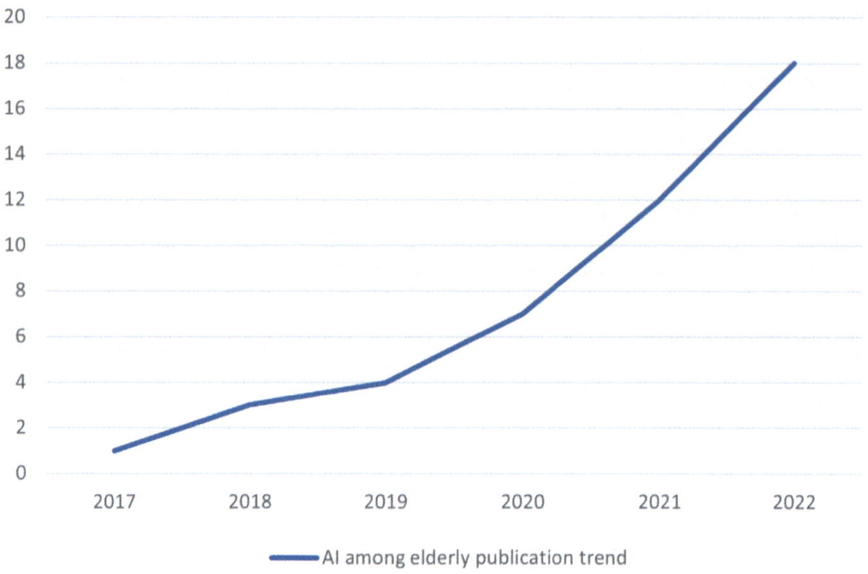

Fig. 6.2 AI among elderly, annual publication trend

Analysing 55 papers on AI and the elderly reveals notable temporal trends (Fig. 6.2). In 2017, one paper was published, indicating initial but limited interest. Interest grew slightly in 2018, with three papers, and continued to increase in subsequent years: four papers in 2019, seven in 2020, 12 in 2021, 18 in 2022, and 10 in 2023. This trend reflects a steady growth in research activity on AI and the elderly over the years.

6.4.2 Qualitative Analysis of the Selected Papers

To address RQ1, an in-depth analysis of 55 documents was conducted using inductive coding [64] to identify AI applications among the elderly. The results, summarized in Table 6.2, highlight key sectors such as "Healthcare and well-being"; "Mental health and well-being"; "Cultural and social activities"; "Health monitoring"; "Services management", along with their respective sub-domains, which are as follows:

1. Home Assistance Technology: concerns the use of technologies to provide assistance to the elderly within their homes, aiming to improve their safety and comfort. The associated sector is "Healthcare and Well-being".
2. Health-related behaviour: focuses on technologies that positively influence the behaviour of the elderly to promote health and well-being. The associated sector is "Mental Health and Well-being".

3. Entertainment: involves the use of technologies to offer entertainment and social stimulation to the elderly, promoting engagement in cultural and social activities. The associated sector is "Cultural and social activities".
4. Monitoring needs: pertains to the use of technologies for diagnosing and monitoring the health of the elderly, often through continuous monitoring of vital parameters. The associated sector is "Health Monitoring".
5. Assistance request: involves the use of technologies to simplify access to services, most of literature focuses on banking and healthcare services and improve the elderly's experience in this sector. The associated sector is "Service management".

These domains illustrate the significant impact of AI on the elderly.

In "Home Assistance Technology" technologies like "Fall Management System" and "Service Robots" were identified, mentioned in 3 and 7 studies, respectively.

The "Health-related Behaviour" domain highlights that technologies like "Social Robots", "Care Robots", "Voice Assistants", and "Chatbots" are crucial tools to address issues like mental health, as reported in the 15 included studies.

Also, in the "Entertainment" domain, an instance of using "Voice Assistants" to support cultural and, importantly, social activities of the elderly was noted.

"Wearable Devices" in the "Monitoring needs" domain were emphasized in 5 studies for real-time monitoring of elderly's physiological parameters.

Finally, in "Assistance request", "Chatbots" facilitate service interaction, noted in 3 studies.

Table 6.2 AI domain applications and use by the Elderly

Sector	Domain of application	Sub-domain of application	Technology	Description	Authors
Healthcare and well-being	Home assistance technology	Guided learning, customisation, real-time monitoring	Fall Management Systems (FMS)	An "FMS" is an AI-based system aimed at preventing and managing falls among the elderly. It utilizes sensors, wearables, and other technologies to detect potential falls or risky situations.	Aidemark and Askenäs [20]; Yoo and Oh [59]; O'Connor et al. [35].
		Real-time monitoring, customisation, intuitive interaction.	Service robots	Service robots are designed to provide assistance and support to the elderly within their home environment	Belk [19]; Hertog et al. [52]; Ho et al. [47]; Kipnis et al. [29]; Shareef et al. [58]; Westerlund [38]; Grandinetti et al. [39].

(continued)

Table 6.2 (continued)

Sector	Domain of application	Sub-domain of application	Technology	Description	Authors
Mental health and well-being	Health-related behaviour	Emotional interactions, customization, social learning, variable morphology.	Social robots	Social robots are physically embodied autonomous agents capable of emotional communication and interaction with humans. They operate in personal environments serving as both emotional companions and practical assistants.	Poulsen et al. [30]; Kalisz et al. [21]; Westerlund [38]; Blaurock et al. [32]; Poulsen and Burmeister [63]; Shareef et al. [6].
		Practical assistance, enhancement of well-being.	Care robots	Care robots can assist in personal care needs, administer medications, and even provide basic medical assistance.	Tsiotsou and Boukis [37]; Rubeis et al. [10]; Kanumuri et al. [57].
		Smart home integration, voice recognition.	Voice assistant	Voice assistants provide an intelligent environment through voice recognition systems, enabling elderly individuals to effortlessly manage home appliances via connected smart devices.	Alshudukhi [56]; Tsiotsou and Boukis [37]; Park and Kim [60].
		24/7 availability, continuous evolution, automated interaction.	Chatbot	A chatbot is a computer program that simulates a conversation with users through text or voice interaction.	Bhattacharya and Sinha [61]; Cheng et al. [62]; Ho et al. [47].

(continued)

Table 6.2 (continued)

Sector	Domain of application	Sub-domain of application	Technology	Description	Authors
Cultural and social activities	Entertainment	Smart home integration, voice recognition.	Voice assistant	Voice assistants offer a wide range of functionalities, including checking weather conditions, making phone calls, playing music, sending messages, scheduling appointments and searching for information.	Buhalis and Moldavska [27].
Health monitoring	Monitoring needs	Real-time monitoring, personalized health insights, integration with IoT.	Wearable devices	Wearable devices support regular, non-invasive medical assessments by monitoring real-time biometric data.	Katsaliaki and Kumar [40]; Srivastava et al. [36]; Tang et al. [53]; Xing et al. [31]; Rubeis et al. [10].
Services management	Assistance request	24/7 availability, continuous evolution, automated interaction.	Chatbot	Chatbot in online banking provides 24/7 support for assistance and recommendations.	Cheng et al. [62]; Alt and Ibolya [43]; Rajaobelina and Ricard [51].

These findings underscore AI's increasing adoption to enhance elderly lives across autonomy, entertainment, mental and physical health monitoring.

6.4.3 Factors of AI Adoption by the Elderly

Several theoretical frameworks and constructs have been applied to explore AI adoption among the elderly (Table 6.3). Prominent models include the Technology Acceptance Model (TAM) and its extensions, such as UTAUT and UTAUT2, which are widely used across different contexts. Additionally, theories like Uses & Gratification Theory, Signalling Theory, and Prospect Theory are employed in diverse theoretical frameworks. In general, a number of key constructs emerge that influence the adoption of AI among the elderly:

Table 6.3 Summary of reviewed articles about AI adoption

Article	Context	Leading framework or theory	Methodology	Dimension and construct of the analysis	Key findings
Belanche et al. [44]	Robo-advisors	Decision theory, SOR	Online survey in USA	Perceived ease of use (PEOU), security (SE), perceived benefits (PB), perceived risk (PR), WOM, loyalty, behaviour intention (BI)	SE and PEOU affect PB positively and reduce PR. PB increases willingness to use and recommend the service, PR weakens the relationship between PB and BI.
Cheng et al. [62]	Financial artificial intelligent customer service (FAICS)	TAM, ISS	Interviews and questionnaires	Perceived intelligence (PI), perceived virtual identity (PVI), perceived anthropomorphism (PA), perceived usefulness (PU), PEOU, satisfaction (SA), cognitive age (CA), usage intention (UI).	PA and PVI positively affect PEOU. Similarly, PI positively influences PEOU, SA, and PU. CA can moderate the effects of PU and SA on UI of FAICS.
Dawar et al. [45]	Chatbot	TAM, UTAUT	Questionnaire and online survey in India	User experience (UE), performance expectancy (PE), effort expectancy (EE), trust (TR), BI, Chatbot usage (CU).	EE and TR strongly influence CU, while positive UE boosts customer intention to use chatbot services. Additionally, PE positively impacts BI.
Hameed et al. [46]	AI in healthcare (AIH)	UATUT2	Cross-sectional survey	PE, EE, BI, social influence (SI), initial trust (IT), personal innovativeness (PIIT), technological characteristics (TC), task complexity (TAC), perceived substitution crisis (PSC).	PE, EE, and IT positively influence healthcare providers' BI to use AIH. PIIT positively impacts EE, while TAC and TC also contribute positively to EE for AIH.
Ho et al. [47]	Emotional AI (robots, chatbots, diagnostic software)	TAM	Survey in Japan	PU, PEOU, self-related familiarity, knowledge with AI technologies and ethical concerns, socio-demographic factors; participation in community activities and mobility.	Perceptions and adoption of emotional AI applications in healthcare are influenced by ethical, demographic, and social factors. Familiarity with AI positively correlates with favourable perceptions of emotional AI applications. However, elderly patients tend to perceive emotional AI technologies more negatively. Community activities and mobile internet access positively impact attitudes towards emotional AI applications.

Lee et al. [48]	Service robot	Extended TAM	Online questionnaire	Hedonistically motivated consumer innovation (hMCI), functionally motivated consumer innovation (fMCI), socially motivated consumer innovation (sMCI), PU, TR, perceived enjoyment (PerE), intention to use (InT).	fMCI, hMCI, and sMCI all had positive effects on InT and PerE. TR positively influences InT.
Liang et al. [49]	Fashion AI product-Echo look	TAM	Questionnaire in USA	PU, PEOU, perceived risk (PR), technology attitudes (TechAtt), attitude towards AI, purchase intention (PurI).	PU and PEOU positively predict consumer attitudes towards AI. PR negatively predicts attitudes; if the product fails to meet expectations, satisfaction decreases, leading to negative attitudes towards AI.
Maroufkhani et al. [50]	Interactive voice assistants	UGT, signalling theory, prospect theory	Online survey in China	Overall perceived value (OPV), utility features (UTF), hedonic features (HDF), social presence (SCP), perceived privacy risk (PPR), irritation (IRR), voice assistant continued usage intention (VCU), brand loyalty (BRL), perceived anthropomorphism (PAP), perceived intelligence (PCI), attractiveness (ATR), trustworthiness (TRW), expertise (XPR)	PPR hinders the OPV of voice assistants. BRL moderates this relationship. PCI and PAP positively influence VCU.
Shareef et al. [6]	Autonomous homecare systems	ATM	Experiment and survey in Canada	Perceived cognitive complexity (PCC), perceived psychological complexity (PPC), perceived cognitive risk (PCR), perceived psychological risk (PPR), perceived cognitive performance (PCP), perceived psychological performance (PPP), personal ability and control (PAC), technological uncertainty and reliability (TUR), personal benefit and accomplishment (PBA), empathetic cooperation and social interaction (ECSI), TR, self-concept and behavioural attitude (SCBA), BI.	PAC and ECSI significantly impact TR in autonomous systems. TUR and PBA do not significantly contribute to TR. PAC and ECSI play crucial roles in TR.

(continued)

Table 6.3 (continued)

Article	Context	Leading framework or theory	Methodology	Dimension and construct of the analysis	Key findings
Shareef et al. [58]	Automated assistance systems	ETM4MA	Survey in Bangladesh	Expected cognitive complexity (ECC), expected psychological complexity (EPC), expected cognitive risk (ECR), expected psychological risk (EPR), expected cognitive performance (ECP), expected psychological performance (EPP), expected personal ability and control (EPAC), expected technological uncertainty (ETU), family benefit and accomplishment (FBA), expected empathetic cooperation and social interaction (EECSI), self-concept and personality & image (SCPI), TR, BI.	Elderly acceptance of ambient intelligence-based automation as a substitute for human care support depends on concerns about emotional belonging, EECSI, and technology proficiency. TR is crucial in forming the intention to accept machine autonomy.
Xing et al. [31]	Wearable medical devices (WMDs)	Theories on barriers and issues influencing large-scale deployment of AI-enabled WMDs	Focus group and thematic analysis in China	Technological, managerial, clinical, financial, legal, and personal barriers	The study identifies 16 significant barriers impacting the adoption of technologies among the elderly, including concerns about device performance, data privacy, functional deficiencies, and psychological resistance.

- Perception of benefits and risks: Perception of benefits of AI use, such as ease of use and safety, has a positive impact on intention to use, while perception of risks has a negative impact.
- Ease of use and satisfaction: Ease of use of AI is a key factor affecting satisfaction and perceived usefulness of AI.
- Trust: is an important factor influencing intention to use and adoption of AI.
- Personal capability expectations: reflects expectations regarding personal ability to use and benefit from AI and can play a significant role in influencing attitudes towards the adoption of advanced technologies such as AI.

More than others, trust and personal capability expectations are identified as crucial factors influencing AI acceptance among the elderly, impacting behavioural intentions towards machine autonomy.

6.5 Discussion and Implications

The systematic literature review underscores the rising significance of AI adoption among the elderly across various domains like home assistance, health monitoring, and entertainment. Tools like interactive voice assistants, wearable health monitors, social robots, and chatbots play crucial roles in enhancing the quality of life for elderly individuals [47]. These technologies alleviate isolation and promote autonomy among older adults [58]. AI optimizes daily tasks, fosters independence, and revolutionizes proactive health management through personalized experiences and meaningful interactions [19, 21]. Voice assistants add entertainment value, while real-time assistance and safety monitoring enhance care quality and strengthen caregiver relationships [40, 60]. The introduction of chatbots in the domain of service management and service requests simplifies request procedures, enhancing accessibility and responsiveness [62]. Key factors influencing AI adoption include trust, perceived benefits, ease of use, and privacy concerns [44, 48, 49]. However, emotions and personal perceptions also play pivotal roles [47, 62] and overcoming privacy, financial, and social barriers is crucial for successful adoption [47, 50]. Prioritizing trust, ease of use, and privacy measures can enhance AI acceptance among the elderly, enabling active aging. Further research is imperative to understand the long-term effects of AI on elderly care, guiding theoretical and managerial advancements. Particular attention should be given to longitudinal studies that monitor how the elderly adapt to continued use of these technologies. Additionally, considering human–machine interaction, it is essential to investigate the elderly's perceptions of these technologies in order to better tailor them to their needs and preferences. Studies are necessary, taking into account differences among various groups of elderly individuals, including age, generational cohort, education, and health. Accessibility is a crucial issue, so exploring ways to make these technologies more intuitive and accessible to a wider audience—such as through simpler interfaces and learning support—is essential for improving the user experience for elderly customers and their caregivers. Future research could also focus on overcoming remaining obstacles related to digital literacy and due to different types of technologies and related pathologies and ageing difficulties, economics, and psychosocial factors to enhance accessibility. Lastly, given the continuous innovations in AI technologies, future research should remain at the forefront and promote interdisciplinary collaboration between AI experts, health-care professionals, and innovation processes. Healthcare professionals and technology companies should provide flexible, personalized solutions tailored to individual needs.

6.6 Conclusion and Discussion

The rising applications of AI in elderly care, combined with the challenges of an ageing population, underscore the need for structured studies to identify key variables influencing AI development and adoption. The dynamic landscape of AI in elderly care presents opportunities where technology trends intersect with market dynamics, offering insights for companies and entrepreneurs. However, this systematic literature review has limitations. The chosen keywords and the rapid evolution of AI technologies may have limited the scope of the analysis. The heterogeneity of studies and the exclusion of non-English articles may have affected comparability. Future research should expand keyword searches and consider a broader range of contributions, including conference papers and books.

Acknowledgement The project was also realized thanks to the Medium Funding Project, a.a. 2022: "Il Sistema della Salute in ottica sistemico - vitale. L'impatto della digitalizzazione sugli anziani di oggi e di domani. - n. protocollo RM1221816C56680F" of Sapienza University.

References

1. Alabed, A., Javornik, A., Gregory-Smith, D.: AI anthropomorphism and its effect on users' self-congruence and self–AI integration. Technol. Forecast. Soc. Chang. **182**, 121786 (2022)
2. Alcover, C.M., Guglielmi, D., Depolo, M., Mazzetti, G.: "Aging-and-tech job vulnerability": a proposed framework on the dual impact of aging and AI, robotics, and automation among older workers. Organ. Psychol. Rev. **11**(2), 175–201 (2021)
3. Armour, J., Sako, M.: AI-enabled business models in legal services: from traditional law firms to next-generation law companies? J. Prof. Organ. **7**(1), 27–46 (2020)
4. Lehto, X.Y., Park, S., Mohamed, M.E., Lehto, M.R.: Traveler attitudes toward biometric data-enabled hotel services. Cornell Hosp. Q. **64**(1), 74–94 (2023)
5. Pentina, I., Xie, T., Hancock, T., Bailey, A.: Consumer–machine relationships in the age of artificial intelligence. Psychol. Mark. **40**(8), 1593–1614 (2023)
6. Shareef, M.A., Kumar, V., Dwivedi, Y.K., et al.: A new health care system enabled by machine intelligence: elderly people's trust or losing self control. Technol. Forecast. Soc. Chang. **162**, 120334 (2021)

7. WHO: Ageing and Health. https://www.who.int/news-room/fact-sheets/detail/ageing-and-health (2023). Accessed 10 Sept. 2023
8. Malik, M.A., Singh, S.P., Pattanaik, F.: Economics of healthy aging in India: a multidimensional perspective. IZA J. Labor Policy. **12**(1), 1–22 (2022)
9. Angioni, M., Musso, F.: New perspectives from technology adoption in senior cohousing facilities. TQM J. **32**(4), 761–777 (2020)
10. Rubeis, G., Fang, M.L., Sixsmith, A.: Equity in AgeTech for ageing well in technology-driven places. Sci. Eng. Ethics. **28**(6), 49 (2022)
11. Zhu, Y., Yang, Q., Mao, X.: Global trends in the study of smart healthcare systems for the elderly. Int. J. Comput. Intell. Syst. **16**, 105 (2023)
12. Tranfield, D., Denyer, D., Smart, P.: Towards a methodology for developing evidence-informed management. Br. J. Manag. **14**, 207–222 (2003)
13. Palmatier, R.W., Houston, M.B., Hulland, J.: Review articles: purpose, process, and structure. J. Acad. Mark. Sci. **46**, 1–5 (2018)
14. Stamova, I., Draganov, M.: Artificial intelligence in the digital age. In: International Scientific Conference "Digital Transformation on Manufacturing, Infrastructure and Service", vol. 940, pp. 1–9. Russian Federation, St. Petersburg (2020)
15. Einola, K., Khoreva, V.: Best friend or broken tool? Exploring the co-existence of humans and artificial intelligence in the workplace ecosystem. Hum. Resour. Manag. **62**(1), 117–135 (2023)
16. Booyse, D., Scheepers, C.B.: Barriers to adopting automated organisational decision-making through the use of artificial intelligence. Manag. Res. Rev. **47**(1), 64–85 (2023)
17. Cusack, M., Quintos, C.: A review of the methods, applications, and challenges of adopting artificial intelligence in the property assessment office. J. Prop. Tax Assess. Admin. **19**(1), 21–80 (2022)
18. Liu, Y.: Application analysis of artificial intelligence in computer network technology under the age of big data. In: 4th International Conference on Education, Management and Information Technology (ICEMIT 2018), vol. 14, pp. 1133–1136, Changchun (2018)
19. Belk, R.: Ethical issues in service robotics and artificial intelligence. Serv. Ind. J. **41**(13–14), 860–876 (2021)
20. Aidemark, J., Askenäs, L.: Fall prevention as personal learning and changing behaviors. Procedia Comput. Sci. **164**, 498–507 (2019)
21. Kalisz, D.E., Khelladi, I., Castellano, S., Sorio, R.: The adoption, diffusion & categorical ambiguity trifecta of social robots in e-health. Futures. **129**, 102743 (2021)
22. Pilotto, A., Boi, R., Petermans, J.: Technology in geriatrics. Age Ageing. **47**, 771–774 (2018)
23. Nevi, G., Gogiali, G., Dezi, L.: Artificial intelligence for elderly intelligence. In: 5th IEEE International Conference on Technology Management, Operations and Decisions (IEEE ICTMOD), pp. 1–6. IEEE, Rabat, Marrakesh (2023)
24. Kraus, S., Breier, M., Lim, W.M., et al.: Literature reviews as independent studies: guidelines for academic practice. Rev. Manag. Sci. **16**, 2577–2595 (2022)
25. Paul, J., Barari, M.: Meta-analysis and traditional systematic literature reviews—what, why, when, where, and how? Psychol. Mark. **39**, 1099–1115 (2022)
26. Boell, S.K., Cecez-Kecmanovic, D.: On being 'systematic' in literature reviews. Formulating Research Methods for Information Systems, **2**, 48–78 (2015)
27. Buhalis, D., Moldavska, I.: Voice assistants in hospitality: using AI for customer service. J. Hosp. Tour. Technol. **13**(3), 386–403 (2022)
28. Keeler, L.W., Bernstein, M.J.: The future of aging in smart environments: four scenarios of the United States in 2050. Futures. **133**, 102830 (2021)
29. Kipnis, E., McLeay, F., Grimes, A., et al.: Service robots in long-term care: a consumer-centric view. J. Serv. Res. **25**(4), 667–685 (2022)
30. Poulsen, A., Fosch-Villaronga, E., Burmeister, O.K.: Cybersecurity, value sensing robots for LGBTIQ+ elderly, and the need for revised codes of conduct. Australas. J. Inf. Syst. **24** (2020)

31. Xing, F., Peng, G., Zhang, B., et al.: Socio-technical barriers affecting large-scale deployment of AI-enabled wearable medical devices among the ageing population in China. Technol. Forecast. Soc. Chang. **166**, 120609 (2021)
32. Blaurock, M., Čaić, M., Okan, M., Henkel, A.P.: A transdisciplinary review and framework of consumer interactions with embodied social robots. Int. J. Consum. Stud. **46**(5), 1877–1899 (2022)
33. Gao, Y., Liu, H.: Artificial intelligence-enabled personalization in interactive marketing: a customer journey perspective. J. Res. Interact. Mark. **17**(5), 1–18 (2022)
34. Kumar, V., Ramachandran, D., Kumar, B.: Influence of new-age technologies on marketing. J. Bus. Res. **125**, 864–877 (2021)
35. O'Connor, S., Gasteiger, N., Stanmore, E., et al.: Artificial intelligence for falls management in older adult care. J. Nurs. Manag. **30**(8), 3787–3801 (2022)
36. Srivastava, S., Pant, M., Agarwal, R.: Role of AI techniques and deep learning in analyzing the critical health conditions. Int. J. Syst. Assur. Eng. Manag. **11**(2), 350–365 (2020)
37. Tsiotsou, R.H., Boukis, A.: In-home service consumption: a systematic review, integrative framework and future research agenda. J. Bus. Res. **145**, 49–64 (2022)
38. Westerlund, M.: An ethical framework for smart robots. Technol. Innov. Manag. Rev. **10**(1), 35–44 (2020)
39. Grandinetti, R., Bettiol, M., Di Maria, E.: Contexts of consumption and their evolution in the digital age. Adm. Sci. **12**(4), 121 (2022)
40. Katsaliaki, K., Kumar, S.: The past, present, and future of the healthcare delivery system through digitalization. IEEE Eng. Manag. Rev. **50**(4), 21–33 (2022)
41. Nedić, B.: Gartner's top strategic technology trends. In: 13th International Quality Conference, Proceedings on Engineering Sciences, vol. 1(2), pp. 433–442, Kragujevac (2019)
42. Wan, H., Chin, K.: Exploring internet of healthcare things for establishing an integrated care link system in the healthcare industry. Int. J. Eng. Bus. Manag. **13**, 18479790211019526 (2021)
43. Alt, M.A., Ibolya, V.: Identifying relevant segments of potential banking Chatbot users based on technology adoption behavior. Market-Tržište. **33**(2), 165–183 (2021)
44. Belanche, D., Casaló, L.V., Flavián, M., Loureiro, S.M.C.: Benefit versus risk: A behavioral model for using robo-advisors. Serv. Ind. J. **45**, 1–28 (2023)
45. Dawar, S., Savita, S.P., Dhaka, S., Kudal, P.: Antecedents and role of trust in chatbot use intentions. Mark. Manag. Innov. **13**(4), 198–206 (2022)
46. Hameed, B.Z., Naik, N., Ibrahim, S., et al.: Breaking barriers: unveiling factors influencing the adoption of artificial intelligence by healthcare providers. Big Data Cogn. Comput. **7**(2), 105 (2023)
47. Ho, M.-T., Le, N.T.B., Mantello, P., Ghotbi, N.: Understanding the acceptance of emotional artificial intelligence in Japanese healthcare system. Technol. Soc. **72**, 102166 (2023)
48. Lee, J., Kwak, M.-K., Cha, S.-S.: The effect of motivated consumer innovativeness on perceived value and intention to use for senior customers at AI food service store. J. Distrib. Sci. **19**(9), 91–100 (2021)
49. Liang, Y., Lee, S.H., Workman, J.E.: Implementation of artificial intelligence in fashion. Cloth. Text. Res. J. **38**(1), 3–18 (2020)
50. Maroufkhani, P., Asadi, S., Ghobakhloo, M., et al.: How do interactive voice assistants build brands' loyalty? Technol. Forecast. Soc. Chang. **183**, 121870 (2022)
51. Rajaobelina, L., Ricard, L.: Classifying potential users of live chat services and chatbots. J. Financ. Serv. Mark. **26**(2), 81–94 (2021)
52. Hertog, E., Fukuda, S., Matsukura, R., et al.: The future of unpaid work: estimating the effects of automation on time spent on housework and care work in Japan and UK. Technol. Forecast. Soc. Chang. **191**, 122443 (2023)
53. Tang, V., Lam, H.Y., Wu, C.H., Ho, G.T.S.: A two-echelon responsive health analytic model for triggering care plan revision in geriatric care management. J. Organ. End User Comput. **34**(4), 1–29 (2021)

54. Thakur, N., Han, C.Y.: Indoor localization for personalized ambient assisted living of multiple users in multi-floor smart environments. Big Data Cogn. Comput. **5**(3), 42 (2021)
55. Wong, B., Ho, G.T.S., Tsui, E.: Development of an intelligent e-healthcare system for the domestic care industry. Ind. Manag. Data Syst. **117**(7), 1426–1445 (2017)
56. Alshudukhi, J.S.: Smart and interactive healthcare system based on speech recognition using soft margin formulation and kernel trick. Int. J. Syst. Assur. Eng. Manag. **15**(1), 324–333 (2022)
57. Kanumuri, C.R., Reddy, D.S., Swamy, A.M.: Deep learning based assistive robot. Int. J. Recent Technol. Eng. **8**(1S3), 154–155 (2019)
58. Shareef, M.A., Ahmed, J.U., Giannakis, M., et al.: Machine autonomy for rehabilitation of elderly people. J. Bus. Res. **164**, 113961 (2023)
59. Yoo, S., Oh, D.: An artificial neural network–based fall detection. Int. J. Eng. Bus. Manag. **10**, 1–7 (2018)
60. Park, S., Kim, B.: The impact of everyday AI-based smart speaker uses on the Well-being of older adults living alone. Technol. Soc. **71**, 102133 (2022)
61. Bhattacharya, C., Sinha, M.: The role of artificial intelligence in banking for leveraging customer experience. Australas. Bus. Account. Finance J. **16**(5), 89–105 (2022)
62. Cheng, X., Qiao, L., Yang, B., Li, Z.: An investigation on the influencing factors of elderly people's intention to use financial AI customer service. Internet Res. **34**(3), 690–717 (2023)
63. Poulsen, A., Burmeister, O.K.: Overcoming carer shortages with care robots. Australas. J. Inf. Syst. **23** (2019)
64. Gioia, D.A., Corley, K.G., Hamilton, A.L.: Seeking qualitative rigor in inductive research. Organ. Res. Methods. **16**(1), 15–31 (2013)

Chapter 7
Enhancing Value Co-creation Through AI in the Cultural Sector: An Exploration in Italian Museums

Laura Clemente and Francesco Bifulco

Abstract In the rapidly evolving landscape of Artificial Intelligence (AI), a growing number of industries are revolutionizing the way services are developed and value co-creation processes are nurtured. The potential of AI has been explored in numerous studies, although the benefits of its use in the cultural sector are still an underexplored topic. This article is based on an exploratory analysis to investigate the implementation of AI in museums for cultural and participatory value co-creation. The research is developed from the analysis of case studies of the Italian cultural sector and the information obtained is systematized through the use of the co-creation framework (CCF), which allowed to demonstrate how the use of AI, in its different declinations and forms, enhances the interaction between consumers and cultural resources for the co-creation of value, allowing museums to exploit data on their audiences to identify and build a customized offer on them.

Keywords Artificial Intelligence · Value Co-creation · Service-dominant logic · Cultural sector · Museum · Digital transformation

7.1 Introduction

Digital transformation has pervaded every aspect of contemporary society and imposed a paradigm shift in the ways firms collaborate to create and capture value [1, 2]. The infusion of technology into service delivery has helped to provide companies with better opportunities for meaningful engagement and co-creation of unique experiences for their customers. Among the enabling technologies driving technological innovation, Artificial Intelligence is emerging as a disruptive force,

L. Clemente (✉)
Sapienza University, Rome, Italy

F. Bifulco
University of Naples Federico II, Naples, Italy

F. Schiavone et al. (eds.), *Advanced Perspectives and Trends in Digital Transformation of Firms, Networks, and Society*, Springer Proceedings in Business and Economics, https://doi.org/10.1007/978-3-031-80692-6_7

due to its advanced capabilities in data processing, service personalization, and communication with customers through a human-like language. Its continuous development in an increasing number of fields has led scholars and practitioners to reflect on its application in museum context, a field in which it can be considered a valuable tool to be implemented in visitor engagement strategies to renew traditional communication trajectories [3, 4].

In particular, the application of AI can provide new coordinates within which to develop innovative services and value co-creation processes supported by interaction mechanisms that can be perceptual or reactive [5]. Drawing on these assumptions, this study aims to contribute to the debate on value co-creation with reference to the museum sector, analyzing the ways in which AI can support such processes and highlighting the managerial challenges associated with the implementation of this technology, which requires synergies between internal and external resources, fostering collaboration between organizations and stakeholders.

7.2 Objective

Artificial intelligence—meant as the ability attributed to machines to perform tasks usually performed by humans [6]—has stimulated growing interest among scholars, emerging as the most revolutionary technological force of the last decades [7]. The advent of artificial intelligence, and more recently, generative artificial intelligence (GAI), has made it possible to offer increasingly efficient and customizable services and has led to a transformation in the way companies interact with customers, seamlessly engaging them in physical and virtual environments [8, 9]. AI therefore represents a turning point for several industries [10], paving the way for the development of innovative services and new value co-creation processes resulting from the interaction between people and material entities (e.g., devices), supported by digital technologies [11].

Value co-creation is a key concept of the service-dominant logic (SDL) that emphasizes collaboration and reciprocity between producers and consumers at every stage of the value creation process [12]. In a service-based perspective [13], the customer is perceived as an active resource that must be involved in the value creation process and is able to influence the available organizational resources to create innovative products and services [14]. Other stakeholders, such as technology service providers, also play an active role in finding opportunities and supporting the co-creation of value together with customers [1, 15].

The cultural sector is experiencing a major change thanks to the new digital paradigm, which affects the dynamics of value co-creation [16, 17] and the ways in which visitors are engaged [18]. Several scholars have emphasized the effects of new technologies on the visitor experience, hence fostering managerial studies focused on the relationships between cultural organizations and their stakeholders [19, 20]. Despite this, the understanding of the role of AI in value co-creation and its effects in the cultural context is limited [21, 22] and empirical studies on these issues are still at an early stage compared to practical advances. In accordance with these preliminary considerations, this research work aims to conduct an exploratory analysis to investigate the implementation of AI in museums for the co-creation of cultural and participatory value. To meet this purpose, the following research question was identified:

RQ. *How can AI enhance value co-creation in museums?*

Upon answering this question lies the originality of this research that highlights the transformative potential of AI in reshaping visitor engagement practices in museums, while emphasizing the importance of strategic partnerships that drive co-creation of value and contribute to the improvement of the overall museum experience.

7.3 Methodology

The empirical analysis conducted in this study aims to offer a new perspective in the conceptualization of generative AI as an enabler of value co-creation in museums. In line with the exploratory purpose of the present research, the authors adopted a qualitative method based on an interpretive research approach [23] and chose a comparative case study method. This approach allowed moving beyond the specifics of a single case to identify similarities and differences through careful abstraction [24], providing a coherent framework to answer the research question. The cases were selected as representative of the most recent AI-based projects in Italian museums, while information was obtained through the analysis of official publicly available documents and secondary sources such as interviews, website content, social media pages, and reports [25]. More specifically, the selected cases are: the National Archaeological Museum of Naples with the "ArtWare Caronte" app; the National Cinema Museum of Turin with the project "The future of cinema, the cinema of the future"; the MAXXI Museum of Rome; the Colosseum Museum and Archaeological Park with the "Nerone" chatbot and the I-MUSE app, developed with the collaboration of eight museums of Turin.

The Co-Creation Framework (CCF), developed by Ramaswamy and Ozcan [11], has been used to qualitatively analyze the approach adopted by museums to implement AI in value co-creation, while the four thematic categories into which the framework is divided, namely artifacts, processes, interfaces, and people, made it possible to construct the classification of empirical data. Starting from these elements, Ramaswamy and Ozcan [11] introduce the concept of "interactive value creation," shifting the locus of value from activities to interactions. From a cultural heritage perspective, artifacts are related to the museum collection and include physical and digitized things in the form of text, images, audio and video; processes identify how visitors interact with artifacts; interfaces represent the tools through which interaction is created; people include all the individuals who interact with the cultural experience, in their different roles.

7.4 Research Setting

It should be noted that the five cases considered, although all belonging to the museum sector, have specific characteristics that highlight the heterogeneity of the museum types found in Italy, which include archaeological museums, museums of modern or contemporary art, and museums of cinema and performing arts.

The I-Muse project was born out of the collaboration between the University of Turin, the Foundation Compagnia di San Paolo and eight of Turin's major museums, including the Reggia di Venaria Reale, the Egyptian Museum, Palazzo Madama, Galleria Civica d'Arte Moderna e Contemporanea, Museo d'Arte Orientale, the National Cinema Museum, the National Automobile Museum, and the Pinacoteca Giovanni e Marella Agnelli. The primary objective of the project is the integration of GAI to improve the experience of visiting museums. Through an app, visitors can create customized thematic itineraries and set their own preferences, adding the exhibited works by framing the appropriate Qrcode, while the GAI integration allows for the automatic creation of inter-museum connections.

The National Cinema Museum of Turin, in partnership with Synesthesia, carried out an experimental project lasting approximately 3 months, entitled "The future of cinema, the cinema of the future," which offered museum visitors the opportunity to become authors of a film script through the use of generative artificial intelligence, allowing visitors to become prosumers, co-creating their own personalized visitor experience. Through an interactive touch screen totem and the integration of OpenAI and machine learning-based technologies, visitors were able to create their own tailor-made film subject and script based on their preferences.

"ArtWare Caronte" is the app of the National Archaeological Museum of Naples to improve visitor fruition of exhibition and digital content with the aim of involvement, inclusiveness, and integration between the physical and digital environment. The app is configured as a hybrid software where different technologies are put in synergy to engage visitors by providing information in an innovative way. In particular, AI is used for the exploration of the MANN Campania Romana section and makes it possible to bring sculptures to life by combining the artworks with new forms of museum fruition. In addition, the connection with the "Archimedes" system makes it possible to acquire data and information on the public, which, analyzed by means of Machine Learning models, makes it possible to develop personalized and proactive strategies toward visitors.

The MAXXI Museum in Rome has implemented a virtual assistant that, exploiting AI, communicates with visitors, anticipates their curiosities and tests their knowledge through *gamification*. In order to increase the interactivity of the visitor experience and customer loyalty, engagement and information, while maintaining a continuous connection with each visitor in the before-and-after visit phases, the museum management collaborated with Engineering Interactive to design an interactive guide that accompanies visitors through works, exhibitions, and events, with the possibility of winning special prizes. The system is integrated with several components, including Google Dialogue Flow, to offer the public natural language interaction, and the museum's website, to allow the chatbot to automatically update itself with available information, such as new exhibitions or events.

In early 2023, the Colosseum Museum and Archaeological Park launched Nerone, a chatbot powered by AI and designed to engage a wider audience and provide information about the cultural site and its services. It is the result of an international collaboration with the Italian company Machineria srl and the French Ask Mona, both active in the production of innovative services for cultural enterprises. The study of the most frequently asked questions made it possible to intercept recurring themes and potential critical issues encountered by users, and this information was used to improve the chatbot's performance with more detailed answers in line with visitors' expectations and to turn problematic aspects into opportunities for value creation.

7.5 Findings

This section details how co-creation of value can be effectively achieved by implementing AI in the cultural activities offered by the different museum analyzed, generating new interactions through the relationship between the four different elements resulted from the CCF. In order to achieve a systematization of this information, the following summary table (Table 7.1) illustrates the key elements for each of the five case studies, labeled with a number that will identify them in the body of the text.

Table 7.1 Key items in selected cases

Label	Name of the project or museum	Artifacts	Processes	Interfaces	People
1	I-Muse	Collections of the involved museums	Exploration	Mobile Device	Visitors, online users, University of Turin, Foundation Compagnia di San Paolo
2	National Cinema Museum of Turin	Customized scripts	Exploration and content creation	Interactive totem	Visitors, Synesthesia
3	ArtWare Caronte MANN	Collection and exhibition environment	Storytelling and exploration	Mobile Device	Visitors, online users, University of Naples Federico II
4	MAXXI Museum	Museum collection and exhibitions	Exploration and gamification	Mobile Device	Visitors, online users, Engineering Interactive
5	Nerone Chatbot Colosseum Museum and Archaeological Park	Archaeological relics and exhibition environment	Exploration and information	Mobile Device or PC	Visitor, online users, Machineria srl, Ask Mona

Source: Author's elaboration

Artifacts are the first category and include not only the works belonging to museum collections or archaeological park, but also their related content, consisting of stories, meanings, anecdotes, information, explanations, interpretations, conveyed by digital interfaces. In line with what Marini and Agostino [16] state, the use of AI leads to the development of personal content, identified as that content that is owned by visitors because it is generated by them, as Case 2 shows. The implementation of AI also makes it possible to display content that would otherwise not be accessible or available, as in case 3, in which AI conveys additional content based on the needs and preferences of individual visitors.

The ways in which this content is shared with the audience represent the processes, the second element of the framework, which can be divided into storytelling, exploration, information, and gaming. The cases analyzed show the use of a strongly narrative and conversational approach, made possible by AI's ability to generate human-like content. In case 1, generative AI enables an exploration process for visitors, stimulating them to find hidden connections between artworks and suggesting new itineraries. Case 5 uses AI to provide answers to users' questions and needs, while recording information to improve service quality. Case 4 bases its processes on interactions with users and activates an innovative communication and interaction channel using the tool of gamification to achieve greater involvement even after the visit. A central theme, particularly relevant in the cultural context, is the importance of providing the user with the idea that the interaction is not conducted by a technological system [26] but by a human being, capable of guiding the visitor along a learning path with the same degree of involvement that an expert human guide can provide [27].

Processes are mediated by interfaces, which constitute the third element of the framework. Although they may come in different forms, such as apps, interactive totems and web pages, they represent the technical means through which processes can manifest themselves and enable the co-creation of value. A key component that determines the likelihood of the visitor benefiting from the process of value creation through AI is the user's competence or level of high-tech experience [28]. Therefore, visitors may not be fully engaged if they are not familiar with the more sophisticated aspects and skills related to different technological interfaces [29].

The last element refers to people, a term used in the cultural sector to refer not only to visitors, but also to non-visitors, managers, and other stakeholders gravitating in the orbit of museums, such as universities, research centers, and tech partners involved in value co-creation processes. In this regard, it should be noted that all the projects analyzed involved the participation of actors from outside the museum system, whose contribution was decisive in the development of interfaces and performance monitoring.

7.6 Implications

The exploratory analysis conducted aims to provide an empirical contribution to managerial studies concerning the implementation of new technologies in the museum sector and the consequent ability to co-create value in a service-dominant logic (SDL). The need to observe, monitor, and investigate the practical use of AI by cultural enterprises plays a prominent role in management studies, in line with what was stated in Sfodera et al. [30]. Museums are rethinking their relationship with visitors and users, through digital technologies such as AI, redesigning the elements of interaction (artifacts, content, processes, people, and interfaces), and the type of experience to be offered, which is moving toward an increasingly marked digitization.

From a theoretical perspective, this study makes its contribution to SDL studies by emphasizing the shift from an asset-centered to a service-centered value creation through the integration of AI technologies in specific contexts, such as the museum sector [28]. Furthermore, by emphasizing visitor engagement and the definition of personalized experiences, this research reinforces the knowledge generation on customer engagement conveyed by new technologies in the context of cultural institutions. In particular, the analysis of the cases shows that the implementation of AI, in its different declinations and forms, enhances the interaction between consumers and cultural resources for the co-creation of value, highlighting the use of a mostly perceptual mechanism [5], which allows museums to understand customers' needs and provide further opportunities to support them in a meaningful way, identifying and building an offer focused on them [1].

From a managerial point of view, the proposed analysis highlights how AI technologies imply specific know-how and resources that often have to be sought outside the boundaries of the cultural sector. It is therefore crucial, at least in the initial stages of technology implementation, to establish collaborations and partnerships with third parties capable of handling the different levels of complexity of AI. Lastly, a relevant aspect to be considered by museum managers is the use of a customer-centered approach, in a way that interaction with AI can bring out the individuality of visitors, who can thus personalize their visit experience according to their choices and preferences.

References

1. Leone, D., Schiavone, F., Appio, F.P., Chiao, B.: How does artificial intelligence enable and enhance value co-creation in industrial markets? An exploratory case study in the healthcare ecosystem. J. Bus. Res. **129**, 849–859 (2021)
2. Teece, D.J.: Profiting from innovation in the digital economy: enabling technologies, standards, and licensing models in the wireless world. Res. Policy. **47**(8), 1367–1387 (2018)

3. Boiano, S., Borda, A., Gaia, G., Rossi, S., Cuomo, P.: Chatbots and new audience opportunities for museums and heritage organisations. In: Electronic Visualisation and the Arts, pp. 164–171. BCS Learning & Development (2018)
4. Huang, M.H., Rust, R.T.: Engaged to a robot? The role of AI in service. J. Serv. Res. **24**(1), 30–41 (2021)
5. Lenka, S., Parida, V., Wincent, J.: Digitalization capabilities as enablers of value co-creation in servitizing firms. Psychol. Mark. **34**(1), 92–100 (2017)
6. Jarrahi, M.H.: Artificial intelligence and the future of work: human-AI symbiosis in organizational decision making. Bus. Horiz. **61**(4), 577–586 (2018)
7. Raj, M., Seamans, R.: Primer on artificial intelligence and robotics. J. Organ. Des. **8**, 1–14 (2019)
8. Dwivedi, Y.K., Pandey, N., Currie, W., Micu, A.: Leveraging ChatGPT and other generative artificial intelligence (AI)-based applications in the hospitality and tourism industry: practices, challenges and research agenda. Int. J. Contemp. Hosp. Manag. **36**(1), 1–12 (2023)
9. Mondal, S., Das, S., Vrana, V.G.: How to bell the cat? A theoretical review of generative artificial intelligence towards digital disruption in all walks of life. Technologies. **11**(2), 44 (2023)
10. Raisch, S., Krakowski, S.: Artificial intelligence and management: the automation–augmentation paradox. Acad. Manag. Rev. **46**(1), 192–210 (2021)
11. Ramaswamy, V., Ozcan, K.: What is co-creation? An interactional creation framework and its implications for value creation. J. Bus. Res. **84**, 196–205 (2018)
12. Vargo, S.L., Lusch, R.F.: Institutions and axioms: an extension and update of service-dominant logic. J. Acad. Mark. Sci. **44**, 5–23 (2016)
13. Vargo, S.L., Lusch, R.F.: Evolving to a new dominant logic for marketing. J. Mark. **68**(1), 1–17 (2004)
14. Solakis, K., Katsoni, V., Mahmoud, A.B., Grigoriou, N.: Factors affecting value co-creation through artificial intelligence in tourism: a general literature review. J. Tourism Futures. **10**(1), 116–130 (2022)
15. Grönroos, C., Voima, P.: Critical service logic: making sense of value creation and co-creation. J. Acad. Mark. Sci. **41**, 133–150 (2013)
16. Marini, C., Agostino, D.: Humanized museums? How digital technologies become relational tools. Mus. Manag. Curatorsh. **37**(6), 598–615 (2022)
17. Sacco, P.L., Ferilli, G., Tavano Blessi, G.: From culture 1.0 to culture 3.0: three socio-technical regimes of social and economic value creation through culture, and their impact on European cohesion policies. Sustainability. **10**(11), 3923 (2018)
18. Taormina, F., Baraldi, S.B.: Museums and digital technology: a literature review on organizational issues. Eur. Plan. Stud. **30**(9), 1676–1694 (2022)
19. Carignani, F., Clemente, L., Iodice, G., Bifulco, F.: Digital marketing in cultural heritage: an approach to Metaverse. In: Cultural Marketing and Metaverse for Consumer Engagement, pp. 142–163. IGI Global (2023)
20. Orea-Giner, A., Muñoz-Mazón, A., Villacé-Molinero, T., Fuentes-Moraleda, L.: Cultural tourist and user experience with artificial intelligence: a holistic perspective from the industry 5.0 approach. J. Tourism Futures. (2022)
21. Barile, S., Bassano, C., Piciocchi, P., Saviano, M., Spohrer, J.C.: Empowering value co-creation in the digital age. J. Bus. Ind. Mark. **39**(6), 1130–1143 (2021)
22. Clemente, L., Iodice, G., Carignani, F., Greco, F., Bifulco, F.: Phygital approach to value co-creation in international museums. Meas. Bus. Excell. **28**(2), 209–221 (2024)
23. Eisenhardt, K.M., Graebner, M.E.: Theory building from cases: opportunities and challenges. Acad. Manag. J. **50**(1), 25–32 (2007)
24. Yin, R.K.: Case Study Research: Design and Methods, vol. 5. Sage (2009)
25. Creswell, J.W., Creswell, J.D.: Research Design: Qualitative, Quantitative, and Mixed Methods Approaches. Sage Publications (2017)
26. Casillo, M., De Santo, M., Mosca, R., Santaniello, D.: An ontology-based Chatbot to enhance experiential learning in a cultural heritage scenario. Front. Artif. Intell. **5**, 808281 (2022)

27. Iodice, G., Clemente, L., Bifulco, F.: The role of Chatbot Technology in the cultural value proposition: a managerial perspective in human-like interaction. In: The Use of Artificial Intelligence in Digital Marketing: Competitive Strategies and Tactics, pp. 166–187. IGI Global (2024)
28. Shaw, G., Bailey, A., Williams, A.: Aspects of service-dominant logic and its implications for tourism management: examples from the hotel industry. Tour. Manag. **32**(2), 207–214 (2011)
29. Kirova, V.: Value co-creation and value co-destruction through interactive technology in tourism: the case of 'La Cité du Vin'wine museum, Bordeaux, France. Curr. Issues Tourism. **24**(5), 637–650 (2021)
30. Sfodera, F., Mingo, I., Mattiacci, A., Colurcio, M.: Night at the museum: technology enables visitor experiences. Sinergie Italian J. Manag. **38**(1), 231–250 (2020)

Chapter 8
The Impact of Digital Transformation on Family Security

Atheer Binsalman, Khalid Salman, and Maha Alqohmani

Abstract The research aims to determine digital transportation's impact on family security. The research sample amounted to 300 families with children at school stages, a non-random (intentional) non-probability sample. Research approach: descriptive-analytical approach. Research tools: the digital transformation questionnaire and the family security questionnaire. The research resulted in there is a statistically significant discrepancy between the dimensions of the digital transformation questionnaire (public sector, economy, society, and digital transformation (as a whole)) depending on the variable of the place of residence, where the value of (f), respectively, was statistically significant at the level of 0.05. There is a direct correlation statistically between the digital transformation (as a whole), each of its dimensions, family security (as a whole), and each of its dimensions, where the value of "r" reached its statistical significance at the level of (0.01). There is a difference in the percentage of participation of demographic variables affecting the dimensions of the family security questionnaire, where the multiple linear regression reached a substantial value of F (0.591) at a moral level (0.05). The father's educational level variable ranked first, and the father's age ranked last. In light of this, the researchers recommended urging the inclusion of training programs for the employees of the Ministry of Education to spread the culture of digital transformation (its mechanisms, strategies, goals, and importance) to contribute to spreading digital awareness and stimulating researchers to produce scientific research specialized in the aspect of family security.

Keywords Digital transformation · Digitization · Security · Family security

A. Binsalman · M. Alqohmani
Umm al-Qura University, Makkah, Saudi Arabia
e-mail: mhgohmani@uqu.edu.sa

K. Salman (✉)
Research Center, King Abdullah Medical City, Makkah Healthcare Cluster, Makkah, Saudi Arabia
e-mail: salman.k@kamc.med.sa

F. Schiavone et al. (eds.), *Advanced Perspectives and Trends in Digital Transformation of Firms, Networks, and Society*, Springer Proceedings in Business and Economics, https://doi.org/10.1007/978-3-031-80692-6_8

8.1 Introduction and Background

At the beginning of the last century, there was talk about the emergence of an electronic space and a virtual world caused by the repercussions of modern technology. This transformation quickly became a feature of the era, but it was he who drew the shape and dimensions of the era in which we live and enabled the contemporary communications revolution and Information Development to penetrate modern society, which has become one of the urgent needs and important necessities in our current time. This is what has made the world witness a global digital revolution today that has made tremendous leaps represented by multiple forms of modern technology [1], digital technologies, and the internet as the main engine that leads society toward advancement, development, and prosperity, in addition to being the main gateway for society to enter this new revolution and keep up with it, which has been called digital transformation [2].

The increasing reliance on digitization in modern society's economic and social system has created indispensable pillars with which the importance of applying digital transformation has increased. Virtual, informational, and interactive capabilities have increased with it, where this revolution was able to attract a large segment of society [3], one of the most affected of which was the family, as it is located among the most important social institutions, perhaps this is what emphasized the importance of its role in caring for its members and the need to raise their awareness and insight into their need to adapt to the changes of the times and take on new roles imposed by the nature of the digital revolution [4]. However, the gap between technological development, on the one hand, and the inability of family members to deal with this technology, on the other, created a kind of threat to the family, especially the security aspect, in the face of this acceleration of knowledge, the qualitative leap in the information civilization and the unlimited openness of the world to different customs, traditions, cultures, and the enormous prevalence of virtual communities, the family has become difficult to fulfill its role and maintain its security [3].

Therefore, the role played by the family is increasing in this era, as these developments are an easy way to affect its security. Family security is a crucial requirement that the family must provide to its members. Therefore, the family must perform its role to the fullest to achieve this purpose, starting with physiological needs and reaching other needs, such as self-realization, appreciation, and respect [5]. Since the family is the central nucleus in building any society, with which society can progress, prosper, and develop, attention has been focused on how to take care of the family, protect it, and the need to support its security because the family is considered the backbone in shaping the main features of the individual, so the security of the family and making it an Incubator to ensure the security of the whole society is the first function assigned to all groups of society collectively and individuals from this point of view [6], many scientists and specialists in various fields paid attention to the issue of family security, and it became the priorities in which family members should direct their care and efforts [7].

When we see what we are experiencing from the technical developments around us, we find their reflection on our lives significantly, that accelerated digital transformation necessitates the family to try to keep up with and in line with it, which requires the support of the family for its security and stability, which makes the family at stake in front of that digital transformation and the change in the components of family security, and the study of Haya Alharbi emphasizes on the importance of supporting family security among the members of the research sample and the need to provide electronic devices as a requirement to keep pace with digital transformation [8], as the qualitative shift from traditional to digitization requires acceptance of digital transformation within one of the pillars of modern society, so the family should focus on maintaining its security and safety, and this is one of the most critical roles in light of this transformation, and in light of the above, this study attempts to reveal the impact of digital transformation on family security.

8.2 Research Exploration

8.2.1 Research Questions

RQ1. To what extent does digital transformation affect family security, depending on demographic changes?

RQ2. What is the relationship between the impact of digital transformation on family security depending on demographic variables?

RQ3. What are the most demographic variables affecting the dimensions of digital transformation?

RQ4. What are the priorities of demographic variables affecting the dimensions of family security?

8.2.2 Research Importance

Scientific Significance First, we present information and knowledge in the field of specialization, as the studies have not been subjected to studying the impact of digital transformation on family security (as far as the researchers are aware) in what was discussed, put forward, published, and circulated.

Second, the current research provides information and knowledge related to digital transformation to understand digital developments and provide knowledge that improves the family's ability to make plans and overcome obstacles to digital transformation, supporting the pillars of family security.

Third, we will focus on presenting concepts related to family security and theories related to digital transformation data and identify the factors most affected and influencing them in consolidating the stable fixed rules of the family in light of the surrounding developments.

Practical Importance The research highlights a topic of importance in the lives of individuals, namely family security. Shedding light on the reality of digital transformation and its acceleration in the Kingdom of Saudi Arabia by wide leaps. The study contributes to the family's insight into its essential role in stabilizing the pillars of family security in light of the digital transformation characteristic of the era and benefits from the results of the current research by placing families as the focus of attention of government departments and family associations during their training and awareness programs in enriching families with ways and methods to contribute to raising the level of family security.

The current research is a new addition to the studies that dealt with the two research axes of digital transformation and family security as well as a new reference to be added to research centers, scientific institutions, and knowledge libraries.

8.2.3 Research Objectives

Finding statistically significant differences between the dimensions of the digital transformation questionnaire and the dimensions of the family security questionnaire, depending on the different demographic characteristics. Finding the statistically significant correlation between the dimensions of the digital transformation questionnaire in its various dimensions and between the dimensions of the family security questionnaire in its various dimensions. Finding the most variables and the most influential dimensions on the dimensions of the digital transformation questionnaire and finding the most variables and the most influential dimensions on the dimensions of the family security questionnaire.

8.2.4 Research Assignments

Assessment if there are significant statistical differences between the dimensions of the digital transformation questionnaire (public, economic, and community sector) and the dimensions of the family security questionnaire (intellectual, cognitive, and socioeconomic) depending on the different demographic characteristics (place of residence, father's educational level, mother's educational level, father's age, mother's age, father's job, mother's job, income level, number of children and children's school stages). Also, it should assess any statistically significant correlation between digital transformation (as a whole), its dimensions, family security (as a whole), and each dimension.

8.2.5 Research Terms

Digital Transformation Nawal Al-balushiya et al. define it as: "a government project that includes all the services of various institutions and sectors in the country, and consists in transforming vital and basic services related to the service of individuals, institutions, and various investments, from their traditional form to the smart electronic form, relying on modern and advanced technologies." [9]

However, the researchers define it procedurally as: "changing the traditional working and living lifestyle to a modern digital one through the use of the latest modern and digital technologies in various fields (government, public and private sectors, religious, health, educational, economic, banking, community, cultural and entertainment) and others to facilitate procedures and services in various sectors to save time, effort and money and keep abreast of developments."

Family Security Whereas Fahad Al-Tayyar defines it as: "reassuring the individual and the family, freeing them from fear and satisfying their psychological, social, spiritual, physical, and mental needs by achieving family cohesion and social control within the family, which accordingly achieves more security and stability for their society." [10]

But the researchers define it procedurally as: "the security of the family internally among its members, and externally among those around it, such as relatives, neighbors, friends, and acquaintances, in terms of (intellectual, cognitive, social, and economic), in light of contemporary conditions and widespread digital transformations, through the performance of its members of their responsibilities to preserve its entity and prevent any defect infiltrating it that could threaten it is being a single interconnected unit."

8.3 Research Method

8.3.1 Methodology

The researchers in this study followed the descriptive approach, defined by Ibrahim Abu Aqeel as "the descriptive approach means a study that deals with the study of reality and phenomena by describing them accurately and expressing this description either quantitatively by giving the phenomenon or its size and degrees of correlation with other phenomena a numerical description, or how to describe the phenomenon in writing with specifying its characteristics, to understand these phenomena and their relationship with other phenomena and the degree of their correlation" [11].

8.3.2 Research Variables

These are the research variables that have been studied to find relationships between them, process them statistically, and come up with publishable results, as the research included the following variables: **Independent variable** is the one whose impact on the dependent variable will be measured, and in this research, the independent variable is digital transformation. **The dependent variable** is the one whose influence will be measured by the independent variable, and in this research, the dependent variable is family security. **Demographic variables** are the variables related to the description of the sample members, which affect the dependent and independent research variables and work with them and from them (place of residence, the educational level of the father, the educational level of the mother, the father's age, the mother's age, the father's job, the mother's job, income level, the number of children, the educational stages of the children, whether the children live with their parents or not and the children own digital devices).

8.3.3 Research Tool

The research relied on the questionnaire as a data collection tool. It designed and built this tool starting with the research topic and its questions and ending with the formulation of research hypotheses after reading the research literature available to the researchers in order to achieve the hypotheses and reach the results.

Procedures Review the theoretical framework and previous studies related to the research topic to benefit from them. Identification of the research variable (the impact of digital transformation on family security). Preparation and construction of the research tool: the questionnaire was the tool used in this research, it included the following axes: General data form, Special data form, Digital transformation questionnaire, and Family security questionnaire.

8.3.4 Research Limitation

Boundaries The researchers chose the Makkah region, which includes three major cities within its administrative scope (Makkah, Jeddah, and Taif). Still, they limited it to the main cities without their villages so the researchers could read those cities geographically to determine the influences surrounding the sample.

Human The sample of the study was selected by the method of non-random (intentional) sampling, also called the non-probability sample, as it included 300 families with children and required the answer of one of the parents or their representatives who have a son/daughter or more at study stages, the researchers have selected a group of members of the research sample in Makkah governorate in each of the cities. The researchers expanded the access to them through an electronic link to the questionnaire (Google Form), in addition to converting it into an electronic Code (barcode). Then, the researchers used a set of channels to obtain the responses of the members of the research sample.

8.4 Results and Discussion

- There is a statistical difference between the dimensions of the digital transformation questionnaire (public sector, economy, society, and digital transformation (as a whole)) depending on the variable of the place of residence, where the value of (f), respectively, was statistically significant at the level of 0.05.
- There are statistical differences between the dimensions of the family security questionnaire (intellectual, social, and family security (as a whole)) depending on the variable of the place of residence, where the value of (f), respectively, was statistically significant at the indicative level (0.05).
- There are no statistical differences between the dimensions of the digital transformation questionnaire (economy, society, and digital transformation (as a whole)) depending on the variable of the father's educational level, where the value (f) was, respectively, non-statistically significant.
- There is a statistical difference between the dimensions of the family security questionnaire (cognitive, social, economic, and family security (as a whole)) depending on the variable of the father's educational level, where the value of (f) was, respectively, statistically significant at an indicative level (0.05).
- There is a statistical difference between the dimensions of the digital transformation questionnaire (as a whole) depending on the variable of the mother's educational level, where the value of (f) was, respectively, statistically significant at the indicative level (0.05).
- There are statistical differences between the dimensions of the family security questionnaire (intellectual, economic, (and family security as a whole)) depending on the variable of the mother's educational level, where the value of (f) was, respectively, statistically significant at an indicative level (0.05).
- There are no statistical differences between the dimensions of the digital transformation questionnaire (public sector, economy, society, and digital transformation (as a whole)) depending on the variable of the father's age, where the value of (f) was, respectively, non-statistically significant.
- There are no statistical differences between the dimensions of the family security questionnaire (intellectual, cognitive, social, economic, and family security (as a whole)) depending on the variable of the father's age, where the value (f) was, respectively, non-statistically significant.
- There are no statistical differences between the dimensions of the digital transformation questionnaire (public sector, economy, society, and digital

transformation (as a whole)) depending on the variable of the mother's age, where the value of (f) was, respectively, non-statistically significant.

- There are no statistical differences between the dimensions of the family security questionnaire (intellectual, cognitive, social, economic, and family security (as a whole) depending on the variable of the mother's age, where the value of (f) was, respectively, non-statistically significant.
- There are no statistical differences between the dimensions of the digital transformation questionnaire (public sector, economy, society, and digital transformation (as a whole)) depending on the variable of the father's function, where the value (f) was, respectively, non-statistically significant.
- There are no statistical differences between the dimensions of the family security questionnaire (intellectual, cognitive, social, economic, and family security (as a whole) depending on the father's job variable, where the value of (f) was, respectively, non-statistically significant.
- There are no differences between the dimensions of the digital transformation questionnaire (public sector, economy, society, and digital transformation (as a whole)) depending on the variable of the parent function, where the value of (f) was, respectively, non-statistically significant.
- There are no differences between the dimensions of the family security questionnaire (intellectual, cognitive, social, economic, and family security (as a whole) depending on the variable of the mother's function, where the value of (f) was, respectively, statistically significant.
- There are no statistical differences between the dimensions of the digital transformation questionnaire (public sector, economy, society, and digital transformation (as a whole)) depending on the income level variable, where the value of (f) was, respectively, non-statistically significant.
- There are no statistical differences between the dimensions of the family security questionnaire (intellectual, cognitive, social, economic, and family security (as a whole) depending on the income level variable, where the value (f) was, respectively, statistically significant.
- There is a direct correlation statistically between the digital transformation (as a whole) and each of its dimensions, where the value of "r" reached its statistical significance at the level of (0.01).
- There is a difference in the percentage of participation of demographic variables affecting the dimensions of the digital transformation questionnaire, where the multiple linear regression reached a substantial value of F (0.428) at a significant level (0.05), and the variable of the mother's educational level ranked first and the place of residence ranked last.
- There is a difference in the percentage of participation of demographic variables affecting the dimensions of the family security questionnaire, where the multiple linear regression reached a substantial value of F (0.591) at a moral level (0.05). The father's educational level variable ranked first, and the father's age ranked last.

8.5 Conclusion and Recommendation

The negative effect findings of a few digital transformation variables on family security direct the competent authorities in the Ministries of Education and Ministries of Informatics to provide community partnerships with heads of families within the framework of inter-programs that help them to improve their knowledge levels and spreading the culture of digital transformation among members of society to increase the awareness of family members of the current era and its changes.

Also, to work on hackathons, events, and conferences to support innovative ideas and present them on sound scientific bases, which aims to reduce cultural conflicts between generations and contributes to achieving family security by focusing on discussing topics related to the family and all that supports its security and stability as well as stimulate research centers and researchers to produce scientific research specialized in family security.

Acknowledgments I want to thank Umm al-Qura University, Makkah, Saudi Arabia, for the Master's degree and my mentor for directing me to use better methodologies to reach the optimal results from this study while writing this research. I also thank my family for supporting me materially and morally through my education journey, and special thanks to my husband for standing with me until the end of this achievement. I appreciate the opportunity to present and publish this scientific research from the Department of Management & Quantitative Studies, University of Naples Parthenope, Italy, and the Management Department, Paris School of Business, France.

References

1. Break, M.H.: The reality of the digital transformation of NGOs and a proposed vision from the perspective of social planning to activate it. J. Fut. Soc. Sci. **11**(1), 245–287 (2022)
2. Eid, B., Eid, Y.: The role of digital transformation as a mechanism to activate the performance of the educational process in university education. J. Humanit. Lit. Stud. **27** (2022)
3. Zgheina, N.: Family security in the shadow of the virtual community. In: 1st International Conference: Family Security Reality and Challenges, Turkey (2018)
4. Saleem, H.: Educational security for the Egyptian family in light of the Corona pandemic, from the educational Elite's perspective (a proposed scenario). J. Educ. Sohag Univ., 307–402 (2021)
5. Wahba, S.: Family security and its relationship to school bullying in middle school students. J. Res. Field Spec. Educ. **7**(36), 1–68 (2021)
6. Salmi, L., Boukacha, S.: Family security in the Islamic West through Fiqh principles. Master's Thesis. Yahya Fares University in Medea, Faculty of Humanities and Social Sciences, Department of Humanities, Algeria (2021)
7. Al-Gohary, S.A.: Marital disputes management strategies for battered wife and their relationship with family security for children. J. Home Econ. **36**(2), 35–84 (2020)
8. Al-Harbi, H.: Determinants of family security among Saudi university students. A descriptive study was applied to female students of King Faisal University in Al-Hasa. J. Stud. Soc. Work Humanit., 7–37 (2014)
9. Al-Balushi, N.A., Al-Harasi, N., Al-Awfi, A., Ali, A.: The reality of digital transformation in Omani institutions. J. Inf. Stud. Technol. **1**(2), 14–16 (2020)

10. Al-Tayyar, F.: The interactive relationship in socialization between parents and children and its relationship to family security. Arab J. Secur. Stud., 29–58 (2013)
11. Abu-Aqeel, I.: Scientific Research: its Methods, Designs, and Statistical Methods, 1st edn. Dar Al-Ayyam for Publishing and Distribution, Amman (2020)

Chapter 9
Digital Transformation and Innovation in Services: An Exploratory Study in Brazilian Museums

Eduardo Raupp de Vargas

Abstract Digital transformation presents itself as a new paradigm in the pace and direction of innovations. In services, the potential of digitalization replacing interpersonal relationships challenges the very foundations of the co-creation of value while enabling previously unthinkable forms of interaction in new service ecosystems. This scenario is exciting in museums that have taken on new purposes and have been affirmed as effervescent spaces of sociocultural debates. Digital technologies have challenged them to develop innovation strategies that allow them to play a leading role in this reconfiguration of services. The objective of the ongoing research and exploratory studies presented here is to analyze how managing innovation in services takes place and the context of digital transformations in the ecosystem of services of museums. Two cases were selected in this study: the development of the website Brasiliana Museus by the Brazilian Institute of Museums and innovation initiatives developed within the National Museum linked to the Federal University of Rio de Janeiro. The results contribute to the debate on the technology-services relationship in the context of digital transformation, aiming to consolidate an integrative approach to innovation in museum services.

Keywords Digital transformation · Innovation in services · Museums

9.1 Introduction

Digital transformation presents itself as a new paradigm in the pace and direction of innovations. In services, the potential of digitalization replacing interpersonal relationships challenges the very foundations of the co-creation of value while enabling previously unthinkable forms of interaction in new service ecosystems.

E. R. de Vargas (✉)
COPPEAD, Universidade Federal do Rio de Janeiro, Rio de Janeiro, Brazil
e-mail: eduardo.raupp@coppead.ufrj.br

F. Schiavone et al. (eds.), *Advanced Perspectives and Trends in Digital Transformation of Firms, Networks, and Society*, Springer Proceedings in Business and Economics, https://doi.org/10.1007/978-3-031-80692-6_9

This scenario is especially exciting in museums. Viewed by most as stable organizations, museums have taken on new purposes and have been affirmed as effervescent spaces of sociocultural debates. Digital technologies have challenged them to develop innovation strategies that allow them to play a leading role in this reconfiguration of services. The Musée d'Orsay recently announced that it broke its attendance record when it launched an interactive exhibit about the work produced by Van Gogh in his last months of life spent in France. In the exhibit, artificial intelligence trained with the texts of a hundred letters written by the painter at that time directly answered questions from the audience about the life and work of the painter as if he were there personally [1]. The success of the exhibit in the Parisian Museum illustrates how digital innovations can add new services, expanding audiences and invigorating museums and cultural services of all kinds [2]. However, it is necessary to go beyond the discussion of the impact of these technologies and understand how the management of these organizations can explore digitalization, innovate in their services, and not lose sight of their mission [3]. As Simone, Cerquetti, and La Sala [4] pointed out, the physical and virtual are merged together in a new configuration of how we view museums. We maintain that it is through managing innovation in services that museums will escape the technicist trap and develop integrative strategies that place digitalization as an enhancement of potential in their activities and not a substitute.

The role of technology has always been at the heart in the various approaches to innovation in services when analyzing its determinants. In the beginning the predominated vision was that the dynamics of the service sector could be explained directly by the technological diffusion. This resulted in what came to be known as a technicist bias toward innovation in services [5, 6]. Given the undeniable growth of the sector, the research on innovation in services surpassed this bias by building an integrative approach [7]. The interactivity in the provision of services and the intangibility of many of these activities have taken up a symmetrical space in the analysis of the factors that compete for innovation [8], as well as the development of the notion of service logic. The pervasive character of the digital transformations, however, has again brought to the forefront the theoretical debate on the determinants of innovation in services [9, 10]. Could we say that we are on the threshold of the triumph of the technicist approach?

We understand that this perspective remains insufficient for analyzing innovation in services and this is confirmed when examining different branches of services [11]. Digital transformation has enabled forms of interaction that generate opportunities for co-creating value [12]. More than that, it has expanded the extension and intensity of service relations even in consolidated activities.

In cultural services, which is the case of museums, literature recognizes the profusion of innovative services that have become more consolidated [13]. Digitization, even in its early stages, allows administrative services to be modernized. The relationship between museums and the public has been renewed [14–18]. New digital technologies allow the relationship with visitors to begin before they actually arrive at the museum in order to create innovative services during the visit and allow this relationship to extend over time [16, 19]. Online visits and social networks break barriers of space and time and generate new forms of public engagement with museums [20, 21]. The digitization of collections and remote educational tools establish a new reach for scientific and educational outreach activities. Accessibility and inclusion allow new audiences to be integrated as digital media reach the masses [22]. New forms of narrative allow us to unveil different attractions for the existing collections [23] and collaborate for reassessments that advance what has been called the decolonization of museums [24, 25]. New technologies revolutionize research and development that allow the advancement of knowledge disseminated across different types of museums [4, 26]. The transformation of museums was accelerated even more during and after the Covid-19 pandemic, as several examples show [18, 27, 28].

These transformations, however, are rarely analyzed from the point of view of managing innovation in services [29–31]. This diminishes the perception of the role of museums in these processes and the recognition of the transformative dynamics taking place in the ecosystem in which their services are inserted. Russo-Spena et al. [3] mentioned the need and relevance of a management vision regarding these processes, and they moved forward with the proposal to analyze the changes of business models in the area of cultural heritage. Along these same lines, we seek to focus the role of innovation management in this context and investigate how managing innovation in services in museums takes place in the context of digital transformations.

We assume that an integrative approach that considers the technological impact but does not neglect the other actors of the service ecosystem, especially their teams, managers, and visitors who co-create the value offered by museums, will allow identifying new logic and forms of innovation in museums [7, 32–34]. Here we will present the initial results of two exploratory studies: one in a Brazilian museum and another in the governmental institute set up for the museums in the country.

9.2 Objective

The objective of the ongoing research and exploratory studies presented here is to analyze how managing innovation in services takes place and the context of digital transformations in the ecosystem of services of museums. Our focus includes both classic and virtual museums where we seek to identify the degree of digitalization, the innovation strategies by museums, and to analyze innovation cases with a look at their management.

We understand that by achieving these objectives we can provide support for improving innovation management in museums and substantiate public policies enabling digital transformation advancement. Specifically in the cases discussed here, we seek to identify the degree of digitization that exists and the strategic perspective to innovate in services in the scenario where such aspects are at the heart of the new configuration of what we understand to be a museum.

9.3 Methods

Understanding the dynamics of innovations in services in museums requires a deepening of the ongoing initiatives. The case study method lends itself exactly to this, offering the possibility of a detailed description of the innovation process with eyes focused on its management. Two cases were selected in this study: the development of the website Brasiliana Museus by the Brazilian Institute of Museums and innovation initiatives developed within the National Museum linked to the Federal University of Rio de Janeiro. The case of the Brasiliana website was studied based on documents, scientific articles, and visits and use of this portal. The research initiatives with the National Museum, on the other hand, were based on interviews with its managers. The evidence collected was submitted to a content analysis considering the following categories: description of innovation, determinants of innovation, actors involved, resources employed, and main results.

9.4 Findings

The Brasiliana Museus Portal [35] was developed by the Brazilian Institute of Museums (IBRAM) in partnership with a research laboratory from the University of Brasilia [36]. Based on the application of Tainacan software and inspired by websites such as Europeana and Google Arts & Culture, the portal brings together the digitized collections of 20 Brazilian museums that voluntarily joined the initiative. It should be noted that IBRAM currently has 3519 museums on record in Brazil, 27 of which are directly linked to the agency with the others being linked to the federal government, state governments, municipalities, in addition to the private initiative. In our analysis, the digitization of collections can be considered one of the initial steps in the process of digital transformation but is still distant from other stages where it gives rise to new services [37]. Currently the portal has an important development, which is the construction of an analytical panel (https://cadastro.museus.gov.br/painel-analitico/) that makes it possible to search and filter information from the 3905 museums registered with IBRAM.

The second case study was carried out at the National Museum that is linked to the Federal University of Rio de Janeiro, a public federal university. Founded in 1818 by the King of Portugal Dom João VI, the museum initially had the donation of its workforce that belonged to the royal family and its alleged intent was to collaborate toward the cultural and economic progress of the country, a colony at the time (www.museunacional.ufrj.br). It is a university museum with a scientific vocation, and it sponsors teachers and researchers for developing graduate study activities in Social Anthropology, Archeology, Botany, Linguistics, Indigenous Languages, Zoology, and Geosciences. Destroyed by a fire in 2018, the museum today has developed a robust reconstruction program with support from UNESCO, the federal government, and the private initiative, in what is known as Museu

Nacional Vive, meaning that the National Museum continues on. This Museum was chosen so as to understand how the trauma from the fire is influencing the digitalization strategy. With a reopening scheduled for 2026, the museum has maintained virtual exhibits such as the "First Brazilians," which had its physical collection preserved because it was in a temporary exhibit at the National Congress in Brasilia (https://museunacional.ufrj.br/destaques/os_primeiros_brasileiros_exposicao.html). In addition, an invitation had been made prior to the fire to visit the main highlights of the collection in partnership with Google (https://artsandculture.google.com/project/museu-nacional-brasil?hl=pt).

Several initiatives were identified such as the digitization of scientific collections, the creation of a virtual center for visitors, and initiatives related to gamification and the metaverse. The National Museum, therefore, has seen advances in digitalization that go beyond the digitalization of collections and involves initiatives that seek to change the relationship with its audience. It is true that these initiatives have a certain imbalance because there are advances in certain areas but not yet in others. It can be noticed that there is no integrated innovation strategy, but that the Museum advances in proposals that use digital technologies while preparing for its physical reopening scheduled for 2026.

9.5 Implications

We understand that the results of this exploratory study allow to contribute to the debate on the technology-services relationship in the context of digital transformation with a view to consolidating an integrative approach on innovation in services in museums. From a practical point of view, mapping innovation management practices allows identifying the current stage of the changes taking place in Brazilian museums and highlighting the debate on innovation management in these services. In the case of the Brasiliana Museus portal, for example, we see the fulfillment of the digitization of collections in participating museums. The goal now is to involve more museums and move toward other technological solutions. In the case of the National Museum, the challenge is to develop a strategy of innovation in services that allows the Museum to reopen already counting on the potential that a digital experience can bring. Further results of this research are to allow a critical analysis of the Brazilian public policies and their relationship with the purposes that characterize these organizations contemporaneously.

References

1. Binswanger, J.: Musée d'Orsay breaks attendance records with interactive Vincent van Gogh exhibition. Smithsonian Magazine. https://www.smithsonianmag.com/smart-news/musee-dorsay-breaks-attendance-records-with-immersive-ai-van-gogh-show-180983789/ (2024)

2. Najda-Janoszka, M., Sawczuk, M.: Exploring engagement in value creation–a multi-stakeholder perspective in the museum context. Mus. Manag. Curatorsh. **38**(1), 92–109 (2023). https://doi.org/10.1080/09647775.2022.2158910
3. Russo-Spena, T., Tregua, M., D'Auria, A., Bifulco, F.: A digital business model: an illustrated framework from the cultural heritage business. Int. J. Entrep. Behav. Res. **28**(8), 2000–2023 (2022). https://doi.org/10.1108/IJEBR-01-2021-0088
4. Simone, C., Cerquetti, M., La Sala, A.: Museums in the Infosphere: reshaping value creation. Mus. Manag. Curatorsh. **36**(4), 322–341 (2021). https://doi.org/10.1080/09647775.2021.1914140
5. Gallouj, F.: Innovating in reverse: services and the reverse product cycle. Eur. J. Innov. Manag. **1**(3), 123–138 (1998). https://doi.org/10.1108/14601069810230207
6. de Vargas, E.R., Zawislak, P.A.: Inovação em serviços no paradigma da economia do aprendizado: a pertinência de uma dimensão espacial na abordagem dos sistemas de inovação. Revista de Administração Contemporânea. **10**(1), 139–159 (2006)
7. Gallouj, F.: Innovation in the Service Economy: The New Wealth of Nations. Edward Elgar (2002)
8. Desmarchelier, B., Djellal, F., Gallouj, F.: Users' involvement in value co-creation: the more the better? Eur. Manag. Rev. **17**(2), 439–448 (2019). https://doi.org/10.1111/emre.12365
9. Helmer, J., Huynh, T.M.T., Lobacz, K., Kör, B., Wakkee, I.: Innovating digitally for services: a review of innovation process literature focused on digital innovation and service innovation. Procedia Comput. Sci. **192**, 2797–2806 (2021). https://doi.org/10.1016/j.procs.2021.09.050
10. Kitsios, F., Kamariotou, M.: Service innovation process digitization: areas for exploitation and exploration. J. Hosp. Tour. Technol. **12**(1), 4–18). Emerald Group Holdings Ltd (2021). https://doi.org/10.1108/JHTT-02-2019-0041
11. Park, H., Lee, M., Back, K.J.: A critical review of technology-driven service innovation in hospitality and tourism: current discussions and future research agendas. Int. J. Contemp. Hosp. Manag. **35**(12), 4502–4534 (2023). https://doi.org/10.1108/IJCHM-07-2022-0875
12. Dalle Nogare, C., Murzyn-Kupisz, M.: Do museums foster innovation through engagement with the cultural and creative industries? J. Cult. Econ. **45**(4), 671–704 (2021). https://doi.org/10.1007/s10824-021-09418-3
13. Sundbo, J., Rubalcaba, L., Gallouj, F.: Servitization in the creative and cultural industries. Int. J. Qual. Serv. Sci. **14**(1), 65–85 (2022). https://doi.org/10.1108/IJQSS-01-2021-0017
14. Evrard, Y., Krebs, A.: The authenticity of the museum experience in the digital age: the case of the louvre. J. Cult. Econ. **42**(3), 353–363 (2018). https://doi.org/10.1007/s10824-017-9309-x
15. Navarrete, T.: Digital heritage tourism: innovations in museums. World Leis. J. **61**(3), 200–214 (2019). https://doi.org/10.1080/16078055.2019.1639920
16. Raimo, N., De Turi, I., Ricciardelli, A., Vitolla, F.: Digitalization in the cultural industry: evidence from Italian museums. Inter. J. Entrep. Behav. Res. **28**(8), 1962–1974 (2022). https://doi.org/10.1108/IJEBR-01-2021-0082
17. Tham, A., Liu, Y., Loo, P.T.: Transforming museums with technology and digital innovations: a scoping review of research literature. Tour. Rev. (2023). https://doi.org/10.1108/TR-02-2023-0112
18. Trunfio, M., Lucia, M.D., Campana, S., Magnelli, A.: Innovating the cultural heritage museum service model through virtual reality and augmented reality: the effects on the overall visitor experience and satisfaction. J. Herit. Tour. **17**(1), 1–19 (2022). https://doi.org/10.1080/1743873X.2020.1850742
19. Borowiecki, K.J., Navarrete, T.: Digitization of heritage collections as indicator of innovation. Econ. Innov. New Technol. **26**(3), 227–246 (2017). https://doi.org/10.1080/10438599.2016.1164488
20. Romolini, A., Fissi, S., Gori, E.: Visitors engagement and social media in museums: evidence from Italy. Int. J. Digit. Cult. Electron. Tour. **3**(1), 36–53 (2020)

21. Zollo, L., Rialti, R., Marrucci, A., Ciappei, C.: How do museums foster loyalty in tech-savvy visitors? The role of social media and digital experience. Curr. Issue Tour. **25**(18), 2991–3008 (2021). https://doi.org/10.1080/13683500.2021.1896487
22. Garcia Carrizosa, H., Sheehy, K., Rix, J., Seale, J., Hayhoe, S.: Designing technologies for museums: accessibility and participation issues. J. Enabling Technol. **14**(1), 31–39 (2020). https://doi.org/10.1108/jet-08-2019-0038
23. Carrozzino, M., Bergamasco, M.: Beyond virtual museums: experiencing immersive virtual reality in real museums. J. Cult. Herit. **11**(4), 452–458 (2010). https://doi.org/10.1016/j.culher.2010.04.001
24. Baradaran, A.: Towards a decolonial I in AI: mapping the pervasive effects of artificial intelligence on the art ecosystem. AI & Soc. **39**(1), 7–19 (2023). https://doi.org/10.1007/s00146-023-01771-5
25. Hakopian, M.F.: Art histories from nowhere: on the coloniality of experiments in art and artificial intelligence. AI Soc. **39**(1), 29–41 (2023). https://doi.org/10.1007/s00146-023-01768-0
26. Palumbo, R., Manna, R., Cavallone, M.: The managerialization of museums and art institutions: perspectives from an empirical analysis. Int. J. Organ. Anal. **30**(6), 1397–1418 (2022). https://doi.org/10.1108/IJOA-10-2020-2438
27. Kamariotou, V., Kamariotou, M., Kitsios, F.: Strategic planning for virtual exhibitions and visitors' experience: a multidisciplinary approach for museums in the digital age. Digit. Appl. Archaeol. Cult. Herit. **21**, e00183 (2021). https://doi.org/10.1016/j.daach.2021.e00183
28. Trunfio, M., Campana, S., Magnelli, A.: Measuring the impact of functional and experiential mixed reality elements on a museum visit. Curr. Issue Tour. **23**(16), 1990–2008 (2020). https://doi.org/10.1080/13683500.2019.1703914
29. Angelo, V.D.: The value of management in the digitalization era: evidence from an Italian museum. Int. J. Digit. Cult. Electron. Tour. **3**(1), 94–108 (2020)
30. Bertacchini, E.E., Dalle Nogare, C., Scuderi, R.: Ownership, organization structure and public service provision: the case of museums. J. Cult. Econ. **42**(4), 619–643 (2018). https://doi.org/10.1007/s10824-018-9321-9
31. Lusch, R.F., Nambisan, S.: Service innovation: a service-dominant logic perspective. MIS Q.: Manag. Inf. Syst. **39**(1), 155–175 (2015). https://doi.org/10.25300/MISQ/2015/39.1.07
32. Camarero, C., Garrido, M.J.: Fostering innovation in cultural contexts: market orientation, service orientation, and innovations in museums. J. Serv. Res. **15**(1), 39–58 (2012). https://doi.org/10.1177/1094670511419648
33. Kershaw, A., Bridson, K., Parris, M.A.: The muse with a wandering eye: the influence of public value on coproduction in museums. Int. J. Cult. Policy. **26**(3), 344–364 (2020). https://doi.org/10.1080/10286632.2018.1518980
34. Koskela-Huotari, K., Edvardsson, B., Jonas, J.M., Sörhammar, D., Witell, L.: Innovation in service ecosystems-breaking, making, and maintaining institutionalized rules of resource integration. J. Bus. Res. **69**(8), 2964–2971 (2016). https://doi.org/10.1016/j.jbusres.2016.02.029
35. IBRAM – Instituto Brasileiro de Museus: Brasiliana Museus – Plataforma do Patrimônio Museológico Brasileiro. https://brasiliana.museus.gov.br/ (2024)
36. Siqueira, J., Lopes, D.M.: Acervos Agregados do Instituto Brasileiro de Museus: Desenvolvimento do Painel de Visualização Analítica. Tendências Da Pesquisa Brasileira Em Ciências Da Informação. **15** (2022) https://orcid.org/0000-0003-0246-8443
37. Lu, S.E., Moyle, B., Reid, S., Yang, E., Liu, B.: Technology and museum visitor experiences: a four-stage model of evolution. Inf. Technol. Tour. **25**(2), 151–174). Springer Science and Business Media Deutschland GmbH (2023). https://doi.org/10.1007/s40558-023-00252-1

Chapter 10
Artificial Intelligence in Project Management: An Empirical Study of Project Managers' Appropriation of Generative Chatbots

Marco Greco and Vincenzo Corvello

Abstract The inclusion of AI tools, such as chatbots, in project management is transforming the field. This article investigates how project managers are embracing and customizing these tools, with a specific focus on ChatGPT, to enhance project management practices. Employing Adaptive Structuration Theory, the research examines how project managers are adopting generative AI. It considers factors like Attitude Toward Innovation, Peer Influence, and Alignment of Tasks with Technology, using a survey conducted among project managers in Italy. The methodology used to analyze the data is based on Partial Least Squares—Structural Equation Modeling. The findings validate the importance of the proposed factors in the adoption of AI tools. Attitude Toward Innovation and Peer Influence are found to have a positive impact on the innovative and flexible use of AI in project management. Alignment of Tasks with Technology is crucial for successful integration of AI, influencing both innovative behavior and flexible adoption. The study underscores the significance of a forward-thinking mindset, peer interactions, and task compatibility in the effective utilization of AI tools in project management. It proposes potential avenues for future research, including exploration of cultural and organizational contexts, as well as the rapid advancements in AI technologies.

Keywords Project Managers · Generative Artificial Intelligence · ChatGPT · Appropriation theory · Structural Equation Modeling

M. Greco
Wish Innovation, Rende, Italy

V. Corvello (✉)
Department of Engineering, University of Messina, Messina, Italy
e-mail: vincenzo.corvello@unical.it

F. Schiavone et al. (eds.), *Advanced Perspectives and Trends in Digital Transformation of Firms, Networks, and Society*, Springer Proceedings in Business and Economics, https://doi.org/10.1007/978-3-031-80692-6_10

10.1 Introduction

These sophisticated AI systems, capable of learning, adjusting, and generating content, are reshaping the concept of innovation, enabling businesses to automate intricate creative tasks that previously demanded significant human intelligence and time [1]. The emergence of AI is marking the onset of a new era across various domains, and project management is no exception [2]. These chatbots go beyond simple task automation; they are transforming how project managers engage with team members, stakeholders, and the project landscape, offering fresh avenues for project planning, decision-making, and risk assessment [3]. The incorporation of AI tools into project management signifies a transformative change in the field, particularly through the adoption of generative chatbots by project managers.

Despite the promising integration of AI in project management, there remains a distinct gap in understanding how these technologies, particularly generative chatbots, are being adopted and utilized by project managers in real-world scenarios. Existing literature emphasizes the broad use of AI in project management [4], but lacks a detailed exploration of generative chatbot technology.

This paper presents an empirical investigation into this phenomenon, examining how project managers are not only embracing but also customizing generative AI tools like ChatGPT to enhance project oversight. We argue that technological evolution, traditionally focused on enhancing operational efficiency and productivity, is now extending into cognitive tasks.

In this study, we employ the theoretical framework of Adaptive Structuration Theory (AST) [5] to comprehend how generative AI technologies are structured and adapted within project management, highlighting how the impact of the same technology varies across organizations due to diverse usage in alignment with their work needs. At the core of AST lies the concept of appropriation, which refers to how individuals or groups adopt and adapt technologies to suit their specific requirements or integrate them into their social practices [5]. Utilizing appropriation theory, we delve into how project managers leverage generative AI tools within their unique operational contexts. We particularly focus on Innovation Attitude (IA), Peer Influence (PI), and Task-Technology Fit (TTF) as independent variables impacting the Appropriation of Generative AI tools by Project Managers.

From a theoretical standpoint, this work seeks to deepen the understanding of how project managers adopt and effectively utilize generative AI, emphasizing the pivotal role of project manager capabilities in this process. Preliminary findings suggest that these factors significantly influence how project managers integrate ChatGPT into their workflow, offering new insights into the theoretical understanding of technology adoption in project management.

10.2 Literature Review

The emergence of advanced project management software and collaboration tools has transformed the processes of project planning, monitoring, and control [6]. Technologies are often employed in both expected and unexpected ways, offering flexibility that can result in unexpectedly positive outcomes [7]. The process of adapting technology goes beyond technical integration and involves cultural, procedural, and strategic considerations to ensure alignment with the organization's ethos and objectives [8].

Adaptive Structuration Theory (AST) provides a framework for understanding how groups and organizations adapt and utilize advanced technologies [5]. According to AST, the adoption and use of technology in organizations are shaped by existing structures, while technology use, in turn, modifies these structures, leading to an adaptive structuration process.

The appropriation of technology, particularly Information and Communication Technologies (ICTs), frequently leads to changes in organizational routines and practices [9]. Orlikowski [10] highlights that this change is not always straightforward or predictable, as appropriation can vary significantly among different users and contexts, resulting in diverse and sometimes unexpected outcomes. The central concept of technology appropriation revolves around two key aspects: customization and reinterpretation. Customization involves users adapting a technology to better suit their specific needs [11]. Research indicates that some technologies are more adaptable than others. Unlike rigid technologies with fixed functionalities and limited applicability, malleable technologies can be shaped, modified, and customized by users to meet their specific requirements and contexts [12]. As highlighted by Leonardi [9], malleable Information Technology (IT) enables users to alter and adapt the technology in various ways, resulting in diverse usage patterns and outcomes. Generative AI chatbots, characterized by their ability to learn and evolve from user interactions, exemplify malleable IT. These systems are dynamic, evolving continuously and providing a dynamic interface for user interaction and appropriation [3].

Generative AI, with its ability for ideation and creative problem-solving, offers project managers an opportunity to transcend traditional boundaries and foster an environment conducive to innovation. AI serves as a multifaceted enabler, enhancing project managers' capabilities across various dimensions of their role in a rapidly evolving business environment. Continuous interaction with AI systems facilitates ongoing learning, enabling project managers to stay abreast of the latest technologies and methodologies in project management.

Our aim is to investigate factors influencing the adaptation of Generative AI Chatbots for efficient and effective customized use by project managers, termed Generative AI Appropriation. This appropriation can significantly impact project outcomes, influencing communication, decision-making, and information management practices. Unfaithful appropriation involves using technology in a manner different from its intended purpose [11]. Creative Behavior is derived from Amabile's componential theory of creativity, which suggests that creativity involves expertise, creative-thinking skills, and intrinsic task motivation [13]. The adoption of Generative AI Chatbots by project managers could profoundly influence their creative behavior throughout the project management lifecycle. This integration of AI tools suggests a potential enhancement in the creative capabilities of project managers, leading to more innovative project management approaches [14]. This aspect aligns with the concept of computational creativity, where AI aids in extending human cognitive abilities [15].

These issues have predominantly been addressed in the literature on human–machine interaction [16, 17].

In this study, we integrate the degree of faithfulness and creative behavior to model the appropriation of AI chatbots in project management. This approach aims to underscore the dynamic nature of project management in the digital age, where creative and adaptive technology use plays a pivotal role in achieving project success.

10.3 Research Model and Hypotheses Development

To meet our research objective, we establish and verify the research model outlined in Fig. 10.1. We propose three predictors hypothesized to influence project managers' adoption of Generative AI tools. These predictors include Innovation Attitude (IA), Peer Influence (PI), and Task-Technology Fit (TTF). In this study, appropriation is operationalized through two scales developed to measure the extent of unfaithful appropriation (UFOA) and the level of creative behavior (CB) exhibited by users when adopting new technology. The UFOA scale is scored in a way that higher scores reflect greater levels of appropriation.

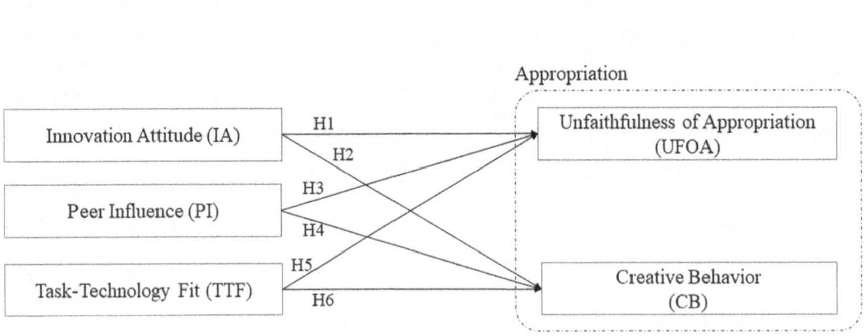

Fig. 10.1 The research model

10.3.1 Innovation Attitude

Innovation attitude encompasses various facets that reflect an individual's inclination toward embracing new ideas, methodologies, or technologies [18].

In the domain of project management and the adoption of AI tools, innovation attitude is a multifaceted concept. It reflects a project manager's readiness to adopt and utilize innovative technologies, particularly AI tools, in project management. This attitude goes beyond mere openness to new technologies; it is closely associated with the ability to think creatively, solve problems innovatively, and demonstrate receptiveness to new experiences and concepts [19].

The correlation between innovation attitude and the appropriation of Information Technology (IT) has been examined [20]. Individuals who possess a strong intrinsic motivation and enthusiasm for exploring novel ways of utilizing AI chatbots persist in overcoming obstacles and challenges inherent in new technologies. Consequently, they are more inclined to adapt these technologies to their work practices [12]. This mindset encourages them to think innovatively and explore unconventional uses of technology.

Innovation attitude is also inherently linked to creative behavior within the realm of project management. This mindset fosters an environment where experimenting with AI tools is encouraged, motivating project managers to think creatively and innovate in their project approaches.

As a consequence, the following hypotheses are formulated:

H1: *Innovation Attitude has a positive impact on Unfaithfulness of Appropriation.*
H2: *Innovation Attitude has a positive impact on Creative Behavior.*

10.3.2 Peer Influence

Peer influence refers to the extent to which an individual's beliefs and behaviors are influenced by colleagues and peers in the industry [21]. This concept closely aligns with the notion of "subjective norm" introduced in the Theory of Planned Behavior [22], which suggests that the stronger the subjective norm favoring a behavior, the greater the intention to engage in that behavior. Similarly, within the Unified Theory of Acceptance and Use of Technology (UTAUT) model, peer influence is encapsulated in the concept of "social influence," one of the four core determinants of technology acceptance and usage [23].

Adaptive Structuration Theory (AST) underscores the significance of existing social structures and their transformation through technology adoption. Within this framework, peer influence serves as a crucial social structure that influences how technology is assimilated in workplace environments [24].

Peer influence fosters experimentation with technology, often leading to innovative and occasionally unconventional uses. This behavior is influenced by observing peers who may utilize the technology in innovative ways, demonstrating the advantages of such applications. Additionally, witnessing **peer**s creatively employ technology inspires individuals to experiment and explore the technology's full potential themselves. This fosters a culture where creative utilization of technology is not only tolerated but actively encouraged, creating an environment conducive to innovative thinking and problem-solving.

Therefore, it can be argued that the utilization of ChatGPT is correlated with an enhancement in creativity [25, 26].

Hence, we propose the following hypotheses:

H3: *Peer Influence has a positive impact on Unfaithfulness of Appropriation.*
H4: *Peer Influence has a positive impact on Creative Behavior.*

10.3.3 Task-Technology Fit

The Task-Technology Fit theory suggests that technology is more likely to have a positive influence on both individual and organizational performance when it is aligned effectively with the tasks being performed [27].

Task-Technology Fit evaluates the extent to which technology aids users in accomplishing their particular tasks [28]. Past research has emphasized how the compatibility between technology functionalities and task demands can enhance technology usage [29]. Technologies that align closely with user tasks frequently offer a strong basis for users to develop innovative solutions. A technology that fits well serves as a catalyst for creative ideas and innovative approaches [30].

H5: *Task-Technology Fit has a positive impact on Unfaithfulness of Appropriation.*
H6: *Task-Technology Fit has a positive impact on Creative Behavior.*

10.4 Methods

10.4.1 Measurement Instrument

From June 2023 to September 2023, the researchers conducted a survey using a custom-designed questionnaire to empirically assess the hypotheses outlined in this study. The questionnaire comprises 22 inquiries and is structured to systematically investigate the constructs of the research model. Each query is accompanied by a 5-point Likert scale, ranging from "strongly disagree" to "strongly agree," allowing project managers to convey their perspectives. Additionally, a specific section of the questionnaire is dedicated to collecting general and demographic information from participating project managers. This information, including gender, age, work experience, and company size, contributes to comprehending the backgrounds of the respondents.

10.4.2 Data Collection

Before commencing the survey, the researchers conducted a validation phase to ensure the questionnaire's reliability. Copies of the questionnaires were distributed to 11 project managers employed across various companies. Following the finalization of the questionnaire, it was distributed to the Project Management Institute—Southern Italy Chapter (PMI-SIC), which comprises 450 members. Approximately 32% of the survey elicited responses, resulting in a total of 144 completed surveys. Prioritizing data quality, the authors reviewed the responses and removed any entries that were repetitive or inconsistent. This data cleansing process yielded 131 valid questionnaires, establishing the fundamental dataset for the research analysis.

10.4.3 Data Analysis

The author utilized Partial Least Squares—Structural Equation Modeling (PLS-SEM) as a statistical technique to empirically examine the proposed research hypotheses [31]. PLS-SEM was chosen due to its suitability for small sample sizes [32], its applicability in exploratory studies [33], and its strong recommendation for datasets with a limited number of indicators per latent variable [34]. For the analysis, SmartPLS4 software from SmartPLS GmbH was employed. The analysis began with assessing the measurement model to determine the reliability of the constructs used in the study. Subsequently, the evaluation of the structural model was conducted to ascertain the hypothesized relationships between the research model constructs and their statistical significance.

10.5 Findings

In this portion, the findings of the study are disclosed. Section 10.5.1 provides an overview of descriptive statistics concerning the demographic details of project managers who took part in the survey. Section 10.5.2 outlines the findings of the measurement model, whereas Sect. 10.5.3 presents the results of the analysis of the structural model.

10.5.1 Descriptive Statistics

Table 10.1 presents the demographic characteristics of the project managers surveyed. The data reveal that a majority of the respondents are male, comprising 66% of the interviewed project managers, while females make up the remaining 34%.

Table 10.1 Project Managers' demographical data

Demographic character		Frequency(n)	Percentage (%)
Gender	Male	86	66
	Female	45	34
Age	< 25 years	3	2
	25–30 years	22	17
	31–40 years	27	21
	41–50 years	40	31
	51–60 years	33	25
	> 60 years	6	5
Work experience	< 5 years	35	27
	5–10 years	31	24
	11–15 years	23	18
	16–20 years	20	15
	21–25 years	13	10
	25 > years	9	7
Company size	< 10 employees	11	8
	10–50 employees	14	11
	51–250 employees	45	34
	> 250 employees	61	47

The age distribution among the project managers is diverse, with those under 25 years representing 2% of respondents. Age groups ranging from 25 to 60 years exhibit varying proportions, with the highest percentage found in the 41 to 50 years category at 31%. Respondents aged over 60 years account for 5% of the total. Regarding work experience, project managers with less than 5 years of experience constitute the largest group at 27%. The distribution then gradually declines across subsequent experience categories, with 7% having over 25 years of experience. Lastly, the distribution of project managers across various company sizes indicates that companies with fewer than ten employees represent a minority, accounting for 8%, while the majority, comprising 47% of the total, belong to companies with over 250 employees.

10.5.2 Measurement Model Results

The evaluation of the measurement model begins with the assessment of the indicator's reliability [35]. In this regard, the calculation of indicator loadings is employed to assess the strength of the relationship between indicators and latent constructs [36, 37]. The results of this analysis, presented in Table 10.2, confirm the model's robustness with all values exceeding the recommended threshold of 0.708 [36].

Progressing in the assessment, the authors turn their attention to the model's internal consistency reliability, employing both the composite reliability coefficient

Table 10.2 Indicator loadings

	CB	UFOA	IA	PI	TTF
CB-1	0.792				
CB-2	0.863				
CB-3	0.881				
CB-4	0.891				
CB-5	0.782				
CB-6	0.753				
UFOA-1		0.892			
UFOA-2		0.922			
UFOA-3		0.812			
IA-1			0.808		
IA-2			0.778		
IA-3			0.721		
IA-4			0.754		
IA-5			0.812		
PI-1				0.847	
PI-2				0.851	
PI-3				0.811	
PI-4				0.755	
TTF-1					0.869
TTF-2					0.903
TTF-3					0.765
TTF-4					0.895

Table 10.3 Internal consistency reliability

	Cronbach's alpha	Composite reliability (rhoC)
CB	0.908	0.913
UFOA	0.848	0.853
IA	0.838	0.863
PI	0.834	0.843
TTF	0.881	0.888

(rho) and Cronbach's alpha [34]. The outcomes of this evaluation are presented in Table 10.3 and adhere to the criteria established by Hair et al. [34], thus further confirming the validity and reliability of the measurement model.

Moving forward, the model's convergent validity has been evaluated through the Average Variance Extracted (AVE) metric. Table 10.4 shows that all constructs exceed the threshold limit of 0.5 [38], confirming the model's validity.

The evaluation of the measurement model concludes with the assessment of its discriminant validity [35] by using metrics such as Cross-loadings [34], Fornell-Larcker [39], and Heterotrait-Monotrait Ratio (HTMT) [40], the authors assess the relationships between the model constructs. The tests have positive outcomes, align with discriminant validity criteria as established by Fornell and Larcker [39] and

Table 10.4 Convergent validity

	Average Variance Extracted (AVE)
CB	0.687
UFOA	0.768
IA	0.601
PI	0.668
TTF	0.739

Table 10.5 Variance Inflation Factors (VIFs)

	VIF
IA - > CB/UFOA	1.154
PI - > CB/UFOA	1.280
TFF - > CB/UFOA	1.448

Hair et al. [35]. The examination further reinforces the validity and robustness of the measurement model.

10.5.3 Structural Model Results

The authors initiated the evaluation of the structural model by investigating potential collinearity concerns. The Variance Inflation Factor (VIF) was employed for this purpose [41]. Table 10.5 displays the VIF values for each construct. None of these values surpass the threshold limit of 5 [41], thereby affirming the absence of collinearity issues in the structural model.

In the subsequent phase, the study evaluates the significance and importance of the path coefficients. Utilizing the PLS-SEM algorithm, the research confirms the positive impacts of the model's independent constructs (IA, PI, and TFF) on the adoption of generative chatbots by project managers (CB and UFOA), as outlined in Table 10.6. Notably, the TTF construct demonstrates the most substantial influence on both CB and UFOA, with respective values of 0.458 and 0.433. Following closely, the IA construct exhibits significant influence with values of 0.388 for UFOA and 0.320 for CB. Although the PI construct also contributes positively to the appropriation of generative chatbots, its impact is relatively lower, as indicated by values of 0.164 (UFOA) and 0.147 (CB). The statistical significance of the path model has been evaluated by employing bootstrapping standard errors to compute the t-values of path coefficients. Table 10.6 illustrates that each construct yields t-values exceeding 1.96, which serves as the lower threshold to confirm the statistical significance of indicator weights [35]. These findings validate all proposed hypotheses in the research model.

Concluding the evaluation of the structural model, the assessment of the model's explanatory power has been carried out through the coefficient of determination R^2 [42]. Table 10.7 reports values approximately around 0.5 for both CB and UFOA

Table 10.6 Path coefficients—Mean, STDEV, T-values, p-values

	Original sample (O)	Sample mean (M)	Standard deviation (STDEV)	T statistics (|O/STDEV|)	P values	Results
IA - > UFOA (H1)	0.388	0.388	0.078	4.947	0.000	Supported
IA - > CB (H2)	0.320	0.322	0.091	3.524	0.000	Supported
PI - > UFOA (H3)	0.164	0.168	0.062	2.637	0.008	Supported
PI - > CB (H4)	0.147	0.148	0.073	2.014	0.044	Supported
TTF - > UFOA (H5)	0.433	0.431	0.083	5.245	0.000	Supported
TTF - > CB (H6)	0.458	0.457	0.101	4.515	0.000	Supported

Table 10.7 R^2 values

	R^2
CB	0.499
UFOA	0.551

constructs, thus affirming a satisfactory level of explanatory power of the proposed model [41].

10.6 Discussion and Implications

The concept of generative AI appropriability is operationalized through the variables of Creative Behavior (CB) and Unfaithfulness of Appropriation (UFOA). The empirical analysis has confirmed all the formulated hypotheses.

In the initial findings, the study establishes a positive correlation between Innovation Attitude (IA) and both CB (H2) and UFOA (H1). Essentially, project managers with a more innovative mindset, as indicated by IA, are more inclined not only to unfaithfully appropriate AI tools but also to exhibit creative behaviors. This positive association can be explained by recognizing that individuals with a strong inclination toward innovation tend to approach new technologies with a proactive and open-minded outlook. These project managers, characterized by an innovative mindset, are predisposed to explore and leverage the full potential of generative AI tools. Regarding CB, their innovative mindset likely encourages them to think creatively and apply AI tools in unconventional ways, pushing the boundaries of traditional project management practices.

Additionally, the study confirms a positive correlation between Peer Influence (PI) and both CB (H4) and UFOA (H3). This indicates that the influence of colleagues or industry peers significantly impacts how project managers adopt and utilize generative AI tools. When project managers observe their peers successfully applying generative AI tools in creative ways, it serves as motivation and inspiration. This positive influence motivates them to not only adopt these tools but also to apply them in innovative and creative ways. Success stories from peers establish a positive social norm, encouraging project managers to explore the full potential of AI tools and think creatively about their applications.

Moreover, the positive relationship between Task-Technology Fit (TTF) and the dependent constructs CB (H6) and UFOA (H5) suggests that when project managers perceive AI tools as well-suited for their project tasks, the likelihood of effective utilization increases. When project managers perceive that generative AI tools align well with the requirements and demands of their project tasks, it fosters synergy between the technology and the tasks. This alignment creates a conducive environment for creative thinking and problem-solving. Project managers, confident in the suitability of AI tools for their tasks, are more inclined to engage in creative behaviors. The positive correlation with UFOA implies that the AI tools are perceived as valuable and relevant to the specific context of project management tasks, leading to unfaithful appropriation.

10.6.1 Implications

The theoretical implications of the study contribute to an enhanced understanding of how project managers adopt and effectively utilize generative AI to improve decision-making within organizations. The confirmed positive association between Innovation Attitude (IA) and both Creative Behavior (CB) and Unfaithfulness of Appropriation (UFOA) enriches theories of innovation adoption. It underscores the pivotal role of a proactive and open-minded approach in exploring and maximizing the potential of emerging technologies.

Moreover, the positive correlation between Peer Influence (PI) and CB, as well as UFOA, underscores the social dynamics in technology adoption. It emphasizes the significant impact of peer success stories in motivating project managers to creatively adopt and unfaithfully use generative AI tools. Additionally, the study's findings regarding the positive relationship between Task-Technology Fit (TTF) and both CB and UFOA contribute to technology adoption theories. They highlight the importance of perceived fit between AI tools and project tasks, aligning with models like the Technology Acceptance Model (TAM) [43].

In terms of practical implications, understanding the influence of an innovative mindset on AI adoption suggests that training programs can focus on enhancing innovation attitudes among project managers. By emphasizing the benefits of open-mindedness and proactivity in approaching new technologies, organizations can enhance project managers' willingness to explore and creatively utilize generative AI tools. Additionally, organizations can leverage the positive influence of peers by promoting a culture that shares success stories and innovative applications of AI tools.

10.7 Conclusion and Limitation

As shown in Table 10.1 the respondents are 66% men and 34% women, 77% are aged between 31 and 60 (21% 31–40, 31% 41–50, 25% 51–60) and in relation to Company Size 34% work for companies with 51–250 employees, 47% for companies with more than 250 employees. The study attempts to advance the theoretical understanding of how project managers adopt and effectively utilize generative AI to refine project management methods and enhance decision-making within organizations [44]. Foremost, the study contributes to the body of knowledge on the use of generative AI in project management by underscoring the essential role of adaptability in project managers for the effective application of these advanced technologies. The findings of the work highlight the importance of an innovative mindset, peer influence, and task-technology alignment in determining how project managers adopt and use generative AI tools. This research remarks on the multifaceted nature of technology adoption, where psychological factors (like innovation attitude), social dynamics (peer influence), and task-related considerations (task-technology fit) play a significant role in how new tools are integrated into professional practices. This exploration provides valuable insights, contributing to an overall understanding of AI adoption in the context of project management.

UFOA plays a crucial role in fostering an environment where CB can thrive. This deviation is not merely a matter of non-compliance but a creative reimagining of the technology's capabilities. In the context of AI chatbots, project managers often find themselves navigating uncharted territories, necessitating a departure from traditional methods and a move toward more innovative practices. CB, on the other hand, is characterized by the ability of users to leverage technology for novel idea generation, problem-solving, and the implementation of innovative solutions. The findings offer project managers guidance on how to adapt and leverage generative AI tools effectively, ensuring they are equipped to meet the evolving demands of modern project management. Finally, the findings guide organizations on leveraging the positive impact of innovation attitude, peer influence, and task-related fit to facilitate the appropriation of AI tools within the project management context.

References

1. Jia, N., Luo, X., Fang, Z., Liao, C.: When and how artificial intelligence augments employee creativity. Acad. Manag. J. **67**(1), 5–32 (2023)
2. Taboada, I., Daneshpajouh, A., Toledo, N., De Vass, T.: Artificial intelligence enabled Project Management: a systematic literature review. Appl. Sci. **13**(8), 5014 (2023)

3. Kaplan, A., Haenlein, M.: Siri, Siri, in my hand: Who's the fairest in the land? On the interpretations, illustrations, and implications of artificial intelligence. Bus. Horiz. **62**(1), 15–25 (2019)
4. Sheoraj, Y., Sungkur, R.K.: Using AI to develop a framework to prevent employees from missing project deadlines in software projects – case study of a global human capital management (HCM) software company. Adv. Eng. Softw. **170**, 103143 (2022)
5. Wang, T., Deng, C.P., Teo, T.S.: Appropriating IT outsourcing for IT alignment: an adaptive structuration theory perspective. Technol. Forecast. Soc. Chang. **192**, 122581 (2023)
6. Barcaui, A., Monat, A.: Who is better in project planning? Generative artificial intelligence or project managers? Project Leadership and Society. **4**, 100101 (2023)
7. Leonardi, P.M., Barley, S.R.: What's under construction here? Social action, materiality, and power in constructivist studies of technology and organizing. Acad. Manag. Ann. **4**(1), 1–51 (2010)
8. Boudreau, M.C., Robey, D.: Enacting integrated information technology: a human agency perspective. Organ. Sci. **16**(1), 3–18 (2005)
9. Leonardi, P.M.: When flexible routines meet flexible technologies: affordance, constraint, and the imbrication of human and material agencies. MIS Q., 147–167 (2011)
10. Orlikowski, W.J.: Using technology and constituting structures: a practice lens for studying technology in organizations. Organ. Sci. **11**(4), 404–428 (2000)
11. Carroll, J.M., Rosson, M.B.: Participatory design in community informatics. Des. Stud. **28**(3), 243–261 (2007)
12. Shao, Z., Li, X.: The influences of three task characteristics on innovative use of malleable IT: an extension of adaptive structuration theory for individuals. Inf. Manag. **59**(3), 103597 (2022)
13. Amabile, T.M.: A model of creativity and innovation in organizations. Res. Organ. Behav. **10**(1), 123–167 (1988)
14. Nevo, S., Nevo, D., Pinsonneault, A.: Exploring the role of IT in the front-end of innovation: an empirical study of IT-enabled creative behavior. Inf. Organ. **30**(4), 100322 (2020)
15. Alimam, H., Mazzuto, G., Tozzi, N., Ciarapica, F.E., Bevilacqua, M.: The resurrection of digital triplet: a cognitive pillar of human-machine integration at the dawn of industry 5.0. J. King Saud Univ. – Comput. Inf. Sci. **35**(10), 101846 (2023)
16. Hoc, J.M.: From human—machine interaction to human—machine cooperation. Ergonomics. **43**(7), 833–843 (2000)
17. Liu, C.: Human-machine Trust interaction: a technical overview. Int. J. Dependable Trust. Inf. Syst. (IJDTIS). **1**(4), 61–74 (2010)
18. Ferraro, G., Iovanella, A.: Revealing correlations between structure and innovation attitude in inter-organisational innovation networks. Int. J. Comput. Econ. Econom. **6**(1), 93–113 (2016)
19. Vartanian, O.: Attention, cognitive flexibility, and creativity. In: Creativity and Reason in Cognitive Development (2016)
20. Bubou, G.M., Job, G.C.: Individual innovativeness, self-efficacy and e-learning readiness of students of Yenagoa study Centre, National Open University of Nigeria. J. Res. Innov. Teach. Learn. **15**(1), 2–22 (2022)
21. Venkatesh, V., Thong, J.Y., Xu, X.: Unified theory of acceptance and use of technology: a synthesis and the road ahead. J. Assoc. Inf. Syst. **17**(5), 328–376 (2016)
22. Ajzen, I.: The theory of planned behavior. Organ. Behav. Hum. Decis. Process. **50**(2), 179–211 (1991)
23. Venkatesh, V., Morris, M. G., Davis, G. B., Davis, F. D.: User acceptance of information technology: toward a unified view. MIS Q., 27(3), 425–478 (2003)
24. Ilie, V., Turel, O.: Manipulating user resistance to large-scale information systems through influence tactics. Inf. Manag. **57**(3), 103178 (2020)
25. Rane, N.: ChatGPT and similar generative Artificial Intelligence (AI) for smart industry: role, challenges and opportunities for Industry 4.0. Industry 5.0 and Society 5.0. Challenges and Opportunities for Industry. **4** (2023). https://doi.org/10.2139/ssrn.4603234
26. Trust, T., Whalen, J., Mouza, C.: ChatGPT: challenges, opportunities, and implications for teacher education. Contemp. Issues Technol. Teach. Educ. **23**(1), 1–23 (2023) Waynesville:

Society for Information Technology & Teacher Education. https://www.learntechlib.org/primary/p/222408/
27. Zigurs, I., Buckland, B.K.: A theory of task/technology fit and group support systems effectiveness. MIS Q. **22**(3), 313–334 (1998)
28. Goodhue, D.L., Thompson, R.L.: Task-technology fit and individual performance. MIS Q. **19**(2), 213–236 (1995)
29. Lu, H.P., Yang, Y.W.: Toward an understanding of the behavioral intention to use a social networking site: an extension of task-technology fit to social-technology fit. Comput. Hum. Behav. **34**, 323–332 (2014)
30. Zhou, J., Shalley, C.E.: Research on employee creativity: a critical review and directions for future research. Res. Pers. Hum. Resour. Manag. **22**, 165–217 (2003)
31. Dijkstra, T.K., Henseler, J.: Consistent partial least squares path modeling. MIS Q. **39**(2), 297–316 (2015)
32. Willaby, H.W., Costa, D.S.J., Burns, B.D., MacCann, C., Roberts, R.D.: Testing complex models with small sample sizes: a historical overview and empirical demonstration of what partial least squares (PLS) can offer differential psychology. Personal. Individ. Differ. **84**, 73–78 (2015). https://doi.org/10.1016/j.paid.2014.09.008
33. Hair, F., Risher, J.J., Sarstedt, M., Ringle, C.M.: When to use and how to report the results of PLS-SEM. Eur. Bus. Rev. **31**(1), 2–24 (2019). https://doi.org/10.1108/EBR-11-2018-0203
34. Hair Jr., J.F., Sarstedt, M., Ringle, C.M., Gudergan, S.P.: Advanced Issues in Partial Least Squares Structural Equation Modeling. Sage Publications (2017)
35. Hair Jr., J.F., Hult, G.T.M., Ringle, C.M., Sarstedt, M., Danks, N.P., Ray, S.: Partial Least Squares Structural Equation Modeling (PLS-SEM) Using R. Springer Publications (2021). https://doi.org/10.1007/978-3-030-80519-7
36. Jöreskog, K.G.: Simultaneous factor analysis in several populations. Psychometrika. **36**(4), 409–426 (1971)
37. Chin, W.W.: Commentary: issues and opinion on structural equation modeling. MIS Q. **22**(1), vii–xvi (1998)
38. Hair, J.F., Hult, T., Ringle, C.M., Sarstedt, M.: A Primer on Partial Least Squares Structural Equation Modeling (PLS-SEM), 3rd edn. Sage, Thousand Oaks (2022). https://doi.org/10.1007/978-3-030-80519-7
39. Fornell, C., Larcker, D.F.: Evaluating structural equation models with unobservable variables and measurement error. J. Mark. Res. **18**(1), 39–50 (1981). https://doi.org/10.1177/002224378101800104
40. Henseler, J., Ringle, C.M., Sarstedt, M.: A new criterion for assessing discriminant validity in variance-based structural equation modeling. J. Acad. Mark. Sci. **43**(1), 115–135 (2015). https://doi.org/10.1007/s11747-014-0403-8
41. Hair, J.F., Ringle, C.M., Sarstedt, M.: PLS-SEM: indeed a silver bullet. J. Mark. Theory Pract. **19**, 139–151 (2011). https://doi.org/10.2753/MTP1069-6679190202
42. Shmueli, G., Koppius, O.R.: Predictive analytics in information systems research. MIS Q. **35**(3), 553–572 (2011). https://doi.org/10.2139/ssrn.1606674
43. Davis, F.D.: Perceived usefulness, perceived ease of use, and user acceptance of information technology. MIS Q. **13**(3), 319–339 (1989)
44. Felicetti, A.M., Cimino, A., Mazzoleni, A., Ammirato, S.: Artificial intelligence and project management: an empirical investigation on the appropriation of generative Chatbots by project managers. J. Innov. Knowl. **9**(3), 100545 (2024)

Chapter 11
A Bibliometric Analysis on the Digital Transformation and Knowledge Management Interrelatedness Pattern

Olga Grieva, Fatima Guadamillas, and Mario J. Donate

Abstract In compliance with the PRISMA 2020 guidelines, this article communicates the results of a bibliometric analysis on the interrelations between Digital Transformation (DT) and Knowledge Management (KM). The operating data were obtained from the Scopus database and cross-validated with the Web of Science (WoS) Core Collection. The literature search was run in June 2022 to detect all the existing evidence from 1990 to 2021 inclusively. To study a sample of 232 English language peer-reviewed publications this paper employed a co-occurrence of keywords technique on the grounds of the science mapping analysis. The network visualization maps in the period and longitudinal dimensions were created by leveraging the SciMAT software. Having shed light on the key theoretical bases and having provided a glimpse of the future of the field, the executed bibliometric analysis offers a generic picture of the current state of research in the intersection of DT and KM.

Keywords Digital transformation · Knowledge management · Bibliometric analysis · Science mapping · Keyword co-occurrence analysis · SciMAT

11.1 Introduction

Digital transformation (hereinafter referred to as DT), the first central concept to explore, can be defined as the combined effects of digital innovations, bringing about novel actors, structures, practices, and values that change, replace, or complement the existing rules of the game within organizations, ecosystems,

O. Grieva (✉) · F. Guadamillas
University of Castilla-La Mancha, Toledo, Spain

M. J. Donate
University of Castilla-La Mancha, Ciudad Real, Spain

F. Schiavone et al. (eds.), *Advanced Perspectives and Trends in Digital Transformation of Firms, Networks, and Society*, Springer Proceedings in Business and Economics, https://doi.org/10.1007/978-3-031-80692-6_11

industries, or fields [1]. The second one, knowledge management (hereinafter referred to as KM), is deemed as an integral process, which connects technologies, techniques, and people, and implies the procedures and practices that allow, on the one hand, exploring knowledge (e.g., creation, acquisition) and, on the other, exploiting it (e.g., transfer, storage, application, protection) to achieve strategic objectives and sustainability goals such as improved competitiveness or performance [2]. Once the persistent technological evolution spurred on the knowledge-based economy, the interconnectedness between DT and KM became discernible [3]. The up-to-date organizational theory [4] elucidates that KM has come to be an effective source for promoting DT by determining a strategic direction and developing competitive advantages within a company from the integration and optimization of the transformational resources, i.e. people, processes, knowledge, and technology. With the aim of improving overall comprehension of the impact jointly induced by DT and KM—which leads to essential societal, organizational, and commercial implications—a broad-ranging investigation is urged to merge the efforts that have dealt with the complexity of ongoing digital disruption thus far. Given these circumstances, to provide valuable information for scholars to advance the development of the research field, our scientific proposal stands for the necessity to conduct a bibliometric analysis on the pattern of interrelations between DT and KM. According to Tranfield, Denyer, and Smart [5], undertaking a literature review is a fundamental research objective for the respective academic and practitioner communities of any discipline. In a similar vein, Iivari, Hirschheim, and Klein [6] agree that knowing the current state of conceptual base of the investigated field and its uncertainties is a substantial prior step for any research project. With these evidence-informed principles in mind, we recognized the need to first gather a consolidated understanding of the approached realm to further formulate an academically sensitive research problem and objectives. Therefore, we reviewed and synthesized relevant literature for the specific inquiry. The inference of such an intention can be drawn up as follows: the current state of knowledge on the pattern of interrelations between the two theoretically sound and methodologically rigorous research domains—DT and KM— is still quite scattered and lacks the foundations on which the onward research could be carried out. Before we proceed to the research objectives and questions being addressed, the sought contribution of the newly conceived bibliometric analysis should be explicitly stated. First and foremost, it regards filling the abovementioned limitations of the existing evidence base by providing significant insights into the unrevealed theoretical foundations and introducing previously non-applied methodological practices, thus comprising the review's originality and value. As far as this study constitutes the first attempt to conduct a detailed, systematic, and objective review of the academic literature in the intersection of DT and KM, the anticipated contribution will be multiplied by methodologically analysing the evolution of the field over time, in addition, by exploring the relationships with other topics and sub-topics to unfold potential streams and future research avenues, capable of producing meaningful scientific advances to strengthen the subject within the academic and business literature. By and large, this bibliometric analysis tends to improve the general understanding to what extent the concepts of DT and KM have been jointly studied in the academic research so far, whether they have been considered within the same context, if so, to what extent, finally, what important inferences can be accentuated from the existing body of literature that deals with these concepts. The pursued contribution determines the main objectives of the research (1) to explore the common knowledge base of DT and KM by analysing the scientific production since the 90s onward through the science mapping of the co-word networks, and (2) to identify the pattern of the examined interrelations by interpreting the SciMAT software generated strategic diagrams with the thematic networks, precisely, by categorizing the research topics within clusters in accordance with their status and studying the connections of these topics with other clusters. The research questions arisen from the accorded agenda are: RQ 1—What is the conceptual structure of the common knowledge base of DT and KM? RQ 2—How has this conceptual structure developed over time? RQ 3—What are the influential research topics and potential directions for future research associated with the particular research line? The rest of this article is organized as follows: Section 11.2 describes the conceptual framework. Section 11.3 explains the methodology and the data collection process. Section 11.4 presents the bibliometric analysis results. Lastly, Sect. 11.5 hypothesizes and discusses conclusions, implications, limitations, and future research avenues.

11.2 Conceptual Framework

11.2.1 Digital Transformation

Whether the etymology might be suffering a considerable change under the circumstances of perpetual transformations, if so, how comparable are the former and latter views? As a starting point for our investigation, we commemorated the definition of DT dated back to the pre-COVID-19 times (See the Introduction). The cited statement of the meaning of the phenomenon is as much universal, so that the essence of the notion continues being relevant throughout history. Despite its absolute inclusiveness [7], we forged ahead and found one of the modernistic approaches to the definition and interpretation of the concept, suggested by [8], which reflects the newly developed research context: "Digital transformation is the profound transformation of business and organizational activities, processes, competencies, and models to fully leverage the changes and opportunities of a mix of digital technologies in a strategic and prioritized way". Meaningfully, the produced metamorphosis in the post-COVID-19 epoch invoked the straightening of the connotation of DT from its literal and primary meaning of the simple application of information technologies and data to reaching out the centre of business strategies.

11.2.2 Knowledge Management

Even though the reflection upon the ubiquitous influence of the COVID-19 on the plausible alteration of the definition of the KM concept in the ante- and post-settings remains appropriate, herein we designate as paramount to observe how the digital economy forced to appoint knowledge as a factor of development and the main competitive advantage. Coherently, to explain the importance of KM for the success of an organization in both the real and digital environments we quote Mizintseva and Gerbina [9], who formulated a groundbreaking in terms of digital humanism conception of KM: "Knowledge management is a system of knowledge, skills, and abilities that provides for critical problems of the enterprise's activity and implements a synergetic combination of data and information processing methods with application of information technologies and creative innovative abilities of people".

11.2.3 Interrelations Between DT and KM: The Need of a Systematic Review

Due to the accelerating pace and expanding scope of technological developments in the modern economy and society, the relevance of research on the DT has shifted to the fore, registering a formidable growth in the number of publications since the last decade [10]. Recently, the impact of the COVID-19 pandemic has strikingly stimulated the investigation related to the adoption of DT in the manufacturing sector and the role of technology in business sustainability. However, the specificities of the digital transformation process implementation and the performance challenges remained superficially addressed [11]. In the meantime, the literature in the field of KM has exponentially increased as well owing to its broad interdisciplinary implications and practical appropriateness for all types of organizations [12]. With the view to identify what our bibliometric analysis may add to the context of existing knowledge, all co-authors worked systematically and independently to profoundly screen the most relevant articles that seem to reflect the focal analysis items ("Bibliometric analysis" OR "Bibliometrics" AND "Digital transformation" AND "Knowledge management"), highlighting the key points of the published research outcomes. The conclusions were subsequently compared and summarized in Table 11.1 to fully understand the rationales behind each study. The existing body of knowledge on the pattern of interrelations between DT and KM points out at a certain level of theoretical and empirical consistency of each research domain on one's own with some respective deficiencies, finding both in a developing and consolidation phase, regarding conceptualization, typologies, frameworks, and applications [18], whereas the investigation of their interrelatedness and future-oriented potential of their synergetic confluence has just recently become a perspective focus of the academic research. At this point, drawing from the introductory overview of the most eligible bibliometrics and systematic literature reviews

Table 11.1 The frame of existing knowledge

Authors	Title	Scope	Focus	Design	Findings
Manesh et al. [13]	**"Knowledge management in the fourth industrial revolution: mapping the literature and scoping future avenues"**	Knowledge management (KM), Industry 4.0, bibliometrics	The intellectual structure and trends of KM in Industry 4.0	Bibliometric analysis and systematic literature review on a total of 90 relevant articles	Six clusters related to different research areas emerging from studies of KM and Industry 4.0
Di Vaio et al. [14]	**"Digital innovation and disruptive technologies in the intellectual capital (IC) and knowledge management systems (KMS) disclosure: a bibliometric analysis"**	Digital innovation, disruptive technologies, intellectual capital (IC), knowledge management systems (KMS), integrated reporting (IR), integrated thinking (IT)	The literary corpus on the impact of disruptive technologies on intangible factors	Bibliometric analysis of 115 publications in English between 2006 and 2020	The study outlines the importance of implementing IC diffusion models through KMS in organizations oriented to digital innovation
Di Vaio et al. [15]	**"Artificial intelligence and business models in the sustainable development goals perspective: a systematic literature review"**	Artificial intelligence (AI), machine learning, sustainability, cultural drift, sustainable business models (SBMs), knowledge management system (KMS)	The literary corpus on the role of Artificial Intelligence (AI) in the construction of sustainable business models (SBMs)	Bibliometric analysis of 73 publications in English with publication dates from 1990 to 2019	The paper highlights the key underlying relationship between AI and SBMs, identifying a research gap regarding KMS through AI, and the implications of AI, concerning SDG #12
De Bernardi et al. [16]	**Digital Transformation in SMEs: a systematic review and future directions**	Digitalization, SMEs, structured literature review, bibliometric analysis, innovation, information systems, knowledge management, decision making	The literary corpus on the interplays existing between digital transformation processes in the context of small and medium-sized enterprises (SMEs)	A two-step structured literature review (Bibliometrics and content analysis) of articles indexed in the timespan 2010–2019	Three major research streams: (1) innovation, (2) information and knowledge management systems, and (3) decision making

(continued)

Table 11.1 (continued)

Authors	Title	Scope	Focus	Design	Findings
Alvino et al. [17]	**"The role of digital innovation in knowledge management systems: a systematic literature review"**	Digital transformation (DT), knowledge management (KM), business model (BM), sustainable performance (SP)	The literary corpus on digital innovation in knowledge management systems (KMS)	Bibliometric analysis of 46 articles written in English and published in the last three decades (1990–2020)	The findings stress the links between innovation and sustainability, revealing that digital transformation tools contribute to the value creation process over the long-term

(hereinafter referred to as SLRs), the relevance of developing a bibliometric analysis on the DT and KM interrelatedness pattern acquires even more importance for the forward-looking implications and becomes entirely rationalized, since no studies were found on this particular inquiry, besides a firm request for enriching the research on the interconnectedness between KM and DT-congenerous constructs has come to the fore.

11.3 Methodology

11.3.1 Bibliometric Analysis

Bibliometric analysis, also referred to as bibliometrics, is defined as an analytical technique widely used to explore the knowledge structure and development of research fields by analysing the related publications [19–21]. According to M. J. Cobo, López-Herrera, Herrera-Viedma, and Herrera [22, 23] there are two general methods of bibliometric analysis applied to visualize the qualitative, quantitative, and structural changes in a specific research area: performance analysis and science mapping. The performance analysis employs a wide range of bibliometric indicators to evaluate the productivity of authors, journals, institutions, and countries, likewise, the impact of their contributions to research fields [24]. By comparison, the science mapping is used to establish conceptual, intellectual, and social linkages among different units of analysis (e.g., documents, citations, authors, and keywords), thus leveraging the structural mapping of a scientific field [25]. Since the science mapping analysis represents a combination of bibliometric analysis with graphical visualization, depending on the unit of analysis, it may perform various techniques such as co-citation analysis, bibliographic coupling, or keyword co-occurrence analysis [4]. Among manifold analyses of networks available, the most appropriate technique to uncover the theoretical foundations and key concepts that

Fig. 11.1 Stages of the research. Source: [27]

underpin the interrelations between DT and KM, as well to reproduce their evolution and the agenda for further development of the research area is the keyword co-occurrence analysis, also known as co-word analysis [26]. The application of this bibliometric research technique promises to enable a highly robust, structured, and comprehensive review of the rapidly expanding research domain, as long as it is the only method that uses the actual content of the documents to construct a similarity measure, while others connect documents indirectly through citations or co-authorships [27]. There are hardly any transparent and reproducible guidelines for conducting an SLR with the bibliometric methods. One of the few, Zupic and Čater [27] provided efficient workflow guidance for bibliometric analysis in the field of Management and Organization, whose conceptualized principles constitute the current research management model. Specifically, the steps, recommended for the science mapping analysis, were applied to successfully accomplish the research process.

Drawing from this methodological pattern (Fig. 11.1), we elaborated the following research structure: research design comes as the preliminary step that precedes the research process and entails determining the eligibility criteria, identifying the information source(s) and explaining the search strategy (Step 1). The next step comprises data selection: database search, refinement process, qualitative assessment, data export, and final core document set composition (Step 2). In the third step, the bibliometric analysis is rolled out (Step 3). The ultimate step involves the interpretation and reporting of results (Step 4). The methods and results at each stage of the bibliometric analysis should be reported in sufficient detail to allow users to assess the trustworthiness and applicability of the review findings. Hence, in the frame of the present research, the bibliometric analysis is undertaken adhering to the PRISMA 2020 guidelines[1], proposed by Page, Moher et al. [28]. However, as this approach is quite extensive, it might require further adaptation to the research project.

11.3.2 *Research Design*

Eligibility Criteria

To decide whether or not a retrieved document is suitable for inclusion in the final review, we set the specific eligibility criteria: the language of publication (English-language literature only) and the study orientation (if the focus of the study is out of

[1] http://www.prisma-statement.org/.

the effect produced from the synergy between DT and KM, an analysed item will be excluded).

Information Sources
Bibliographic data may be obtained from numerous databases, including Scopus, Web of Science (WoS), Google Scholar, etc. [29]. Since the validity of research depends on an adequate selection of the bibliographic database [30], we decide to work with the Elsevier solution—Scopus, which contains the largest multidisciplinary abstract and citation database of a peer-reviewed literature, covering scientific journals, conference proceedings, and books [31].

Search Strategy: Setting Search Criteria
Search strategy is an indispensable component of any SLR, and the PRISMA-S, an extension to the updated PRISMA 2020 Statement, is the most budding reporting guidance for that [32]. To verify that each stage of the search is completely reported and, therefore, replicable, we apply this interdisciplinary guidance tool.

1) *Keyword(s) and/or search term(s) selection.* We performed a preliminary review of the most relevant terms, associated with the "digital transformation" and "knowledge management" concepts, within the Business and Management literature. The former "digital transformation" in front of the definitional ambiguity [10] has been misleadingly compiled with the digit* terms, such as "digitization" and "digitalization" [33], asserting its exceptional interpretation in relation to its specific subject. The second term "knowledge management" hasn't shown a multi-syntax character.
2) *Source document selection.* Since academic journals might lag behind due to the time needed for the publication process, to embrace the timely information in addition to journal articles other sources like conference proceedings, management consulting, industry reports, and technological maturity surveys were included [34].

11.3.3 Selection Process

So far as an input into the bibliometric analysis regards an affirmative sample of the selected documents, on which to apply the technique, in the first phase, to identify and systematize the main contributions of scientific research favourable to our investigation, a search in the Scopus database was performed within article title, abstract, keywords with the query: TITLE-ABS-KEY ("Digital Transformation" AND "Knowledge Management"). The retrieval was further refined to the publications written in English only (See the Eligibility criteria), unveiling two hundred and seven (n = 207) records. From our perspective, the activated filter is not critical and does not produce any serious effects on the research results. Meanwhile, we hypothesized that Scopus might not include all relevant documents for analysis, so, we did a manual search for other coherent contributions in Web of Science (WoS)

Core Collection using the same parameters. As a result, one hundred and two (n = 102) publications were additionally recovered. The information sources were browsed in June 2022. Three hundred and nine (n = 309) records identified in both searches were coded as potentially eligible and, subsequently, were exported into SciMAT for its further pre-processing (the Scopus documents in a RIS format and the WoS knowledge base in an ISIWoS format), giving a start to the second phase, in which we systematically ascertained the content of each record to ensure consistency with the research theme and questions by screening the titles, abstracts, and keywords. After removing duplicate entries (n = 62) and discarding publications of an unfinished 2022-year (n = 15), two hundred and thirty two (n = 232) documents constituted the final knowledge dataset for the consecutive analysis. Since the PRISMA Statement calls for a high level of reporting detail in systematic reviews and meta-analyses, we summarized the study selection process with an application of the PRISMA 2020 flow diagram template[2] (See Fig. 11.2):

11.4 Bibliometric Analysis

11.4.1 Elaboration of Scientific Maps

Taking into consideration that SciMAT[3] is the only open-source software tool incorporating methods, algorithms, and measures for all steps of the general science mapping workflow—from data loading to the interpretation of results (see Fig. 11.3)—we continued working with this program to delineate the conceptual structure of DT and KM as a joint research domain.

After emending the knowledge base by eradicating misspelling in titles and author names (i.e., a consequence of text format conversion) and normalizing the keywords by unifying similar terms under a single concept (i.e., initially one thousand five hundred and twenty five items (n = 1525) were manually allocated into three hundred and forty three working groups of keywords (n = 343), applying the investigators' personal criteria), we proceeded from the pre-processing stage to the design and execution of the bibliometric analysis by following through eleven steps configured to envisage and create networks and clusters through a graphical representation [22, 23]:

Step 1 (Selecting the periods)[4]—there are two evolution periods analysed: the first starts with the publication date of the premier theme-related document (2006) and continues until 2019 and the second covers the term from 2020 up to 2022. Such

[2] PRISMA2020 | Evidence Synthesis Hackathon (eshackathon.org)—the link provides access to the free-to-use web-based tool run by DiagrammeR R package and ShinyApp to produce systematic review flow charts that conform to the latest update of the PRISMA Statement.

[3] https://sci2s.ugr.es/scimat/.

[4] To have periods for choosing for the analysis in Step 1, at the pre-processing stage these should be set up by passing to "Knowledge base" module and working with "Periods manager" function.

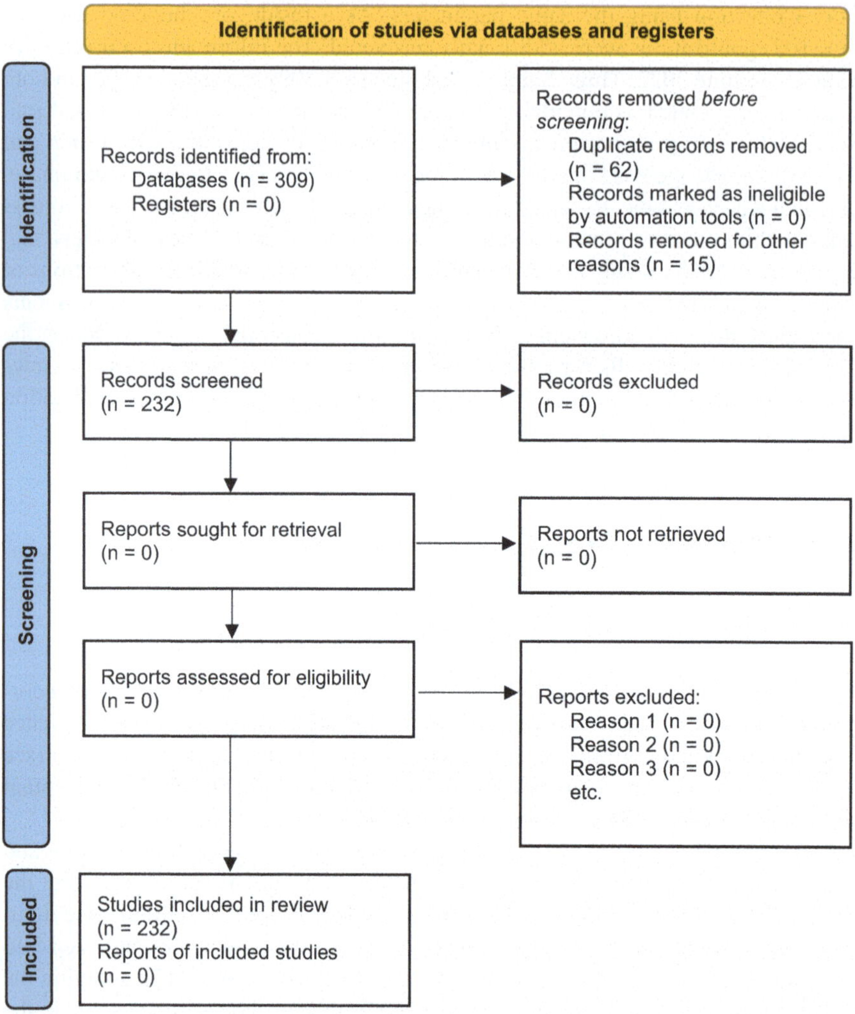

Fig. 11.2 The PRISMA 2020 flow diagram. Source: Authors' elaboration based on eshackathon.org [35]

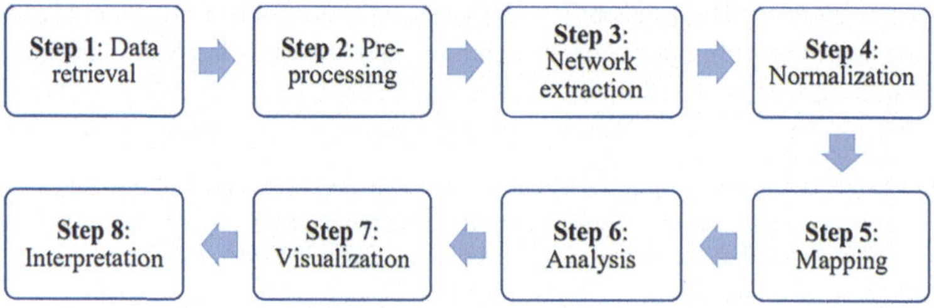

Fig. 11.3 The general workflow of science mapping analysis. Source: (Manuel J Cobo, López-Herrera, Herrera-Viedma, & Herrera, [22, 23, 36])

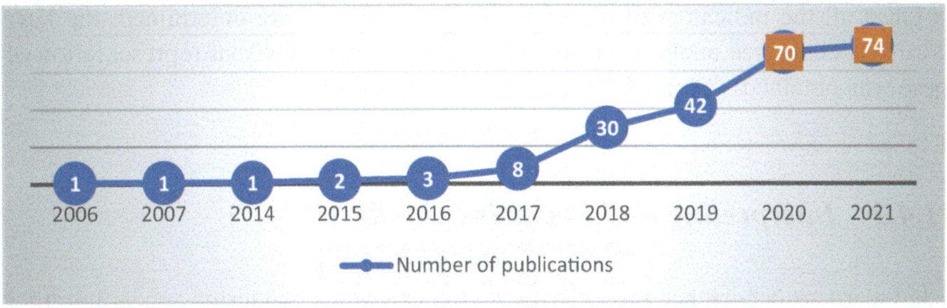

Fig. 11.4 Research evolution in DT & KM over the period 2006–2021. Source: Authors' elaboration

a division is due to a heterogeneous tendency in research evolution, reflected in the number of publications registered annually over 15 years: a piecemeal progression from 2006 to 2019, with 88 publications, followed by strengthened growth in scientific production in the common knowledge base of DT and KM over the last 2 years, nearly doubling to a total of 144 documents (See Fig. 11.4).

Step 2 (Selecting the units of analysis)—to unfold the fundamental topics and their linkages that reflect the conceptual landscape of the studied research field, a series of keywords (author, source, added) are to be explored.

Step 3 (Selecting the data reduction methods)—the minimum number of documents in which a keyword must appear is established with the value 2, since it is the inferior limit recommended by the software.

Step 4 (Selecting the kind of matrix)—to lay out the pattern of interrelationships within the common DT and KM conceptual base, we appeal to the keyword co-occurrence analysis.

Step 5 (Selecting the network reduction method)—the minimum value that allows reducing the network extracted from the analysis equals to 2.

Step 6 (Selecting the normalization measure)—the Equivalence index was selected, since it is the most adequate similarity measure for network normalization [37].

Step 7 (Selecting a clustering algorithm)—the Simple centres algorithm [38] was selected, with a minimum value of 3 clusters and a maximum value of 12 clusters, as recommended by [36].

Step 8 (Selecting the document mappers)—in this study the core mapper and the union mapper are used.

Step 9 (Selecting the quality measures)—h-index and the sum citations were selected.

Step 10 (Selecting the measures for the longitudinal map)—to attain the evolution map of the concepts we opted for the Jaccard index, which detects the transition of clusters between different periods. Meanwhile, to gain the overlapping map, the Inclusion index was used, which measures the degree of overlap of the clusters in various periods.

Once all the indicators of the aforementioned steps were determined, the keyword co-occurrence analysis results were generated and a visual representation of these was obtained—**Step 11**.

11.4.2 Interpreting and Reporting the Results

A keyword co-occurrence refers to the number of times two terms appear together in a set of documents [37]. Drawing from the presence of these co-occurrences and their frequency in the common knowledge base of DT and KM, on the one hand, we get a series of strategic diagrams for each analysed period that allow observing the relationships between clusters of keywords, where each cluster is considered a research theme, while, on the other hand, a longitudinal view is gained that allows knowing the evolution of the topics throughout the analysed periods.

Period View

Period 2006–2019

At the first stage of development of the academic and business research in the intersection of DT and KM, 12 thematic clusters appeared to be prevailing (See Fig. 11.5).

Each research theme disclosed in this process is characterized by two parameters: y-axis "density" and x-axis "centrality". Density tends to evaluate the internal strength of the network, while the centrality is responsible for measuring the degree of interaction of a network with others. Both median and mean values for density and centrality are used in classifying the themes into four groups [36]:

- **Motor clusters** (themes in the upper-right quadrant that present strong centrality and high density): "knowledge management", "knowledge-based system", "business", "computers" and "innovation" are mature topics with well-developed associations both internally within the cluster network and externally with the rest of the clusters.
- **Highly developed and isolated clusters** (themes in the upper-left quadrant that have fully developed internal cohesion within the thematic network, but nonessential external links with the rest of the research topics): "application programs" and "modelling" can be categorized as a specialized but the least significant cluster network for the key research field.
- **Emerging or declining clusters** (themes in the lower-left quadrant that present weak centrality and low density): integration of the underdeveloped "big data" concept into the persistent "human resource management" and "workers" network placed their thematic cluster at the incipient stage with an exemplary scarce internal cohesion, moreover, insulated external relations with other topics of the central research area.

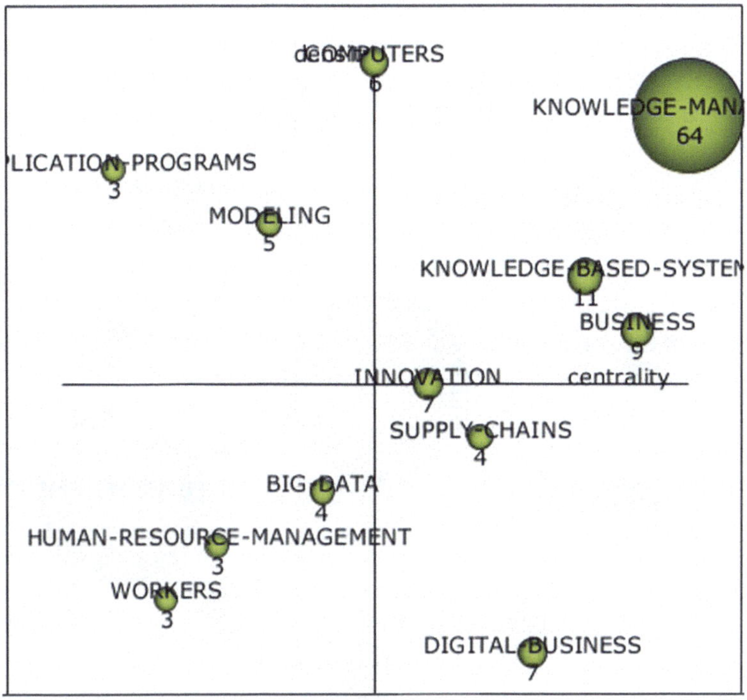

Fig. 11.5 Strategic diagram for the period 2006–2019. Source: SciMAT

- **Basic and transversal clusters** (themes in the lower-right quadrant that lack sufficiently developed internal cohesion within the cluster network, whereas opt for a high relevance for the targeted research field due to their rigorous external linkages): "supply chains" and "digital business". As shown in the strategic diagram, one of the primary concepts of this bibliometric analysis, "knowledge management" was the first to consolidate into the most significant construct for structuring the studied research field. Meantime, the concept of "digital transformation" seems to have undergone a thorough while to reinforce in joint research with KM, what can be learned from the knowledge management cluster's network (See Fig. 11.6). Another important outcome, deriving from this scientific map, leads to an explicit understanding of the foundations for the DT and KM common knowledge base, laid out on the basis of the unravelled tripartite relationships between KM, DT and each of the detected conceptual elements: "Industry 4.0 technologies", "knowledge process", "information systems and technologies", "information management", "decision-making process", "data asset management", and others.

Period 2020–2021

In recent years, the situation shaped by the COVID-19 propelled each organization to digitally transform to adapt to the disrupting circumstances and leverage the new

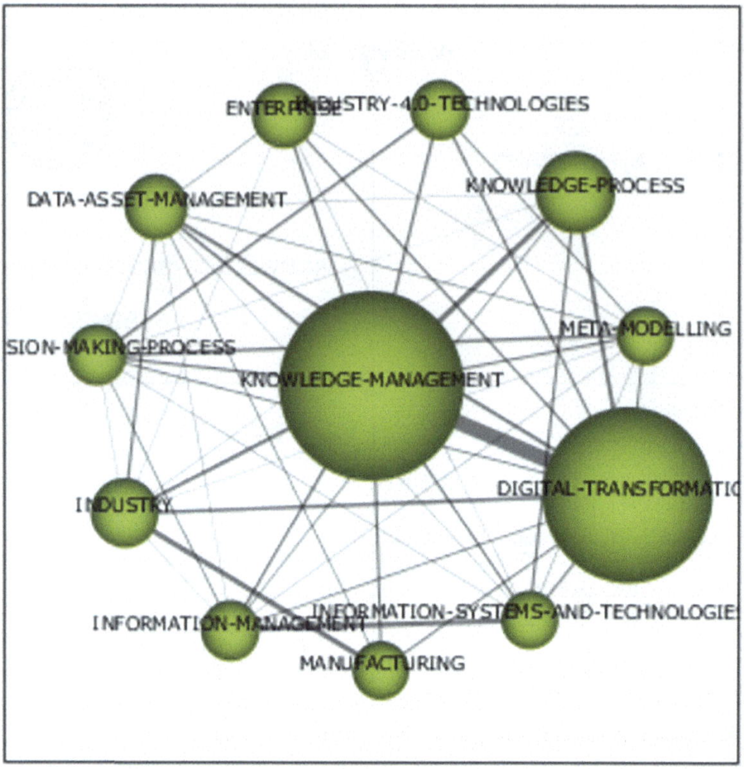

Fig. 11.6 "Knowledge management" cluster composition in the period 2006–2019. Source: SciMAT

business opportunities [39]. To approach this newly configured context academic and business research was exposed to prioritize novel investigative orientations in the Business and Management field, such as "risk management", "technological change", "production process", "healthcare industry", "Industry 4.0", "digitalization", and others, what is literally reflected in the strategic diagram, which visualizes the common knowledge base of DT and KM in the period 2020–2021 (See Fig. 11.7).

Apart from the foreknown evolutionary changes (i.e., significant increase in scientific production, extended thematic scope for research, consolidation of the motor, emerging, and transversal themes), the conceptual panorama in the intersection of DT and KM has been noticeably modified, reverberating the new paradigm. As a result, the renewed grouping of research themes was assorted into [36]:

- **Motor clusters** (themes in the upper-right quadrant that present strong centrality and high density): when compared to the ex-period, "knowledge management" cluster is seen to have enhanced its centrality, breaking free greater interaction capabilities with the topics from distinct cluster networks, while internally the

Fig. 11.7 Strategic diagram for the period 2020–2021. Source: SciMAT

bond with the new associate elements, "education" and "research action", is found rather less tight.

- **Highly developed and isolated clusters** (themes in the upper-left quadrant that have fully developed internal cohesion within the thematic network, but nonessential external links with the rest of the research topics): "risk management", "production process", "organizational aspects", "technological change", and "educational institutions" imply the existence of particular conditions, when these fully-fledged research streams show higher internal bondage, whereas bringing less importance for the central research field.
- **Emerging or declining clusters** (themes in the lower-left quadrant that present weak centrality and low density): "software" and "healthcare industry" figure rather like emerging research lines for multiple areas of investigation activity, including the DT and KM joint research, that are in attempt to nurture cohesion within the cluster network itself, likewise, intensify external relations with other topics of the focal research area.
- **Basic and transversal clusters** (themes in the lower-right quadrant that lack sufficiently developed internal cohesion within the cluster network, whereas opt for a high relevance for the targeted research field due to their rigorous external linkages): among "Industry 4.0", "digitalization", "decision-making process",

and "methodology", the first three constructs come out with the greatest strength to align DT and KM, while keeping an ample capacity for solidification of its interconnectedness.

The "knowledge management" cluster composition in the period 2020–2021 embodies rather scrutinized, aligned, and advanced pattern of interrelationships within the common knowledge base of DT and KM, expressed in conceptualizing deficient correlations between the foundational building blocks (i.e., "industry"—"Industry 4.0 technologies"—"knowledge process"; "knowledge process"—"meta modelling"), whereas embedding groundbreaking untested conceptual items (i.e., "innovation", "workers", "transfer activities", and "digital technology") (See Fig. 11.8). At this point, the up-to-date conceptual structure of DT and KM as a joint research domain performs a number of lines optative for further investigation that have been marginally considered in the academic and business literature so far (See Fig. 11.9), for instance, the interrelation Digitalization-KM through the mediating role of Strategic Approach, Business Model and Human Resource Management; the interrelation Technological Change-KM from the sustainable development

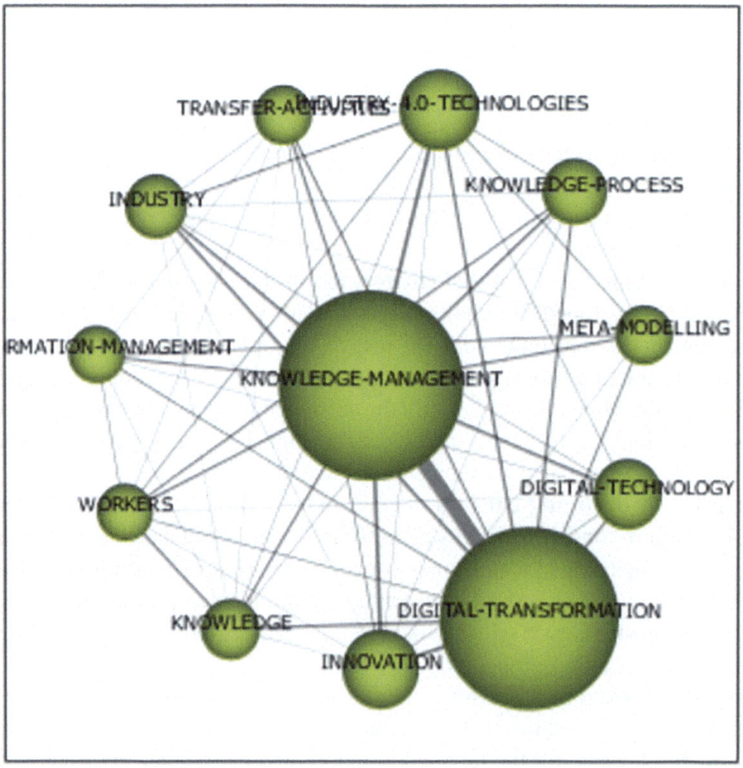

Fig. 11.8 "Knowledge management" cluster composition in the period 2020–2021. Source: SciMAT

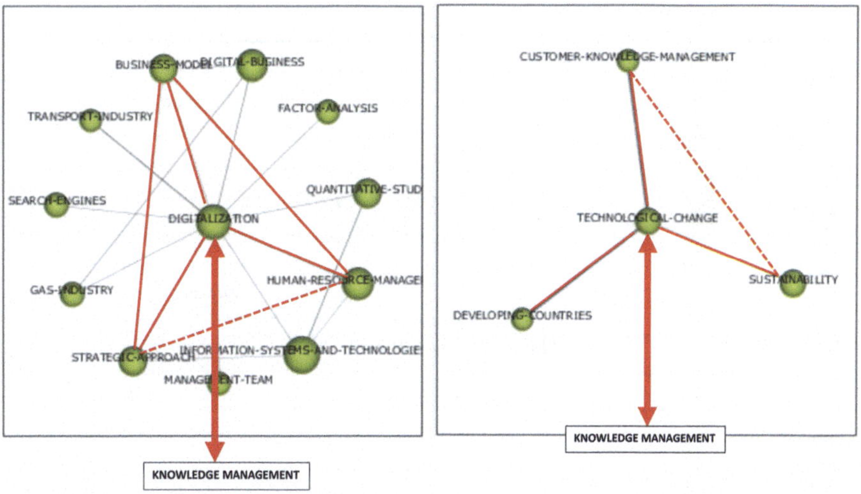

Fig. 11.9 "Digitalization" and "technological change" clusters' composition in the period 2020–2021. Source: SciMAT (solid red lines—tested correlations; dotted red lines—hypothesized correlations)

perspective or the interrelation Technological Change-KM mediated by Customer Knowledge Management and Sustainability. However, these specializations call for reviews such as ours to establish a common ground across studies that integrate different disciplinary perspectives to prevent fragmentation of the field as the literature grows.

Longitudinal View

Once a static outlook of the conceptual structure of the common knowledge base of DT and KM was performed, to study the evolution of the research topics, identified at the preceding stage, a dynamic perspective was applied. For these needs, the science mapping in longitudinal framework generates an evolution map that depicts the transition of the thematic clusters between periods, otherwise stated, tracks the structural development of a research field. In this regard, Fig. 11.10 combines several graphic illustrations: map (I.) is the SciMAT unprocessed version of the evolution representation, and the other two are the pre-processed maps, elaborated with the end to facilitate a comprehensive grouping of the pivotal research themes, distinguishing self-sufficient and influential clusters (II.), as well as prospective thematic streams (III.).

To interpret the given maps properly, it is worth clarifying that each column of the research topics refers to a pre-determined period, that is, the first column corresponds to the term 2006–2019, while the second one reflects the interval 2020–2021. Another principal aspect to understanding these graphics are the lines, which are responsible for underlining the relationships between different themes:

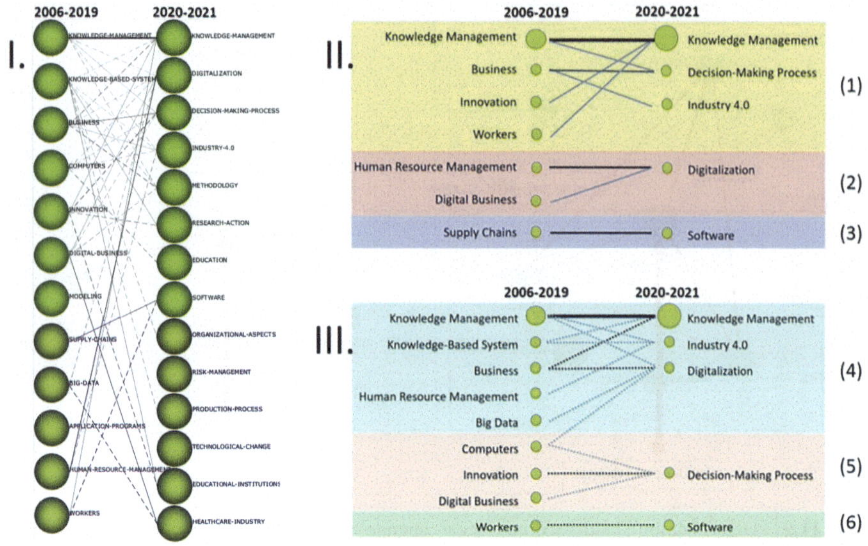

Fig. 11.10 Evolution map. Source: (I.) SciMAT; (II.) and (III.) authors' elaboration

particularly, the solid lines entail substantial relations between the linked themes, whereas a dotted line means an implicit connectedness. The gradient of colour of the edge from less to more saturated tone indicates density of characterized linkages, and the volume of the spheres is proportional to the number of published documents of each theme [22, 23]. Having explained the legend of the evolution maps, the composition criteria for the tailor-created graphs (II.) and (III.) should be clearly stated. First, the pre-processing measures refined the original scope (See map (I.)) to three cluster groups: motor, basic, and transversal, and emerging or declining, with a remark for the last one "all the themes of this cluster are forthcoming". On this basis, as soon as there were no evolutionary paths determined with "application programs" and "modelling" from the first column and "risk management" and "technological change" from the second column, these items were excluded from the analysis. Moreover, "methodology", "research action", "education", and "healthcare industry" were labelled as very particularized research lines and, therefore, disregarded too. Eventually, "organizational aspects", "production process", and "educational institutions", classified as isolated themes, were discarded. The conceptual evolution is shown through the coloured groupings of thematic areas (See maps (II.) and (III.)), deriving from the process of manual reordering of the conceptual elements (See map (I.)) with the main purpose to unravel a multitude of lines crossed in various directions. Giving an insight into each grouping of thematic areas will clear the way for framing the agenda of future research consistent with the developmental course. The second map provides a picture of the sets of influential research topics of the common knowledge base of DT and KM, featuring strong and enduring longitudinal relationships:

(1) *Knowledge management—Business—Innovation—Workers—Decision-making process—Industry 4.0.*

As the bibliometric analysis exhibits, KM is a driving force to the development of the joint research domain of DT and KM, and the rationales for that are not exclusively the highest centrality and density values, but also a constant and strengthening presence over time unrevealed from the longitudinal view. In point of fact, the inferences of the existing core contributions highlight a crucial role of knowledge management systems (hereinafter referred to as KMSs) in ensuring the optimization of resources and processes to achieve an improved performance through the fulfilment of knowledge-sharing strategies and the concurrent adoption of tools [40]. The adoption of technological solutions for the development of new processes, products, and services nurtures the firm's innovation capacity [41]. Innovation is configured as a governance issue, which influences the business model, pushing entrepreneurs to launch intervention strategies capable of satisfying the contingencies of an increasingly globalized and liberalized market [42]. The function of DT in this case becomes to provide an integrated strategic solution that supports managers in their decision-making process and operationally guides the company towards innovative and sustainable business models [40]. Thereby, while KM drives business processes towards innovation and the use of disruptive technologies, DT promotes new business models through the optimization of knowledge. Further exploration of the connections between these scientific fields could be a beneficial contribution to the management sciences.

(2) *Human Resource Management—Digital business—Digitalization.*

Di Vaio, Hassan et al. [14, 15] reassure that the workers' experience and skills, expressed in intellectual capital and knowledge, in conjunction with the technological component contribute to the organizational success. The integration of knowledge and learning processes in DT offers enormous potentials for the intangible value of the company (i.e. a higher level of market intelligence, a successful realization of the business goals, better product development, and improved customer orientation), and reversely, digitalization of the business processes is a favourable chance for developing strategies for KM, as well as digital and organizational learning [43]. The latest research findings in the common knowledge base of DT and KM point out the increasing demand for a new type of employee—one who can work in unstructured and unpredictable circumstances that are often complex and involve continual change. Indeed, the digital future requires a more flexible workforce with the ability to change and adapt within a sophisticated global economic environment [44]. Pavlekovskaya, Urintsov, Staroverova, and Nefedov [45] deduced that the technological foundation creates a multiplier effect for the price and value of intellectual capital and human potential, this might be exemplified by an innovative and entrepreneurial spirit, combined with an agile mindset of lifelong learning [46]. Consequently, the workforce proficient in new technologies is of a great request, and the skill sets that equate to a combination of certain cognitive (critical thinking, problem-solving, persuasion, and negotiation) and socio-emotional skills (creativity, emotional intelligence, leadership, and social influence) will likewise retain or increase in prominence [47].

(3) *Supply chains—Software.*

Information systems (hereinafter referred to as ISs) are nowadays vitally important to any organization to exercise its internal operations and its interactions with the external environment [48]. ISs can be defined as "software systems for business management, encompassing the modules that serve organizational functions such as planning, manufacturing, sales, marketing, distribution, accounting, financial, human resources management, project management, inventory management, service and maintenance, transportation and e-business supported by computers, software, people, processes, and data" [49]. Open-source software (hereinafter referred to as OSS) platforms are an example of highly responsive and innovative development models in supply chain management. Earlier, Barbaroux and Le Texier [50] identified that the modularity in OSS architecture establishes effective modes of KM and innovation. By dint of the third map the investigation releases an assembly of the most conducive research directions found at the inceptive stage of formation and acknowledgement.

(4) *Knowledge Management—Knowledge-Based System—Business—Human Resource Management—Big Data—Industry 4.0—Digitalization.*

From the compilation with the first grouping of thematic areas, the fourth combination appears to deploy the early evolutionary alterations consisting in posing a greater emphasis on investigating the role of the knowledge-based system, human resource management, Big Data and digitalization, the concepts that suggest enforcing the density and centrality of DT in the common knowledge base with KM. Influenced by an increasingly competitive environment, modern companies are seen to rely more on knowledge and its exploitation to sustain a long-term advantage. However, since the advent of the new millennium, the KM processes have been deeply affected by the Industry 4.0 [13]. The opportunities arisen for the radical improvement in data processing and knowledge mining included the Big Data systems [51]. Nevertheless, the current knowledge-based systems (hereinafter referred to as KBSs) confront with an unprecedented amount of data, emanating from the Big Data sources [52]. As a result, a new generation of KBSs for exploring and exploiting Big Data could become a major focus for the upcoming research.

(5) *Computers—Innovation—Digital Business—Decision-Making Process.*

In conditions of a strong reliance of employees on tacit and embedded knowledge in their decision-making, knowledge-driven decision support systems have been identified as an important cornerstone of the onward digitalization [53]. This research inference can be firmly endorsed adding that the KM of technological innovations substantially improves the business competitiveness by supporting processes and decisions through risk reduction for value creation [54]. To open seminal research avenues, the other accordant issues might need more attention and development in the future investigation.

(6) *Workers—Software.*

The DT forms the latest information space of the business and opens it up widely for the organization's employees. Thanks to the smart means of processing and analysis of data (e.g., software), employees receive virtually unlimited knowledge to solve their tasks [55]. Nonetheless, in the settings of soaring digitization, there is still plenty of room for debate on the role of cutting-edge technologies in the business performance improvement.

11.5 Discussion and Conclusions

The concluding remarks of the study begin with an evidence-supported discussion of the main research findings and terminate with the theoretical and practical implications, analysis limitations, and avenues for future research. Having efficiently articulated the set of tactics, destined to achieve the programmed research objectives, and respond to the arisen questions, the telic contribution was completely ensured. That is to say, the divulgence of the concealed conceptual strands of the common knowledge base of DT and KM, its evolutionary dynamics and interrelatedness pattern facilitated the improvement of the general understanding to what extent and in which context/s the concepts of DT and KM have been jointly studied by the academy so far. At this point, if we appeal to the strategic diagrams of the period view analysis for both terms (2006–2019) and (2020–2021), an inquisitive phenomenon can be observed, in particular, the pervasiveness of the concept of KM against the implicitness of the concept of DT, namely, its invisibility throughout the common knowledge structure grids. On the face of it, a penetrating glance at the KM cluster composition turned out to be of the utmost importance to disclose the disguised interplay between the examined analysis items—DT and KM. Not long ago the DT and KM joint research remarkably gained on scholarly dynamism and interdisciplinarity [12, 33]. However, the state of affairs corresponding to the earlier times (2006–2019) denotes the infirmity and unsteadiness of the DT as a sovereign research domain to oppose against the level of density and centrality of a group of associated research themes like "computers", "innovation", "digital business", and "big data" in the contextual relationships with the KM theory. Nevertheless, we can break it down further and identify from a sectional drawing of the KM cluster composition in the period 2006–2019 a conspicuous tight connection between the primary analysis items—DT and KM. The observation of the proximate period (2020–2021), along with the longitudinal analysis, significantly contributes to the ongoing discussion by shedding light on the current state of research at the intersection of DT and KM. Nonetheless, regardless of the transformative power of the most recent environmental turbulences (e.g., COVID-19), the DT construct remains tacitly related to the concept of KM, yielding to "technological change", "Industry 4.0", and "digitalization". In the meantime, stemming from the KM cluster composition in the period 2020–2021, the linkage between DT and KM is perceived rather significative. On these premises we can unhesitatingly conclude that the academic and practitioner communities have been persistently striving on the way to pave a theoretically and empirically consistent foundation for the DT and KM joint research thus far. Moreover, lately, concerning the escalating knowledge-based orientation and digitalization of the contemporary economy and society [45], they have been trying to foster rather a single setting in terms of which the joint DT and KM impact could be understood and assessed than an uncoupled one. From the authors' standpoint these state-of-the-art insights could be of a major interest for those who anticipate the breakthroughs in the development of the joint research domain of DT and KM.

11.5.1 Implications

In the process of substantiating the relevance of developing a bibliometric analysis on the DT and KM interrelatedness, we came to find out five comparable scientific contributions (See Table 11.1) that throughout the whole course served us as a reference point in creating an absolutely original paper of its kind, while addressing the acknowledged drawbacks and scholarly prescribed recommendations. Successively, the conducted bibliometric analysis distinctly improved on the antecedent research design practices, having reviewed a larger set of documents and, in some cases, of a longer time span. Added to everything else, initially this bibliometric study might seem to have much in common with the newly published structured literature review authored by de Bem Machado et al. [33]. However, first, the scope of this research was narrowed to the first thematic cluster resultant from the KM and DT joint research domain for Industry 4.0. Consequentially, these bibliometric analysis results complement prior findings and continue the inaugurated debate on the topic. Second, different research tools were tested for a similar methodology, including distinct research guidelines, information sources, and software, in the rest the named article appears as an inspiration. In this way, having justified the originality and authenticity of the present research work, from a theoretical point of view, the generated results represent an important academic value to make progress in further investigation on the role of the DT as one of the most preeminent KM processes' enablers, placing a greater emphasis on the organizational component of transformations. What regards an empirical applicability, the created knowledge is suggested to assist practitioners and policy makers in informing their decisions to respond to the strategic challenges that may predict a forthcoming organizational change. As a result, all target audiences can benefit from the produced outcome, once it enhances a better understanding of the analysed phenomenon.

11.5.2 Limitations

This bibliometric analysis remained subjected to certain limitations. In the first place, there should be mentioned the shortcoming related to the performed Scopus and Web of Science searches. In essence, the used search string was exceptionally aimed at publications from the joint knowledge base, however, there are many pertaining either to the field of DT or KM. In this respect, to potentiate rather a holistic literature review, the proposition could be to apply a single search query to each subject area and then connect the two concepts. Secondly, when abiding by the "investigators' personal criteria" in normalizing the keywords range, there is no way to make the pending research results exempt from the authors' bias. The manner to resolve it questions what the exodus would be if we used the raw selection of keywords (1525 instead of 343). Next to this, the inclusion of documents basing on the keywords in title, abstract, and keywords might represent another aspect for consideration, since it allowed focusing on those studies, in which DT and KM are the backbone, meanwhile, the studies, in which these keywords play a less central and, to some extent, perplexing role, might have been excluded. This could serve the novel investigators as a well-outlined threshold for capturing hidden and unexpected connections between diverse literature streams.

11.5.3 Future Research Avenues

Previously, the development of a well-grounded framework for the future research agenda has reached the culmination at the stage of interpreting the bibliometric analysis results (See Sect. 11.4). In this part, the narrative will consider the essentiality of an exhaustive SLR to move forward in forging the theory of an extensive research area such as the intersection of DT and KM, and in second place, while improving on the methodological diversity, it might appear as an interesting context in the realm of the academic research to ponder the interconnection between KM and DT through the lenses of the execution of knowledge management strategies by means of the technological infrastructure. Adamczewski [56] advocates that Information and Communication Technology (hereinafter referred to as ICT) Systems constitute the foundation of the modern organizations. Zbuchea and Vidu [12] went ahead stating that the ensuing digital transformations and intelligent devices not only support knowledge storage and sharing, but also enable co-creativity and innovation for more complex exploitation of knowledge for the benefit of both companies and its stakeholders. That is to say, influencing continuously the knowledge management processes, the advanced ICT infrastructure turns out to be an indispensable condition for the effective KM.

References

1. Hinings, B., Gegenhuber, T., Greenwood, R.: Digital innovation and transformation: an institutional perspective. Inf. Organ. **28**(1), 52–61 (2018)
2. Bhatt, G.D.: Knowledge management in organizations: examining the interaction between technologies, techniques, and people. J. Knowl. Manag. **5**(1), 68–75 (2001)
3. Sarina, T.: Enhancing knowledge management (KM) in the fourth industrial revolution era: the role of human resource systems. In: The Palgrave Handbook of Knowledge Management, pp. 411–435. Springer (2018)
4. Gaviria-Marin, M., Merigó, J.M., Baier-Fuentes, H.: Knowledge management: a global examination based on bibliometric analysis. Technol. Forecast. Soc. Chang. **140**, 194–220 (2019)
5. Tranfield, D., Denyer, D., Smart, P.: Towards a methodology for developing evidence-informed management knowledge by means of systematic review. Br. J. Manag. **14**(3), 207–222 (2003)
6. Iivari, J., Hirschheim, R., Klein, H.K.: Towards a distinctive body of knowledge for information systems experts: coding ISD process knowledge in two IS journals. Inf. Syst. J. **14**(4), 313–342 (2004)
7. Schaller, A., Vatananan-Thesenvitz, R., Schaller, A.-A.: Assessing relations between sustainable business models and digital transformation: A bibliometric analysis. Paper Presented at the PICMET'21 Conference Proceedings, Daejeon City, South Korea (2021)
8. Kelly, P., Cole, G.A.: Management Theory and Practice, 9th edn. CENGAGE, Singapore (2020)
9. Mizintseva, M.F., Gerbina, T.V.: Knowledge management: a tool for implementing the digital economy. Sci. Tech. Inf. Process. **45**(1), 40–48 (2018). https://doi.org/10.3103/S0147688218010094
10. Weinmann, A.: An Overview of the Current State of Research on Digital Transformation/ Author Alexandra Weinmann, BSc. Universität Linz (2019)
11. Bin, M., Hui, G.: A systematic review of factors influencing digital transformation of SMEs. Turkish Journal of Computer and Mathematics Education (TURCOMAT). **12**(11), 1673–1686 (2021)
12. Zbuchea, A., Vidu, C.: Knowledge management in the digital era. Paper Presented at the 6th International Academic Conference on Strategica – Challenging the Status Quo in Management and Economics, Bucharest, ROMANIA, October 11–12 (2018)
13. Manesh, M.F., Pellegrini, M.M., Marzi, G., Dabic, M.: Knowledge Management in the Fourth Industrial Revolution: mapping the literature and scoping future avenues. IEEE Trans. Eng. Manag. **68**(1), 289–300 (2021). https://doi.org/10.1109/tem.2019.2963489
14. Di Vaio, A., Hassan, R., Palladino, R.: Digital innovation and disruptive technologies in the "intellectual capital (IC) and knowledge management systems (KMS) disclosure": a bibliometric analysis. Paper Presented at the 2020 IEEE International Conference on Technology Management, Operations and Decisions (ICTMOD) (2020a)
15. Di Vaio, A., Palladino, R., Hassan, R., Escobar, O.: Artificial intelligence and business models in the sustainable development goals perspective: A systematic literature review. J. Bus. Res. **121**, 283–314 (2020b)
16. De Bernardi, P., Forliano, C., Bertello, A.: Digital Transformation in SMEs: A Systematic Review and Future Directions. Paper presented at the 13th Annual Conference of the EuroMed Academy of Business (2020)
17. Alvino, F., Di Vaio, A., Hassan, R., & Palladino, R. (2020). Intellectual capital and sustainable development: A systematic literature review. J. Intellect. Capital, **22**, 76
18. Cuc, J.E.: Trends of business model research: a bibliometric analysis. J. Bus. Model. **7**(5), 1–24 (2019)
19. Ahmi, A., Elbardan, H., Raja Mohd Ali, R.H.: Bibliometric analysis of published literature on industry 4.0. Paper presented at the 18th International Conference on Electronics, Information, and Communication, ICEIC 2019 (2019)

20. Lu, K., Liao, H., Zavadskas, E.K.: An overview of fuzzy techniques in supply chain management: Bibliometrics, methodologies, applications and future directions. Technol. Econ. Dev. Econ. **27**(2), 402–458 (2021)
21. Pritchard, A.: Statistical bibliography or bibliometrics. J. Doc. **25**(4), 348–349 (1969)
22. Cobo, M.J., López-Herrera, A.G., Herrera-Viedma, E., Herrera, F.: An approach for detecting, quantifying, and visualizing the evolution of a research field: a practical application to the fuzzy sets theory field. J. Informet. **5**(1), 146–166 (2011a). https://doi.org/10.1016/j.joi.2010.10.002
23. Cobo, M.J., López-Herrera, A.G., Herrera-Viedma, E., Herrera, F.: Science mapping software tools: review, analysis, and cooperative study among tools. J. Am. Soc. Inf. Sci. Technol. **62**(7), 1382–1402 (2011b)
24. Durieux, V., Gevenois, P.A.: Bibliometric indicators: quality measurements of scientific publication. Radiology. **255**(2), 342–351 (2010)
25. Agostini, L., Nosella, A., Sarala, R., Spender, J.C., Wegner, D.: Tracing the evolution of the literature on knowledge management in inter-organizational contexts: a bibliometric analysis. J. Knowl. Manag. **24**(2), 463–490 (2020). https://doi.org/10.1108/JKM-07-2019-0382
26. Benavides-Velasco, C.A., Quintana-García, C., Guzmán-Parra, V.F.: Trends in family business research. Small Bus. Econ. **40**(1), 41–57 (2011). https://doi.org/10.1007/s11187-011-9362-3
27. Zupic, I., Čater, T.: Bibliometric methods in management and organization. Organ. Res. Methods. **18**(3), 429–472 (2015)
28. Page, M.J., Moher, D., Bossuyt, P.M., Boutron, I., Hoffmann, T.C., Mulrow, C.D., et al.: PRISMA 2020 explanation and elaboration: updated guidance and exemplars for reporting systematic reviews. BMJ. **372**, n160 (2021b). https://doi.org/10.1136/bmj.n160
29. Paoloni, M., Coluccia, D., Fontana, S., Solimene, S.: Knowledge management, intellectual capital and entrepreneurship: a structured literature review. J. Knowl. Manag. **24**(8), 1797–1818 (2020). https://doi.org/10.1108/JKM-01-2020-0052
30. Granda-Orive, J.I., Alonso-Arroyo, A., García-Río, F., Solano-Reina, S., Jiménez-Ruiz, C.A., Aleixandre-Benavent, R.: Ciertas ventajas de Scopus sobre Web of Science en un análisis bibliométrico sobre tabaquismo. Revista española de documentación científica. **36**(2), e011–e011 (2013)
31. Borregan-Alvarado, J., Alvarez-Meaza, I., Cilleruelo-Carrasco, E., Garechana-Anacabe, G.: A bibliometric analysis in industry 4.0 and advanced manufacturing: what about the sustainable supply chain? Sustainability (Switzerland). **12**(19), 7840 (2020). https://doi.org/10.3390/SU12197840
32. Rethlefsen, M.L., Kirtley, S., Waffenschmidt, S., Ayala, A.P., Moher, D., Page, M.J., et al.: PRISMA-S: an extension to the PRISMA statement for reporting literature searches in systematic reviews. Syst. Rev. **10**(1), 39 (2021). https://doi.org/10.1186/s13643-020-01542-z
33. De Bem Machado, A., Secinaro, S., Calandra, D., Lanzalonga, F.: Knowledge management and digital transformation for industry 4.0: a structured literature review. Knowl. Manag. Res. Pract. **20**(2), 320–338 (2022). https://doi.org/10.1080/14778238.2021.2015261
34. Culot, G., Nassimbeni, G., Orzes, G., Sartor, M.: Behind the definition of industry 4.0: analysis and open questions. Int. J. Prod. Econ. **226**, 107617 (2020)
35. Page, M.J., McKenzie, J.E., Bossuyt, P.M., Boutron, I., Hoffmann, T.C., Mulrow, C.D., et al.: The PRISMA 2020 statement: an updated guideline for reporting systematic reviews. The BMJ. **372**, n71 (2021a). https://doi.org/10.1136/bmj.n71
36. Cobo, M.J., López-Herrera, A.G., Herrera-Viedma, E., Herrera, F.: SciMAT: a new science mapping analysis software tool. J. Am. Soc. Inf. Sci. Technol. **63**(8), 1609–1630 (2012)
37. Callon, M., Courtial, J.-P., Turner, W.A., Bauin, S.: From translations to problematic networks: an introduction to co-word analysis. Soc. Sci. Inf. **22**(2), 191–235 (1983)
38. Coulter, N., Monarch, I., Konda, S.: Software engineering as seen through its research literature: a study in co-word analysis. J. Am. Soc. Inf. Sci. **49**(13), 1206–1223 (1998)
39. Feroz, A.K., Zo, H., Chiravuri, A.: Digital transformation and environmental sustainability: a review and research agenda. Sustainability. **13**(3), 1530 (2021)

40. Di Vaio, A., Palladino, R., Pezzi, A., Kalisz, D.E.: The role of digital innovation in knowledge management systems: A systematic literature review. J. Bus. Res. **123**, 220–231 (2021)
41. Gil-Gomez, H., Guerola-Navarro, V., Oltra-Badenes, R., Lozano-Quilis, J.A.: Customer relationship management: digital transformation and sustainable business model innovation. Econ. Res.-Ekon. istraž. **33**(1), 2733–2750 (2020)
42. Ghezzi, A., Cavallo, A.: Agile business model innovation in digital entrepreneurship: lean startup approaches. J. Bus. Res. **110**, 519–537 (2020)
43. Schumann, C.A., Tittmann, C.: Digital business transformation in the context of knowledge management. Paper presented at the Proceedings of the 16th European Conference on Knowledge Management, ECKM 2015 (2015)
44. Van Den Berg, C.: 21st century learning: changes to knowledge acquisition in a digital world. Paper presented at the 15th International Conference on Intellectual Capital, Knowledge Management and Organisational Learning, ICICKM 2018 (2018)
45. Pavlekovskaya, I., Urintsov, A., Staroverova, O., Nefedov, Y.: The impact of digital transformation of the Russian economy on knowledge management processes. Paper presented at the 19th European Conference on Knowledge Management, ECKM 2018 (2018)
46. Philbeck, T., Davis, N., Larsen, A.M.E.: Values, Ethics and Innovation: Rethinking Technological Development in the Fourth Industrial Revolution. World Economic Forum (2018)
47. Angrist, N., Djankov, S., Goldberg, P., Patrinos, H.A.: Measuring human capital. Available at SSRN 3339416 (2019)
48. Romero, D., Vernadat, F.: Enterprise information systems state of the art: past, present and future trends. Comput Ind. **79**, 3–13 (2016)
49. Moreira, F., Ferreira, M.J., Seruca, I.: Enterprise 4.0 – the emerging digital transformed enterprise? Paper presented at the International Conference on ENTERprise Information Systems / International Conference on Project MANagement / International Conference on Health and Social Care Information Systems and Technologies, CENTERIS/ProjMAN/HCist 2018 (2018)
50. Barbaroux, P., Le Texier, T.: Managing knowledge within military open-source software (MOSS) projects. Paper Presented at the 4th International Conference on Intellectual Capital Knowledge Management and Organisational Learning, Univ Stellenbosch Business Sch, Stellenbosch, SOUTH AFRICA, October 15–16 (2007)
51. Afanasev, M., Dneprovskaya, N., Kliachin, M., Demidko, D.: Digital transformation of the knowledge management process. Paper presented at the 19th European Conference on Knowledge Management, ECKM 2018 (2018)
52. Le Dinh, T., Phan, T.C., Bui, T.: Towards an architecture for big data-driven knowledge management systems. Paper presented at the 22nd Americas conference on information systems: surfing the IT innovation wave, AMCIS 2016 (2016)
53. Vössing, M., Wolff, C., Reinerth, V.: Digitalization of field service planning: the role of organizational knowledge and decision support systems. In: 9th International Conference on Exploring Service Science, IESS 2018, vol. 331, pp. 138–150. Springer Verlag (2018)
54. Alberti-Alhtaybat, L.V., Al-Htaybat, K., Hutaibat, K.: A knowledge management and sharing business model for dealing with disruption: the case of Aramex. J. Bus. Res. **94**, 400–407 (2019)
55. Ananyin, V.I., Zimin, K.V., Lugachev, M.I., Gimranov, R.D., Skripkin, K.G.: Digital organization: transformation into the new reality. Bus. Inform. **2**, 45–54 (2018). https://doi.org/10.17323/1998-0663.2018.2.45.54
56. Adamczewski, P.: Knowledge Management of Intelligent Organizations in Turbulent Environment. Paper Presented at the 28th International Scientific Conference on Economic and Social Development (ESD), Paris, FRANCE, April 19–20 (2018)

Chapter 12
The Impact of Digital Transformation in the Italian Insurance Sector

Antonio Coviello

Abstract Leading global insurance companies have been implementing digital transformation projects for some years now, in order to effectively respond to renewed customer needs and, at the same time, seize the opportunities offered by new technologies in terms of optimisation of operational processes, automation and cost reduction, and progress in the integration of innovative channels with traditional networks. The objective of this paper will be to prepare an analysis of the diffusion of the digital transformation process, assessing its impact on the organisational processes and management of insurance companies with a view to maximising business opportunities and reducing risks. The analysis, therefore, of the possible challenges that Italian insurance companies will be called to face will be conducted with reference to three levels: innovation of the service model, increase in customer knowledge, and redefinition of the organisational structure. The insurance company, operating in a network structure, is called upon to continuously interact and develop a complex system of communications with numerous other players. These stakeholders include not only agents, brokers, banks and financial networks, but also service companies, other reinsurance companies (with which it manages co-insurance, reinsurance, retrocession) and, more generally, other financial intermediaries. Given the circumstance of the intangibility of the product and the very connection that it poses between product and process, it appears necessary to analyse the internal and external processes of the insurance company that are fully subject to innovation that takes place continuously over time or in "additional" terms. It appears, then, with all the evidence that insurance companies, with the intention of responding in a targeted manner to the growing needs of the insured, tend to vary the structure of the insurance product over time both in the risk coverage component and in the complementary services.

A. Coviello (✉)
National Research Council of Italy, Institute for Research on Innovation and Services for Development, Naples, Italy
e-mail: antonio.coviello@cnr.it

F. Schiavone et al. (eds.), *Advanced Perspectives and Trends in Digital Transformation of Firms, Networks, and Society*, Springer Proceedings in Business and Economics, https://doi.org/10.1007/978-3-031-80692-6_12

Keywords Digital transformation · Insurance · Insurance services innovation

12.1 Introduction

Digital transformation in the insurance sector finds its genesis in changes in the environment and the market. The consequences that this change is generating represent real challenges for the Italian Insurance Company against which it must take a reactive and proactive attitude: detect the evolution of the external context and anticipate change. The aim of this paper is to investigate this phenomenon in order to respond to the change thrusts of the external context and to strengthen the relationship of trust between the Insurance Company and its reference stakeholders, which increasingly requires anticipation of possible future scenarios. The main innovations are mostly related to digitalisation and the organisation of information now available in large quantities, to multi-channelling and to the definition of new organisational models. For insurance companies, therefore, digital transformation and innovation is linked to the translation of new ideas into new products and/or new processes, aimed at changing the positioning of the insurance company on the market with the objective of generating, defending, and/or increasing its competitive advantage [1–3].

Also for Insurance Companies, innovation is to be linked to the translation of new ideas into new products and/or new processes, aimed at modifying the positioning of the insurance company on the market in order to generate, defend, and/or increase its competitive advantage. For these companies, and in particular for their management peculiarities, innovation must be correctly framed in order to grasp its characteristic features [4, 5].

The methodological approach mainly used in the definition of the working hypotheses is of a qualitative-quantitative type and substantially based on the analysis of case studies and best practices. In particular, by means of the widespread use of focus groups, the research work intended to focus mainly on the structure and organisational climate of Italian insurance companies. The study is conducted by examining a number of contributions, selected from a sample of journals, that deal with non-competitive relationships in the firm (internal environment) and between firms (external environment). The articles examined are placed in an analysis scheme that relates the purpose of the work to the approach used, where the purpose and the approach represent the chosen interpretive variables.

The research work aims to contribute to the advancement of theory and practice on insurance service innovation through the definition of models suitable for analysing and understanding the specificities of innovation in insurance services and the identification of risk factors that can affect the success of development strategies and innovation processes. The proposed research approach also attempts to provide a response to the external pressures on insurance companies arising from increasingly high demand expectations and regulatory constraints.

12.2 Innovation in Insurance Services: Characteristics and Considerations

Insurance innovation finds its genesis in changes in the environment and the market. The consequences of this change represent real challenges for the Insurance Company against which they take a reactive and proactive attitude: to detect the evolution of the external context and to anticipate change. The need to Innovate in order to respond to the pushes of change in the external context and to strengthen the relationship of trust between the Insurance Company and its reference stakeholders increasingly requires an anticipation of possible future scenarios [3, 6]. The main innovations are mostly related to the digitalisation and organisation of information now available in large quantities, to multi-channel and the definition of new organisational models. Therefore, for insurance companies, innovation is linked to the translation of new ideas into new products and/or new processes, aimed at changing the positioning of the insurance company on the market with a view to generating, defending, and/or increasing its competitive advantage.

Given the circumstance of the intangibility of the insurance service/product and of the very connection that it places between product and process, it is necessary to analyse the internal and external processes of the insurance company which are fully subject to an innovation that takes place continuously over time or in "additional" terms. It is clear, therefore, that insurance companies, with the intention of responding in a targeted manner to the growing needs of the insured, tend to vary the structure of the insurance product/service over time, both in the risk coverage component and in the complementary services [4, 7].

Innovation is, therefore, one of the strategic priorities for insurance companies, and it is precisely in this context that corporate objectives aim to outline themes of innovation dependent on technological resources, or rather on the use of new ways of carrying out insurance processes, not only in terms of the development of IT and telematics solutions, but also technical and legal ones. An example of this is the construction in recent decades of new insurance policies that have not been contemplated until now, such as those covering cyber risks or those due to climate change risks [1, 8].

The technological revolution is fully affecting the insurance world and is calling on all players in the sector, companies and brokers, to deal with the coverage needs of an increasingly evolved, aware and demanding client, who is changing his purchasing behaviour by exploiting technology as an enabling factor.

The challenge facing operators is to move from selling policies to offering insurance services that enable the end customer to achieve a customer experience in line with expectations [9]. The evolution of behaviours, on the other hand, requires a radical change that puts the customer in a position to have full usability of services in a multi-access logic, both in the purchase phase and in the subsequent management of coverage.

The insurance industry is dynamic by nature and is always responding to threats and incremental changes by providing an indispensable service to its customers. In fact, insurance plays a great social role and is a vehicle to help customers obtain real solutions for their changing needs.

Technological changes are a fact of life and bring opportunities to improve the quality of insurance service, products, and distribution methods [4].

Consequently, the insurance industry must necessarily confront technological progress, promoting research in order to innovate, above all, distribution channels in order to offer a wide range of benefits to customers and society as a whole [8].

Technology, in fact, acts as a catalyst for renewing the approach to how we address a wide range of challenges such as economic, socio-political, and environmental issues, new products, and distribution channels.

The insurance industry, therefore, has a proven track record of dealing with incremental technological progress; in recent years, the pace of radical technology has impacted insurance products and distribution channels precisely because of innovations. This has drastically accelerated the way consumers want their products distributed, causing the insurance industry to lag behind others, creating an ambient atmosphere for insurers to seize this opportunity by offering their services through "InsurTech" [5, 10].

In this context, the insurance industry is currently in a dilemma and is witnessing major challenges caused by the rapid pace of technological innovations. These innovations are reshaping all the activities of the insurance industry, through the use of Internet of Things (IOT), blockchain and artificial intelligence (AI), etc.

In the recent past, radical technological progress has accelerated, and by virtue of new technologies, another macro-trend has emerged that has revolutionised life globally. This revolution has had an impact on the global insurance industry.

New technologies, in fact, such as Big Data and the emergence of smart devices and applications, have increased the frequency and quality of communications between people and things; social media platforms and sharing economy concepts are transforming the ways and methods of innovation, changing insurance business models, and intensifying the demand for new products and distribution channels [1].

New partnerships pioneered with large Internet giants (such as Google, Facebook, Apple, and Amazon) and any other future innovations will culminate the way the distribution channels of the future will handle the sale of insurance products in the new millennium. Through these partnerships, opportunities are created for the insurance industry to access large pools of user data: selling insurance products to a large customer base, devising new marketing strategies, and further diversifying products and distribution methods [8]. Due to rapid technological progress, some market analysts predict that traditional insurers will be forced out of business and face the same fate as "Kodak" (who famously failed to predict future trends in their industry and, as a result, this led them to file for bankruptcy in 2012, ed).

However, most insurers recognise the transformative nature of technology as they still have time to adapt to the "customer-centric" approach and utilise accelerated technological advances and integrate them into their distribution channels. Moreover, by embracing smart innovations, they will have the opportunity to strengthen their relevance to current and future customers.

12.3 Innovation and Digital Disruption for Insurance Services: Discussion and Implication

The insurance industry is experiencing a moment of transition. What we today call "insurtech", referring to all technology-driven innovation in the insurance industry, will increasingly permeate the entire sector and eventually render obsolete even the terminological distinction between insurance and insurtech [5, 10].

The capacity for innovation is a complex phenomenon that involves many subjects and institutions and is the result of human intelligence, application, and resources that in many ways are allocated to research.

Insurance innovation finds its genesis in changes in the environment and the market. The consequences of this change represent real challenges for the Insurance Company, against which they take a reactive and proactive attitude: to detect the evolution of the external context and to anticipate change (a) Artificial intelligence, (b) blockchain, (c) Internet of Things, (d) big data, (e) sharing economy, (f) customer experience, and more. Digital technologies are driving innovation in the insurance industry [1, 3, 4].

1) Artificial Intelligence is the most important technology for the insurance world today, as it is the one founding and enabling a number of other technologies, such as driverless car, robo-advisors, chatbots, search engines, analytics, etc.
2) Blockchain technology has the power to make insurance take a leap into a new era, starting right from the new models of micro insurance, P2P (peer-to-peer o person-to-person), parametric insurance. The power of blockchain in insurance lies in its ability to power new ways of financial transactions, improve existing insurance processes, and keep track of documents. Blockchain-based digital currencies can support many new insurance models, particularly micro insurance and P2P.
3) The Internet of Things represents one of the most innovative technological contexts, closely linked to other technologies such as artificial intelligence and big data analytics, and even the most conservative estimates predict double-digit annual growth rates. For the insurance world it is a great opportunity as it enables new business models (also in terms of preventive insurance) and opens a scenario of new possible services that can be offered at lower costs, especially in the health and smart home sectors.
4) When we talk about big data we refer to a wealth of information of high volume, great speed and variety, which requires innovative and efficient forms of data interpretation in order to make a valid contribution to decision-making and process automation. For insurance companies, which have always relied on the use of data, advanced analytics and data-driven management are the key to promoting more and more "preventive insurance" models.
5) The sharing economy is one of the drivers of change that is having the greatest impact on the insurance industry and is currently posing a real challenge, as the entire value chain of companies is threatened by new business models linked to

the sharing economy and digitalisation. Products, marketing, distribution, prices are transformed. The virtuous circle between insurance and sharing economy has just begun and the growth of peer-to-peer insurance is a manifestation of this.

6) The term customer experience refers to the complex of elements that represent, positively or negatively, the relationship between a customer and a brand, a product or a service. It therefore refers both to the emotional relationship that is created with a company (in the case of insurance, it is, in particular, the relationship of trust) but also to the more practical, physical experience, the so-called "customer journey" that the customer faces at various times of interaction with a company. In the case of insurance, the customer journey can start today from the web, from an app, but also from a broker and embrace the different moments until the claim. In today's insurances, which increasingly put the customer at the centre, the customer experience is obviously fundamental, it is the continuous reference theme that pushes towards digitalisation and the experimentation of new technologies (such as artificial intelligence or blockchain) and new business models (see, for example, micro-insurances, cat-bonds, p2p insurances).

Also for Insurance Companies, innovation is to be linked to the translation of new ideas into new products and/or new processes, aimed at modifying the positioning of the insurance company on the market in order to generate, defend, and/or increase its competitive advantage. For these companies, and in particular for their management peculiarities, innovation must be correctly framed in order to grasp its characteristic features [3, 5].

For insurance companies, the strategic priority is to be reviewed in the innovation of products/services, and the areas in which the main forms of innovation are concentrated—as repeatedly stated in the literature—are in particular:

- Risk Intelligence for the analysis and reduction of risk exposures;
- The Improvement of the operational efficiency of Enterprises;
- User Experience to put the Customer at the centre of the strategy;
- Commercial and Marketing strategies.

In the current insurance panorama, distribution, as a strategic lever of the Companies, assumes an increasingly fundamental role due to several factors:

1) competition and competitiveness of other subjects not typically insurance companies;
2) internal structure of the Companies (the offices in charge of hiring and management activities) rationalised and much reduced in Human Resources, contrary to the strong strengthening of the commercial management inspection networks in order to communicate, assist, and control the external sales networks;
3) evolution of the client's profile in the sense of a positive dynamic change towards the insurance culture and therefore in his ability to identify his own needs/requirements and link them to the most suitable characteristics of the product/policy proposed by the distribution networks.

The insurance company, operating in a network structure, is called upon to continuously interact and develop a complex system of communications with numerous other players. These stakeholders include not only agents, brokers, banks and financial networks, but also service companies, other reinsurance companies (with which it manages co-insurance, reinsurance, retrocession) and, more generally, other financial intermediaries [2, 6].

Insurance Innovation requires a number of conditions (internal and external) and at its base presupposes the generation of new ideas; however, it is the internal organisational context of these companies that effectively contributes to the creative process of new ideas.

Leading global insurance companies have been engaged for some years in the implementation of numerous digital transformation projects, which allow them both to keep pace with the big digital players and meet the renewed needs of customers, and to seize the opportunities offered by new technologies in terms of optimisation of operational processes, automation and cost reduction, and progress in the integration of innovative channels with traditional networks [7, 8, 10, 11].

In recent years, the main driver of change has been cost reduction, which has today made the claims area one of the most evolved, but for the immediate future, much of the investment will be concentrated in the "business" area with particular attention to the issues of omnichannel and relationship management in order to develop distinctive customer experiences able to fully satisfy customer needs.

The need to innovate in order to respond to the pressure of change in the external context and to strengthen the relationship of trust between the insurance company and its stakeholders increasingly requires an anticipation of possible future scenarios [8].

It is therefore essential for Digital Transformation to have a set of high quality data, validated and verified according to specific insurance needs and updated dynamically.

Whether it's geo-referenced images for the management of car claims, or Machine Learning processes for the real-time updating of risk profiles resulting from Climate Change, the quality and accuracy of the data is essential to support the process of change and innovation of Insurance Companies. Therefore, in order to respond in a targeted manner to the growing needs of insured parties, insurance companies are called upon to change the structure of the insurance product over time, both in the risk coverage component and in the complementary services [4].

The main insurance companies at global level have been engaged for some years in the implementation of numerous digital transformation projects, which allow them both to keep up with the big digital players and meet the renewed needs of customers, and to seize the opportunities offered by new technologies in terms of optimisation of operational processes, automation and cost reduction and progress in the integration of innovative channels with traditional networks [3, 5]. The need to Innovate in order to respond to the pushes of change in the external context and to strengthen the relationship of trust between the Insurance Company and its stakeholders increasingly requires an anticipation of possible future scenarios.

12.4 Conclusion and Limitations

Also for Insurance Companies, innovation is to be linked to the translation of new ideas into new products and/or new processes, aimed at modifying the positioning of the insurance company on the market in order to generate, defend, and/or increase its competitive advantage. For these companies, and in particular for their management peculiarities, innovation must be correctly framed in order to grasp its characteristic features.

It is therefore essential for Digital Transformation to have a set of high quality data, validated and verified according to specific insurance needs and updated dynamically [3, 5]. Whether it's geo-referenced images for the management of car claims, or Machine Learning processes for the real-time updating of risk profiles resulting from Climate Change, the quality and accuracy of the data is essential to support the process of change and innovation of Insurance Companies [1, 10]. Therefore, in order to respond in a targeted manner to the growing needs of insured parties, insurance companies are called upon to change the structure of the insurance product over time, both in the risk coverage component and in the complementary services.

The study was able to highlight that for Italian insurance companies, the strategic priority is product/service innovation; the organisational areas in which the main innovations are concentrated are known to be Sales, Marketing, and IT [1]. In the course of the research activities, it was found that the main innovations can be traced back to multi-channelling, digitalisation and the definition of new organisational models. It was thus possible to outline that, for insurance companies, Innovation is a strategic priority; furthermore, although it is implicit in all company activities (even if not formalised), the role of Communication must be defined in the operational plans of insurance companies. However, the design of Innovation and Change in Insurance certainly represents an ambitious programme that requires a careful focus on the critical issues and innovation needs of the insurance sector, which cannot be separated from the development of an action plan for change management.

The study activities outlined the development of a new corporate culture for Italian insurance companies, leading to the establishment of a set of "shared values" among the people who operate in different roles within the organisations, facilitating creativity and the exchange of new interactions. The evaluations and elaborations conducted have thus been able to contribute to outlining a new role of human resources engaged in the production of Value through synergies that can be realised between persons invested with different roles in the organisational structure of Insurance Companies. It was thus possible to identify an open organisational climate capable of stimulating participation and cooperation between the various components of the organisation of Italian insurance companies.

References

1. Coviello, A.: In: Gallouj, F., et al. (eds.) Innovation in Insurance Services, in Elgar Encyclopedia of Service. Edward Elgar Publishing, Cheltenham (2023). https://doi.org/10.4337/9781802202595
2. Isidoro, S.: The Key to Unlock Innovation in the Insurance Industry. e-book Truewind (2020)
3. Kumar, P., Taneja, S., Mukul, E.Ö.: In: Sood, K., Grima, S., Young, P., Ozen, E., Balusamy, B. (eds.) Digital Transformation of the Insurance Industry (2023). https://doi.org/10.1002/9781394167944.ch6
4. Holland, C.P., Kavuri, A.: Artificial intelligence and digital transformation of insurance markets. Journal of Financial Transformation 21. https://www.capco.com/Capco-Institute/Journal-54-Insurance/Artificial-Intelligence-And-Digital-Transformation-In-Insurance-Markets (2021)
5. VanderLinden Sabine, L.B., Millie, S.M., Anderson, N., Chishti, S.: *The Insurtech Book: The Insurance Technology Handbook for Investors*, Entrepreneurs and Fin Tech Visionaries, 1st edn. Wiley Editor (2018)
6. Nicoletti, B.: Digital Insurance, Business Innovation in the Post-Crisis Era. Springer Link (2016)
7. Coviello, A.: *The impact of ICT in the insurance industry: the role of Customer Relationship Management,* IX Conference of the Italian Chapter of AIS "Organization change and Information Systems: Working and living together in new ways", Università Cattolica del Sacro Cuore, September 28th – 29th, 2012, Roma, Italy – Proceedings. http://www.cersi.it/itais2012-EdizioniITHUMSrl (ISBN: 978-88-6685-085-4) (2012)
8. Wiesböck, F., Li, L., Matt, C., Hess, T., Richter, A.: *How Management in the German Insurance Industry Can Handle Digital Transformation* (Management Reports des Instituts für Wirtschaftsinformatik und Neue Medien). LMU München. https://boris.unibe.ch/105693/ (2017)
9. Coviello, A., Di Trapani, G.: *The customer satisfaction in the insurance industry*, Electronic copy. http://ssrn.com/abstract=2144684/ (2014)
10. Lunesu, V.: Insurtech: il modello cinese. Rivista di Diritto Economia e Finanza delle Assicurazioni Private. (3) https://air.unimi.it/handle/2434/892350 (2021)
11. Coviello, A.: L'utilizzo delle tecnologie avanzate nel settore assicurativo. L'e-insurance, Giappichelli Editore, Torino (2009)

Chapter 13
Public Service Delivery in the Digital Era: A Bibliometric Review

Francesco Bellini, Laura Leonori, and Pina Murè

Abstract The study aims to understand through the review of the literature the areas of study and competence related to digital transformation in the public sector, adopting a specific focus on the user perspective. To achieve this, we conducted a bibliometric review which provides an overview of the literature published on Scopus, from 2009 to 2024, and we highlight key trends, influential authors, and potential areas for future research.

The inclusion criteria for the study encompass open-access, peer-reviewed documents in English and listed as 2 stars or more in the ABS list 2021. The findings reveal a significant increase in research on digital transformation and user interaction in public administration since 2009, with a particular emphasis on the period following the Tallinn Declaration of 2017.

We perform a co-authorship network and co-occurrence keywords analysis to identify productive authors and research themes. The Journal of Business Research was found to be the most productive source, with a focus on digitalization, smart cities, and customer engagement. The analysis also revealed the dominance of male authors and UK institutions in the dataset. The co-authorship analysis identified collaboration patterns and clusters of authors mainly focused on European institutions. Finally, the study presented the most cited articles on user orientation in the public sector, emphasizing the importance of understanding digital technologies and their impact on public service delivery.

Keywords Bibliometric analysis · Public service delivery · User orientation · Digital transformation

---

F. Bellini · L. Leonori (✉) · P. Murè
Department of Management, Sapienza University of Rome, Rome, Italy
e-mail: francesco.bellini@uniroma1.it; laura.leonori@uniroma1.it; pina.mure@uniroma1.it

F. Schiavone et al. (eds.), *Advanced Perspectives and Trends in Digital Transformation of Firms, Networks, and Society*, Springer Proceedings in Business and Economics, https://doi.org/10.1007/978-3-031-80692-6_13

13.1 Introduction

Digital Transformation applies to services, processes, and organizational structures through the use of information technology (IT), information systems (IS), and web-based applications; therefore, the link with management is so vital as companies need to establish management practices to govern these complex transformations [1]. The same reasoning can be translated to the public sector. The main objective of the digitization of public services is to provide the population with high-quality administrative services electronically. In particular, digital public services refer to the provision of public services using digital technologies, where the interaction with the public sector organization is mediated by the technological system [2].

The digital transformation of government involves simplifying transactions for citizens, businesses, and government agencies using data analysis to create information which supports and improves decision-making. This makes it possible to simplify transactions for citizens and, thanks to the acquisition and processing of data, to simplify decision-making processes [3].

In the context of contemporary governance, the digitalization of public administration services has a pivotal role in enhancing service delivery, promoting citizen engagement, and fostering administrative efficiency. The European Union has been at the forefront of this digital transformation, embarking on a strategic journey delineated by key policy documents and declarations. The Malmö Declaration of 2009 marks a significant milestone in this trajectory, articulating a vision to harness Information and Communication Technologies (ICTs) to make public services more accessible, efficient, and citizen-centric. Built upon this foundation, the EU e-government Action Plan 2016–2020 underscores objectives to modernize public administrations and mitigate fragmentation in digital modernization efforts. Subsequently, the Tallinn Declaration of 2017 foresees the development of User-Centricity Principles and delineates a roadmap to accelerate the digital transformation of the public sector.

As stated by Porter [4], an industry with demanding customers is likely to receive valuable information for product or service development or improvement. This concept can also be applied to the public service sector. As concerns public service delivery, information sharing plays a crucial role in supporting new models of service delivery [5]. Additionally, the continuous improvement process (CIP) of public service systems relies on customer feedback to drive improvements and ensure customer satisfaction [6]. Furthermore, a customer-centric approach to product and service development in the public sector recognizes the active and creative role of customers, emphasizing the value that can be derived from both products and processes [7]. Therefore, like their counterparts in the private sector, demanding customers within the public service domain can provide valuable information for the development and improvement of products and services. It is crucial for public entities to move beyond rigid bureaucratic procedures and to adopt a comprehensive approach, which entails incorporating knowledge from various fields like service marketing and digital marketing [8].

This paper aims to support existing literature, which is mainly focused on investigating specific fields of analysis related to user orientation, each according to the perspective of the study of the authors, by providing a wide-ranging analysis of user orientation in public administrations within the context of digital transformation. We aim to identify the macro areas that deal with the topic and investigate whether it is possible to assess synergy for a more systematic understanding and development of user orientation in the public sector.

To assess this knowledge gap, we examined the structure and dynamics of scholarly production regarding the public service delivery to users in the digital era through bibliometric analysis [9], investigating the trend in research production over time and the domain structure of the analysis.

The study is based on the collection of 2231 records from the Scopus bibliographic database spanning from 2009 to 2024 and employs the bibliometric software VOSviewer. We analysed the scholarly production over time and identified the most influential journals and popular articles. We analysed the interests of scholars and we investigated the thematic structure by performing a co-occurrence keyword analysis.

The findings indicate a broad spectrum of research interests in user orientation. The interdisciplinary nature of this field emphasizes the multifaceted approaches required to address challenges and opportunities in public sector innovation. The study highlights a concerning disparity in authorship, with male dominance prevailing, and UK institutions dominating the network of collaborative research efforts.

Future research could broaden its scope by expanding data collection methods and, given rapid technological advancements, exploring the societal and organizational impacts of human-centric digital technology, deepening insights into public sector dynamics.

13.2 Methodology and Data Sample

The work adopts the bibliometric literature review with the objective of delineating the fundamental concepts that form the basis of a research area on user orientation in public services and identifying the primary sources and types of evidence available to determine the main lines of research.

The analysis is conducted following the framework outlined by Grant et al. (2009) [10], termed the SALSA (Search, Appraisal, Synthesis, and Analysis) framework (Fig. 13.1).

The research question focuses attention on the structure and dynamics of scholarly production within the domain of public service delivery for users in the digital era. During the planning of the data collection phase, we take the following strategic decisions for enhancing the sample adequacy. The document search was conducted on the Scopus database using three different search strings (see Table 13.1).

The search resulted in a total of 97.641 records. After we applied exclusion and inclusion criteria to assess the suitability of specific search results, we obtained a

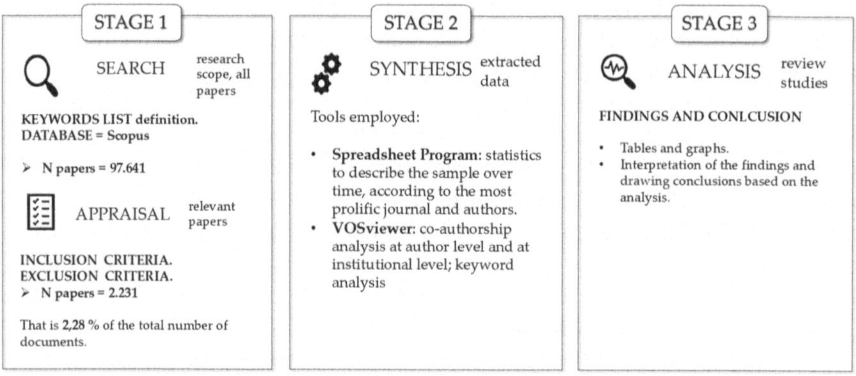

Fig. 13.1 Research protocol adopted according to the SALSA framework stage

Table 13.1 Search strings applied to Scopus database

String no.	Search string
S1	((user AND orient*) AND ((public AND administration) OR (public AND service)))
S2	((citizen AND centric) OR (user AND centric) AND ((public AND sector) OR (public AND service)))
S3	((citizen OR experience) OR (user AND experience)) AND (((public AND service* AND design) OR (digital AND public AND service*))) AND (digital AND transformation))

total number of records equal to 2505. This represents 2.57% of the total number of documents.

The inclusion criteria developed to guide the collection of studies relevant to this review are the following:

- Only open-access documents have been selected, meaning the materials are peer-reviewed, accessible online, and available without restrictions.
- Studies published in the period 2009–2024.
- Studies published in the Scopus databases by departments located in European countries.
- Studies published in the English language.
- Studies published by journals listed in the ABS list 2021, released by the Chartered Association of Business School, with two stars or more.

To strengthen the search process when deciding on the relevance of each selected article, the following exclusion criteria were used:

- Reports, master's or doctoral theses, duplicate notes.
- Studies from non-EU countries.

We performed the analysis using the spreadsheet program Excel, to describe the sample over time and the most prolific journal and authors, and the bibliometric software VOSviewer (see www.vosviewer.com, [11]) for co-authorship analysis at the author and institutional level, and keyword analysis.

13.3 Findings

13.3.1 The Production of Scholarly Articles over Time

The sample consists of 2231 articles. From 2009, the research on digital transformation in public administration related to user orientation has grown rapidly.

Precisely, we carried out a polynomial interpolation with a quadratic degree to better study the growth of the number of publications during the years, providing a visual representation of the increase in production over time (Fig. 13.2). The analysis suggests that every year the number of publications may increase exponentially, which represents a signal of growing interests in the topic. Since 2017, there has been a notable increase in publications: about 87% of the articles in the dataset were published after 2017. This growth is in line with the Tallinn Declaration on e-Government of 2017, when all EU Member States and European Free Trade Association (EFTA) Countries signed a political commitment to guarantee the provision of high-quality, user-centric digital public services for citizens, alongside facilitating seamless cross-border public services for businesses.

To show the publications' performance [9] more comprehensively, the analysis proceeds to study the most productive journals, the articles most cited, the authors

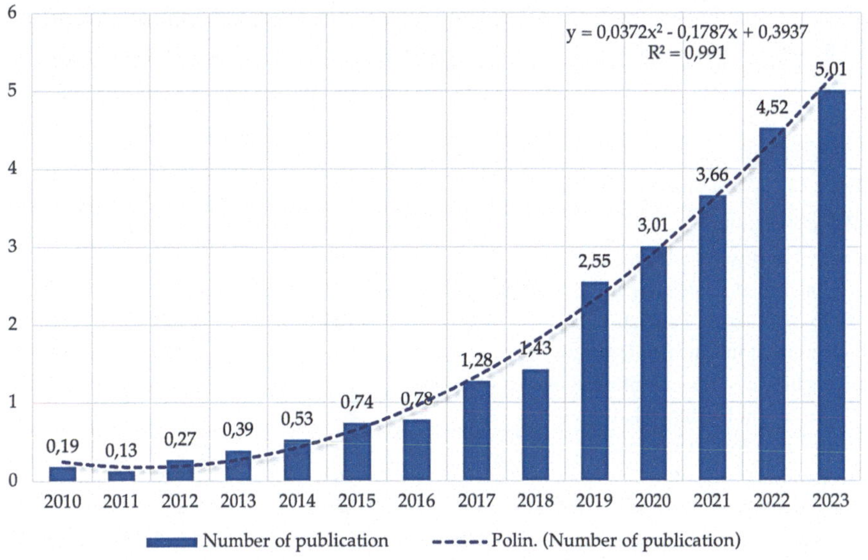

Fig. 13.2 Polynomial interpolation of the number of publications with time

most proficient making a substantial contribution to the rising trend in scholarly research and the co-occurrence analysis of keywords.

13.3.2 Influential Journal and Articles

The sample analysed includes 94 publication titles. Approximately 40% of the articles are published by the top ten most productive journals listed in Table 13.2.

The *Journal of Business Research* holds the top position with 199 articles, followed by *Technological Forecasting and Social Change* with 134. *The Journal of Cleaner Production* and *Government Information Quarterly* follow closely with 105 and 99 articles, respectively. Together these four journals include over a fifth of the sample's articles.

The distribution of records across different journals reflects the diverse focus areas within the fields covered by these publications. The presence of journals like *Government Information Quarterly* and *Research Policy* suggests a focus on governance, policy, and research methodologies; although there is a strong interest in business-related research topics, technology and innovation, information management and systems; sustainability and environmental management.

The papers with the highest number of citations are listed in Table 13.3. The first most cited article (1756 citations) is entitled 'Developing a framework for responsible innovation' [12]: it is jointly written by three authors, affiliated with three institutions from two countries, the University of Exeter Business School and the Durham University in UK, and the State University of Campinas in Brazil; it was published in 2013 by Research Policy and discusses the development of a framework for responsible innovation, focusing on the dimensions of anticipation, inclusion, and responsiveness in the governance of emerging science and technology.

Table 13.2 Most productive journal

Publication titles	Record count	Overall percentage
Journal of Business Research	199	8.92%
Technological Forecasting and Social Change	134	6.01%
Journal of Cleaner Production	105	4.71%
Government Information Quarterly	99	4.44%
International Journal of Information Management	67	3.00%
Computers in Human Behaviour	65	2.91%
Information Systems Frontiers	63	2.82%
Technovation	56	2.51%
Research Policy	54	2.42%
Transportation Research Part A: Policy and Practice	52	2.33%

Table 13.3 Most cited articles

Authors	Article title	Source title	Number of citations
Stilgoe et al. (2013) [12]	Developing a framework for responsible innovation	Research Policy	1756
Neirotti et al. (2014) [13]	Current trends in Smart City initiatives: Some stylized facts	Cities	1539
Voorberg et al. (2015) [14]	A systematic review of co-creation and co-production: Embarking on the social innovation journey	Public Management Review	1276
Verhoef et al. (2021) [15]	Digital transformation: A multidisciplinary reflection and research agenda	Journal of Business Research	1247
Sivarajah et al. (2017) [16]	Critical analysis of big data challenges and analytical methods	Journal of Business Research	1192
Gawer (2014) [17]	Bridging differing perspectives on technological platforms: Towards an integrative framework	Research Policy	1001
Dwivedi et al. (2021) [18]	Artificial intelligence (AI): Multidisciplinary perspectives on emerging challenges, opportunities, and agenda for research, practice and policy	International Journal of Management	943
Nambisan et al. (2019) [19]	The digital transformation of innovation and entrepreneurship: Progress, challenges, and key themes	Research Policy	906
Mollen and Wilson (2010) [20]	Engagement, telepresence, and interactivity in online consumer experience: Reconciling scholastic and managerial perspectives	Journal of Business Research	894
Jaakkola and Alexander (2014) [21]	The role of customer engagement behaviour in value co-creation: A service system perspective	Journal of Service Research	829

13.3.3 Collaborative Ties Between Prolific Authors

The software VOSviewer was used to perform a co-authorship analysis: we analysed the collaboration pattern between authors who reached the minimum requirement of having published at least five articles and it resulted that just a few numbers of authors (80) have five articles or more (Fig. 13.3). Further, we mapped the strength of links between authors based on direct collaboration through joint publications.

A connection is identified for 45 prolific authors grouped in seven connected clusters, each representing distinct collaboration patterns. As concern the remaining scholars, they are grouped in isolated clusters: the most densely populated is the yellow cluster composed of 6 authors; while eight clusters are composed of two authors per cluster; lastly no connections were identified for 13 authors.

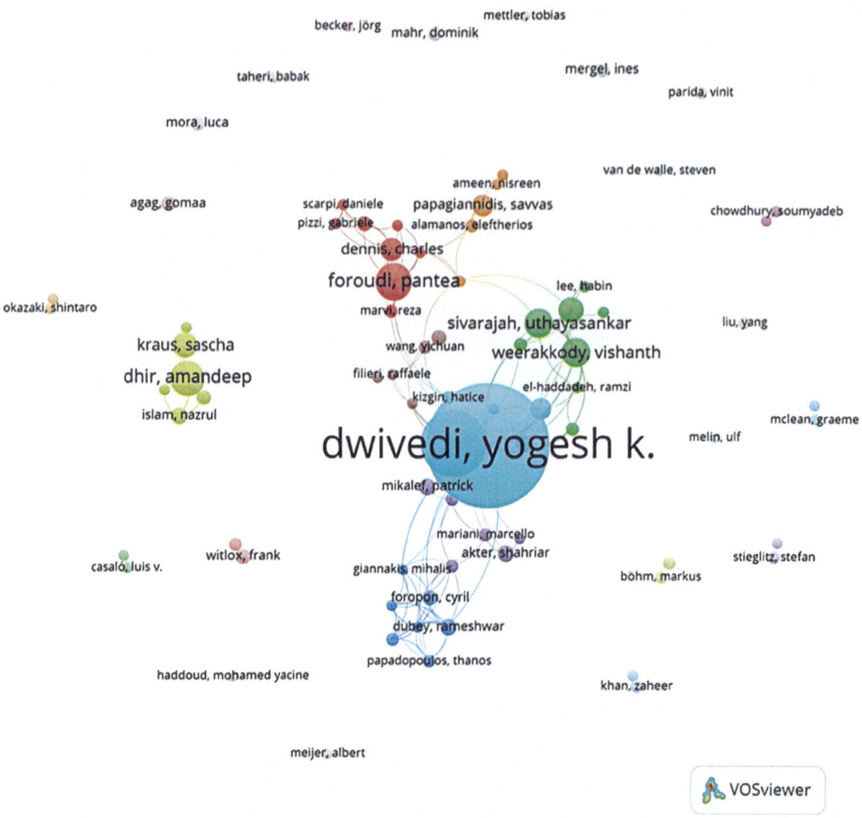

Fig. 13.3 Co-authorship analysis at author level

We identified the most important link strength in the sky-blue cluster which comprises six authors. Some of them are linked with a geographically defined institutional context: Yogesh K. Dwivedi and Michael D. Williams are affiliated with Swansea University (UK); Marijn Janssen and Hatice Kizgin are both affiliated with institutions located in the Netherlands, respectively, with Delft University of Technology and with the University of Twente; Marcello M. Mariani is affiliated with the Alma Mater Studiorum—Università di Bologna (Italy); Nripendra P. Rana is currently associated with the College of Business and Economics at Qatar University (Qatar). The highest link strength, which indicates how many publications two researchers have co-authored, identifiable within the cluster is between Dwivedi and Rana for a value at least of 25: they collaborated on papers covering subjects as the discussion on the use of the Unified Theory of Acceptance and Use of Technology (UTAUT) [22] and adoption of an electronic government [23].

The second cluster (green) is composed of eight authors, half of them are affiliated with British institutions (University of Bradford and Brunel University London), while there is quite heterogeneity between the other affiliations: North

South University (Bangladesh), Qatar University (Qatar), Universidade Nova de Lisboa (Portugal) and NEOMA Business School (France). As concerns the research area, the more prominent themes are digital government, information systems, and energy innovative technology. For instance, the study co-authored by Weerakkody and El-Haddadeh reveals how empowerment, usefulness, and privacy significantly shape citizens' perceived value of IoT in public service engagement, informing future policy and implementation strategies. [24].

The second most populated cluster is the red one and it is composed of eight authors, six of them are affiliated with British institutions (Middlesex University, Aston Business School, Brunel University London) while the other two are associated with the Italian Alma Mater Studiorum—Università di Bologna. Pantea Faroudi has the strongest link connection albeit her cluster relationships do not involve the Italian scholars. The main research interests of the group deal with consumer behaviour focusing on several research streams including corporate brand design and identity, SDGs, technology, emotional analytics, and customer experience in the service context. The article co-authored by Foroudi and Marvi, the two scholars with the highest connection, examines how co-creation behaviour affects peer commitment in P2P platforms [25].

Another sizable cluster is the blue one and includes seven authors from two French institutions, respectively, the Montpellier Business School (Rameshwar Dubey, Cyril Foropon, David Roubaud) and the Audencia Business School (Mihalis Giannakis); from two British universities, the Plymouth Business School (Stephen J. Childe) and the University of Kent (Thanos Papadopoulos); from Penn State Harrisburg in the USA (Angappa Gunasekaran). Within this cluster, the highest total link strength of 29 is attributed to R. Dubey, who collaborated with A. Gunasekaran, S.J. Childe, and T. Papadopoulos to co-author a paper published in the British Journal of Management, which investigates the impact of external institutional pressures on organizational resources and big data culture, and how these factors influence the adoption of big data and predictive analytics (BDPA) and how these capabilities may play a role for gaining a competitive advantage [26]. The most common research topics of the cluster deal with supply chain management, management information systems, and sustainable and humanitarian operations management.

The subsequent cluster analysed is the purple one, comprising six authors from diverse institutions, including Toulouse Business School in France, Norwegian University of Science and Technology in Norway, University of Wollongong in Australia, NIIT University in India, and Alma Mater Studiorum—Università di Bologna in Italy; scholars overarching topics are digital transformation and analytics in business [27].

The fifth most populated cluster is the orange, and it is made up of five researchers all affiliated with British institutions, respectively, Suraksha Gupta, Savvas Papagiannidis, Eleftherios Alamanos are associated with the Newcastle University Business School, Jyoti Choudrie with the University of Hertfordshire, and Nisreen Ameen with the Royal Holloway Business School. The predominant focus among the authors revolves around digital marketing, technological innovation, and the intersection of sustainability and social impact issues [28].

Elvira Ismagilova, Yichuan Wang, Nick Hajli, and Giampaolo Viglia, affiliated with British universities, respectively, with the University of Bradford, University of Sheffield, Loughborough University, University of Portsmouth and Raffaele Filieri affiliated with the Audencia Business School in France are part of the sixth cluster. The cluster mainly produces research on digital marketing and consumer behaviour; marketing and management, and technology and innovation [29].

Finally, the yellow cluster has no connection even if comprehends six scholars affiliated with institutions from different countries such as Italy, Finland, Norway, the UK, and India. Researches mainly insist on business strategy, entrepreneurship, and innovation, digitalization and technology, consumer behaviour and tourism, organizational dynamics, and management also in the public sector [30].

Lastly, the analysis of the gender distribution among the top 80 most proficient authors found that male authors predominate, constituting 86% of the total, while female authors comprise only 14%. The ten most cited authors, who predominantly come from the United Kingdom, represent the same distribution: the majority are male and only one female author is represented among the top cited (Table 13.4).

Table 13.4 Top 10 most influential scholars (ranked by citation)

Ranking	Author	Documents	Citations	Total link strength	Institution	Country	Gender
1	Dwivedi, Yogesh K.	64	7475	71	Swansea University	UK	M
2	Rana, Nripendra P.	33	5155	44	Qatar University	Qatar	M
3	Weerakkody, Vishanth	14	2220	24	University of Bradford	UK	M
4	Sivarajah, Uthayasankar	14	2000	19	University of Bradford	UK	M
5	Irani, Zahir	12	1910	18	University of Bradford	UK	M
6	Dubey, Rameshwar	8	1592	29	Montpellier Business School	France	M
7	Gunasekaran, Angappa	6	1465	23	Penn State Harrisburg	USA	M
8	Childe, Stephen J.	5	1404	23	Plymouth Business School	UK	M
9	Janssen, Marijn	10	1387	8	Delft University of Technology	The Netherlands	M
10	Ismagilova, Elvira	5	1220	11	University of Bradford	UK	F

13.3.4 *Co-Authorship Analysis at Institutional Level*

The organizational level is the next point of examination of the co-authorship network analysis. From the 5.772 institutions identified within the dataset, 70 institutions met the established criterion of having published a minimum of three papers over the past 16 years. The largest set of connected items consists of 37 items.

Concerning sample identification and data cleaning we noticed that institutions are often recorded differently, primarily due to the inclusion or omission of the address or the research centre causing variations in the identifier strings. Therefore, out of the 5.772 institutions recorded, there might be inconsistencies in the total count due to the software's inability to recognize all possible variants of organization names belonging to the same entity.

For this reason, we opted for a broad approach and have set a threshold of 3 documents per organization. This resulted in the identification of 115 organizations. The subsequent step involves the manual analysis of the resultant list. Considering the total number of institutions identified by VOSviewer (115), the manual scraping of the list has revealed the presence of 73 near-duplicates, indicating pairs of strings that can be traced back to the same entity. This process leads to the creation of the thesaurus file for merging various variants of organization names. After, the file was uploaded to VOSviewer. This resulted in a discrepancy of 46 units in the total count of organizations, reducing the count to 70 organizations.

The largest set of connected items comprises 37 items (Fig. 13.4). More than half of the institutions are based in the UK, while the others are distributed between France, India, Sweden, the Netherlands, Belgium, Australia, Saudi Arabia, and Italy.

The largest and most productive cluster, in red colour, comprises eight institutions. Swansea University in the UK leads in terms of documents with 28, while the Department of Management at Symbiosis Institute of Business Management, Pune, and Symbiosis International (Deemed University) in India hold the highest total link strength of 14. Scholars from these organizations collaborate with different institutions based in France (3), such as Audencia Business School and the Rennes School of Business which belong to the same cluster; UK (3), the Leeds University Business School; India (2), the Guru Nanak dev University and the Management Development Institute of Gurgaon; Italy (1); and Saudi Arabia (1) as Department of Management Information Systems of The King Abdulaziz University.

The green cluster is the second per documents co-authored and consists of five British institutions, namely the Newcastle Business School, the Aston Business

Fig. 13.4 Co-authorship network at institution level

School, the Nottingham University Business School, the Essex Business School and Durham University, and the University of Sadat City based in Egypt.

The third cluster (blue) is composed of five universities, based in different countries, such as the University of Bologna in Italy, Middlesex University and University of Reading in the UK, Luleå University of Technology in Sweden and Tbs Business School in France.

Cluster 4 (yellow) is composed only of British institutions, precisely the University of Manchester, the King's College London, Bournemouth University, University of Surrey, which are mainly connected to other UK-based organizations. Similarly, the sky-blue cluster comprises solely British organizations, namely Coventry University, University of Hull, University of Kent, University of Strathclyde.

The purple cluster is constituted by organizations located in the Netherlands, Radboud University and Utrecht University; Belgium, Public Governance Institute; Loughborough University in the UK.

The smallest connected clusters describe a relationship between French and British institutions, respectively, between the Montpellier Business School and the Plymouth University (orange cluster); the Neoma Business School and the University of Bradford (brown circles).

Finally, the pink cluster groups together the Linköping University in Sweden and the University of Queensland in Australia.

13.3.5 Themes Connected to User Orientation in the Public Sector

To examine the main topics related to user orientation in public services we performed a co-occurrence analysis of keywords in VOSviewer. For data cleaning, we created a thesaurus file to merge similar words to foster the accuracy of the analysis by consolidating related terms. This allows for a standardization of term variations.

The dataset comprises 11.915 keywords, setting a minimum number of co-occurrences of 7 results in a subset of 517 words. Notably, recurrent keywords were organized into seven distinct clusters, visually represented in Fig. 13.5.

A significant portion of the research focuses on the adoption of digital technologies, processes, and platforms across various sectors, reflecting the ongoing digital transformation and innovation efforts in businesses, governments, and public services. The red cluster comprises 97 keywords such as 'innovation', which has the highest total link strength of 809, 'digital transformation', 'value creation', 'e-government', 'user experience', 'public services'.

The second cluster, highlighted in green, encompasses 74 keywords including 'human', 'social media', 'healthcare', 'policy-making', 'consumer', 'citizen participation'.

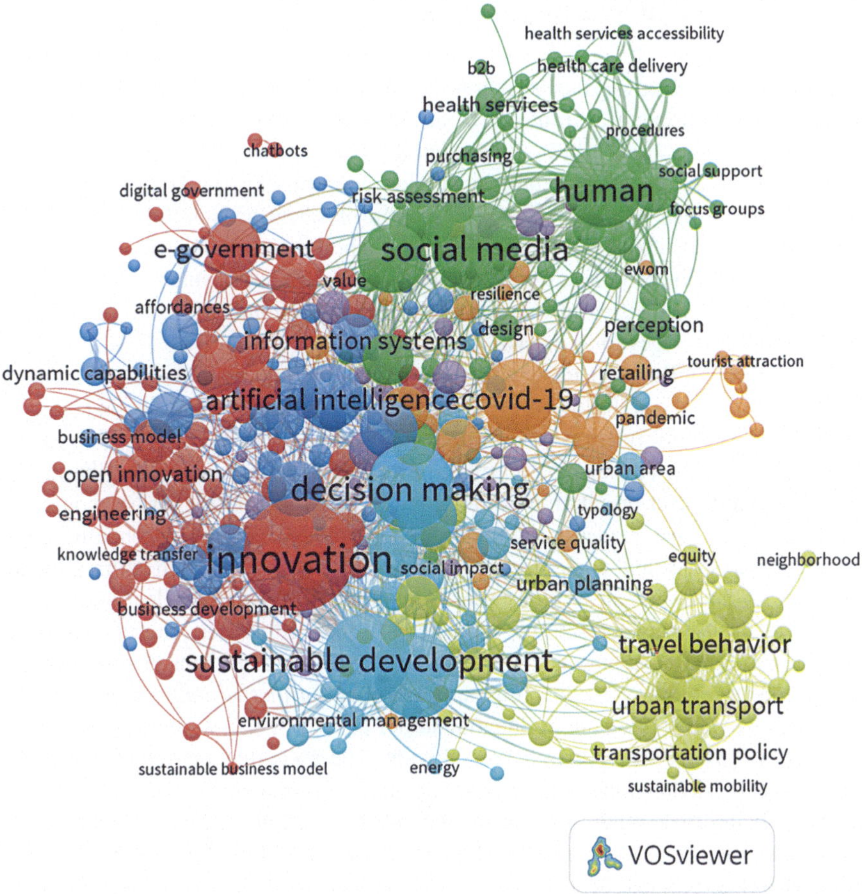

Fig. 13.5 Co-occurrence keywords analysis

The third cluster (blue) consists of 69 closely related keywords indicating a significant interest in 'artificial intelligence', 'big data', 'machine learning', 'technology adoption'.

The yellow cluster, made of 65 items, is related to transportation issues regarding 'public transport', 'mobility', 'travel behaviour', and 'climate change'.

The themes of the purple cluster, consisting of 39 items are cross-cutting and deal with the 'economic and social effects' of 'technology adoption' and development, 'industry 4.0', and 'crowdsourcing'.

The sky-blue cluster, comprising 38 items, focuses on 'decision-making', 'sustainable development', 'technological development', 'SDGs' and 'smart cities'.

Finally, the seventh cluster (orange) is the smallest, including 34 words, and deals with 'COVID-19' related topics such as 'ICT', 'consumption behaviour', 'tourism management', and 'metaverse'.

13.4 Discussion and Implication

This bibliometric analysis has been conducted on 2.231 articles written by scholars affiliated with European institutions from 2009 to 2024 on user orientation in the public sector. This study shows a substantial growth in the literature and suggests that the commitment of the European Union countries to providing user-centric public services has encouraged the production of scientific literature. Out of the total publications, about 87% have been published since 2017, after the Tallinn Declaration which promotes user-centricity principles in public service delivery and increased attention on user involvement, reflecting a commitment to providing high-quality digital services for citizens.

The Journal of Business Research emerges as the most productive and influential source in this analysis, followed by Technological Forecasting and Social Change, and Government Information Quarterly. Notably, almost half of all articles originate from the top 10 most productive journals, indicating their significant contribution to the discourse in user orientation within governmental and public service contexts. As concern the most cited articles, there is a clear focus on digitalization, other topics like smart cities and customer engagement are also present, indicating a broad range of research interests within user orientation. However, the selection of journals from the Scopus dataset may have been influenced by our criteria—we only considered ABS-listed journals with 2 stars or more. This approach may have excluded some sub-areas, impacting our assessment of journal influence [31].

The most contributing and influential author is Yogesh K. Dwivedi., followed by Nripendra P. Rana and Vishanth Weerakkody. The co-authorship analysis reveals that the most cited authors in the dataset primarily originate from the UK, with the majority being male, and only one female author is featured among the top cited. Men dominate social science authorship, accounting for 70% of published research [32]. This disparity can be attributed to several factors, such as women submitting their work to prestigious journals less frequently, all-male teams being more common in academic publishing, and certain research topics being male-dominated [33]. Furthermore, men tend to use sensationalistic terms in titles or abstracts, which often lead to more citations [34]. These disparities exacerbate gender gaps in authorship within social science fields.

As concerns the top authors' affiliated institutions, the largest set of connected items comprises 37 items. UK institutions dominate the network, while the others are distributed between France, India, Sweden, the Netherlands, Belgium Australia, Saudi Arabia, and Italy. The findings provide insights into research collaboration patterns and potential areas for future partnerships. British universities have a strong presence in social science and management, thanks to the support of two organizations: the British Academy of Management (BAM) and the Association of Business Schools (ABS). This has led to increased legitimacy and recognition of management education, along with the historical legacy of British universities. Strong international networks and collaborations with institutions worldwide have also facilitated knowledge exchange, research collaboration, and funding opportunities.

This is confirmed by QS Ranking 2024 by Subject, which shows that out of 236 European institutions in social science, 26% are in the UK and eight of the top ten institutions in Europe are in the UK.

The co-occurrence analysis of keywords reveals seven principal areas of focus. Of notable prominence are themes encompassing innovation, value creation, e-government, digital transformation, and user experience. It is an interdisciplinary field, intersecting with theories of co-creation, public value management, collaborative governance, and strategic management [35]. All these aspects have a role in shaping public service delivery and have the potential to create a positive impact on society. Research on smart cities is relevant because it enhances public services such as smart living, mobility, and environmental sustainability. COVID-19 and related terms emerge for their relevance to the public sector, stimulating innovation in government services and healthcare. Overall, one core concept for all issues related to public sector innovation is the human element which highlights the importance of citizens' needs and experiences as sources of innovation.

13.5 Conclusion and Limitation

This paper provides an overview of the research areas that influence the concept of user orientation in the public sector using comprehensive bibliometric analysis. Literature has grown significantly since 2017, reflecting a commitment to high-quality digital services and an increased focus on user involvement in public services.

Our findings emphasize the critical importance of relating multiple elements from different realms, such as marketing, business, and technology, to form coherent and organic knowledge. This comprehension is pivotal in navigating the transition from the established Industry 4.0 to the emergent Industry 5.0 paradigm. Industry 5.0 is characterized by the dynamic interplay between user and technological functions, and being proficient in these areas requires an alignment with the principles of public management knowledge. Additionally, this represents a top-down change that should be assessed and systematized at all levels of public administration, and it implies the recruitment of highly qualified personnel and the cultivation of a shared and coherent vision of delivering public services.

Future investigations could adopt a wider approach by expanding the database for data collection. And, due to the high degree of technological evolution, the analysis could be expanded in the future by investigating the impact at the societal and organizational level of developing human-centric digital technology.

Funding Information *Funded by the European Union – NextGenerationEU, Mission 4, Component 1, CUP B83C22003100006.*

References

1. Reis, J., Amorim, M., Melão, N., Matos, P.: Digital transformation: a literature review and guidelines for future research. In: Rocha, Á., Adeli, H., Reis, L.P., Costanzo, S. (eds.) Trends and Advances in Information Systems and Technologies, pp. 411–421. Springer International Publishing, Cham (2018)
2. Lynn, T., Rosati, P., Conway, E., Curran, D., Fox, G., O'Gorman, C.: Digital public services. In: Lynn, T., Rosati, P., Conway, E., Curran, D., Fox, G., O'Gorman, C. (eds.) Digital Towns: Accelerating and Measuring the Digital Transformation of Rural Societies and Economies, pp. 49–68. Springer International Publishing, Cham (2022)
3. Battisti, D.: The digital transformation of Italy's public sector: government cannot be left behind! JeDEM. **12**, 25–39 (2020). https://doi.org/10.29379/jedem.v12i1.591
4. Porter, M.E.: The competitive advantage of nations. Harvard Business Review. **68**(2), 73–93 (2001)
5. Curtis, S., Edwards, J.: Improving public services by sharing the right information. Public Money Manag. **39**, 355–358 (2019). https://doi.org/10.1080/09540962.2019.1611238
6. Schweitzer, E., Aurich, J.C.: Continuous improvement of industrial product-service systems. CIRP J. Manuf. Sci. Technol. **3**, 158–164 (2010). https://doi.org/10.1016/j.cirpj.2010.04.002
7. Fukuda, S.: Product and service development with customers. In: Pokojski, J., Fukuda, S., Salwiński, J. (eds.) New World Situation: New Directions in Concurrent Engineering, pp. 635–641. Springer, London (2010)
8. Saxena, D., Muzellec, L., McDonagh, J.: From bureaucracy to citizen-centricity: how the citizen-journey should inform the digital transformation of public services. Int. J. Electron. Gov. Res. **18**, 1–17 (2022). https://doi.org/10.4018/IJEGR.305230
9. Donthu, N., Kumar, S., Mukherjee, D., Pandey, N., Lim, W.M.: How to conduct a bibliometric analysis: an overview and guidelines. J. Bus. Res. **133**, 285–296 (2021). https://doi.org/10.1016/j.jbusres.2021.04.070
10. Grant, M.J., Booth, A.: A typology of reviews: an analysis of 14 review types and associated methodologies. Health Inf. Libr. J. **26**, 91–108 (2009). https://doi.org/10.1111/j.1471-1842.2009.00848.x
11. van Eck, N.J., Waltman, L.: Software survey: VOSviewer, a computer program for bibliometric mapping. Scientometrics. **84**, 523–538 (2010). https://doi.org/10.1007/s11192-009-0146-3
12. Stilgoe, J., Owen, R., Macnaghten, P.: Developing a framework for responsible innovation. Res. Policy. **42**, 1568–1580 (2013). https://doi.org/10.1016/j.respol.2013.05.008
13. Neirotti, P., De Marco, A., Cagliano, A.C., Mangano, G., Scorrano, F.: Current trends in Smart City initiatives: some stylised facts. Cities. **38**, 25–36 (2014). https://doi.org/10.1016/j.cities.2013.12.010
14. Voorberg, W.H., Bekkers, V.J.J.M., Tummers, L.G.: A systematic review of co-creation and co-production: embarking on the social innovation journey. Public Manag. Rev. **17**, 1333–1357 (2015). https://doi.org/10.1080/14719037.2014.930505
15. Verhoef, P.C., Broekhuizen, T., Bart, Y., Bhattacharya, A., Qi Dong, J., Fabian, N., Haenlein, M.: Digital transformation: a multidisciplinary reflection and research agenda. J. Bus. Res. **122**, 889–901 (2021). https://doi.org/10.1016/j.jbusres.2019.09.022
16. Sivarajah, U., Kamal, M.M., Irani, Z., Weerakkody, V.: Critical analysis of Big Data challenges and analytical methods. J. Bus. Res. **70**, 263–286 (2017). https://doi.org/10.1016/j.jbusres.2016.08.001
17. Gawer, A.: Bridging differing perspectives on technological platforms: toward an integrative framework. Res. Policy. **43**, 1239–1249 (2014). https://doi.org/10.1016/j.respol.2014.03.006
18. Dwivedi, Y.K., Hughes, L., Ismagilova, E., Aarts, G., Coombs, C., Crick, T., Duan, Y., Dwivedi, R., Edwards, J., Eirug, A., Galanos, V., Ilavarasan, P.V., Janssen, M., Jones, P., Kar, A.K., Kizgin, H., Kronemann, B., Lal, B., Lucini, B., Medaglia, R., Le Meunier-Fitz Hugh, K., Le Meunier-FitzHugh, L.C., Misra, S., Mogaji, E., Sharma, S.K., Singh, J.B., Raghavan, V., Raman, R., Rana, N.P., Samothrakis, S., Spencer, J., Tamilmani, K., Tubadji, A., Walton,

P., Williams, M.D.: Artificial Intelligence (AI): multidisciplinary perspectives on emerging challenges, opportunities, and agenda for research, practice and policy. Int. J. Inf. Manag. **57**, 101994 (2021). https://doi.org/10.1016/j.ijinfomgt.2019.08.002
19. Nambisan, S., Wright, M., Feldman, M.: The digital transformation of innovation and entrepreneurship: progress, challenges and key themes. Res. Policy. **48**, 103773 (2019). https://doi.org/10.1016/j.respol.2019.03.018
20. Mollen, A., Wilson, H.: Engagement, telepresence and interactivity in online consumer experience: Reconciling scholastic and managerial perspectives. J. Bus. Res. **63**, 919–925 (2010). https://doi.org/10.1016/j.jbusres.2009.05.014
21. Jaakkola, E., Alexander, M.: The role of customer engagement behavior in value co-creation: a service system perspective. J. Serv. Res. **17**, 247–261 (2014). https://doi.org/10.1177/1094670514529187
22. Dwivedi, Y.K., Rana, N.P., Jeyaraj, A., Clement, M., Williams, M.D.: Re-examining the Unified Theory of Acceptance and Use of Technology (UTAUT): towards a revised theoretical model. Inf. Syst. Front. **21**, 719–734 (2019). https://doi.org/10.1007/s10796-017-9774-y
23. Rana, N.P., Dwivedi, Y.K., Lal, B., Williams, M.D., Clement, M.: Citizens' adoption of an electronic government system: towards a unified view. Inf. Syst. Front. **19**, 549–568 (2017). https://doi.org/10.1007/s10796-015-9613-y
24. El-Haddadeh, R., Weerakkody, V., Osmani, M., Thakker, D., Kapoor, K.K.: Examining citizens' perceived value of internet of things technologies in facilitating public sector services engagement. Gov. Inf. Q. **36**, 310–320 (2019). https://doi.org/10.1016/j.giq.2018.09.009
25. Foroudi, P., Marvi, R., Colmekcioglu, N.: Antecedents and consequences of co-creation value with a resolution of complex P2P relationships. Int. J. Contemp. Hosp. Manag. **34**, 4355–4388 (2022). https://doi.org/10.1108/IJCHM-10-2021-1278
26. Dubey, R., Gunasekaran, A., Childe, S.J., Blome, C., Papadopoulos, T.: Big Data and predictive analytics and manufacturing performance: integrating institutional theory, resource-based view and Big Data culture. Br. J. Manag. **30**, 341–361 (2019). https://doi.org/10.1111/1467-8551.12355
27. Sultana, S., Akter, S., Kyriazis, E., Wamba, S.F.: Architecting and developing big data-driven innovation (ddi) in the digital economy. J. Global Inf. Manag. **29**, 165–187 (2021). https://doi.org/10.4018/JGIM.2021050107
28. Davlembayeva, D., Papagiannidis, S., Alamanos, E.: Sharing economy: studying the social and psychological factors and the outcomes of social exchange. Technol. Forecast. Soc. Change. **158**, 120143 (2020). https://doi.org/10.1016/j.techfore.2020.120143
29. Wang, Y., Hajli, N.: Exploring the path to big data analytics success in healthcare. J. Bus. Res. **70**, 287–299 (2017). https://doi.org/10.1016/j.jbusres.2016.08.002
30. Khan, A., Krishnan, S., Dhir, A.: Electronic government and corruption: Systematic literature review, framework, and agenda for future research. Technol. Forecast. Soc. Change. **167**, 120737 (2021). https://doi.org/10.1016/j.techfore.2021.120737
31. Van Campenhout, G., Van Caneghem, T., Uytbergen, S.: A comparison of overall and sub-area journal influence: The case of the accounting literature. Budapest Scientometrics. **77**, 61–90 (2008). https://doi.org/10.1007/s11192-007-1907-5
32. Lockhart, J.W., King, M.M., Munsch, C.: Who authors social science? Demographics and the production of knowledge. https://osf.io/dsqcp (2023)
33. Bartosch, J., Lohmeyer, N., Auschra, C., Jackson, G.: Gendered publication patterns in Socio-Economic Review. Socio-Econ. Rev. **21**, 1273–1289 (2023). https://doi.org/10.1093/ser/mwad041
34. Lerchenmueller, M.J., Sorenson, O., Jena, A.B.: Gender differences in how scientists present the importance of their research: observational study. BMJ. **367**, l6573 (2019). https://doi.org/10.1136/bmj.l6573
35. Torfing, J., Ferlie, E., Jukić, T., Ongaro, E.: A theoretical framework for studying the co-creation of innovative solutions and public value. Policy & Politics. **49**, 189–209 (2021). https://doi.org/10.1332/030557321X16108172803520

Part II
Healthcare Digital Innovations

Chapter 14
Analyzing the Drivers of Digital Innovation in Healthcare: A Panel Analysis from European Countries

Alessia Spignese, Federica Zeuli, and Caterina Cantone

Abstract Digital health represents a significant innovation in healthcare that can enhance patient engagement and improve health outcomes. Various studies have highlighted the benefits of these technologies in healthcare settings, the improved diagnostics, personalized care, and real-time patient monitoring. Recent research emphasizes the importance of understanding both the internal and external drivers of digital innovation in healthcare, particularly in diverse geographical contexts. Our study aims to address the gap in the literature by investigating the macroeconomic variables influencing digital healthcare innovation, focusing on the disparities across European countries. By exploring these factors, we seek to provide insights into the strategies that can facilitate effective digital transformation in healthcare organizations. We used the quantitative methodology of the regression model to investigate the drivers of digital innovation in the healthcare industry. Our research contributes to identifying the drivers of digital innovation in the healthcare sector. In particular, the analysis results highlight additional factors that influence digital innovation in the healthcare industry that the researchers have not empirically investigated yet.

Keywords Digital innovation · Healthcare · Financial development · Medical technology

A. Spignese (✉) · C. Cantone
Department of Economics, Luigi Vanvitelli University, Capua, CE, Italy
e-mail: alessia.spignese@unicampania.it

F. Zeuli
University of Naples Parthenope, Naples, Italy

F. Schiavone et al. (eds.), *Advanced Perspectives and Trends in Digital Transformation of Firms, Networks, and Society*, Springer Proceedings in Business and Economics, https://doi.org/10.1007/978-3-031-80692-6_14

14.1 Introduction

Digital health represents a type of innovation in healthcare wherein platforms leverage digital technologies to engage patients in their healthcare journey. Various authors concur that digital health encompasses using different technologies to administer health services to improve patient's well-being [1, 2]. The integration of digital technologies in hospitals has gained widespread acceptance, facilitated by their reduced size, increased accessibility, and enhanced capacity for handling extensive amounts of data [1, 3, 4, 20]. Therefore, digital healthcare innovation involves adopting advanced technologies to augment and streamline healthcare services [5]. Marques and Ferreira [3] pointed out that the last two decades have seen a sharp increase in the number of studies on digital innovation in healthcare. However, the authors point out that most of these studies have mainly focused on identifying the different digital technologies implemented, the benefits and advantages of digital technology implementation, and the barriers and challenges associated with implementing these technologies. Among the digital technologies in health, biomedical and digital sensors are those referred to by most academic studies [4, 6–8]. Digital healthcare technologies also include mobile devices [9], apps, cloud-based solutions [6], artificial intelligence, and data analytics. The benefits of implementing these digital technologies in health care are considerable. They include improving diagnostic capabilities, providing personalized healthcare, and monitoring patients in real time [2, 4, 6, 10]. Finally, the researchers underscored various challenges in integrating digital technologies within healthcare institutions. These challenges encompass issues such as information security vulnerabilities, patent-related concerns, limited collaborative partnerships, and a scarcity of expertise in information and communication technology (ICT) [4, 5, 8, 11, 22, 12]. Only recently, scholars have expressed a keen interest in investigating the potential catalysts for digital healthcare innovation [4, 8]. The factors driving innovation can vary across organizations, sectors, and nations. Consequently, innovation management is contingent on internal and external factors in the healthcare domain. As posited by Bedir [9] and Visconti et al. [13], variables such as per capita income and the state of the local economy can influence healthcare organizations' strategies and effectiveness. Conversely, Tortorella et al. [4] asserted that factors like the number of beds and the functionality and experience of employees play a role in shaping the dynamics of healthcare organizations. A limited number of studies have explored the factors influencing the extent of digital innovation in healthcare organizations. Some of these studies have exclusively focused on emerging economies [4] and African countries [8]. Furthermore, a recent study specifically delves into the internal drivers of digital innovation within the healthcare industry, investigating their impact on the digital transformation of Italian hospitals [14]. While these studies are valuable for comprehending the factors shaping digital innovation in the healthcare sector, there is a clear imperative for additional empirical research to investigate additional drivers of digital innovation in healthcare organizations and to analyze this phenomenon in different geographical contexts. In the world, different situations require innovation development in the health sector. In particular, in Europe, there are notable differences among countries in terms of development and innovation in healthcare [12]. For the reasons set out above, our study aims to fill the gap in the literature that answers the following research question (RQ): what variables drive innovation in healthcare through the macro-economic lens?

Our study aims to contribute to the existing literature that explores the drivers of digital innovation in the healthcare industry. This research seeks to fill the gap in the literature about innovation driving in the health field [15]. Some studies support this research topic, but there is still a need to understand better the factors that drive digital innovations, especially from a macroeconomic lens and different geographic aspects. Beyond identifying additional drivers for this phenomenon, our analysis seeks to provide a macroeconomic perspective by examining multiple countries within the European context. The countries under analysis include Italy, France, Germany, Spain, and Sweden. This selection is justified by the notable similarity in their national healthcare systems, with only subtle variations that add an exciting dimension to the analysis. Specifically, all these countries share a universalistic health system, ensuring access for all citizens. They rely on taxes, social contributions, and public funding as the primary financing sources for their national health systems. However, differences emerge in the decentralization of healthcare systems (Italy, Spain, and Sweden) or their centralization (France and Germany). Additionally, variations exist in the role played by the private sector, particularly the insurance sector in France and Germany [12]. To the best of our knowledge, our study represents the first attempt to investigate the presence of supplementary drivers of digital innovation in healthcare within the European context.

14.2 Method

The study is structured into three main phases. In the first step, we collected data for each country from 2011 until 2021 for the analysis. Two criteria were used to select the time period. First, we considered digital healthcare innovation a recent phenomenon, attracting academic interest in the last decade [6, 23, 7, 16] (See Table 14.1 above).

Table 14.1 This table describes the variables and the official source used

Variable	Description	Source
Healthcare digital innovation	Number of patent applications to EPO.	OECD database
Medical technology	Technological and digital hospital equipment units per 1,000,000 inhabitants.	OECD database
R&D expenditure	R&D expenditure (% GDP).	EUROSTAT database
Size	Numbers of available beds x 100,000 inhabitants.	OECD database
Specialized	Number of skilled workers in the health sector x 100,000 inhabitants.	OECD database
Financial development	Financial development index: Financial markets and financial institutions.	World Bank database

Source: Authors' elaboration

Second, we had to consider the availability of data. In the second step, we proposed the regression model (1) to investigate the drivers of digital innovation in the healthcare industry (See Table 14.1 above).

$$\begin{aligned} Healthcare\ DIi,t = & \ \beta 0 + \beta 1(Medical\ Technology)i,t \\ & + \beta 2(R\,\&\,D\ Expenditure)i,t + \beta 3(Size)i,t \\ & + \beta 4(Specialised)i,t + \beta 5(Financial\ Development)i,t \end{aligned}$$

Finally, we performed a GLS regression, after which the Hausman test was applied. Generalized least squares (GLS) estimation, an extension of ordinary least squares (OLS), has proven advantageous for fitting linear models to data sets affected by heteroscedasticity problems [17, 18].

14.3 Findings

By providing a broader perspective, our research contributes to identifying the drivers of digital innovation in the healthcare sector. The analysis results highlight additional factors influencing digital innovation in the healthcare industry that, to the best of our knowledge, have not been empirically investigated. Specifically, variables such as Medical technology, reflecting the presence of technological and digital devices in hospitals, the development of the financial system, and the availability of high-skilled workers emerge as noteworthy drivers. Recognizing the potential enhancement of healthcare quality and operational efficiency through digital technologies [16] should guide the integration of these technologies and promote digital innovation in the sector. The financial system's development also plays a crucial role, providing essential resources for medical research, advancing health technologies, and implementing sophisticated medical practices [19]. Additionally, a highly specialized healthcare workforce can drive digital innovation, although a negative relationship suggests that specialized knowledge may only sometimes translate into digital innovation benefits. Some professionals may need help adopting new digital technologies due to unfamiliarity or concerns about patient privacy, potentially slowing down the digital innovation process in healthcare [15].

14.4 Discussion and Implication

This study contributes to the academic literature in several ways. First, it identifies additional drivers of digital innovation in healthcare. Second, the analysis is not limited to one geographical context. It covers five different European countries with very similar national healthcare systems. Third, the study has a new methodology (GLS) for measuring the level of digital innovation in the healthcare industry, which can be used by future research. Finally, this study contributes to the knowledge of the drivers of the adoption and implementation of digital solutions by healthcare organizations, showing the role of medical technology, financial development, and specialized workforce. The results obtained also offer important implications for policymakers. Recognizing that the digital innovation process is necessary to promote the complete digitization of healthcare services [23], policymakers need to encourage digital innovation in healthcare through regulations to promote the adoption and implementation of specific digital technologies in European hospitals so that the digital innovation process continues. Furthermore, policy interventions should address initiatives to develop the financial system and its involvement in financing innovative healthcare investment projects. Finally, to support greater uptake of digital health technologies and ensure an improvement in the health and well-being of the population, policymakers should also pay attention to the e-skills of the healthcare workforce.

References

1. Ancarani, A., Di Mauro, C., Gitto, S., Mancuso, P., Ayach, A.: Technology acquisition and efficiency in Dubai hospitals. Technol. Forecast. Soc. Chang. **113**, 475–485 (2016)
2. Shivers, C.M., Sonnier-Netto, L., Lee, G.K.: Needs and experiences of family caregivers of individuals with autism spectrum disorders across the lifespan. J. Policy Practice Intell. Disab. **16**(1), 21–29 (2019)
3. Garai, Á., Péntek, I., Adamkó, A., Németh, Á.: A clinical system integration methodology for bio-sensory technology with cloud architecture. Acta Cybern. **23**(2), 513–536 (2017)
4. Tortorella, G.L., Fogliatto, F.S., Espôsto, K.F., Vergara, A.M.C., Vassolo, R., Mendoza, D.T., Narayanamurthy, G.: Effects of contingencies on healthcare 4.0 technologies adoption and barriers in emerging economies. Technol. Forecast. Soc. Chang. **156**, 120048 (2020)
5. Lupton, D.: Health promotion in the digital era: a critical commentary. Health Promot. Int. **30**(1), 174–183 (2014)
6. Elhoseny, M., Abdelaziz, A., Salama, A.S., Riad, A.M., Muhammad, K., Sangaiah, A.K.: A hybrid model of internet of things and cloud computing to manage big data in health services applications. Futur. Gener. Comput. Syst. **86**, 1383–1394 (2018)
7. Hamidi, H.: An approach to develop the smart health using Internet of Things and authentication based on biometric technology. Futur. Gener. Comput. Syst. **91**, 434–449 (2019)
8. Preko, M., Boateng, R.: Assessing healthcare digitalisation in Ghana: a critical realist's approach. Health Policy Technol. **9**(2), 255–262 (2020)
9. Bedir, S.: Healthcare expenditure and economic growth in developing countries. Adv. Econ. Bus. **4**(2), 76–86 (2016)
10. Munzer, B.W., Khan, M.M., Shipman, B., Mahajan, P.: Augmented reality in emergency medicine: a scoping review. J. Med. Internet Res. **21**(4), e12368 (2019)
11. Aceto, G., Persico, V., Pescapé, A.: The role of Information and Communication Technologies in healthcare: taxonomies, perspectives, and challenges. J. Netw. Comput. Appl. **107**, 125–154 (2018)
12. AGENAS, 2022_Quaderni_Monitor_Cure_Primarie_Europa_lettura_Web: https://www.agenas.gov.it/images/agenas/In%20primo%20piano/CURE_PRIMARIE_IN_EUROPA/Quaderni_Monitor_Cure_Primarie_Europa_lettura_web.pdf
13. Visconti, R.M., Dos, A., Gurgun, A.P.: Public–private partnerships for sustainable healthcare in emerging economies. In: The Emerald Handbook of Public–Private Partnerships in Developing and Emerging Economies, pp. 407–437. Emerald Publishing Limited, Bingley (2017)

14. Raimo, N., De Turi, I., Albergo, F., Vitolla, F.: The drivers of the digital transformation in the healthcare industry: an empirical analysis in Italian hospitals. Technovation. **121**, 102558 (2023)
15. Jacob, C., Sanchez-Vazquez, A., Ivory, C.: Social, organizational, and technological factors impacting clinicians' adoption of mobile health tools: systematic literature review. JMIR Mhealth Uhealth. **8**(2), e15935 (2020)
16. Kraus, S., Schiavone, F., Pluzhnikova, A., Invernizzi, A.C.: Digital transformation in healthcare: analyzing the current state-of-research. J. Bus. Res. **123**, 557–567 (2021)
17. Menke, W.: Review of the generalized least squares method. Surv. Geophys. **36**(1), 1–25 (2015)
18. Meuleman, B., Loosveldt, G., Emonds, V.: Regression analysis: assumptions and diagnostics. In: The SAGE Handbook of Regression Analysis and Causal Inference, pp. 83–110 (2015)
19. World Health Organization: Research and development to meet health needs in developing countries: strengthening global financing and coordination: report of the consultative expert working group on research and development: financing and coordination (2012)
20. Prieto González, L., Jaedicke, C., Schubert, J., Stantchev, V.: Fog computing architectures for healthcare: wireless performance and semantic opportunities. J. Inf. Commun. Ethics Soc. **14**(4), 334–349 (2016)
21. Sannino, G., De Falco, I., De Pietro, G.: A continuous noninvasive arterial pressure (CNAP) approach for health 4.0 systems. IEEE Trans. Industr. Inform. **15**(1), 498–506 (2018)
22. Pan, J., Ding, S., Wu, D., Yang, S., Yang, J.: Exploring behavioural intentions toward smart healthcare services among medical practitioners: a technology transfer perspective. Int. J. Prod. Res. **57**(18), 5801–5820 (2019)
23. Toni, M., Mattia, G., Pratesi, C.A.: What's next in the healthcare system? The contribution of digital innovation in achieving patient-centricity. Futures. **156**, 103304 (2024)

Chapter 15
Mapping Technology Acceptance in Healthcare: Insights from a Systematic Review

Nohaila Kartti and Annaluce Mandiello

Abstract This study investigates, through a systematic review of literature, the acceptance of digital health solutions. The digitalization of the healthcare system leads to several benefits increasing quality of care, efficiency of the processes, and reducing unnecessary costs. However, most of digital technologies need to be accepted by the actors of a system to be effective because someone's behavior toward an innovation can affect its capability to provide value through outputs and outcomes. This work facilitates the understanding of the mechanisms below innovation acceptance considering different contexts, technologies, and actors and highlighting the specific needs and behaviors that regulate them. This analysis is expected to show actors' reaction to digital health solutions across the years, using a multifaceted approach. The implications of this research include the identification of tailored and effective strategies that must be implemented by policy-makers and practitioners to foster the entrance of digital health technologies in the healthcare.

Keywords Digital health · Technology acceptance · Systematic literature review

15.1 Introduction

15.1.1 Overview

Nowadays, the main innovations that are revolutionizing society at individual, organizational, and societal levels are the ones resulting from the impactful phenomenon generally known as "Digital Transformation" (DT) [1, 2] which is, indeed, "the combined effects of several digital innovations bringing about novel actors (and actor constellations), structures, practices, values, and beliefs that change, threaten, replace or complement existing rules of the game within organizations, ecosystems,

N. Kartti (✉) · A. Mandiello
University of Naples "Parthenope", Naples, Italy

F. Schiavone et al. (eds.), *Advanced Perspectives and Trends in Digital Transformation of Firms, Networks, and Society*, Springer Proceedings in Business and Economics, https://doi.org/10.1007/978-3-031-80692-6_15

industries or fields" [3]. In this regard, digital health (DH) is the meeting point where the innovative solutions of DT encounter the healthcare (HC) system to "improve human health, healthcare services, and wellness for individuals and across population" [4] becoming an essential part of it. However, despite the important role of the DT in the improvement of value co-creation processes within HC and the implementation of more cost-effective ways to provide personalized care services [5], "the HC industry is lagging behind other sectors in its DT" [6] because of several reasons.

Generally, in order to understand how and to what extent an innovation can truly transform the aimed market, it must be understood if the market "wants" to be innovated. For example, according to Kumari et al. [7], the users' resistance behavior is one of the crucial reasons for technological innovations' failure. Hence, different concerns come with an innovation especially when it must be inserted within a sector grounded on delicate and significant human needs like health. In the case of DH, resistance is mostly related to "privacy concerns about personal health information, immaturity of the technology being used, organizational readiness, and incompatible stakeholder views about the technology used" [5].

Actually, what has been widely and mostly shown in the literature is the importance of focusing on the acceptance of technology in order to understand the main variables that influence the actors' behavioral intention. These last, indeed, can be used as strategic levers through which the introduction of an innovation within a market is facilitated. It can be stated, hence, that analyzing the technology adoption first can also help the avoidance of resistant behaviors and the related consequences. This is the reason why several studies aim at identifying also the resistance variables through acceptance models. More specifically, several different models investigating the acceptance of a technology have been adapted to the healthcare, like the "Technology Acceptance Model" (TAM) [8], the "Unified Theory of Acceptance and Use of Technology (UTAUT)" [9] and their evolutions. Moreover, the focus of the management literature on the technology acceptance investigated through these models applied to the healthcare has been increasing in recent years (e.g. [10–13]). Thus, it can be helpful to understand more in depth this topic through an analysis of the literature in order to provide an overview of what is truly happening within healthcare in relation to the DH, putting the basis for future research.

15.2 Objective

15.2.1 Purpose of the Work

Through a systematic review of the literature, the aim of this research is to analyze and systematize all the studies and research conducted on the acceptance of digital health solutions, from the beginning of the millennium to the present day since it represents the period in which the study on digital health started. The importance of this topic is significantly supported by the ever-growing trend of publications on the acceptance of digital health solutions, which increased during the pandemic years and peaked in 2023 (Scopus). This growing trend in literature reveals the increasing interest of the scientific community on this topic which is making the utility of a systematic literature review on it even more crucial and necessary. For these reasons, this systematic literature review on the acceptance of digital solutions in healthcare addresses the following questions:

RQ1: How is the acceptance of digital health solutions investigated in management literature?

RQ2: Why do multiple stakeholders accept digital-health innovations across various healthcare contexts?

To answer the research questions abovementioned the authors defined the specific research contexts based on the perspectives of the following stakeholders: patients and healthcare personnel, whose acceptance of existing digital health solutions has been generally investigated in last years [11].

15.3 Methodology

15.3.1 Systematic Literature Review

To fulfill the aims of this study and address the research questions posed, a comprehensive review of the literature will be conducted utilizing a widely recognized method, namely, a Systematic Literature Review (SLR) (e.g., [14]). Systematic reviews became increasingly important in all fields being also adopted in interdisciplinary areas such as IT, digitalization, and healthcare sectors [15]. Professionals and academics in the IT, digitalization and healthcare industries rely on systematic reviews to stay updated in their fields, often utilizing them as a starting point for developing technological guidelines [16] for adoption in various domains, including healthcare.

Conducting an SLR, indeed, is advantageous as it allows for a thorough and systematic examination of existing research findings, as emphasized by Kaur et al. [17]. Additionally, a systematic review serves as a means to identify, evaluate, and interpret all available studies relevant to a particular research question, topic, or phenomenon of interest [18]. It is defined as a methodology that summarizes the process of collecting, arranging, and assessing literature in a review domain [19]. It adopts a replicable, scientific, and transparent process in order to avoid or minimize bias through exhaustive literature research studies [20].

A systematic review was considered suitable for this study given its aim to discern key findings in the management research and offer recommendations for future studies [21, 22]. Furthermore, it can significantly contribute to the understanding of the research area by identifying gaps and suggesting forthcoming research topics [23]. The scholars, indeed, may be able to develop a series of "proven and field-based technological rules" [24], which can be useful not only to academics but to professionals and managers as well [25].

This systematic review will follow the structured approach proposed by Watson [26], which clearly delineates the procedures and stages involved in literature exploration. The search procedure will be based on planning, execution, and reporting activities. Due to the evolving literature in the realms of digitalization and healthcare, it is expected to issue a challenge for professionals and healthcare experts to truly consider the literature as a relevant tool for evidence-based practices [27]. In both practical and academic settings, indeed, strong empirical results are fundamental to guide future actions and practices. Evans [28] suggests that employing a systematic review is a prominent method to support evidence-based practices in IT, digitalization, and healthcare. Boell and Cecez-Kecmanovic [29] argue that adhering strictly to a predetermined protocol and a well-defined research process enhances the efficiency of systematic reviews.

While Watson acknowledges the importance of efficiency in research, he contends that effectiveness is also crucial and it is possible to achieve it by "synthesizing the literature and revealing the depth of knowledge on an area's critical key concepts and the relationships between these concepts" ([26], p. 185). The authors of this study, hence, consider the adherence to the rules and guidelines of systematic reviews as an imperative way to adopt this approach [18]. Moreover, particular attention will be paid to the need to start the review process with a protocol for identifying, selecting, and evaluating pertinent literature since the fact that it will improve the effectiveness of the systematic review [30]. The systematic approach should be replicable, impartial, transparent, fair, and thorough [29].

Based on the above, this study will be conducted in three stages, as proposed by Watson [26]. Furthermore, several collective rules and guidelines will be applied for the different steps of this systematic review, as identified by Kitchenham and Charters [18] and Ali et al. [31–33]. Rules and guidelines applied during the planning stage (1) include the identification of the gap to fulfill with the systematic review, definition of a classification framework, definition of research questions, and formulation of research strategies. The execution stage (2) entails techniques such as keyword search, filter application, title, and abstract reading, full article reading, backward snowballing, and quality assessment. In the reporting stage (3), the research will involve classifying selected articles and discussing the results and future implications.

15.4 Expected Findings and Future Implications

15.4.1 Expected Findings and Next Steps

This study contributes to the academic debate on digital health acceptance and innovation management through a comprehensive analysis of the existing literature. Facilitating the understanding of the mechanisms below innovation acceptance that truly operate within different contexts and highlighting the specific needs that regulate them. However, the complexity of the sector analyzed may led the authors extending the area of research toward the identification of the additional factors that, interacting, can affect and moderate the acceptance of an innovation (such as cultural values).

In a broader sense, this analysis is expected to show the different actors' reactions to the new DH technologies using a multifaceted approach. This last, indeed, is based on the several points of view resulting from the analyses grounded on various technology acceptance models implemented in the management literature. In this manner, it will be easier to comprehend which are the tailored and effective strategies that must be implemented to foster the entrance of an innovation in the HC sector.

References

1. Verhoef, P.C., Broekhuizen, T., Bart, Y., Bhattacharya, A., Dong, J.Q., Fabian, N., Haenlein, M.: Digital transformation: a multidisciplinary reflection and research agenda. J. Bus. Res. **122**, 889–901 (2021)
2. Vial, G.: Understanding digital transformation: a review and a research agenda. Manag. Digit. Transform., 13–66 (2021)
3. Leone, D., Schiavone, F., Appio, F.P., Chiao, B.: How does artificial intelligence enable and enhance value co-creation in industrial markets? An exploratory case study in the healthcare ecosystem. J. Bus. Res. **129**, 849–859 (2021)
4. Kostkova, P.: Grand challenges in digital health. Front. Public Health. **3**, 134 (2015)
5. Balta, M., Valsecchi, R., Papadopoulos, T., Bourne, D.J.: Digitalization and co-creation of healthcare value: a case study in occupational health. Technol. Forecast. Soc. Chang. **168**, 120785 (2021)
6. Kraus, S., Schiavone, F., Pluzhnikova, A., Invernizzi, A.C.: Digital transformation in healthcare: analyzing the current state-of-research. J. Bus. Res. **123**, 557–567 (2021)
7. Kumari, P., Shankar, A., Behl, A., Pereira, V., Yahiaoui, D., Laker, B., Arya, V.: Investigating the barriers towards adoption and implementation of open innovation in healthcare. Technol. Forecast. Soc. Chang. **200**, 123100 (2024)
8. Davis, F.D.: Perceived usefulness, perceived ease of use, and user acceptance of information technology. MIS Q., 319–340 (1989)
9. Venkatesh, V., Morris, M.G., Davis, G.B., Davis, F.D.: User acceptance of information technology: toward a unified view. MIS Q., 425–478 (2003)
10. Baudier, P., Kondrateva, G., Ammi, C., Chang, V., Schiavone, F.: Patients' perceptions of teleconsultation during COVID-19: a cross-national study. Technol. Forecast. Soc. Chang. **163**, 120510 (2021)

11. Edo, O.C., Ang, D., Etu, E.E., Tenebe, I., Edo, S., Diekola, O.A.: Why do healthcare workers adopt digital health technologies-A cross-sectional study integrating the TAM and UTAUT model in a developing economy. Int. J. Inform. Manag. Data Insights. **3**(2), 100186 (2023)
12. Talukder, M., Aroos-Sheriffdeen, S., Khan, M.I., Quazi, A., Abdullah, A.B.M.: Usage behavior of mHealth service users in Australia: do user demographics matter? J. Serv. Mark. (2023)
13. Wei, X., Peng, X., Prybutok, V.: Consumer behavioral intention of adopting emerging healthcare technology. IEEE Trans. Eng. Manag. (2022)
14. Christofi, M., Vrontis, D., Cadogan, J.W.: Micro-foundational ambidexterity and multinational enterprises: a systematic review and a conceptual framework. Int. Bus. Rev. **30**, 101625 (2021)
15. Kamboj, S., Rahman, Z.: Marketing capabilities and firm performance: literature review and future research agenda. Int. J. Product. Perform. Manag. **64**(8), 1041–1067 (2015)
16. Moher, D., Liberati, A., Tetzlaff, J., Altman, D.G.: Preferred reporting items for systematic reviews and meta-analyses: the PRISMA statement. BMJ. **339**, b2535 (2009). https://doi.org/10.1136/bmj.b2535
17. Kaur, A., Garg, R., Gupta, P.: Challenges facing AI and big data for resource poor healthcare system. In: The 2nd International Conference on Electronics and Sustainable Communication Systems, pp. 1426–1433 (2021)
18. Kitchenham, B., Charters, S.: Guidelines for Performing Systematic Literature Reviews in Software Engineering. Keele University, UK (2007)
19. Dabic, M., Vlacic, B., Paul, J., Dana, L.P., Sahasranamam, S., Glinka, B.: Immigrant entrepreneurship: a review and research agenda. J. Bus. Res. **113**, 25–38 (2020)
20. Cook, D.J., Mulrow, C.D., Haynes, R.B.: Systematic reviews: synthesis of best evidence for clinical decisions. Ann. Intern. Med. **126**(5) (1997). https://doi.org/10.7326/0003-4819-126-5-199703010-00006
21. Eteokleous, P.P., Leonidou, L.C., Katsikeas, C.S.: Corporate social responsibility in international marketing: review, assessment, and future research. Int. Mark. Rev. **33**(4), 580–624 (2016)
22. Paul, J., Benito, R.G.: A review of research on outward foreign direct investment from emerging countries, including China: what do we know, how do we know and where should we be heading? Asia Pac. Bus. Rev. **24**(1) (2018)
23. Khatoon, S., Rehman, V.: Negative emotions in consumer brand relationship: a review and future research agenda. Int. J. Consum. Stud. **45**, 719–749 (2021)
24. Van Aken, J.E.: Mode 2 Knowledge Production in the Field of Management. Eindhoven Centre for Innovation Studies. Working Paper, 1 (2001)
25. Paul, J., Criado, A.R.: The art of writing literature review: what do we know and what do we need to know? Int. Bus. Rev. **29**(4), 101717 (2020)
26. Watson, R.T.: Beyond being systematic in literature reviews in IS. J. Inf. Technol. **30**(2), 185–187 (2015)
27. Bastian, H., Glasziou, P., Chalmers, I.: Seventy-five trials and eleven systematic reviews a day: how will we ever keep up? PLoS Med. **7**(9), 1000326 (2010)
28. Evans, D.: Hierarchy of evidence: a framework for ranking evidence evaluating healthcare interventions. J. Clin. Nurs. **12**(1), 77–84 (2003)
29. Boell, S.K., Cecez-Kecmanovic, D.: On being 'systematic' in literature reviews in IS. J. Inf. Technol. **30**(2), 161–173 (2015)
30. Tranfield, D., Denyer, D., Smart, P.: Towards a methodology for developing evidence-informed management knowledge by means of systematic review. Br. J. Manag. **14**(3), 207–222 (2003)
31. Ali, O., Ally, M., Clutterbuck, P., Dwivedi, Y.: The state of play of blockchain technology in financial services sector: a systematic review. Int. J. Inf. Manag. **54** (2020). https://doi.org/10.1016/j.ijinfomgt.2020.102199
32. Ali, O., Jaradat, A., Kulakli, A., Abuhalimeh, A.: A comparative study: blockchain technology utilization benefits, challenges, and functionalities. IEEE Access. **9**, 12730–12749 (2021)
33. Ali, O., Shrestha, A., Soar, J., Wamba, S.F.: Cloud computing-enabled healthcare opportunities, issues, and applications: a systematic review. Int. J. Inf. Manag. **43**, 146–158 (2018)

Chapter 16
Reducing the Environmental Impact of an Operating Room: A Digital Twin Approach

Marco Ferretti, Gerardo Luisi, Stefano Riemma, Francesco Schiavone, and Federica Zeuli

Abstract Environmental pollution is a critical issue that needs attention from policymakers, particularly considering the healthcare sector's substantial carbon footprint. This study explores the environmental impact of healthcare activities, highlighting the significant contribution of hospital operations that increase air pollution through energy consumption and waste generation. Our analysis is on the production of anesthetic gases and intensive water usage in hospital facilities are identified as primary pollutants. To address these challenges, the research examines the potential strategy of digital twins (DT) and life cycle assessment (LCA) methodologies to optimize resource management and reduce emissions in healthcare settings. In our study through agent-based modeling (ABM), we simulated different scenarios to evaluate the environmental consequences of different strategies. This research aims to provide insights for healthcare organizations to implement more sustainable practices and technologies, and for mitigating the environmental impact of the operations in hospital rooms. The research highlights that there are alternatives that are more environmentally friendly even though it is necessary to make initial investments.

Keywords Environmental strategy · Healthcare · Agent-based model · Hospital room

---

M. Ferretti · G. Luisi (✉) · F. Schiavone · F. Zeuli
University of Naples Parthenope, Naples, Italy
e-mail: gerardo.luisi001@studenti.uniparthenope.it

S. Riemma
Department of Economics, University of Salerno, Fisciano, SA, Italy

F. Schiavone et al. (eds.), *Advanced Perspectives and Trends in Digital Transformation of Firms, Networks, and Society*, Springer Proceedings in Business and Economics, https://doi.org/10.1007/978-3-031-80692-6_16

16.1 Introduction

Environmental pollution is a significant concern for public policymakers, who strive to implement effective strategies to achieve environmental goals. Air pollution, particularly in urban areas, is a primary contributor to declining environmental quality and has adverse effects on public health [1, 2]. The healthcare sector, as one of the largest service industries, has a notable carbon footprint [3, 4]. Studies by Eckelman and Sherman [5] have shown that a considerable fraction of national air pollution emissions can be attributed to the healthcare sector.

In 2022, there was a slight increase in hospitalizations [5], leading to higher energy consumption and greenhouse gas emissions from healthcare services [6]. Operating and resuscitation rooms, in particular, use substantial amounts of energy and generate significant waste [7]. Anesthetic gases are identified as primary contributors to pollution in operating rooms [8]. Additionally, extensive water use in hospital facilities for cleaning, sterilization, and plant operation can lead to water management problems [9, 10].

Healthcare management can mitigate pollution by adopting sustainable practices, such as using environmentally friendly technologies, optimizing waste management, and improving energy efficiency in hospitals. Various methodologies, such as life cycle assessment (LCA), can help reduce emissions in operating rooms [11, 12]. The concept of Digital Twin (DT), introduced by Professor Grieves, is a digital model that integrates technologies like IoT, AI, and AR to enhance efficiency and management across sectors [13]. DT allows real-time monitoring and optimization of physical systems, which can be valuable for environmental management in healthcare [14].

LCA is a comprehensive method used to evaluate the environmental impacts associated with a product, process, or service throughout its life cycle [12]. Combined with agent-based modeling (ABM), it can improve resource management, efficiency, and cost reduction in healthcare [15]. The aim of this research is to examine potential ongoing strategies that can be adopted to mitigate pollution resulting from healthcare activities. Our study focuses on identifying strategies, through possible simulations, to reduce polluting emissions from an operating room. The main focus is applying concepts such as "digital twins" and life cycle assessments.

In our research, we will develop a discrete event simulation model to evaluate different scenarios of the sterilization processes and observe and perform life cycle assessments for the emissions resulting from the three activities analyzed. The sterilization process is a highly energy-intensive activity that produces a significant environmental impact [11], different logistic and productive scenarios can be analyzed through a digital twin to evaluate the environmental and effectiveness benefits without affecting the real operating room process. In addition, the digital twin provides accurate data to perform LCA for the different scenarios, including analysis of the use of different anesthetic gases and management systems for surgical fluids.

The research emphasizes that environmentally friendly alternatives exist, often necessitating increased focus on environmental issues or, in some cases, initial investments. In this research, [4], use the model that we proposed in our research but only to improve the management of the resources not for reducing the impact on the environment. On the previous statements, this research tries to fill out the gap in research answering at following research question (RQ): How the ABM can be implemented for reducing the environmental pollution in the hospital room?

16.2 Method

The work focuses on three activities in particular: fluid waste management, inhalation anesthesia, and sterilization services. An ABM [12, 14] was developed that would allow the studied environment to be replicated and an evaluation of different scenarios to be conducted, especially about sterilization processes. It was based on the logic shown in Fig. 16.1. Furthermore, with the data present in the model, it was possible to conduct life cycle assessments for the emissions deriving from the three analyzed activities, with the possibility of evaluating the environmental consequences of different application alternatives.

16.3 Findings

For the sterilization process activities, it was possible to highlight that through various organizational choices, it is possible to reduce emissions and, therefore, the related impacts by up to approximately 50%, without in any way affecting the efficiency of the perioperative processes. Regarding the use of the closed management system for surgical fluids, however, advantages emerged from every point of view, but significant in terms of faster recovery times in the operating room, excellent safety for the staff, and, as highlighted from research, also in terms of minor environmental consequences. The issue of inhalation anesthetic agents has already been widely discussed in the literature; the model wanted to report this information for an even more accessible and simpler understanding of the problem: for the same interventions, the environmental impact generated using desflurane is approximately 20 times higher than that obtained using sevoflurane. Xenon represents an even more advantageous solution from an environmental point of view, but for a realistic application, it requires a gas recovery system due to its high cost. The research highlights that there are alternatives that are more environmentally friendly; sometimes, they simply require greater attention to the environmental problem, and other times, even initial investments.

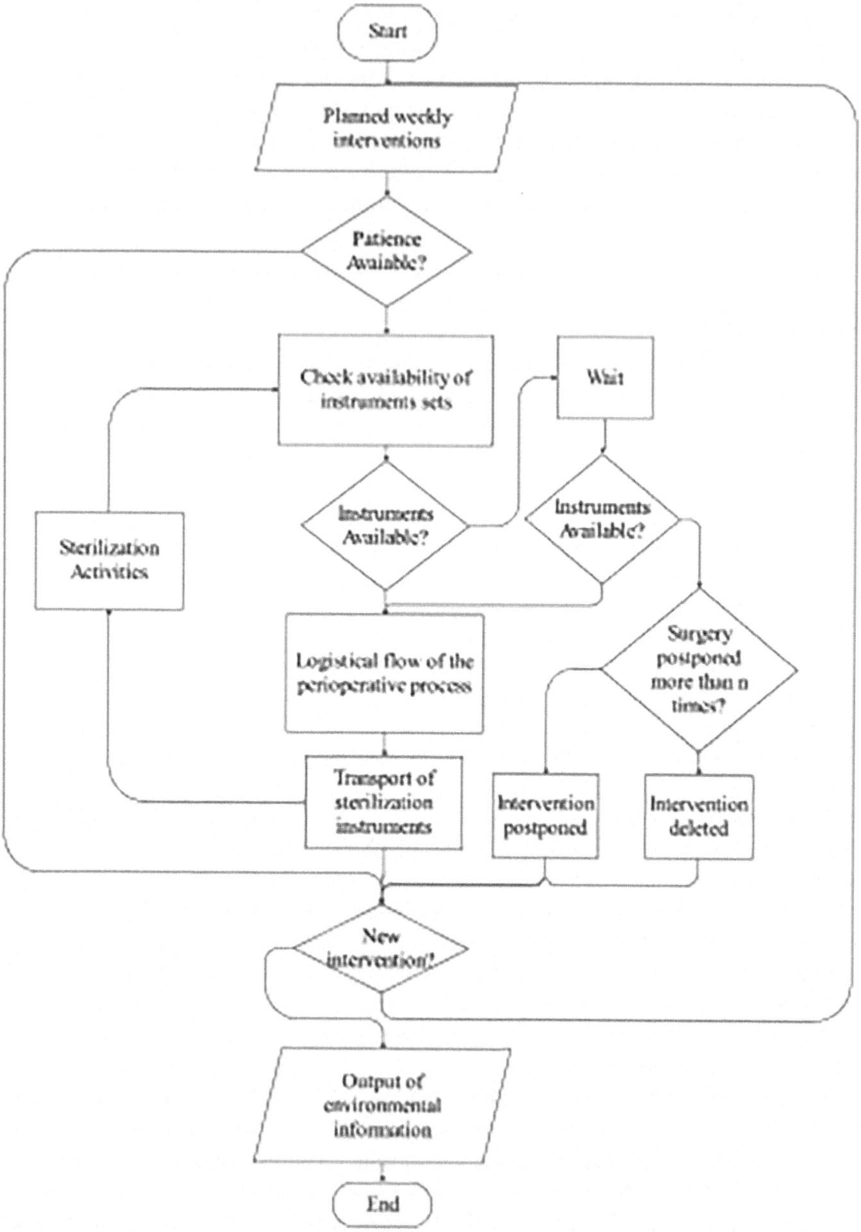

Fig. 16.1 Operating room agent-based model logic. Source. Authors' elaboration

16.4 Discussion and Implication

The development of a simulation model of the logistical flows of an operating theatre represents a tool capable of providing information relating to several different scenarios in a very immediate way. It has proven particularly useful for the analysis of sterilization processes; in relation to them, it was possible to highlight that through various organizational choices, it is possible to reduce emissions and, therefore, the related impacts by up to approximately 50%, without in any way affecting the efficiency of the peri-operational processes. From the information the model provides, it was also possible to carry out environmental assessments for the management system of surgical suction fluids and the anesthetics used by inhalation. The research highlights that there are alternatives that are friendlier to the environment; sometimes, they simply require greater attention to the environmental problem and, at other times, initial investments. The closed management system for surgical fluids represents advantages from every point of view: faster recovery times in the operating room, greater safety for staff, and, as highlighted by the research, also lower environmental consequences. The issue of inhalation anesthetic agents has already been widely discussed in the literature; the model wanted to report this information for an even easier and simpler understanding of the problem: for the same interventions, the environmental impact generated using desflurane is approximately 20 times higher than that obtained using sevoflurane. Xenon represents an even more advantageous solution from an environmental point of view, but for a realistic application, it requires a gas recovery system due to its high cost.

The theoretical contribution based on the findings of this research consists of evidence of how an ABM can be useful for developing a digital twin of an operating room that allows the simulation of different process scenarios and selecting the best option in the comparison, in this way it is possible to improve the efficiency of the process and reduce the environmental impact.

References

1. Attaran, S., Attaran, M., Celik, B.G.: Digital Twins and Industrial Internet of Things: uncovering operational intelligence in industry 4.0. Dec. Anal. J. **10**, 100398 (2024)
2. Vlachokostas, C., Achillas, C., Moussiopoulos, N., Banias, G.: Multicriteria methodological approach to manage urban air pollution. Atmos. Environ. **45**(25), 4160–4169 (2011)
3. Beloeil, H., Albaladejo, P.: Initiatives to broaden safety concerns in anaesthetic practice: the green operating room. Best Pract. Res. Clin. Anaesthesiol. **35**(1), 83–91 (2021)
4. Bhosekar, A., Ekşioğlu, S., Işık, T., Allen, R.: A discrete event simulation model for coordinating inventory management and material handling in hospitals. Ann. Oper. Res., 1–28 (2023)
5. Eckelman, M.J., Sherman, J.: Environmental impacts of the U.S. health care system and effects on public health. PLoS One. **11**(6), e0157014 (2016). https://doi.org/10.1371/journal.pone.0157014

6. Gordon, M., Patricio, M., Horne, L., Muston, A., Alston, S.R., Pammi, M., et al.: Developments in medical education in response to the COVID-19 pandemic: a rapid BEME systematic review: BEME Guide No. 63. Med. Teach. **42**(11), 1202–1215 (2020)
7. Lenzen, M., Malik, A., Li, M., Fry, J., Weisz, H., Pichler, P.P., et al.: The environmental footprint of health care: a global assessment. Lancet Planet. Health. **4**(7), e271–e279 (2020)
8. Lu, Q., Chen, L., Li, S., Pitt, M.: Semi-automatic geometric digital twinning for existing buildings based on images and CAD drawings. Autom. Constr. **115**, 103183 (2020)
9. Luhar, I., Luhar, S., Abdullah, M.M.A.B.: Challenges and impacts of COVID-19 pandemic on global waste management systems: a review. J. Compos. Sci. **6**(9), 271 (2022)
10. D'Alessandro, D., Tedesco, P., Rebecchi, A., Capolongo, S.: Water use and water saving in Italian hospitals. A preliminary investigation. Annali dell'Istituto superiore di sanita. **52**(1), 56–62 (2016)
11. McGain, F., Muret, J., Lawson, C., Sherman, J.D.: Environmental sustainability in anaesthesia and critical care. Br. J. Anaesth. **125**(5), 680–692 (2020)
12. Pearce, B., Hosseini, N., Taaffe, K., Huynh, N., Harris, S.: Modeling interruptions and patient flow in a preoperative hospital environment. In: Proceedings of the 2010 Winter Simulation Conference, pp. 2261–2270. IEEE (2010, December)
13. Sapuan, S.M., Ilyas, R.A., Asyraf, M.R.M., Sapuan, S.M., Ilyas, R.A., Asyraf, M.R.M.: Carbon footprint in healthcare. Saf. Health Compos. Ind., 115–137 (2022)
14. Taaffe, K., Fredendall, L., Huynh, N., Franklin, J.: Computer simulation shows the effect of communication on day of surgery patient flow. AORN J. **102**, 60.e1–60.e14 (2015). https://doi.org/10.1016/j.aorn.2015.05.005
15. Upadhyayula, V.K., Meyer, D.E., Curran, M.A., Gonzalez, M.A.: Life cycle assessment as a tool to enhance the environmental performance of carbon nanotube products: a review. J. Clean. Prod. **26**, 37–47 (2012)

Chapter 17
Gamification Enhancing Patient-Centricity Through Digital Therapeutics

Annaluce Mandiello and Giuseppe Franzese

Abstract This chapter explores the role of gamification in digital therapeutics (DTx) to implement and improve the patient-centered approach in healthcare. DTx, non-pharmacological software solutions, integrate game mechanics to increase patient adherence and engagement in the therapeutic pathway. Through a preliminary analysis of case studies and secondary data, the chapter identifies the main gamification mechanisms used in DTx, such as levels, points and rewards, and their impact on active patient engagement. The methodological approach combines a multiple case analysis with primary and secondary data to explore how gamification can effectively personalize treatment and improve the therapeutic efficacy of DTx. The theoretical and practical implications suggest that integrating gamification elements could promote more effective and patient-centered digital solutions in the contemporary healthcare setting.

Keywords Gamification · Digital therapeutics · DTx · Patient-centricity · Digital health

17.1 Introduction

17.1.1 Research Context

Digital transformation (DT) emerges as a global and disruptive phenomenon which continually reshapes, through the use of digital technologies, the schemes on which business models, organizations, and society at large are built [1, 2]. Among the different sectors, healthcare (HC), which is a deeply intricate ecosystem, opened its doors to DT. In fact, in this specific sector, the aim is to implement high-quality care

A. Mandiello (✉) · G. Franzese
Department of Management and Quantitative Studies, University of Naples Parthenope, Naples, Italy

F. Schiavone et al. (eds.), *Advanced Perspectives and Trends in Digital Transformation of Firms, Networks, and Society*, Springer Proceedings in Business and Economics, https://doi.org/10.1007/978-3-031-80692-6_17

delivery and customer-centric solutions to improve patients' well-being [3, 4] and the value co-creation process in general [5], preventing it from inefficiencies, medical errors, and, thus, unnecessary costs. With the expression "digital health" (DH), scholars draw attention to the impact of digital technologies which revolutionize the whole HC system and its rules [6].

It can be stated that digital innovation enables the implementation of a patient-centric approach to improve individuals' quality of life through "different solutions, non-pharmaceutical options or a new way of delivering care" [7]. In this vein, a digital health innovation called "digital therapeutics" (DTx) entered the HC sector in the last few years and is designed consistently with the previously mentioned changes and trends. DTx are non-pharmaceutical digital solutions (combined with medications, if required) specially designed to help patients along their therapeutic path through the promotion of behavioral and emotional changes. The aim is to encourage people to be proactive in treating and managing their own diseases and eventually in improving a specific health function (Digital Therapeutics Alliance 2023).

According to ISO/TR 11147:2023 "Health informatics—Personalized digital health—Digital therapeutics health software systems," DTx are "health software intended to treat or alleviate a disease, disorder, condition, or injury by generating and delivering a medical intervention that has a demonstrable positive therapeutic impact on a patient's health." Moreover, the digital capabilities in terms of DTx are the ones which are currently changing the world of healthcare (with its services and products), like "data tracking, predictive analytics and algorithms, artificial intelligence (AI) and machine learning (ML), sensor and location services, communications and social media, and gaming and virtual reality (VR)" (Deloitte 2019). DTx are software that generate and provide evidence-based therapeutic interventions (Digital Therapeutics Alliance 2023) through three types of solutions depending on the product's primary purpose. In this regard, DTx can aim at preventing, managing, or improving a health function to take care of patients' medical disorders or diseases in the following fields: "cardiovascular, central nervous system (CNS), endocrine, immunology, oncology, and respiratory" [8]. In synthesis, DTx are regulated by national agencies that operate in the health and pharmacological field (e.g., FDA, LEA) and must always be clinically validated; they are, indeed, defined as "software as a medical device" (SaMD) according to the Food and Drug Administration (FDA). Nonetheless, DTx could be (prescription DTx) / not be (non-prescription DTx) cleared by those agencies [9], and currently, they are not globally available. For example, Italy is one of the countries that is still trying to investigate the compatibility of DTx with its national healthcare system, as Deputy Loizzo's bill presented in 2023 "Disposizioni in materia di terapie digitali (Loizzo et al. 2023) shows. Different studies support the adoption of DTx as medical devices; Jeon et al. [8] brought them together and identified six benefits of DTx over traditional therapeutics.

In summary, the authors [8] state that DTx could bring advantages over traditional therapeutics at the organizational level (lower expense, consistent quality of treatments compared to face-to-face care provided by therapists), the social level (expandability for undertreated stigma-associated disorders such as venereal diseases and mental health disorders), and, above all, the patient level (direct access to patients, more personalized treatment adapted to patients' individual needs and lifestyles, access by patients anytime and anywhere). Therefore, the strengths of this type of DH solutions lie in the ability to engage patients within their care pathway, which is consistent with the primary purpose of the HC: the individual's well-being. It is, indeed, its raison d'être.

DTx that are intended to increase the level of patient engagement in their own journey using unconventional solutions are the gamified ones [10]. In fact, using game mechanics in non-game contexts is essential in improving individuals' engagement and guiding their behaviors [11]. This is due to the fact that commonplace events may be more exciting and fulfilling thanks to game-like aspects like challenges, rewards, stories, and scoring [12]. The gamification process involves creating a game system that provides incentives for completing activities to stimulate high levels of player engagement [13]. This might involve establishing goals, allocating levels or points, and obtaining virtual prizes or badges [14]. These are the main reasons why gamification has become widely accepted in various industries, including marketing, education, and the growing area of healthcare [10, 15, 16]. Nonetheless, the general interest in gamified DTx is still at its early stage in the management field. In this regard, even though managerial literature is currently lacking in evidence-based studies on this topic [10], it can be reasonable to think that by analyzing DTx which use gamified mechanics, more insights could be provided on the DTx role in the patient-centric approach to health.

17.2 Objective

17.2.1 Purpose of the Work

Even though gamification in healthcare is emerging as an innovative strategy to engage and motivate individuals in managing their own health [16] and several scholars are studying this topic, the full potential that gamified solutions can have in the healthcare sector remains unexplored [10, 17]. This is particularly true when it comes to recent technologies like DTx which tendentially use gamified components. The diffusion of this particular type of digital health solution has given rise to various research opportunities for scholars [9]. Nonetheless, there are still many gaps concerning the relationship between gamification in DTx and patient-centricity [10, 18]. In particular, it has not been deeply investigated yet how DTx built on gamified mechanics fit into the patient-centric approach to health and, consequently, it is unclear how gamification enhances the performance of digital therapeutics in terms of patient-centricity and personalization of the care pathway.

However, what is generally known is that gamification is a critical component of DTx in increasing treatment efficacy and patient adherence [19]. By incorporating gamified components into the activities offered by DTx, the therapeutic process becomes more exciting and inspiring [20]. For instance, games with targeted motions or exercises may benefit people using DTx apps for pain management or rehabilitation. In fact, gamification has the potential to enhance patients' enjoyment of therapeutic activities and motivate each individual to pursue treatment goals more assiduously. Additionally, gamification features may be used to track and monitor patients' improvements, providing this information to the doctors [21]. Moreover, the gamification experience offers a motivating component that might help patients to adhere to the recommended therapy, improving clinical outcomes. Ultimately, gamification is a potent catalyst that may increase the effectiveness, accessibility, and satisfaction of medical treatments above all else [22]. In this respect, this work aims to explore the following research question (RQ): "How does gamification enhance patient-centricity in DTx?"

17.3 Methodology

17.3.1 Multiple Case Study Methodology

In order to give an answer to the research question and to study the potential role of DTx gamified solutions in the promotion of a patient-centric approach, the authors conducted a case study analysis because of the novelty of the topic in the literature [23] which is still in an emerging phase [24]. Indeed, a multiple case methodology provides a more accurate and broadly applicable analysis than a single case one, making the conclusions easier to replicate [25]. The data obtained from the case study analysis were then integrated with secondary data retrieved from the web, company reports, and other sources. Considering the exploratory nature of the study, the authors are developing an interview protocol that can be used to collect primary data useful for the research, which will be later incorporated with those from the analysis already carried out, so as to achieve data triangulation [26]. This work, hence, is still in progress but some interesting insights are emerging from the secondary data analysis grounded on public declarations found on the web related to each case.

The reference sample comprises six companies that offer DTx products with gamified mechanics in it. Based on information found on the web, 11 cases were initially selected. The selection criteria that have been used were mainly 2: the digital solutions developed had to fall within the spectrum of DTx and, simultaneously, had to have gamified elements.

Specifically, about the first selection criterion the authors consider, according to the Digital Therapeutics Alliance (2018), "digital therapeutics" all the products that satisfy the following principles: "1) Incorporate design, manufacture, and quality best practices; 2) Engage end users in product development and usability processes; 3) Incorporate patient privacy and security protections; 4) Apply product deployment, management, and maintenance best practices; 5) Publish trial results inclusive of clinically-meaningful outcomes in peer-reviewed journals; 6) Be reviewed and cleared or certified by regulatory bodies as required to support product claims of risk, efficacy, and intended use; 7) Make claims appropriate to clinical evaluation and regulatory status; 8) Collect, analyze, and apply real-world evidence and/or product performance data."

Consequently, 5 cases were discarded because (1) they did not meet the requirements to be considered digital therapeutics or (2) they did not use gamified elements. Despite this, six instances are inside the range of cases that are thought to help produce persuasive results when replicated [23].

Furthermore, the data collected and those subsequently obtained through direct interviews will be subject to thematic content analysis (TCA). Through TCA, it will be possible to organize and describe data sets in detail, identifying, analyzing, and extracting themes from the data [27].

In conclusion, the main purpose of this chapter is to understand the role of gamified mechanics of digital therapeutics in increasing patient-centric processes within a care pathway and to achieve this goal the authors adopted an abductive approach [28]. This systematically combines the deductive approach, which involves developing theories from existing literature, with the inductive one which aims to establish a universal law from individual cases. This chapter is based indeed on a large and recognized body of literature regarding existing theories of digital transformation and patient-centricity in health, but it explores and connects two elements—gamified digital therapeutics and patient-centric approach, which is a new research interest since the effects of this connection are still subject to unclear points.

17.4 Results

17.4.1 Preliminary Findings

The authors conducted a preliminary analysis across three steps. Specifically, the first step aimed at the identification and subdivision for each DTx of the following information: the type of device on which the software is implemented, the geographical development area, the disease treated, and the age category of the patient around which the therapeutic has been designed (child, adult, or generic). The goal of this phase was to provide an overview of the current gamified DTx market state. Then, through the second step of analysis, it has been assigned for each DTx one (or more) primary purpose pursued according to the ones presented by Digital Therapeutics Alliance (2018): treat a disease, manage a disease, and improve a health function (including preventing a disease). In the end, to put the basis of the next step of this work which investigates the role of gamification in patient-centricity, it has been fundamental to identify the gamification mechanics implemented in each DTx. However, it must be stated that this third step is still at its embryonic stage.

Thus, the hypothesis that new mechanics could emerge from a deeper analysis of the cases must not be excluded.

To ensure the anonymity of the companies and DTx analyzed—as the authors do not currently hold the authorizations for the use of the name—the cases studied are identified with a code established by the authors (e.g., Dtx1, DTx2).

Table 17.1 shows the primary information that the authors have seized from the chosen sample. Specifically, this table shows the devices through which the DTx are administered, the areas in which the solutions are developed and diffused, the diseases that these DH technologies are projected to treat, and the final users, essentially, the patients. This first stage of analysis shows how most of the DTx solutions (DTx1, DTx3, DTx4, DITx6) are developed to be provided through digital devices grounded on m-health: smartphones and tablets. It can be observed, though, that it happens for all the products built around children's health diseases, in particular ADHD and ASD. This type of final purpose and patient are the pillars of most DTx inserted in the global health market. However, when the category of final user is not specified because the DTx could be addressed to both adults and children just depending on their health condition (DTx2, DTx5), the technology is designed and provided, also including the use of wearable devices alone (DTx2) or combined with smartphones and tablets (DTx5).

There is no specific development area regarding the geographic origin of these solutions. The DTx phenomenon is developing all over the globe. In this sample, two solutions were developed in Europe, specifically Switzerland and Iceland, while the others were developed worldwide: Korea, the United States, Singapore, and Australia.

Table 17.2 shows the main gamified mechanics used in the cases analyzed. It is evident how most of the selected DTx use digital levels and badges. Half of the sample relies on experience points and achievements. Only a tiny portion of the sample uses quests, and only one DTx uses leaderboards. In addition to this, two or more gamified mechanics are used in almost all DTx. In fact, combining different mechanics improves the result [29]. Specifically, the table shows that the most used techniques are badges and levels, followed by experience points and achievement, and quest and leaderboards. Relative to leaderboards, it is possible to see that this

Table 17.1 DTx main information

Sample	Device	Diffusion area	Aim	Final users
DTx1	Tablet	Australia	Manage/ treat ADHD/ASD	Children
DTx2	Wearable device	Switzerland	Neurorehabilitation	Generic
DTx3	Tablet/smartphone	USA	Manage/ treat ADHD	Children
DTx4	Tablet	Singapore	Manage/ treat ADHD	Children
DTx5	Smartphone/ tablet/ wearable device	Iceland	Manage/treat diabetes, heart disease, mental health, obesity, chronic respiratory diseases	Generic
DTx6	Tablet/smartphone	Korea	Manage/treat ASD	Children

Source: Authors' elaboration

Table 17.2 Most used gamified mechanisms in DTx

Sample	Level	Badge	Leaderboard	Quest	XP Point	Achievement
DTx1	x	x				x
DTx2			x	x	x	
DTx3	x	x			x	x
DTx4					x	
DTx5	x	x		x		x
DTx6	x	x				

Source: Authors' elaboration

mechanic is used in a DTx that is intended primarily for adults, even though it is generically available for all categories of patients. Using a ranking system increases adult competition, improving engagement and results [30]. This from a health perspective translates as a greater commitment to healing and managing one's health. On the contrary, it is inferred that children are more enticed by games with levels and rankings, which confirms what Chow et al. [31] said earlier. Furthermore, solutions concerning children very often use the same mechanics. People may prefer different features of gamification-based products depending on their age. This result aligns with what Yuan and Guo [32] stated and shows that through gamified techniques, it is possible to personalize the experience not only according to the disease treated but also to the patient's personal characteristics.

In conclusion, this preliminary analysis shows, on the one hand, the main characteristics of some of the current most widely used DTx and their primary purposes. On the other hand, it shows some important advantages deriving from the integration of gamification elements in digital therapeutics, as follows:

1) Gamified components significantly increase patients' engagement, encouraging them to participate in their care pathway by assuming proactive behaviors.
2) Gamified elements like points, levels, and awards might facilitate patients' adherence to the recommended therapeutics activities.
3) Gamification in digital health may be more effective thanks to the customization guaranteed by the game mechanics related to each patient's unique preferences and requirements, increasing the therapeutic efficacy of these interventions.

In summary, according to these preliminary results, gamification applied to DTx enhances the implementation and effects of the patient-centric approach in healthcare.

17.5 Implications

17.5.1 Expected Implications

From a strictly theoretical perspective, this study contributes to expanding the literature regarding gamification and digital therapeutics, focusing specifically on the relationship between these phenomena and the patient-centric approach. This topic correlation is highly novel in the literature and thus unexplored. Through a qualitative methodology, it is possible to capture this reality in all its facets [33], focusing in particular on the ways through which gamified mechanics are implemented in digital therapeutics to provide and enhance patient-centric solutions. This work, indeed, can offer an overview of the current link between gamification and a patient-centric approach, helping managers improve the effectiveness of their products through gamified mechanics. It also highlights the importance of adopting digital therapeutics as valid non-pharmaceutical interventions designed to reach high levels of user engagement, adherence to therapy, and, hence, quality of care. The analysis of the future work will be structured to demonstrate if, how, and to what extent a gamified experience supports the individuals' therapeutic pathway, according to the different needs and preferences related to each of the categories of patients identified. In summary, a study so designed is intended to be a demonstration of how digital health technologies acknowledge individuals' role in their own health simply by letting them play. Given the above, this research is aimed at providing valuable theoretical, practical, and managerial insights to decision-makers, policymakers, health staff members, and, moreover, development companies that implement gamified solutions in health. In practical terms, this study can provide some suggestions and guidelines on which gamification mechanics must be implemented within the development process of a DTx to provide more impactful and adequate patient-centric digital health products. In the end, it is believed that such results could support healthcare organizations in avoiding technological investments not centered on the final user, translated into unnecessary costs, ineffective strategies, and general discontent among the stakeholders.

References

1. Balta, M., Valsecchi, R., Papadopoulos, T., Bourne, D.J.: Digitalization and co-creation of healthcare value: a case study in occupational health. Technol. Forecast. Soc. Chang. **168**, 120785 (2021)
2. Vial, G.: Understanding digital transformation: a review and a research agenda. Managing Digital Transformation, 13–66 (2021)
3. Dahl, A.J., Peltier, J.W., Milne, G.R.: Reducing information asymmetry and increasing health value co-creation in a rural healthcare context. J. Consum. Aff. **56**(2), 512–535 (2022)
4. Dahl, A.J., Peltier, J.W., Swan, E.L.: Anticipatory value-in-use in early-stage digital health service transformations: how consumers assess value propositions before and after abrupt, exogenous shocks. J. Bus. Res. **163**, 113910 (2023)

5. Swan, E.L., Peltier, J.W., Dahl, A.J.: Artificial intelligence in healthcare: the value co-creation process and influence of other digital health transformations. J. Res. Interact. Mark. **18**(1), 109–126. (ahead-of-print) (2023)
6. Kraus, S., Schiavone, F., Pluzhnikova, A., Invernizzi, A.C.: Digital transformation in healthcare: analyzing the current state-of-research. J. Bus. Res. **123**, 557–567 (2021)
7. Toni, M., Mattia, G., Pratesi, C.A.: What's next in the healthcare system? The contribution of digital innovation in achieving patient-centricity. Futures. **156**, 103304 (2024)
8. Jeon, E., Yoon, N., Sohn, S.Y.: Exploring new digital therapeutics technologies for psychiatric disorders using BERTopic and PatentSBERTa. Technol. Forecast. Soc. Chang. **186**, 122130 (2023)
9. Hong, J.S., Wasden, C., Han, D.H.: Introduction of digital therapeutics. Comput. Methods Prog. Biomed. **209**, 106319 (2021)
10. Sestino, A., Kahlawi, A., De Mauro, A.: Decoding the data economy: a literature review of its impact on business, society and digital transformation. Eur. J. Innov. Manag. **28**(2), 298–323 (2023)
11. Deterding, S., Dixon, D., Khaled, R., Nacke, L.: From game design elements to gamefulness: defining "gamification". In: Proceedings of the 15th International Academic MindTrek Conference: Envisioning Future Media Environments, pp. 9–15. ACM, Tampere (2011)
12. Yilmaz, Y., Monteiro, S., Chan, T., Chandross, D., Lal, S., Munford, V.: I would rather be playing: enhancing faculty development through gamification. In: Monteiro, S., Kumar, A. (eds.) Innovative Digital Practices and Globalization in Higher Education, pp. 137–159. IGI Global, Hershey (2023)
13. Xiao, R., Wu, Z., Hamari, J.: Internet-of-gamification: a review of literature on IoT-enabled gamification for user engagement. Int. J. Hum.- Comput. Interact. **38**(12), 1113–1137 (2022)
14. Xu, F., Tian, F., Buhalis, D., Weber, J., Zhang, H.: Tourists as mobile gamers: gamification for tourism marketing. In: Buhalis, D., Egger, R. (eds.) Future of Tourism Marketing, pp. 96–114. Routledge, London (2021)
15. Al-Rayes, S., Al Yaqoub, F.A., Alfayez, A., Alsalman, D., Alanezi, F., Alyousef, S., Alanzi, T.M.: Gaming elements, applications, and challenges of gamification in healthcare. Inform. Med. Unlocked. **31**, 100974 (2022)
16. Gajardo Sánchez, A.D., Murillo-Zamorano, L.R., López-Sánchez, J., Bueno-Muñoz, C.: Gamification in health care management: systematic review of the literature and research agenda. SAGE Open. **13**(4), 21582440231218834 (2023)
17. Xie, H.: A scoping review of gamification for mental health in children: uncovering its key features and impact. Arch. Psychiatr. Nurs. **41**, 132–143 (2022)
18. Bos, E., Preller, K.H., Kaur, G., Malhotra, P., Kharawala, S., Motti, D.: Challenges with the use of digital sham: systematic review and recommendations. J. Med. Internet Res. **25**, e44764 (2023)
19. Lutz, J., Offidani, E., Taraboanta, L., Lakhan, S.E., Campellone, T.R.: Appropriate controls for digital therapeutic clinical trials: a narrative review of control conditions in clinical trials of digital therapeutics (DTx) deploying psychosocial, cognitive, or behavioral content. Front. Digit. Health. **4**, 823977 (2022)
20. Woodley, S.J., Moller, B., Clark, A.R., Bussey, M.D., Sangelaji, B., Perry, M., Kruger, J.: Digital technologies for women's pelvic floor muscle training to manage urinary incontinence across their life course: scoping review. JMIR Mhealth Uhealth. **11**, e44929 (2023)
21. Huang, X., Xiang, X., Liu, Y., Wang, Z., Jiang, Z., Huang, L.: The use of gamification in the self-management of patients with chronic diseases: scoping review. JMIR Serious Games. **11**, e39019 (2023)
22. AlSaad, F.M., Durugbo, C.M.: Gamification-as-innovation: a review. Int. J. Innov. Technol. Manag. **18**(05), 2130002 (2021)
23. Eisenhardt, K.M.: Making fast strategic decisions in high-velocity environments. Acad. Manag. J. **32**(3), 543–576 (1989)
24. Yin, R.K.: Designing case studies. Qual. Res. Meth. **5**(14), 359–386 (2003)

25. Greene, D., David, J.L.: A research design for generalizing from multiple case studies. Eval. Program Plann. **7**(1), 73–85 (1984)
26. Stake, R.E.: The Art of Case Study Research. Sage, Thousand Oaks (1995)
27. Braun, V., Clarke, V.: Using thematic analysis in psychology. Qual. Res. Psychol. **3**(2), 77–101 (2006)
28. Awuzie, B., McDermott, P.: An abductive approach to qualitative built environment research: a viable system methodological exposé. Qual. Res. J. **17**(4), 356–372 (2017)
29. Hervás, R., Ruiz-Carrasco, D., Mondéjar, T., Bravo, J.: Gamification mechanics for behavioral change: a systematic review and proposed taxonomy. In: Proceedings of the 11th EAI International Conference on Pervasive Computing Technologies for Healthcare, pp. 395–404. ACM, Barcelona (2017)
30. Chang, T.P., Doughty, C.B., Mitchell, D., Rutledge, C., Auerbach, M.A., Frisell, K., Walsh, C.M.: Leveraging quick response code technology to facilitate simulation-based leaderboard competition. Simul. Healthc. **13**(1), 64–71 (2018)
31. Chow, C.Y., Riantiningtyas, R.R., Kanstrup, M.B., Papavasileiou, M., Liem, G.D., Olsen, A.: Can games change children's eating behaviour? A review of gamification and serious games. Food Qual. Prefer. **80**, 103823 (2020)
32. Yuan, T., Guo, Y.: Gamification design of health apps for the elderly based on the Kano model and conjoint analysis method. In: Human-Computer Interaction. Theory, Methods and Tools, pp. 176–190. Springer, Cham (2021)
33. Aspers, P., Corte, U.: What is qualitative in qualitative research. Qual. Sociol. **42**, 139–160 (2019)

Sitography

Deloitte: Digital Therapeutics Improving patient outcomes through convergence. (2019). https://www2.deloitte.com/content/dam/Deloitte/us/Documents/life-sciences-health-care/us-lshc-digital-therapeutics.pdf. Accessed 28 Feb 2024

DTA Alliance: Digital Therapeutics Alliance – New DTx Definition. https://dtxalliance.org/wp-content/uploads/2023/06/DTA_FS_New-DTx-Definition.pdf. Accessed 28 Feb 2024

Loizzo S., Molinari R., Andreuzza G., Bergamini D., Cavandoli L., Pierro A., Zinzi G., (2023), Camera dei Deputati: Legge n. 126 del 2019. https://documenti.camera.it/leg19/pdl/pdf/leg.19.pdl.camera.1208.19PDL0040170.pdf. Accessed 28 Feb 2024

Chapter 18
The Effect of Personal Innovativeness on Digital Therapeutics Acceptance: An Italian Empirical Study

Annaluce Mandiello and Anna Pia Di Iorio

Abstract Digital transformation is revolutionizing various sectors, including healthcare, by integrating digital technologies such as AI, IoT, and cloud computing in products and processes. This chapter is focused on a new digital health solution, "Digital Therapeutics": a software implemented in different products, with the aim of providing medical interventions with proven therapeutic benefits. In particular, the goal of this chapter is to understand the intention to use "Digital Therapeutics" among Italians exploring the role of a variable defined "personal innovativeness" on it. Italian Government does not permit to prescribe and pay back the use of Digital Therapeutics, yet. The moderating effects of Digital Self-Efficacy and Health Belief are also investigated to understand if there is a possibility to increase the level of intention to use, investing in those variables. The authors, investigating the Italians' intentional behavior through the Unified Theory of Acceptance and Use of Technology (UTAUT) model, expect that personal innovativeness positively influences the intention to use and that this correlation is moderated by Digital Self-Efficacy and Health Belief. Data will be collected through a questionnaire and analyzed through the Structural Equation Modeling (SEM) method. The findings are expected to extend the literature on innovation management and technology acceptance, also providing suggestions to Italian policy-makers and decision-makers of the healthcare system.

Keywords Digital Therapeutics · Technology acceptance · Intention to use · Personal innovativeness · Digital Self-Efficacy · Health Belief

---

A. Mandiello · A. P. Di Iorio (✉)
University of Naples "Parthenope", Napoli, Italy
e-mail: annaluce.mandiello001@studenti.uniparthenope.it;
annapia.diiorio001@studenti.uniparthenope.it

F. Schiavone et al. (eds.), *Advanced Perspectives and Trends in Digital Transformation of Firms, Networks, and Society*, Springer Proceedings in Business and Economics, https://doi.org/10.1007/978-3-031-80692-6_18

18.1 Introduction

18.1.1 Research Context

Digital transformation (DT) is a phenomenon that involves "the efficient integration and utilization of knowledge, information, and data through digital technologies. This transformation leads to changes in the conventional combination of production resources and technologies" [1]. Nowadays, every sector is affected by DT since it enables the implementation of new business models and new business logics to create and capture value [2] consistently with the specificities of the context taken under analysis. In particular, for what concerns the healthcare (HC) system, DT is driving the current wind of change within it [3] creating many opportunities for all the actors of the HC industry [4]. Nowadays, indeed, digital health (DH) is building its solutions on transformative digital technologies like "artificial intelligence (AI), cloud computing, Internet of things (IoT), virtual and augmented reality, blockchains, and connectivity technologies" [5]. In this regard, one of the most disruptive digital health innovations of the last few years is designed taking alternatively into account the majority of these technologies: it is called "Digital Therapeutics."

Digital Therapeutics (DTx) are "health software intended to treat or alleviate a disease, disorder, condition, or injury by generating and delivering a medical intervention that has a demonstrable positive therapeutic impact on a patient's health" [6]. In summary, DTx are both prescription and non-prescription "digital-medical hybrids that had proven therapeutic value and could stand up to rigorous clinical testing and regulatory approval" [7]. DTx also bring several managerial benefits from the lower expense for the healthcare system and the continuously data-collection process measuring the progress of the therapy [8] to the development of several technologies like videogames, chatbot, and wearable devices totally designed in accordance with the needs of patients [9]. In this way, DTx help the treatment and management of both physical diseases and mental disorders (Digital Therapeutics Alliance 2024). Consequently, it can be reasonable to think that unnecessary costs deriving from the low patient engagement within the care pathway and the low effectiveness of the latter can be avoided in a certain way.

However, not all the contexts easily embrace digital health technologies even when their benefits are scientifically and practically proven; it depends on different factors like individuals' level of trust, age, gender, and nationality [10, 11]. Actually, this is essentially the reason why, especially in recent years, scholars implemented technology acceptance theories within managerial studies aimed at identifying the factors guiding actors' intention to adopt DH solutions. The goal is to provide tailored enter strategies, remove barriers, and enhance positive users' behaviors. Nonetheless, due to their novelty and the fact that they are not available within many countries (including Italy), very few managerial studies have been conducted on this topic by the academic literature. For example, Sestino et al. [12] studied the users' acceptance of DTx integrating TAM with other variables like trust and customization, but they focused only on the case of DTx developed as mobile apps grounded on gamification mechanics. What is still missing, thus, is a more generic overview of individuals' intention to adopt a DTx and all the managerial implications that a study so designed could bring.

18.2 Objective

18.2.1 Purpose of the Work

Understanding people's needs, fears, and intentions is the right way to pursue a patient-centric approach, one of the key factors ensuring the effectiveness of a HC system [13]. It can be stated, hence, that measuring the level of users' technology acceptance related to DTx enables and facilitates its entrance within the aimed market. In particular, the Italian government showed its interest in accelerating this process with Deputy Loizzo's bill presented in 2023: Disposizioni in materia di terapie digitali (Loizzo et al. 2023).

For these reasons, this chapter aims to investigate the intention to use DTx among Italians. The goal of this work, in fact, is to understand if Italian people can be interested in the introduction of DTx within their healthcare national system. The authors believe, indeed, that, by understanding actors' behavioral intentions toward DTx, the developers of this type of technologies and the HC decision-makers can use these results to understand which solutions and strategies must be implemented to foster the entrance of DTx within the national healthcare system. In particular, this chapter aims to explore the effect of personal innovativeness (PI) [14] on the intention to use (ITU) [15] DTx among Italians. Personal innovativeness is "an extent to which an individual is innovative by his or her perception of new ideas or technology" [16]. It must be considered, indeed, that there are people who "adopt IT innovations earlier than others" [14] and that they are also the key actors enabling the introduction of new technologies [17, 18].

Moreover, since the goal of this chapter is to suggest the levers that must be used to facilitate the entrance DTx mitigating or enhancing the effect of this relationship. The analysis will also measure the effect that two variables reflecting both the worlds involved in DTx (digital and health) have on it: Digital Self-Efficacy (DSE) and Health Belief (HB), respectively. DSE represents the factors that arise from users' abilities to use DH innovations effectively and competently [19]. For what concerns HB, the authors based this study on Zhang et al.'s assumption [20] that the individuals' technology acceptance in the health sector depends on their perception of how much their health is at risk; more specifically, it will be considered the analysis of this specific factor conducted by Gansser and Reich [18]. The authors, indeed, expect that this type of study will help to understand how to enhance the positive impact of PI on ITU and how to mitigate the negative instinctive reaction to a new digital health technology, by working on DL and HB as strategic points. Furthermore, to contextualize those relationships to the specific DH technology considered, the analysis will be built around the UTAUT model's [21, 22] variables: performance expectancy, effort expectancy, social influence, and facilitating conditions. The UTAUT model has been chosen over the extension of its evolution [23] because of the particular nature of DTx which are associated with medication and therapeutic pathways; consequently, some variables like "Habit" or "Price Value" (see [23]) have not been considered relevant for this research. However, the hypothesis of implementing it in future studies must not be totally excluded.

18.3 Methodology

18.3.1 Research Model and Hypotheses

In this section, the model used to study the effect of personal innovativeness push on the intention to use Digital Therapeutics (DTx) along with its theoretical foundations is presented. The study on individuals' adoption of technology represents one of the main focuses in the research field of information systems (IS) [22]. The investigation of behavioral intention and the use of new technologies is predominantly conducted in the literature through the use of the Unified Theory of Acceptance and Use of Technology (UTAUT) [21]. According to what was previously introduced, UTAUT identifies four main factors influencing the adoption and use of technology: perceived usefulness, perceived ease of use, social influence, and facilitating conditions. Venkatesh et al. [21] defined these constructs in their essence, without theoretical elaboration, as follows:

- Performance expectancy (PE) is "the degree to which an individual believes that the use of the system will help him or her to attain gains in job performance."
- Effort expectancy (EE) is" the degree of ease associated with the use of the system."
- Social influence (SI) is" the degree to which an individual perceives that important others believe he or she should use the new system."
- Facilitating conditions (FC) is" the degree to which an individual believes that an organizational and technical infrastructure exists to support the use of the system."

As explained in the section on the objective of this work, the relationship between personal innovativeness and the intention to use a technology is intended to be analyzed. Derived from the innovation diffusion theory (IDT) [24], personal innovativeness is defined as a personality trait that describes" the willingness of an individual to try out any new information technology" [14]. In this context, individuals who are more innovative are more likely to have positive perceptions of DTx, which will impact their intention to use. The authors propose, therefore:

H1a: Personal innovativeness is positively correlated with intention to use.

Previous research has shown that there is a significant impact of consumer innovativeness on perceived usefulness. According to Jackson et al. [25], indeed, users who are more innovative toward a particular technology have positive beliefs regarding its usefulness. The authors propose:

H1b: Personal innovativeness is positively correlated with performance expectancy.
H1c: Personal innovativeness is positively correlated with effort expectancy.
H1d: Personal innovativeness is positively correlated with social influence.
H1e: Personal innovativeness is positively correlated with facilitating conditions.
H2: Performance expectancy is positively correlated with intention to use.
H3: Effort expectancy is positively correlated with intention to use.
H4: Social influence is positively correlated with intention to use.
H5: Facilitating conditions is positively correlated with intention to use.

An additional aim of this study is to investigate the moderating effect of Digital Self-Efficacy and Health Belief factors.

From a literature review, Digital Self-Efficacy assumes a pivotal position in engagement with digital technologies [26]. Compeau and Higgins [27] adapted the concept of self-efficacy from Social Cognitive Theory [28] and used in the context of Information Systems which is generally known as computer self-efficacy. DSE may exert a crucial influence on future behavior [29]: it not only exhibits a positive correlation with learning and the acquisition of new skills, such as digital competencies, but also significantly influences an individual's propensity to utilize digital systems [30]. Therefore, the authors propose:

H6: The effect of (a) performance expectancy, (b) effort expectancy, (c) social influence, and (d) facilitating conditions on intention to use is higher for individuals having a high level of digital self-efficacy compared to individuals having a low level of digital self-efficacy.

The construct of health belief assumes a foundational role in comprehending and examining the inclination toward adopting technology within the healthcare sector [31]. Within the framework of intention to adopt healthcare technologies, numerous factors exert influence; notably, considerations related to health attributes, particularly health belief, emerge as central in shaping the propensity toward embracing such technologies [20]. Consequently, the authors propose:

H7: The effect of (a) performance expectancy, (b) effort expectancy, (c) social influence, and (d) facilitating conditions on intention to use is higher for individuals having a high level of health belief compared to individuals having a low level of health belief.

18.3.2 Data Collection

To empirically test our hypotheses, stratified sampling will be conducted based on the demographic variable of age and educational qualification. Participants are asked to indicate their level of agreement or disagreement with the statements on a 5-point Likert scale, ranging from strongly disagree to strongly agree.

In order to enhance the reliability of the measurements, the majority of the constructs were derived from pre-existing instruments based on prior research, with minor modifications made where necessary to align with the current research context.

Specifically, performance expectancy (PE), effort expectancy (EE), social influence (SI), and facilitating conditions (FC) are adapted from the UTAUT scale [21]. Intention to use is adapted from a scale based on three items [10]. Health Belief (HB) is measured by a scale based on four items [32].

Digital Self-Efficacy (DSE) is measured by a shortened version of a Digital Self-Efficacy scale [33].

Personal innovativeness is measured using a scale based on three items [34].

The questionnaire is completely anonymized and was developed in three sections: the first section concerns socio-demographic information, such as gender, province of residence, years of schooling, and level of computer tool usage; the second section represents the measurement of constructs present in the theoretical framework through items retrieved from the literature.

18.3.3 Data Analysis

The data will be subjected to a data cleaning phase to identify outliers and missing values and to perform an initial analysis of descriptive statistics. The adoption of the structural equation modeling (SEM) framework has been deemed more appropriate not only for evaluating the measurement of latent variables and addressing the objective of testing the hypothesized relationships among them [35], but also for demonstrating potential in healthcare and medical research [36]. The SEM technique of partial least squares [37] was preferred because it is based on an iterative process that maximizes the explained variance of endogenous constructs, operating similarly to a multiple regression analysis [38]. Through the use of the measurement model, the research will evaluate the reliability and validity of the constructs and structural model employed to examine the significance of the path relationship and to test the hypotheses. Subsequently, the structural model and hypotheses will be validated through the examination of path coefficients.

18.4 Conclusions

18.4.1 Expected Findings and Future Implications

This work is intended to be a valid support to the introduction of an emergent digital health technology within a country which has not adopted it yet. In fact, this study can help to understand how personal innovativeness affects the individual's intention to use a technological product and also how to manage and moderate the effects of this relation. From a theoretical point of view, the study will contribute to the stream of literature on individuals' technology acceptance in healthcare and it will also help to better comprehend the potential power and limitations of Digital Therapeutics, considered as an emerging and complex innovation to be managed. With reference to UTAUT, the aim is to explore the impact of the constructs and the role of personal innovativeness in the behavioral intention to use. Based on our hypotheses, the analysis should show a significantly positive impact of personal innovativeness on the adoption of the new DTx technology. It is expected that the effect of the moderating variables, Digital Self-Efficacy and Health Belief, will be significantly positive in their relationships with the intention to use variable. These expected results are significant because they could provide evidence on the relationship between personal innovativeness and technology adoption, as well as on the effectiveness of health education and digital literacy strategies.

References

1. Long, Y., Liu, L., Yang, B.: The effects of enterprise digital transformation on low-carbon urban development: empirical evidence from China. Technol. Forecast. Soc. Chang. **201**, 123259 (2024)
2. Verhoef, P.C., Broekhuizen, T., Bart, Y., Bhattacharya, A., Dong, J.Q., Fabian, N., Haenlein, M.: Digital transformation: a multidisciplinary reflection and research agenda. J. Bus. Res. **122**, 889–901 (2021)
3. Biancone, P., Secinaro, S., Marseglia, R., Calandra, D.: E-health for the future. Managerial perspectives using a multiple case study approach. Technovation. **120**, 102406 (2023)
4. Shaygan, A., Daim, T.: Technology management maturity assessment model in healthcare research centers. Technovation. **120**, 102444 (2023)
5. Pundziene, A., Gerulaitiene, N., Bez, S.M., Georgescu, I., Mathieu, C., Carrabina-Bordoll, J., Vilutiene, L.: Value capture and embeddedness in social-purpose-driven ecosystems. A multiple-case study of European digital healthcare platforms. Technovation. **124**, 102748 (2023)
6. ISO/TR 11147:2023(en): Health informatics—Personalized digital health—Digital therapeutics health software systems. Available at: https://www.iso.org/standard/83767.html (2023). Accessed 28 Feb 2024
7. Geiger, S., Kjellberg, H.: Market mash ups: the process of combinatorial market innovation. J. Bus. Res. **124**, 445–457 (2021)
8. Jeon, E., Yoon, N., Sohn, S.Y.: Exploring new digital therapeutics technologies for psychiatric disorders using BERTopic and PatentSBERTa. Technol. Forecast. Soc. Chang. **186**, 122130 (2023)

9. Lord, S.E., Campbell, A.N., Brunette, M.F., Cubillos, L., Bartels, S.M., Torrey, W.C., Marsch, L.A.: Workshop on implementation science and digital therapeutics for behavioral health. JMIR Ment. Health. **8**(1), e17662 (2021)
10. Baudier, P., Ammi, C., Deboeuf-Rouchon, M.: Smart home: highly-educated students' acceptance. Technol. Forecast. Soc. Chang. **153**, 119355 (2020b)
11. Zhao, Y., Ni, Q., Zhou, R.: What factors influence the mobile health service adoption? A meta-analysis and the moderating role of age. Int. J. Inf. Manag. **43**, 342–350 (2018)
12. Sestino, A., Bernardo, A., Rizzo, C., Bresciani, S.: An explorative analysis of the antecedents and consequents of gamification in the digital therapeutic context. Eur. J. Innov. Manag. (2023)
13. Ciasullo, M.V., Cosimato, S., Douglas, A.: Customer-centric service innovation in health care: findings from a case study. Sinergie Italian J. Manag. **35**(Sep-Dec), 155–176 (2017)
14. Agarwal, R., Prasad, J.: A conceptual and operational definition of personal innovativeness in the domain of information technology. Inf. Syst. Res. **9**(2), 204–215 (1998)
15. Davis, F.D., Bagozzi, R.P., Warshaw, P.R.: User acceptance of computer technology: a comparison of two theoretical models. Manag. Sci. **35**(8), 982 (1989)
16. Baudier, P., Kondrateva, G., Ammi, C., Chang, V., Schiavone, F.: Patients' perceptions of teleconsultation during COVID-19: a cross-national study. Technol. Forecast. Soc. Chang. **163**, 120510 (2021)
17. Baudier, P., Kondrateva, G., Ammi, C.: The future of telemedicine cabin? The case of the French students' acceptability. Futures. **122**, 102595 (2020a)
18. Gansser, O.A., Reich, C.S.: A new acceptance model for artificial intelligence with extensions to UTAUT2: an empirical study in three segments of application. Technol. Soc. **65**, 101535 (2021)
19. Iyanna, S., Kaur, P., Ractham, P., Talwar, S., Islam, A.N.: Digital transformation of healthcare sector. What is impeding adoption and continued usage of technology-driven innovations by end-users? J. Bus. Res. **153**, 150–161 (2022)
20. Zhang, M., Luo, M., Nie, R., Zhang, Y.: Technical attributes, health attribute, consumer attributes and their roles in adoption intention of health-care wearable technology. Int. J. Med. Inform. **108**, 97–109 (2017)
21. Venkatesh, V., Davis, F., Morris, M.G.: Dead or alive? The development, trajectory and future of technology adoption research. J. Assoc. Inf. Syst. **8**(4), 267–286 (2007)
22. Venkatesh, V., Morris, M.G., Davis, G.B., Davis, F.D.: User acceptance of information technology: toward a unified view. MIS Q. **27**, 425–478 (2003)
23. Venkatesh, V., Thong, J.Y., Xu, X.: Consumer acceptance and use of information technology: extending the unified theory of acceptance and use of technology. MIS Q. **36**, 157–178 (2012)
24. Rogers, E.M.: Diffusion of Innovations, 5th edn. Free Press, New York (2003)
25. Jackson, J.D., Yi, M.Y., Park, J.S.: An empirical test of three mediation models for the relationship between personal innovativeness and user acceptance of technology. Inf. Manag. **50**, 154–161 (2013)
26. Ulfert-Blank, A.S., Schmidt, I.: Assessing digital self-efficacy: review and scale development. Comput. Educ. **191**, 104626 (2022)
27. Compeau, D.R., Higgins, C.A.: Computer self-efficacy: development of a measure and initial test. MIS Q. **19**(2), 189–211 (1995)
28. Bandura, A.: Social Foundations of Thought and Action, Englewood Cliffs, NJ (1986)
29. Marakas, G.M., Yi, M.Y., Johnson, R.D.: The multilevel and multifaceted character of computer self-efficacy: toward clarification of the construct and an integrative framework for research. Inf. Syst. Res. **9**(2), 126–163 (1998). https://doi.org/10.1287/isre.9.2.126
30. Hsu, M.H., Chiu, C.M.: Internet self-efficacy and electronic service acceptance. Decis. Support. Syst. **38**, 369–381 (2004)
31. Champion, V.L., Skinner, C.S.: The health belief model. In: Health Behavior and Health Education: Theory, Research, and Practice, vol. 4, pp. 45–65 (2008)
32. Cheung, M.L., Chau, K.Y., Lam, M.H.S., Tse, G., Ho, K.Y., Flint, S.W., Broom, D.R., Tso, E.K.H., Lee, K.Y.: Examining consumers' adoption of wearable healthcare technology: the

role of health attributes. Int. J. Environ. Res. Public Health. **16**, 2257 (2019). https://doi.org/10.3390/ijerph16132257
33. Swan, E.L., Peltier, J.W., Dahl, A.J.: Artificial intelligence in healthcare: the value co-creation process and influence of other digital health transformations. J. Res. Interact. Mark. **18**(1), 109–126 (2023)
34. Kumari, P., Shankar, A., Behl, A., Pereira, V., Yahiaoui, D., Laker, B., Gupta, B.B.: Investigating the barriers towards adoption and implementation of open innovation in healthcare. Technol. Forecast. Soc. Chang. **200**, 123100 (2024)
35. Babin, B.J., Hair, J.F., Boles, J.S.: Publishing research in marketing journals using structural equation modeling. J. Mark. Theory Pract. **16**, 279–286 (2008)
36. Beran, T.N., Violato, C.: Structural equation modeling in medical research: a primer. BMC. Res. Notes. **3**(1), 1–10 (2010)
37. Wold, H.: Systems Analysis by Partial Least Squares. IIASA (1983)
38. Hair, J.F., Ringle, C.M., Sarstedt, M.: PLS-SEM: indeed a silver bullet. J. Mark. Theory Pract. **19**, 139–152 (2011)

Sitography

Digital Therapeutics Alliance: Consulted on 28.02.2024. https://dtxalliance.org (2024)
Disposizioni in materia di terapie digitali (1208):. Consulted on 28.02.2024. https://www.camera.it/leg19/126?tab=&leg=19&idDocumento=1208&sede=&tipo= (2023)

Chapter 19
The Journey of the Italian National Healthcare System Toward Digital Transformation Maturity

Silvia Cosimato, Maria Vincenza Ciasullo, and Silvia Verde

Abstract Digital transformation is a complex and resource-intensive phenomenon that is causing disruptive changes at social, economic, and institutional levels. This has contributed to the mounting need for better assessing digital transformation maturity through reliable models/frameworks. Even though research provides several models, they suffer from a lack of internal and external validity. Thus, this paper aims to contribute to bridging this gap by proposing a tentative model and preliminarily testing it in the healthcare domain. Drawing on the extant literature a tentative digital transformation maturity model has been applied to the Italian National Healthcare System (NHS) to test it and describe through a content analysis the main dimensions of Italian NHS digital transformation and its maturity stage. The analysis underlined that the Italian NHS is still proceeding in the journey toward digital transformation maturity. It emerged that the analyzed sector needs further effort to strategically approach the wide transformative potential of digital technologies, paying greater attention not only to those dimensions strictly related to processes' efficiency, but also to those that imply a deep cultural change, a focus on processes' participation, and their long-term effectiveness. This study offers some interesting implications for enhancing digital transformation maturity models' reliability and for supporting both health managers and policymakers in strategically approaching this phenomenon. However, its exploratory nature and the application to one specific domain somewhat limit it. The originality of this paper lies in the new dimensions attached to previous digital transformation maturity models and in its practical application to specific and complex sectors such as healthcare.

S. Cosimato (✉) · S. Verde
University of Naples "Federico II", Naples, Italy
e-mail: silvia.cosimato@unina.it; silvia.verde@unina.it

M. V. Ciasullo
University of Salerno, Fisciano, Italy
e-mail: mciasullo@unisa.it

F. Schiavone et al. (eds.), *Advanced Perspectives and Trends in Digital Transformation of Firms, Networks, and Society*, Springer Proceedings in Business and Economics, https://doi.org/10.1007/978-3-031-80692-6_19

Keywords Digital transformation · Digital transformation maturity models · Innovation · Healthcare

19.1 Introduction

In recent years, the emergence of new disruptive technologies has deeply and quickly changed both society and business, reshaping most socio-economic processes, practices, and structures. This is the reason why the technology-driven development or the so-called digital transformation has been globally recognized as a megatrend or a long-term driving force "observable today and will likely have transformational potential in the future" [1].

Digital transformation is "a process that aims to improve an entity by triggering significant changes to its properties through combinations of information, computing, communication, and connectivity technologies" [2]. Across industries, digitalization or digital transformation is a current strategic priority for almost all companies [3] because of its potential positive impact on competitiveness. It follows that the digital shift does not affect just the IT sphere of a company, but also its business model, management practices, organizational structure, and even its corporate culture [4].

Focusing on healthcare, research on digital transformation is still in its infancy even though the disruptiveness of the Covid-19 pandemic has pushed the whole industry forward with an unexpected and rapid change [5]. This change is affecting healthcare not only in terms of productivity, efficiency, and cost reduction [6, 7], but also in service quality, patient safety, well-being, and empowerment [8, 9]. However, it is worth noting that healthcare organizations have differently approached digital transformation and its challenges, often lagging behind companies of other more digitalized industries [10]. This led research to call for further investigation into the degree or maturity of healthcare organizations' digital transformation. DTM is a construct coming from the merge of two different but related concepts, "digital transformation" and "digital maturity," which holistically approach technological and managerial elements [11]. It describes "what a company has already achieved regarding its digital transformation efforts" [12], which mobilizes the competencies and skills gained for driving the transformation process [13]. It follows that DTM refers not only to an organization's position in its digital journey but also to the strategic implementation of digital technologies [14]. Even though the disruptiveness of the COVID-19 pandemic accelerated digital transformation, in healthcare the advancement of its maturity remains heterogeneous and still calls for further assessment frameworks and models to support organizations in understanding their digital status quo as well as the development paths to follow and the areas that need for further enhancement effort [15]. Drawing on these considerations and the results of a literature review on the topic, this study aims to contribute to bridging this gap by proposing and implementing a new approach to the assessment of DTM of healthcare organizations [10, 11, 16]. In doing so, the following research question inspired this work, what is the level of digital transformation maturity of the analyzed Italian healthcare organizations?

Being a preliminary study, a tentative framework has been tested applying it to some articles published by Italian institutional sources and specialized magazines. For this purpose, a content analysis has been conducted. The remainder of the paper has been organized as follows. Section 19.2 is dedicated to the analysis of the theoretical background of the study, while Sect. 19.3 describes the implemented methodology. This is followed by Sect. 19.4, in which the achieved results are presented and discussed. The paper concludes with Sect. 19.5 in which some theoretical, managerial, and policy implications as well as possible future research paths are presented.

19.2 Literature Review

19.2.1 Digital Transformation Maturity in Healthcare

In the era of the digital revolution, the capillary spread of new technologies at the core of the Industry 4.0 paradigm is pushing institutions, business organizations, and even people to change the way they are used to approach socio-economic dynamics [11]. This phenomenon has been globally understood as "digital transformation" [17]. Even though in the extant literature there is no consensus about its definition [18], digital transformation seems to be less about technology and more about people because it is related to the diffusion and the accessibility of "data-enabled services into ever more aspects of life" [19].

Focusing on the business domain, digital transformation has been defined as "an evolutionary process that leverages digital capabilities and technologies to enable business models, operational processes, and customer experiences to create value" [20]. The inner transformative nature of this phenomenon is quickly and deeply changing the structure of global competition, customers' behavior and expectations, production processes, service delivery, and even human resources management [21, 22]. The COVID-19 pandemic has recently enhanced this potential, pushing the world to adopt new digital solutions to solve the current challenges and difficulties in business and society.

Healthcare was the sector that mostly suffered from the pandemic; thus, gradually the initial critical pressure turned into an opportunity for change and growing differently [8]. The acceleration of digital transformation in healthcare set some different evolutionary paths, such as (1) the digitalization of services, (2) the enhancement of stakeholder engagement, and (3) the definition of new digital value-creation processes [9]. However, it is worth noting that even before the pandemic, digital transformation offered new business opportunities to the healthcare sector as well as the possibility to better address timely challenges such as the growing costs, the needs of an aging population, and service accessibility and inclusiveness [23]. To achieve these goals advanced services and solutions—such as self-tracking systems, big data and predictive analytics, e-health or mobile health services, e-patient communities, gamification, robotization, and Artificial Intelligence-based diagnostics [24]—have been developed and implemented for enhancing security, safety, and quality of health-related services, products, and procedures [25]. Even with advancements, when compared with other sectors, the adoption rate of digital technologies in healthcare is still low. This seems to be due to the inner complexity and the strict regulation of this sector, which bond together different interacting stakeholders or actors (e.g., patients, physicians, nurses, and public bodies) [26, 27]. The adoption rate of digital technologies is strictly related to the concept of digital maturity, which is more than the digital transformation status of a company or a sector [28] because it holistically refers to technological, managerial, and social elements. It reflects the technological and managerial approach to the changes that digital technologies have triggered in political, social, and economic processes at different levels [12].

Scholars and practitioners are currently engaged in a lively debate about healthcare and its digital maturity; thus, they are trying not only to better conceptualize it but also to find more effective approaches and tools to assess it. In fact, DTM is a complex concept that delves both on an organization's position in its digital journey and on the strategic implementation of digital technologies [14].

In healthcare, DTM refers to the ability of the whole sector as well as of organizations and people to adapt to new and often unexpected digital scenarios [28]. These changes must be intended to make providers able to better respond to societal needs for a more efficient, effective, and high-quality health system [29]. Moreover, DTM is related not only to how specific digital technologies have been implemented, but also to the changes they have boosted in health-related structures, processes, practices, and policies or in a nutshell "the degree to which a hospital can harness technology for a better patient experience" [10]. It follows that DTM is also about technology integration level, data quality, and adherence to governance principles and international/national laws and standards. Digital transformation maturity has also been approached as a learned ability, depicting the evolutionary path a healthcare organization follows according to two essential variables: space and time [10]. These two variables make DTM not static, but an always-changing concept [30]. This points out the importance of evaluating DTM for a better understanding of how efficiently and effectively a health organization (e.g., a hospital, clinic, research/diagnostic lab, etc.) is implementing digital technologies for gaining better performances and a better medical outcome.

19.2.2 Digital Transformation Maturity Assessment: A Proposed Framework/Model

As stated, the extant literature does not provide commonly agreed definitions of digital maturity [31]. This fragmentation makes its assessment not an easy target to hit because it has produced several different models and frameworks often inspired by opposite perspectives and aims. When dealing with the healthcare domain, these models fail to evaluate care paths and the influence that digitalization has had on them over time. In fact, healthcare evolution toward service digitalization is not only about the implementation of the most recent and innovative technologies but also about organizations' digital congruence and their ability to align strategies to patients' outcomes. It follows that to proceed in their journey toward a full digitalization, health organizations must act as learning systems "capable of reducing error and promoting patient safety and quality" [32]. Even though maturity models and frameworks are generally intended to evaluate and report the stage of healthcare organizations' digital transformation journey, they perform this task by implementing different types of assessment protocols [33], in most cases based on the measurement of different dimensions of digital maturity. This is mainly due to the lack of consensus about the most common and effective dimensions to be considered to assess digital maturity in the healthcare sector and to the long-lasting vagueness of the proposed frameworks [34]. To counteract this vagueness, it is interesting to start describing what a maturity model is.

According to Teichert [11], a digital maturity model is a tool, or a framework based on pre-definite dimensions useful for defining a roadmap of "transformation activities to improve the level of digital maturity" of organizations. Such a model is usually based on dimensions and criteria able to define the level or the stage of an organization in its journey toward digital maturity and the areas/elements to be improved [33]. In sum, a digital maturity model points to assessing organizations' ability to always improve specific dimensions to reach the desired level of maturity [35].

Even though there are several well-established maturity models, frameworks, or standards, such as CMMI (Capability Maturity Model Integration), ISO/IEC 3300xx (ISO2015a,2015b,2015c), and ISO/IEC 15504 standard1, few are those dedicated to the assessment of digital maturity transformation, especially when related to healthcare. Deloitte Digital [16] developed a digital transformation assessment framework pointing to assess "the extent to which digital technologies have transformed an organization's processes, talent engagement, and citizen service models" (p. 2). In a similar vein, Liaw and Godinho [35] proposed a maturity assessment toolkit, based on a list of indicators, aimed at describing the digital profile of healthcare national systems and their organization as well as at supporting them in accomplishing national health priorities. More recently, [36] proposed their DTM framework, which bases its theoretical foundations on dynamic capability theory

[37], merged with the Resource-Based View (RBV) of the organization [38, 39]. Starting from the analysis of these models, a tentative framework has been developed to assess and describe how the healthcare sector and its organizations are approaching digital transformation and the level of digital maturity they have achieved over time. Following the findings of the extant research [16, 32, 36] the proposed digital maturity framework is intended to depict the changes occurring at different stages of the digital transformation process, passing from (1) the early stage, when organizations start to recognize possible opportunities coming from digitalization and their current gaps, (2) the developing stage, when organizations have started to approach value-added processes and activities through digital technologies, and (3) the maturing stage, when organizations can use digital technologies for innovating and, therefore, to enact continuous improvement processes (Table 19.1).

Following the Doiltte's model [16], these three stages at the core of DTM journey have also been related to three of the main dimensions of an organization, people, processes, and technology preparedness, which need for specific capabilities (e.g., seizing, sensing, and transforming) to make healthcare sector or its organization able to fully achieve DTM [36]. The partial achievement of these capabilities relegates an organization at the early or intermediate stage of the digital transformation journey (Fig. 19.1).

Table 19.1 Digital Transformation maturity stages

	Early (not mature)	Developing (maturing)	Mature
People	Lacks awareness and skills; user focus absent	Digitally awareness; growing user focus	Digitally sophisticated; user focus central for digital transformation
Processes	Cost reduction; insufficient investments	Customer experience and decision-making; moderate investments	Processes transformation; adequate investment
Preparedness	Adversity to risk; no innovation strategies and collaboration	Tolerance to risk; accommodating innovation and collaboration	Receptive to risk; open to innovation and collaboration

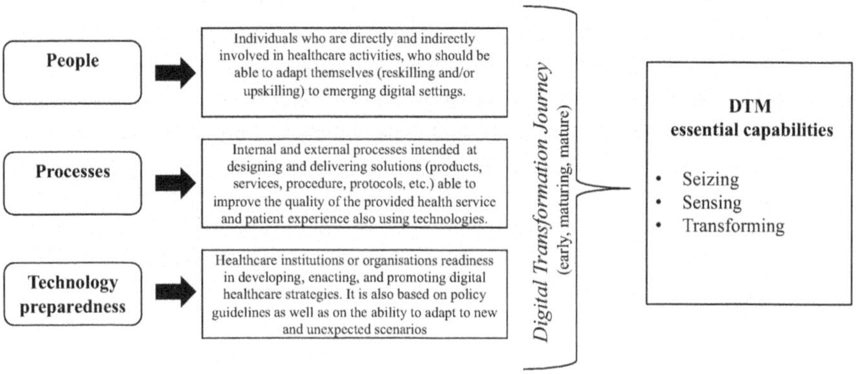

Fig. 19.1 A tentative DTM framework

19.3 Methodology

19.3.1 Digital Transformation Maturity in Healthcare

This study is exploratory in its nature because it aims at better investigating a topic that is not well established in literature, the assessment of digital transformation maturity of healthcare. Due to the fragmentation of the research on the topic, an inductive content analysis—based on discovering if and where findings emerge through researchers' interaction with data [40]—has been performed [41] to test the digital transformation maturity proposed in the previous section and inspired by the extant research. In particular, the proposed framework has been applied to the Italian National Healthcare System (NHS) to understand how digitally mature it is perceived to be.

The content analysis has been conducted on some news published both by health-related institutions and media (e.g., Italian Ministry of Health, and an Italian business magazine), selected according to a convenience sampling strategy [42]. This led to choose the sample units according to their accessibility [43] and checking the following selection criteria, topic, source, publication date (Table 19.2).

The selected 75 articles were red and some of them excluded because not in line with the established sampling criteria because (1) not coherent with the topic under investigation, (2) doubled. Despite these limitations, the achieved results demonstrated a certain degree of generalizability due to their common topic. The final sample consisted of 66 news articles, which were analyzed following the criteria of textual qualitative analysis (Fig. 19.2).

The construct validity, based on suitable operational measures for the analyzed concepts, and the internal validity, based on the appropriateness of the content analysis items and indicators, were ensured by triangulating results with data retrieved from different data sources, such as corporate reports, online news, and social media posts [44].

The content analysis has been based on three main steps: (1) preparation, intended to collect the articles to be analyzed, from the selected database (National Healthcare Ministry, and Sole24Ore database), (2) organization, intended to the definition of a dictionary, containing the 66 keywords retrieved analyzing research on the topic [11, 36, 45] and associated to the coding categories defined after the theoretical analysis, and (3) reporting [46]. Keywords were checked using an English dictionary of synonyms and antonyms and after that added to the dictionary, without articles, auxiliaries, and prepositions (Table 19.3).

Table 19.2 Sampling criteria

Sampling criteria	
Topic	Healthcare digitalization
Source	Italian Ministry of Health, Sole24Ore, health organizations websites
Publication date	March 2020 and June 2023.

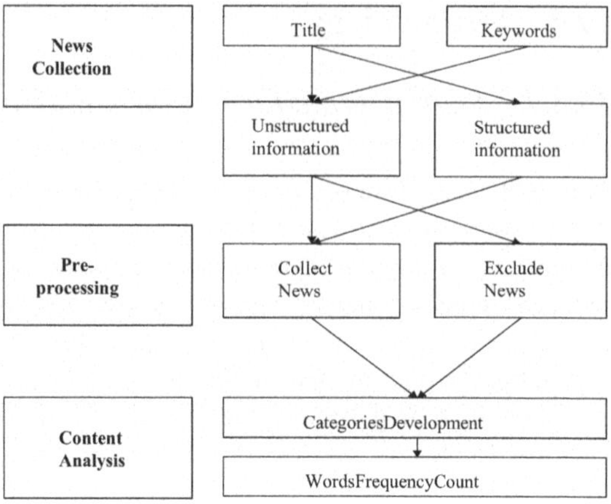

Fig. 19.2 Analysis strategy

Table 19.3 Sampling criteria

Coding categories	Keywords
People	accessib*, adaptability, app*, awareness, benefit, client*, customer centri*, device, digital mindset, digital service*, digital skill*, education, experience*, flexibility, information accessibility, learning, motivation, privacy, safety, *satisfact*.
Processes	agil*, co-creation, collaboration, communication, co-production, cost* reduction, document*, digitalization, digital process*, digital record, engag*, empow*, hybrid channel*, IT, innovation, integration, interoperab*, patient journey, network*, patient centr*, personal*, platform*, technolog*, transform*.
Technology preparedness	analytic*, challenge, compliance, constrain*, digital ecosystem*, digital maturity, digital strateg*, digital culture, efficien*, effect*, innovation strateg*, healthcare 4.0, investment*, law*, leadership, literacy, platform*, quality, risk*, R&D, secur*, software*, telemedicine.

Two researchers independently checked the dictionary consistency to use it for organizing content analysis results. Some content units (or "observations" in inferential statistics) were defined; thus, each sentence of the analyzed news was considered a content unit, which "set the limits on the information to be considered in the description" [47], while the word was set as a unit of analysis. Reporting step was intended to organize and classify the results of the analysis conducted using the MAXQDA software. Results have been organized in a word-count matrix to easily compare them as well as to define patterns or trend.

19.4 Finding Presentation and Discussion

The content analysis demonstrated that the category with the highest number of keywords' occurrences was Technology preparedness (929), follows Processes (869) and People (493) (Fig. 19.3). The most recurrent keywords belong to two of

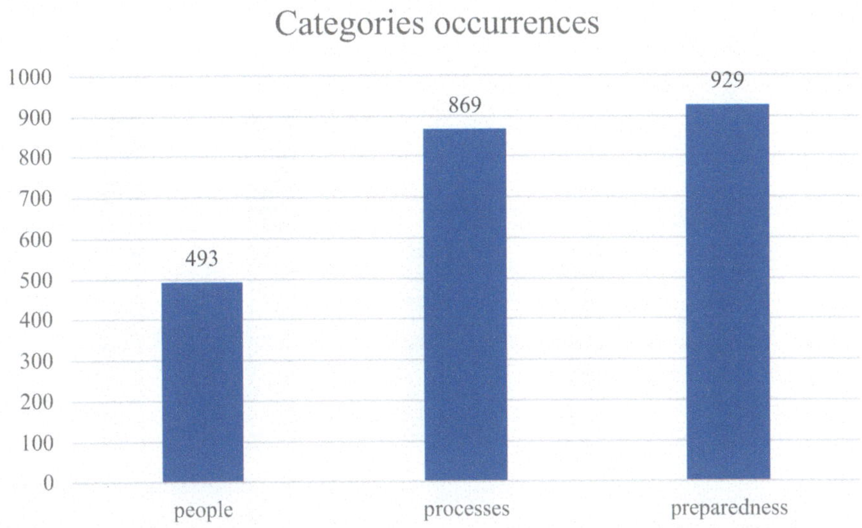

Fig. 19.3 Categories occurrences

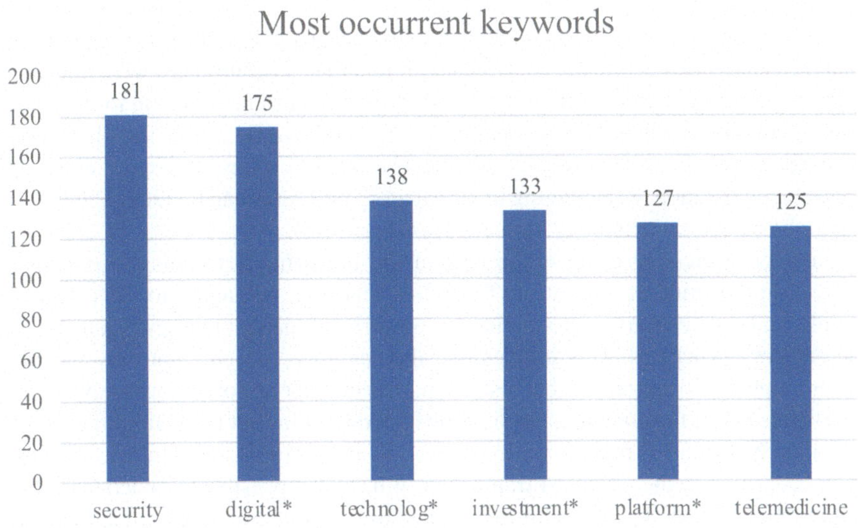

Fig. 19.4 Most occurrent keywords

the three categories: processes and technology preparedness. In fact, security, belonging to the category technology preparedness, was the keyword with the higher number of occurrences (181), follows digitalization (175) and technology (138) belonging to the category processes.

The other keywords with more than one hundred occurrences are platform (127), belonging to the category processes, investment (133) and telemedicine (125) belonging to the category technology preparedness (Fig. 19.4). The occurrences of keywords belonging to the category people do not exceed one hundred occurrences;

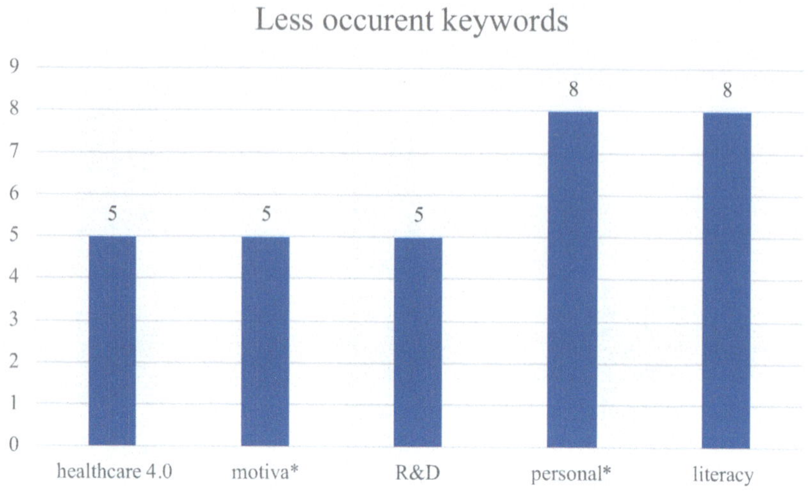

Fig. 19.5 Less occurrent keywords

thus, the most recurrent ones are accessibility (94), patient centricity (70), and app (64).

The content analysis also pointed out the less occurrent keywords for each of the analyzed categories, while the keywords with less than 5 occurrences were not considered; therefore, two over three keywords with 5 occurrences belong to the category technology preparedness, healthcare 4.0 and R&D, while the third one motivation belongs to the category people. The other two less occurrent keywords were personalization (8) belonging to the category processes and literacy (8) belonging to the category technology preparedness (Fig. 19.5).

Another interesting result is the lack of the following keywords, adaptability, device, digital mindset, education, flexibility, learning, belonging to the category people, and co-creation, co-production, empowerment, patient journey belonging to the category processes. This implies that the category "people" not only has the lower number of occurrences, but also the higher number of not retrieved keywords.

The results of the content analysis contributed to point out the main dimensions of Italian NHS digital transformation as well as its maturity stage. In fact, reading them according to the proposed framework, it has been possible to understand that Italian NHS—made up of health-related institutions (e.g., national ministries and local institutions), public and private organizations (e.g., public hospitals, ambulatories, labs, clinics, etc.), and citizens—currently is at an intermediate stage of its DTM journey.

The results of the content analysis contributed to point out the main dimensions of Italian NHS digital transformation as well as its maturity stage. In fact, reading them according to the proposed framework, it has been possible to understand that Italian NHS—made up of health-related institutions (e.g., national ministries and local institutions), public and private organizations (e.g., public hospitals, ambulatories, labs, clinics, etc.), and citizens—currently is at an intermediate stage of its DTM journey. Consequently, it can be defined as "digitally maturing," which means that further effort is needed to fully exploit the technological, economic, and social potential of digital technologies [48]. This finding is in line with the extant literature according to which institutions and public organizations are still at the early stages of the digital transformation and almost none of them have reached the stage of a complete digital maturity [49]. In fact, digital transformation is more than the implementation of as many technologies as possible, but it is rather about their incorporation in organizational culture to concretely change internal and external processes. This is possible engaging different actors and their capabilities, essential for successfully and sustainably implement new healthcare practices [50]. This is in line with the achieved results; thus, they pointed out the Italian NHS is mainly focused on technology preparedness, or the "propensity to embrace and use new technologies to accomplish goals in home life and at work" [51]. This concept is strictly related to organizations willing to invest in innovation, to make new technologies available/accessible to be used to enhance organizational and business processes and in doing so improving corporate performance [52, 53]. This implies an enduring focus of heath institutions and organization on efficiency, flexibility, and business process continuity [51, 54, 55]. This emerges from the achieved results that underline the attention paid to the mobilization of economic resources (investments) for improving the current digital service, for example in terms of security, for both providers and users, service efficiency, effectiveness, accessibility and further develop integrated applications of telemedicine.

Even though technology preparedness is one of the key drivers of digital transformation [56, 57], to reach the full maturity other essential elements or drivers are required. In fact, as stated digital transformation is not only about technology, but also about the capability or the "talent" to exploit it [19] to make processes not only efficient, but also effective, and to address eventual digital skills gaps building or rebuilding the current organizational culture and the corporate capabilities [45]. In this sense, the achieved results demonstrated that even though Italian NHS is currently focused on seizing activities, related to the mobilization of resources to satisfy requirements and value-adding opportunities, more effort is needed to improve sensing activities and, therefore, to the achievement of those resources able to identify, develop, co-develop, and evaluate technological opportunities in relation to customer requirements. This emerges from the achieved results that pointed out a poor attention to patient-oriented initiatives, as findings demonstrated with the absence of keywords such as co-creation, co-production, empowerment, and the few occurrences of the keyword engagement (15). This implies that to proceed into its digital transformation journey, Italian NHS should improve and promote not only citizens/patients' engagement in service design, production, and provisioning [58], but also the collaboration among other actors and the interoperability of digital infrastructures. In this sense, it is worth noting that on the one hand interoperability, based on the sharing of digital protocols and the related information, is also essential for responding to the future evolution of quality standards and requirements of healthcare service [32]. On the other, engagement can effectively enact a co-production logic essential for the ongoing renewal of digital service according to the present and future health-related challenges. This finding is not in line with the extant research, according to which patients or clients are always service co-producers [8] "who must actively become involved in managing their own health." The poor attention dedicated to the issues related to the category people implies that Italian NHS approaches digital transformation not yet as collaborative, open, and people centered process, but rather as a provider-controlled process designed to respond to governmental requirements and corporate goals. This also implies the need of further effort to exploit transformation activities and the related resources, which are at the core of a participative and innovation-based approach to digital healthcare, pointing to be always ready to improve its internal and external processes [36].

19.5 Final Remarks

Even though digital transformation is at the forefront of scholars, practitioners, and policymakers, research on digital transformation maturity as well as on the application of the related assessment models and/or framework is still scarce. Moreover, most of the provided models have been generally designed without a focus on the main characteristics and issues of specific socio-economic domains.

Starting from the analysis of the existent digital transformation maturity models, this paper has been intended at empirically test them to a specific and critical domain such as healthcare. In doing so a personalized tentative model has been tested by the application to the Italian NHS to qualitatively test it, analyzing the perception of the Italian healthcare organizations digital transformation maturity, which is a preliminary assessment for understanding if and how this process affects health service delivery and its quality.

The achieved findings led to address the research question which inspired this work (What is the level of digital transformation maturity of the analyzed Italian healthcare organizations?); thus, the analysis pointed out that even though DT is increasing in Italian healthcare sector, to reach the final stage of the digital transformation maturity journey health organizations need to put further effort toward those processes intended at improving those open and collaborative activities essential for creating and sharing an updated digital know-how and the related skills among different socio-economic actors, making them able to contribute to the core processes of the present and future digital healthcare system.

The analysis offered some interesting implications both for theory and practice. In particular, the theoretical ones delve on the contribution to the international and national research on DTM [11], supporting scholars in better identifying the current gaps and the room for further advancement in extending, prioritizing, and specifying the current incomplete digital maturity models [33]. In fact, these models need for balancing the need for more standardized assessment procedures and the customization of the related activities according to the peculiarities of the application sector. In healthcare sector, digital transformation maturity models should be not only generically focused on the technological potential of this process, but also on digital transformation potential for the quality, continuity, and accessibility of health service [9]. Drawing on this consideration, the proposed tentative model offers a more holistic approach to DTM, focusing on technologies ability to support and enhance a cross-functional collaboration between different health-related actors, not only clinicians and patients, but also families, peers, informal caregivers, other service providers, institutions, etc. However, in doing the analysis led to define the organizational and personal capabilities essential to boost that cultural and contextual change able to digitally transform the coordination mechanisms and the way information are managed and shared to innovate and enhance the provided healthcare service.

In terms of managerial implications, it is worth noting that to complete the digital transformation journey, public and private health managers should place more effort into innovation strategies, inspiring them to patients/users' needs and implementing them according to a collaborative logic based on the direct involvement of patients/citizens [58]. Moreover, the achieved findings also demonstrated that in healthcare digital transformation maturity also calls for an inner organizational change, which pass through the modernization of corporate culture, the development and the implementation of new competences and skills as well as the emergence and the application of new business models [36]. This implies that healthcare organizations must recognize the strategic orientation of this transformation, which requires flexibility and an innovation-based approach to always be ready to face the ever-changing health-related issues. This implies a renewed ability in recognizing the emergent needs of patients and other health-related actors and in involving them in the design of innovative digital services and in the further implementation of the digital technologies that suits the digital transformation process essential to improve the overall patient experience. Moreover, health managers should be more aware about the how DT can support health-related organizations enhancing internal coordination mechanisms and promoting a collaborative approach to decision-making, making the organizations themselves as agile as possible in facing the unpredictability of the current scenarios. However, DT does not come without challenges, which are mainly related to the patients/citizens involvement in digital processes, the service accessibility, and the ability to ensure data privacy and security.

Finally, the analysis offered some interesting policy implication; thus, being the digital transformation journey a collaborative process, it needs for the participation of different healthcare actors, and in particular of patients [53]. To achieve this goal national institutions should enact and promote specific actions of digital literacy, pointing to boost people individual evolution based on a clear knowledge-sharing orientation able to positively affect their technology acceptance [59] and propensity toward using it. In this sense, the definition of a public agenda should support and trigger the transformation of contemporary healthcare system [32]. This will open to the creation of an open collaborative environment, in which new digital, and health-related solution can be developed, tested, and applied for the purpose of enhancing public health and well-being. When dealing with the public dimension of healthcare DT, it is worth noting that its related transformative processes also have implications for healthcare providers, particularly in their ability to reach and offer increased access to services for underserved, remote, or minority communities [60].

Despite its contributions to theory and practice, this study is not without limitations. In fact, on the one hand, the proposed model has been tested applying it to a single sector (healthcare) and needs to be validated also in other sectors or domains. On the other, the analysis is limited to a specific number of secondary data sources, which could be further expanded and supported by primary data collected through interviews with a good number of key informants. Therefore, to ensure a greater validity and generalizability, it needs to be tested using quantitative methods applied to an adequately selected and sized sample. On the other, the model needs for further improvements able to better define the weights of its dimensions, subdimensions, and items to fully describe and evaluate healthcare organizations maturity stage and the related progression in their digital transformation journey.

References

1. Debonne, N., Bürgi, M., Diogo, V., Helfenstein, J., Herzog, F., Levers, C., et al.: The geography of megatrends affecting European agriculture. Glob. Environ. Chang. **75**, 102551 (2022)
2. Vial, G.: Understanding digital transformation: a review and a research agenda. J. Strateg. Inf. Syst. **28**(2), 118–144 (2019)
3. Kozak-Holland, M., Procter, C., Kozak-Holland, M., Procter, C.: The challenge of digital transformation. In: Managing Transformation Projects: Tracing Lessons from the Industrial to the Digital Revolution, pp. 1–11 (2020)
4. Gollhardt, T., Halsbenning, S., Hermann, A., Karsakova, A., Becker, J.: Development of a digital transformation maturity model for IT companies. In: 2020 IEEE 22nd Conference on Business Informatics (CBI), vol. 1, pp. 94–103. IEEE (2020, June)
5. Ghosh, B., Pham, L., Teplova, T., Umar, Z.: COVID-19 and the quantile connectedness between energy and metal markets. Energy Econ. **117**, 106420 (2023)
6. Gopal, G., Suter-Crazzolara, C., Toldo, L., Eberhardt, W.: Digital transformation in healthcare–architectures of present and future information technologies. Clin. Chem. Lab. Med. (CCLM). **57**(3), 328–335 (2019)
7. Tortorella, G.L., Fogliatto, F.S., Espôsto, K.F., Mac Cawley, A.F., Vassolo, R., Tlapa, D., Narayanamurthy, G.: Healthcare costs' reduction through the integration of healthcare 4.0 technologies in developing economies. Total Qual. Manag. Bus. Excell. **33**(3–4), 467–487 (2022)
8. Kraus, S., Schiavone, F., Pluzhnikova, A., Invernizzi, A.C.: Digital transformation in healthcare: analyzing the current state-of-research. J. Bus. Res. **123**, 557–567 (2021)
9. Dal Mas, F., Massaro, M., Rippa, P., Secundo, G.: The challenges of digital transformation in healthcare: an interdisciplinary literature review, framework, and future research agenda. Technovation. **123**, 102716 (2023)
10. Mettler, T., Pinto, R.: Evolutionary paths and influencing factors towards digital maturity: an analysis of the status quo in Swiss hospitals. Technol. Forecast. Soc. Chang. **133**, 104–117 (2018)
11. Teichert, R.: Digital transformation maturity: a systematic review of literature. Acta Univ. Agric. Silvic. Mendel. Brun. **67**(6), 1673–1687 (2019)
12. Chanias, S., Hess, T.: How digital are we? Maturity models for the assessment of a company's status in the digital transformation. LMU Munich Manag. Rep. **2**, 1–14 (2016)

13. Menchini, F., Russo, P.T., Slavov, T.N.B., Souza, R.P.: Strategic capabilities for business model digitalization. Revista de Gestão. **29**(1), 2–16 (2022)
14. Wibowo, Y.F.A., Suwawi, D.D.J., Febriyani, W.: Kerangka Kerja Indikator Pengukuran Tingkat Kematangan Transformasi Digital Perguruan Tinggi. JURIKOM (Jurnal Riset Komputer). **9**(5), 1575–1583 (2022)
15. Thordsen, T., Murawski, M., Bick, M.: How to measure digitalization? A critical evaluation of digital maturity models. In: Responsible Design, Implementation and Use of Information and Communication Technology: 19th IFIP WG 6.11 Conference on e-Business, e-Services, and e-Society, I3E 2020, Skukuza, South Africa, April 6–8, 2020, Proceedings, Part I 19, pp. 358–369. Springer International Publishing (2020)
16. Deloitte Digital: The journey to government's digital transformation Report. Retrieved from DUP_1081_Journey-to-govt-digital-future_MASTER.pdf (deloitte.com) (2019). Accessed 06 Mar 2024
17. Gray, J., Rumpe, B.: Models for the digital transformation. Softw. Syst. Model. **16**, 307–308 (2017)
18. Schallmo, D., Williams, C.A., Boardman, L.: Digital transformation of business models—best practice, enablers, and roadmap. Int. J. Innov. Manag. **21**(08), 1740014 (2017)
19. Frankiewicz, B., Chamorro-Premuzic, T.: Digital transformation is about talent, not technology. Harv. Bus. Rev. **6**(3), 1–6 (2020)
20. Morakanyane, R., Grace, A.A., O'reilly, P.: Conceptualizing digital transformation in business organizations: a systematic review of literature. In: BLED Proceedings, pp. 8–444 (2017)
21. Henriette, E., Feki, M., Boughzala, I.: Digital Transformation Challenges, (2016)
22. Reis, J., Melão, N.: Digital transformation: a meta-review and guidelines for future research. Heliyon. **9**, e12834 (2023)
23. Elton, J., O'Riordan, A.: Healthcare Disrupted: Next Generation Business Models and Strategies. John Wiley & Sons, London (2016)
24. Belliger, A., Krieger, D.J.: The digital transformation of healthcare. In: North, K., Maier, R., Haas, O. (eds.) Knowledge Management in Digital Change: New Findings and Practical Cases. Progress in IS, pp. 311–326. Springer, Cham (2018)
25. Haggerty, E.: Healthcare and digital transformation. Netw. Secur. **8**, 7–11 (2017)
26. Secundo, G., Toma, A., Schiuma, G., Passiante, G.: Knowledge transfer in open innovation: a classification framework for healthcare ecosystems. Bus. Process. Manag. J. **25**(1), 144–163 (2019)
27. Ciasullo, M.V., Lim, W.M., Manesh, M.F., Palumbo, R.: The patient as a prosumer of healthcare: insights from a bibliometric-interpretive review. J. Health Organ. Manag. **36**(9), 133–157 (2022)
28. Kane, G.C., Palmer, D., Phillips, A.N.: Achieving digital maturity. MIT Sloan Manag. Rev. **59**(1) (2017)
29. Marx, E.W., Padmanabhan, P.: Healthcare Digital Transformation: How Consumerism, Technology and Pandemic Are Accelerating the Future. CRC Press (2020)
30. Shahiduzzaman, M., Kowalkiewicz, M.: Digital organisation: A value centric model for digital transformation. In: AOM Specialized Conference: Big Data and Managing in a Digital Economy, pp. 1–1 (2018)
31. Aslanova, I.V., Kulichkina, A.I.: Digital maturity: definition and model. In: 2nd International Scientific and Practical Conference "Modern Management Trends and the Digital Economy: from Regional Development to Global Economic Growth", MTDE 2020, pp. 443–449. Atlantis Press (2020, May)
32. Johnston, D.S.: Digital maturity: are we ready to use technology in the NHS? Future Healthc. J. **4**(3), 189 (2017)
33. Berghaus, S., Back, A.: Stages in digital business transformation: Results of an empirical maturity study. In: MCIS 2016 Proceedings, pp. 1–18 (2016)
34. Canetta, L., Barni, A., Montini, E.J.: Development of a digitalization maturity model for the manufacturing sector. In: 2018 IEE International Conference on Engineering, Technology and Innovation (ICE/ITMC), pp. 1–7. IEEE (2018)

35. Liaw, S.T., Godinho, M.A.: Digital health and capability maturity models—a critical thematic review and conceptual synthesis of the literature. J. Am. Med. Inform. Assoc. **30**(2), 393–406 (2023)
36. Gökalp, E., Martinez, V.: Digital transformation maturity assessment: development of the digital transformation capability maturity model. Int. J. Prod. Res. **60**(20), 6282–6302 (2022)
37. Teece, D.J., Pisano, G., Shuen, A.: Dynamic capabilities and strategic management. Strateg. Manag. J. **18**(7), 509–533 (1997)
38. Barney, J.B.: Resource-based theories of competitive advantage: a ten-year retrospective on the resource-based view. J. Manag. **27**(6), 643–650 (2001)
39. Penrose, E.T.: The Theory of the Growth of the Firm. Oxford University Press, Oxford (2009)
40. Patton, M.Q.: Two decades of developments in qualitative inquiry: a personal, experiential perspective. Qual. Soc. Work. **1**(3), 261–283 (2002)
41. Kohlbacher, F.: The use of qualitative content analysis in case study research. In: Forum Qualitative Sozialforschung/Forum: Qualitative Social Research, vol. 7, No. 1, pp. 1–30. Institut fur Klinische Sychologie and Gemeindesychologie (2006)
42. Etikan, I., Musa, S.A., Alkassim, R.S.: Comparison of convenience sampling and purposive sampling. Am. J. Theor. Appl. Stat. **5**(1), 1–4 (2016)
43. Bryman, A., Cramer, D.: Quantitative Data Analysis with IBM SPSS 17, 18 & 19: a Guide for Social Scientists. Routledge, London (2012)
44. Smith, G.T.: On construct validity: issues of method and measurement. Psychol. Assess. **17**(4), 396 (2005)
45. Eggers, W.D.: Delivering on Digital: The Innovators and Technologies That Are Transforming Government. Rosetta Books (2016)
46. Elo, S., Kyngäs, H.: The qualitative content analysis process. J. Adv. Nurs. **62**(1), 107–115 (2008)
47. Krippendorff, K.: Information of interactions in complex systems, 669–680 (2009)
48. Filke, K.: Erfolgsfaktoren und Handlungsnotwendigkeiten zur Sicherung der Nachhaltigkeit einer kommunalen Digitalisierungsstrategie. Doctoral dissertation, Hochschule für öffentliche Verwaltung und Finanzen (2019)
49. Dener, C., Nii-Aponsah, H., Ghunney, L.E., Johns, K.D.: GovTech Maturity Index: the State of Public Sector Digital Transformation. World Bank Publications (2021)
50. Schwertner, K.: Digital transformation of business. Trakia J. Sci. **15**(1), 388–393 (2017)
51. Parasuraman, A.: Technology readiness index (TRI) a multiple-item scale to measure readiness to embrace new technologies. J. Serv. Res. **2**(4), 307–320 (2000)
52. Jafari-Sadeghi, V., Garcia-Perez, A., Candelo, E., Couturier, J.: Exploring the impact of digital transformation on technology entrepreneurship and technological market expansion: the role of technology readiness, exploration and exploitation. J. Bus. Res. **124**, 100–111 (2021)
53. Ciasullo, M.V., Carli, M., Lim, W.M., Palumbo, R.: An open innovation approach to co-produce scientific knowledge: an examination of citizen science in the healthcare ecosystem. Eur. J. Innov. Manag. **25**(6), 365–392 (2022)
54. Lin, J.S.C., Hsieh, P.L.: The influence of technology readiness on satisfaction and behavioral intentions toward self-service technologies. Comput. Hum. Behav. **23**(3), 1597–1615 (2007)
55. Erdoğmuş, N., Esen, M.: An investigation of the effects of technology readiness on technology acceptance in e-HRM. Procedia Soc. Behav. Sci. **24**, 487–495 (2011)
56. Andriole, S.J., Cox, T., Khin, K.M.: The innovator's Imperative: Rapid Technology Adoption for Digital Transformation. CRC Press (2017)
57. Cosimato, S., Di Paola, N., Vona, R.: Digital social innovation: how healthcare ecosystems face Covid-19 challenges. Tech. Anal. Strat. Manag. **36**(8), 1–16 (2022)
58. Grindell, C., Coates, E., Croot, L., O'Cathain, A.: The use of co-production, co-design and co-creation to mobilise knowledge in the management of health conditions: a systematic review. BMC Health Serv. Res. **22**(1), 877 (2022)

59. Davis, F.D.: Perceived usefulness, perceived ease of use, and user acceptance of information technology. MIS Q. **13**(3), 319–340 (1989)
60. Wong, M.C., Wong, E.L., Huang, J., Cheung, A.W., Law, K., Chong, M.K., et al.: Acceptance of the COVID-19 vaccine based on the health belief model: a population-based survey in Hong Kong. Vaccine. **39**(7), 1148–1156 (2021)

Chapter 20
Toward an Adapted Performance Measurement Model for E-Health Startups: Evidence from Startups in Occitanie, South of France

Alex Hamard and Isabelle Guerrero

Abstract Progress in health information technology has created a huge investment in this field and inventors have funded a lot of startups in this field. However, there is no model to measure the strategic performance of health IT startups to consider unique characteristics of e-health market. This study proposed and tested a hybrid model of strategic performance evaluation to examine the factors that result in financial success and consequently survival of e-health startups.

This paper conducts a quantitative research based on the data collected from French Tech of Montpellier Metropole, which contain an annual directory of companies active in the health IT field in France. We test our hypotheses using data of e-health startups during the 9-month period between September 2020 and June 2021.

His research proposes a comprehensive model of strategic performance evaluation tool, which is of considerable value for e-health entrepreneurs because it is adaptable to specific context of e-health, in which prescribers are the final targets and niche calibration is the greatest determinant of success.

We found that system quality signifying the facility and ease of use of e-health products and service plays a significant role in the adoption of prescribers leading to higher performance. Qualification and skills of founders and employees of e-health startups and also efforts and investments on their training determine their capacity to be innovative and develop patents and this will consequently influence performance of the e-health startup.

Keywords E-health · Startups · Strategic performance · Evaluation · Measurement · Balance Scorecard · DeLone and McLean model

A. Hamard (✉)
CNAM Conservatoire Nationale des Artes et Métiers, Paris, France

I. Guerrero
MRM Management Research Montpellier, University of Montpellier, Montpellier, France

F. Schiavone et al. (eds.), *Advanced Perspectives and Trends in Digital Transformation of Firms, Networks, and Society*, Springer Proceedings in Business and Economics, https://doi.org/10.1007/978-3-031-80692-6_20

20.1 Introduction

The expansion of digital tools has created a transformation in healthcare systems [1, 2, 3] and public accessibility to internet and health IT systems has resulted in a great change in the way healthcare is delivered [4]. The popularity of health applications for smartphones and connected objects is the reason for a great amount of investment in this dynamic sector. Everyday new health IT applications and devices are invented and downloaded and quickly find popularity among the users [5]. New connected objects appear in market in the form of health devices or portables which can be used to measure not only different medical parameters such as heartbeats, glucose, blood pressure, body temperature, etc. but also the parameters related to physical activities and well-being, such as number of steps, number of the burned calories, the quality of sleep, etc. [6]. This trend has caused a huge investment in all domains of information technologies related to health including, mobile applications, connected objects and portables, telemedicine and also electronic health records. This has resulted in the creation of a prosperous market for SMEs and startups to commercialize their technological innovations in form of services and products [4, 7].

Despite increasing prosperity of this market with investments in billions of euros, there is no consent among public authorities, industrialists, representatives of patients and health experts about long-term success of enterprises involved in technologies of health [2, 8–10]. The survival rate of startups is low in general and e-health startups are no exceptions. In France, 39% of startups disappear before the fifth year after their foundation. One of the reasons for high rate of failure among these SMEs and startups might be the doubt on their effectiveness. Based on empiric evidences, some studies show that implementation of technological innovations of health has not been successful [9]. Consequently, certain health professionals and hospital administrators have expressed their doubts about massive investment of time money and energy on projects of health technology [5, 7, 11]. The reason behind uncertainty of success could come from the multitude of stakeholders in the market of health technology with different and often contradictory requirements [2] and the satisfaction of all different needs is very challenging. Due to these complications, the rate of success in this market is quite low and new startups and SMEs show up rapidly in the market and provide the patients with lots of applications, connected objects and services; whereas, with the same rapidity, they fade away and their products and services are abandoned [12–14].

20.1.1 Context

To understand the dynamics of market, it is important to develop a comprehensive and accurate assessment tool for technological innovations of health. However, the creation of this evaluation tool is very challenging for several reasons. Firstly, the speed of change, evolution, and transformation in health technology are very considerable. SMEs and startups quickly create and launch in the market many projects, products, and interventions of health technology [12], which is too fast for researchers to evaluate their potential risks and dangers. Thus, a good measurement tool should be able to catch up with rapid evolution of technological innovations in health. Secondly, a good evaluation tool should take all stakeholders into consideration, namely patients, purveyors, and providers. Patients as the final users of the products and services are direct recipients of medical care, along with their family members or caretakers who serve as a proxy [2]. Providers are another part of stakeholders including hospitals, physicians, and others who assist in direct patient care such as nurses and diagnostic technicians. Purveyors, defined as custodians or keepers of health data [8, 12], constitute the third party of stakeholders. Health IT startups and SMEs are considered to be purveyors because they deal with patient data and make a connection with health professionals (providers). Projects, products, and services of health technology must support the goals of patients (primarily stakeholders) such as improved quality of patient care outcomes, increased coverage of healthcare services, and reduced cost of healthcare [15]. The way to achieve these objectives is a subject of debate and disagreement because each stakeholder has unique concerns and goals that are often in conflict with others' goals. For example, the efforts to increase healthcare quality would increase the costs, which is favorable to some parties of stakeholders but disturbing for others [2].

This context shows the importance of developing a comprehensive evaluation tool to assess the performance of enterprises which supply and launch projects, objects, and services of health technology. If these projects and services are poorly implemented and the evaluation is not accurate, not only huge sums of investments will be wasted [12] but also public health could be jeopardized [2, 16, 17]. Therefore, regarding the critical status of medical issues, there is an urgent need to conduct reliable evaluations with appropriate criteria.

20.1.2 Research Question

So many strategic performance evaluations have been done in healthcare sector and the review of literature reveals so many applications of this model for evaluating the performance of hospitals and public and private healthcare centers [18–21]; however, to the best of our knowledge, there is no application of this model to measure the performance of health IT microenterprises and startups. Furthermore, prior research includes several evaluations in health IT entrepreneurship, but none of the studies in existing literature have provided a comprehensive model of performance measurement to take into account all stakeholders involved; therefore, this study intends to bridge this gap by answering the following research question:

How strategic performance of health IT startups and microenterprises could be measured in a niche market where final users (patients) follow the prescription of healthcare providers? How a conceptual model could be developed to evaluate the performance of health IT startups to take into account the characteristics of health IT innovation?

The objective of this article is to develop a conceptual model that could answer this question. In a review of literature, we are going to investigate other models applied in healthcare context and make a deep reflection on their validity and adaptability to health IT startups and finally propose a pertinent conceptual model.

In order to develop a comprehensive tool for the performance measurement three important elements should be taken into consideration: firstly, the proposed model should be adaptable to the context of technological innovations of health with all its characteristics. Performance models developed for other fields are not necessarily adaptable to health IT. Secondly, different types of enterprises involved in the e-health should be taken into consideration. A performance measurement model that works for a big enterprise is not necessarily suitable for SMEs. Finally, the interpretation of performance should be well defined. In other words, it should be clear exactly which aspect of performance is taken into consideration and the proposed model is designed to measure what aspect of performance.

In the rest of this study, we will investigate the existing literature on performance measurement tools in the field of e-health and argue whether they are adaptable to e-health startups or microenterprises. Afterwards, we present our proposed model of performance measurement in the methodology part and define different types of enterprises based on French standards to clarify our sample enterprises for data collection. We will test our model based on the collected data from French e-health startups and the last part will be discussion about results and conclusion.

20.2 Conceptual Background and Review of Literature

The performance measurement model to be developed to answer our research question should be able to cope with the complexity of e-health startups in terms of stakeholders and the specific requirements of their context. Such a performance measurement model should make a balance between financial and non-financial aspects to follow the performance track e-health microenterprises in the past and provide antecedents that could predict their future performance. That performance measurement model should be compatible with technological complexities of e-health products and services provided by e-health enterprises.

Balanced scorecard (BSC) is a conceptual model measuring performance in a multi-dimensional way. It has received widespread attention and recognition as an important performance measurement system and has been empirically tested in several domains [22]. In the next part, we investigate this conceptual measurement model and then by analyzing its theoretical underpinnings, we will show the relationship with our proposed model design.

20.2.1 Balanced Scorecard

The performance model which has been applied the most in the field e-health is balanced scorecard (BSC) proposed by Kaplan and Norton [23]. Kaplan and Norton developed a comprehensive research model in the 1990s, which inherently makes use of strategic and operational considerations [18, 23, 24]. They argued that the previous models were majorly based on financial accounting measures such as return on investment and that caused an incomplete measurement of a business performance [18, 20, 25, 26]. As shown in Fig. 20.1, they developed a multi-dimensional framework for measuring and managing organizational performance with comprehensive criteria to assess customer, internal process, learning, and innovation [23, 25]. In this line, BSC complements financial measures with operational measures on three dimensions: customer satisfaction, internal processes, and organization's innovation and learning [13, 27]. Therefore, BSC is considered a management system rather than a simple measurement system because it provides strategic guidance toward strategy implementation [21, 25].

Shortly after its release, BSC became a widespread planning and control system tool that helped managers to translate strategy into an integrated set of financial and non-financial measures [25]. There are two great advantages in BSC which made it interesting and commonly used. Firstly, different competitive elements of companies, like customer focus, response time, product quality, work process, and new product cycle time, are all collected in a single management report [20]. Secondly, BSC avoids under-estimation. This management tool attracts the attention of executives to all important performance measures together at the same time, making it possible to see whether the improvement in one area is achieved at the expense of another [19].

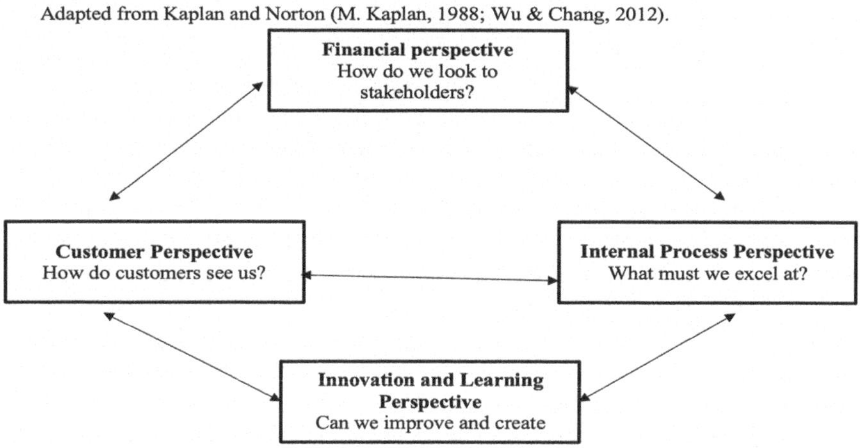

Fig. 20.1 Balanced scorecard

Thanks to these advantages, BSC has been largely adapted in other sectors such as service industries, banking [28], hotel, and also in various business activities, like customer relationship management [29] and supply chain management [29, 30]. Gradually, BSC found its place in healthcare domain and, as Aidemark 2003 puts it, became "meal for today" because consultants more and more recommended this "miraculous treatment" [18, 31]. BSC is now adopted by a broad range of health care organizations, including national health care organizations. [19, 21]. The health organizations that applied BSC very frequently were the hospitals because it could be well adapted to their sophisticated and comprehensive management information systems to achieve their strategic goals [32, 33] and could help them measure and guide their performance [24, 34–36].

In spite of all these applications, there are complications to adapt BSC to healthcare field in general. BSC is a conceptual tool [37], and its four perspectives are not always adaptable to all fields [23]. In particular, due to unique characteristics of healthcare industry, BSC is difficult to get adapted to this field [19]. For example, physicians, as major stakeholders, regard for themselves an absolute authority and superiority in the whole healthcare process [9, 21]. The patients' access to health IT devices and services has empowered them to learn more about their health conditions, take more responsibilities to manage their health, communicate more effectively and efficiently between visits, and ultimately be an actor in the healthcare process [2, 10]. However, physicians tend to stick to traditional models which regard patients as subordinate and passive which makes it difficult to substitute the patient with customer in BSC [19].

In addition to all these problems of adaptability, there are more concerns about our own fieldwork of e-health startups and microenterprises. In this study is we work on "health IT" or "e-health" (not healthcare alone) with its own specifications which are not totally compatible with BSC. Furthermore, we focus on e-health startups and microenterprises while BSC is only applied to big companies and organization of healthcare. There is no evidence of BSC application to e-health startups and applications which could be the sign of BSC incompatibility to our fieldwork. Here we make a reflection on some of the reasons for incompatibility.

Firstly, since its creation, BSC has been mostly applied to big enterprises with lots of ramifications, diverse axes of activity, different departments, a large number of employees and a huge market [38, 39]. As the concept of BSC is more suitable for big enterprises, there are less applications of BSC to SMEs and startups with few employees, lower amounts of turnover, and a limited market [40]. Kaplan and Norton in a later edition of their article suggest that their new framework is applicable to SMEs; however, they do not specifically give a methodology for such an implementation [23]. In fact, Kaplan and Norton have developed implementation methodologies, which are suitable for big and multi-national companies, with no limitation of resource, time, and expertise.

Secondly, in e-health market, final users (patients) are not the ones who decide on the choice of adoption. Although patients are users of health IT products and services, they are not the final target of health IT startups [41, 42]. In different research cases conducted in developing countries the prescribers who prescribe health IT services and products to patients are identified as physicians, regional and local health authorities, public and private hospitals clinics and ambulatories [43]. Kohli et al., referring to different stakeholders involved in health IT, define prescribers are "hospitals, physicians, and others who assist in direct patient care such as nurses and diagnostic technicians" [2].

The final targets of health IT startups are the prescribers of health IT products and services like physicians who ask patients to adopt them [40]. The challenge of applying BSC to health IT entrepreneurship is that, although patients are users of health IT innovations, they are not the ones who decide to adopt these products. Therefore, patients do not independently choose to adopt their health IT connected object or services prescribed by physicians, patient associations, social security interface who act as a mediator between startups and patients [43].

Thirdly, e-health startups have a "niche market" to satisfy prescribers' demands in the context of a healthcare centers like hospitals [2, 40, 44]. In other words, they supply their services and products to health professionals "prescribers" specialized in certain medical fields. As Beaulieu et al. explain in their analysis of health technology startups (HTS), these firms "emphasize their value proposition as novel or performance products while targeting niche markets" [40]. In other words, health IT startups and SMEs are satisfying a distinct set of needs of prescribers "who will pay a premium to the firm that best satisfies their needs, where the niche is not likely to attract other competitors, where the niche firms gain certain economies through specialization" [44]. Stanton et al. [45] define niche as "customer needs (that) are met through tailoring goods and services for small markets" (italics mine) [46].

BSC has been applied to vast markets with different segmentations and it seems to be challenging to apply it to health IT context of niche market in which demand is very specific and corresponds to the needs of a prescriber (physician, patient associations, social security interface, etc.) [2]. E-health lacks variety of segmentations and prescribers look for connected objects, applications, and any other services and products, which can facilitate their care process [47]. As demand is limited and specific, the startups should provide their health IT services and products for a niche market.

Last but not the least, one of the distinctive features of niche markets is "tailor-made mode of production" [44, 46, 48]. Since niche market of prescribers is very specialized and limited, health IT startups use all their expertise and "skill to tailor their resources to meet precisely the unique needs of a prescriber" [44, 48]. To show the mode of production of health IT startups, Beaulieu and Lehoux [40] brought some examples of startups specializing in drug preparation and administration, endoscopic device, medical robotics, telemedicine, surgical tools, imaging device, and medical software. In each field, startups exclusively supplied the demands of a limited number of prescribers [40]. Therefore, tailor-made production is one of the causal characteristics of a niche market structure. Health IT has a customized production context in which startups respond to the needs of prescribers on a one-to-one basis [48]. The startups satisfy specific demands of the prescribers and the context of mass production which is the case of big enterprises is not relevant in health IT market. BSC has been applied to mass production contexts while health IT has a context of tailor-made production in which the number of produced items is limited and corresponds to the narrow demand of the prescribers ([49, 50]). Although BSC is not totally compatible with the context of e-health startups, the theoretical underpinnings of BSC could be used to develop an adaptable performance measurement model.

The theoretical underpinning of BSC is based on this hypothesis that the performance measurement is a multi-dimensional system in which there should be a balance between financial and non-financial performance indicators [22, 23]. The central tenet and underpinning of BSC is based upon four distinct perspectives of financial, customer, internal business process and learning and growth [21, 25, 30]. Therefore, all performance indicators which assess the same perspectives are theoretically grouped into four categories, and that is equal in number to the BSC's four distinct perspectives [22]. That is the reason why, all performance indicators should converge with the performance indicators measuring the same perspective and should discriminate from generic performance indicators measuring the other perspectives [20, 22]. This supposition can be articulated as the theoretical underpinnings of the BSC and we are going to mobilize them for development of a comprehensive diagnostic structure and performance model compatible to highly technology-oriented e-health market, which is the identified gap in the existing literature.

The performance model based on theoretical underpinnings of BSC will not become compatible to e-health context unless it integrates elements of information systems to explain technological aspects of e-health services and products. In the next part, we will argue that, for developing a performance measurement model, the nature of e-health products and services as "tools of information systems" should be taken into consideration and a modern adaptation of IS Success Model of DeLone and McLean in e-health field can be the best response to that concern.

20.2.2 DeLone and McLean IS Success Model

DeLone and McLean have developed IS success model after speculation on so many independent variables, which result in different forms of system success. The model is based on the communication theory, which asserts that accuracy and effectiveness are the most important measures of information system success. In other words, information flows through a series of stages from its production through its use or consumption to its influence on individual and/or organizational performance [51]. Six major dimensions of IS success, shown in Fig. 20.2, are (1) system quality, (2) information quality, (3) use, (4) user satisfaction, (5) individual impact, and (6) organizational impact [51]. Definitions of the constructs are described as follows:

System quality refers to measures of the information processing system itself. System quality is the desired characteristics of the IS itself, which include convenience of access, flexibility of system, integration of systems, response time,

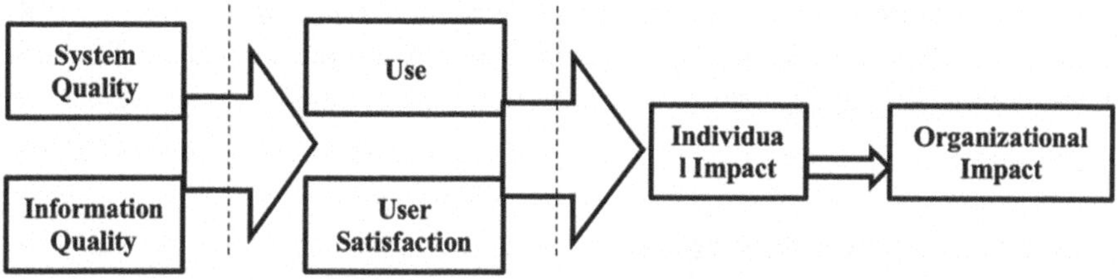

Fig. 20.2 DeLone & McLean IS Success Model

realization of user expectations, reliability, ease of use, ease of learning, usefulness of a technological object.

Information quality refers to the quality of the information in terms of accuracy, precision, currency, reliability, completeness, conciseness, relevance, understandability, meaningfulness, timeliness, comparability, and format.

Use of IS refers to recipient consumption rate of a technological object.

User satisfaction refers to the recipient response to the use of a technological object measured by degree of satisfaction.

Individual impact refers to the effect of information on the behavior of the recipient measured by executive efficiency, higher-quality decisions, communication, and operational control.

Organizational impact refers to the effect of IT on organizational performance measured by profitability, productivity, and financial improvement.

As shown in the model, system quality and information quality, as the elements of IS effectiveness, are antecedents of IS success which consequently influence the use and user satisfaction and finally result in individual and organizational impact. A deep reflection on IS success model reveals that all constructs of BSC are reflected in DeLone & McLean success model except "information quality" and "system quality." Since information technology is the core of all health IT services and products supplied by the startups of the present research, we argue that a measurement tool which ignores this important construct cannot be a complete one; therefore, information quality and system quality constructs should be integrated in our performance measurement model. Since the purpose of this study is to provide a comprehensive model to measure performance of health IT startups and microenterprises, before proposing our model, we should first define performance according to management literature.

Performance Definition

Taking it from management point of view, performance is perceived as valued contribution to reach the goals of a company, enterprise, or organization [24]. Spangenberg elaborates on this point of view and defines performance management as a sequence of activities for (i) planning the value creation, (ii) taking action to control value creation, (iii) measurement of value contribution, and finally (iv) rewarding the value contribution. Then, the task of positioning performance in health IT context will be very challenging. Making a deep reflection in the context of performance management literature, we can adopt the three E's—economy, efficiency, and effectiveness—to position performance in the context of health IT entrepreneurship [24, 52–54].

20.3 Conceptual Model

Since the context of this study is e-health entrepreneurship and e-health startups supply a tool of communication, our model should include certain constructs to evaluate that communication tool and the most adapted constructs relate to Information System literature. Among different concepts and models of information systems, two constructs of information quality and system quality in DeLone & McLean IS success model can measure the effectiveness of a communication tool [51]. We argue that for internal process perspective of our model, two constructs should be integrated into our context-adapted performance measurement model (two constructs of information quality and system quality from a recent adaptation of IS success model in e-health context). We integrate these two constructs from the adaption of IS Success Model by [55] in which they consider four antecedents for information quality and five antecedents for service quality [55]. As it is shown in Fig. 20.3, completeness, accuracy, currency, and transparency are the antecedents of information quality and navigation, delay, accessibility, presentation, and readability are the antecedents of system quality [55]. In learning and innovation perspective, constructs of human capital, patent, and alliance will be measured. Since, in the

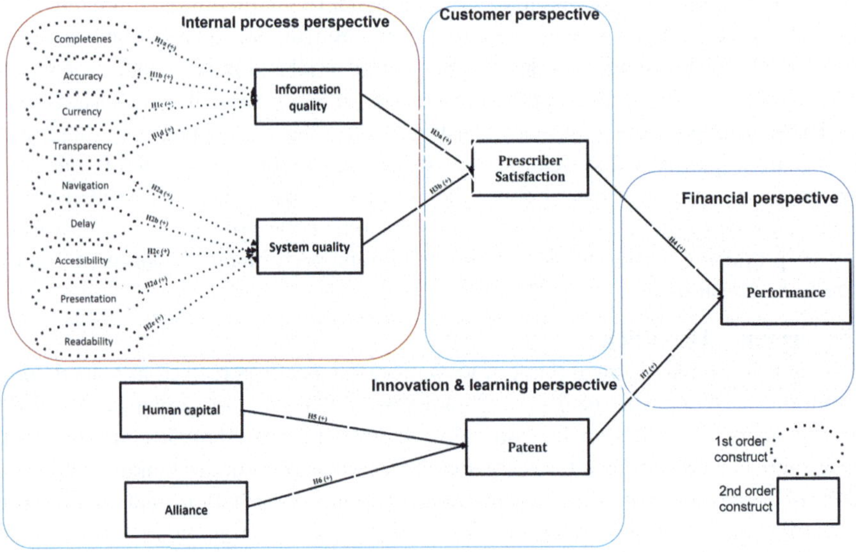

Fig. 20.3 Proposed model

context of e-health, the final user (patients) are not the ones who decide about the adoption of products and services, and health professionals play the role of prescribers, in customer perspective, we rename customer satisfaction as prescriber satisfaction. Our proposed research model is shown in Fig. 20.3.

Internal process perspective, an important underpinning of BSC, is measured by two constructs of information quality and system quality from a recent adaptation of IS success model in e-health context. Two important constructs of IS success model of DeLone & McLean which have been made compatible to e-health context by [55] are information quality and system quality [51].

Information quality construct measures the quality of information circulation in e-health products and services which connect, synchronize, and orchestrate healthcare members together in healthcare process [56]. The objective of this concept is to measure communication quality among healthcare team members and the indicator that we are going to apply for this are communication failures among healthcare team members through health IT product [57]. In recent applications of IS success model in the field of e-health, four antecedents of accuracy, currency, completeness, and transparency have been identified for this construct to measure the quality of information disseminated through e-health services and products [55].

System quality is a central construct of this model because the final product supplied by e-health startups is a communication tool and system quality measures operation flow and smoothness of this final product [57]. By this construct, we intend to measure the system used in communication tool supplied by startups in terms of its effectiveness to reach its fundamental objective of connecting healthcare members together. Five antecedents of accessibility, delay, navigability, presentation, and readability have been identified for this construct to measure the smoothness and effectiveness of the system running e-health services and products [55].

Prescriber satisfaction is an important underpinning of BSC named as customer perspective [23]. The word customer is renamed as prescriber because e-health startups work in a market of prescription. Although patients are the receivers of communication tools supplied by e-health startups, they are not final consumers because they do not choose their e-health connected object or services independently [41]. Relevant e-health products are prescribed by physicians, patient associations, social security interface who act as a mediator between startups and patients. Prescribers are "hospitals, physicians, and others who assist in direct patient care such as nurses and diagnostic technicians" [2]. Prescriber satisfaction construct measures the satisfaction of healthcare providers (prescribers) toward e-health products and services that they have prescribed in terms of improvements and facilitations in healthcare process.

Performance construct, an underpinning of BSC and totally adaptable to e-health context [23], reflects the degree to which an e-health startup is able to generate revenues to satisfy its investors' expectations, survive in market, and finance its future investments. The context of this study is e-health entrepreneurship and the most vital source for the creation and existence of startups is the investment [39, 58, 59]. Therefore, it is critically important for startups to attract investors as inevitable stakeholders who play a central role in their creation. Startups financial outcome can ideally satisfy and ensure investors to continue their support. This construct is very important for investors sponsoring e-health startups such as Business Angels, the funds of capital risk, banks, venture capital mutual funds (FCPR).

Innovation and learning perspective, another important underpinning of BSC ([23]), measures efforts and investments made by e-health startups to innovate a new service or product which could interest potential prescribers [19, 21, 30]. In our proposed model, we have identified three constructs in this perspective: human capital, alliances, and patents. These constructs measure startups efforts on innovation in terms of training its employees to develop new patents and e-health services or products [60, 61].

Human capital construct measures the competence of health IT startup's entire workforce [62]. It deals with education level of startup employees in terms of school and university certificates and their expertise in their startup. It also refers to startup employee training in order to make themselves fit for launching new e-health products and services [62]. This training prepares employees and CEOs to develop, improve, or launch a new e-health product or service [60, 61].

Alliance construct deals with different partnership efforts made by e-health startups. There are two types of alliances. The first type of alliances is made by universities and organizations of scientific and technological expertise. [61] believes that this type of alliance can inspire startups to develop innovative e-health products and services thanks to the knowledge and experience of experts of the field [61]. The second type of alliances, made by pharmaceutical, chemical, and marketing firms, enables the startups expand their network and use the facilities of big enterprises to reach their objectives [40, 60].

Patent construct measures the efforts of an e-health startup in development, registration, and even commercialization of patents. This construct also verifies if patent development can decrease entry barriers and increase legitimacy of an e-health startup facing different public and private stakeholders [63]. As legitimization plays an important role in the success of startups, this construct measures to what extent patent contributes to the legitimization of a startup or microenterprise and signifies the success of technological innovation [64, 65]. In general, this construct measures if there is a meaningful association between patent development and innovation performance of e-health startups and microenterprises.

20.3.1 Hypotheses

After detailed conceptualization of our model constructs, we can now explain the hypothesized relationships. The first part of the proposed model deals with relationships between internal process perspective (represented by information and system quality constructs), customer satisfaction perspective (represented by prescriber satisfaction), and performance and four major hypotheses are made in this part. Our proposed model has been inspired by a recent adaptation of IS success model by

[55] in the field of e-health suggesting that the characteristics of a product or service of e-health determine system and information quality. This recent adaptation of DeLone & McLean IS Success model identifies antecedents that determine system and information quality [66]. These antecedents are compatible with the context of e-health products and services, and in order to make our model comprehensive enough, we include them as first-order constructs to show the components that influence system quality and information quality.

The antecedents of information quality are accuracy, currency, completeness, and transparency. Accuracy refers to the extent to which the information produced in e-health services and products is correct or based on evidence. Currency refers to the extent to which the information produced in e-health services and products is up to date. Completeness refers to the extent to which all necessary information is presented in e-health services and products. Transparency represents the extent to which critical supporting information is provided is produced in e-health services and products. Previous research has shown that accuracy, currency, and completeness are positively related to information quality [55, 66]. Prior research has also found that online users are concerned about transparency. Transparency sheds light on the source of the information by ensuring health professionals that everything is clear behind the information and it determines the information quality.

H1a: Accuracy is positively associated with information quality.
H1b: Currency is positively associated with information quality.
H1c: Completeness is positively associated with information quality.
H1d: Transparency is positively associated with information quality.

The antecedents of system quality include accessibility, delay, navigation, presentation, and readability. Accessibility refers to the extent to which one can access or extract information from e-health services and products [66]. Delay refers to the lag between a user request of certain health information and the system's response to the request. Navigability refers to the system's capability of allowing the user to navigate among different parts of e-health services and products. Presentation refers to the aesthetic format in which information produced in e-health services and products is presented [67]. Readability refers to the extent to which information produced in e-health services and products is visible and understandable for both patients and health professionals. Research has shown that due to some infrastructure problems, there might be some delays for the transfer of information through e-health services and products; therefore, delay is also one of the determinants of system quality. Research has found navigability in e-health services and products to be an important determinant for system quality. There are also two important features in e-health services and products that determine the system quality: presentation and readability. Appropriately presented information can create a pleasant feeling and attract patient attention, while bad presentation can easily destroy their interest. In addition, information produced in e-health services and products must be readable to all patients because there are no physicians beside the patients to explain medical jargons.

H2a: Accessibility is positively associated with system quality.
H2b: Delay is negatively associated with system quality.
H2c: Navigability is positively associated with system quality.
H2d: Presentation is positively associated with system quality.
H2e: Readability is positively associated with system quality.

In e-health literature, there is a considerable emphasis on the quality of information regarding health IT products and services, and the risks of ignoring the quality of information have been warned [10]. Several experts in different healthcare fields have expressed concern if information transferred in e-health devices and services has not the desired standards of exactitude and accuracy, it could endanger final users and patients [68].

At the same time, standard information quality could help contribute to the healthcare process as well as communication among healthcare professionals and patients. The flow of standard information on e-health devices and services could also increase patients' medical knowledge and empower them to perform disease self-management [69]. Therefore, the quality of e-health information can play an important role in determining the satisfaction of the users [3]. In e-health context, the final users (patients) are not the ones who decide about the adoption of e-health devices and services. They are following the prescriptions of healthcare professionals; that is why, we rename customer satisfaction construct in BSC as "prescriber satisfaction." If the information transferred through e-health services and products is exact accurate and correct, health professionals like physicians and nurses can have better results in their treatments and will be satisfied about the e-health services and products that they prescribe making a positive relationship between information quality and prescriber satisfaction.

Another aspect of all communication tools in e-health services and products influences prescriber satisfaction. DeLone and McLean argue that system quality in an IT tools plays a central role and therefore it could cause the satisfaction of users [57]. E-health devices and services with high system quality could make health information more available, accessible, understandable, and enjoyable and, thus, encourage patients to get involved in follow-up and healthcare process [55, 70]. Prior research based on Delone & McLean IS success model has shown positive relationship between system quality and user satisfaction; however, in this research, final user (patient) does not make a decision on the adoption of e-health products and services so we will investigate the impact of system quality on prescriber satisfaction. High system quality creates an impression of satisfaction for prescribers and if e-health service or product has a high quality, it creates a motivation in patients and final users to get involved in treatment leading to better results in healthcare process. That is why we expect a positive relationship between system quality and perceived benefit of e-health prescribers. In other words, if e-health products and services have high quality, health professionals like physicians and nurses can have better results in their treatments and will be satisfied about the e-health services and products that they prescribe creating a positive relationship between system quality and prescriber satisfaction.

H3a: Information quality is positively associated with prescriber satisfaction.
H3b: System quality is positively associated with prescriber satisfaction.

The impact of prescriber satisfaction on performance should be reflected because they play an important role in survival and growth of health IT startups and, according to [59], startups are fragile structure that depends on financial outcomes and funding of investors to a great extent [59]. Some studies prove that prescriber satisfaction has a direct influence on financial aspects of performance [71–75] and others show its strong influence on other aspects of performance [24, 76]. In line with this extensive body of prior research on relationships among these axes, we can hypothesize that prescriber positive attitudes contribute positively to performance of e-health startups. Therefore, we hypothesize that:

Hypothesis 4: Prescriber satisfaction is positively associated with performance.

The second part of the proposed model deals with the relationships between innovation and learning perspective and performance with three major hypotheses. Patents as important indicators of innovation resources at the firm start or during the firm's first 3 years [63] play a central role in showing the relationships between innovation and learning perspective and performance.

A proposition that has been supported frequently by various researchers is that Human Capital has a considerable influence on patent development of a company [77, 78]. Bruns et al. believe that the competence of workforce, to a great extent, determines the output of information quality in enterprises that are based on technological innovations. Afterwards, the same findings on the great influence of human capital on patent development are reflected in other research contexts [79, 80, 81]. This relationship is explained by the impact of qualifications of firm employees and their sense of innovation on the ability of a firm to develop patents that give the firm a competitive advantage [82]. Some indicators like the number of hour spent by employees on training, the amount of money spent by the firm on training of the employees, and the level of education and expertise of employees have been chosen to show the influence of human capital on patent development [83, 84]. Based on these findings, we hypothesize that human capital is positively associated with patent development:

Hypothesis 5: Human capital is positively associated with patent development.

The importance of alliances of startups and microenterprises on patent development is evident in the existent literature. A large number of studies have shown the effect of different types of alliances, whether alliances with universities and research centers or alliances with big multinational companies on patent development [56, 85–87]). The existing literature shows that collaboration of firm and university scientists increases innovative capacity of startups and microenterprises. Shane believes that today universities are concentrating on a type of research that can bring commercial benefits through licensing. The universities are not specialized to sell their research while startups and microenterprises have good capacity and knowledge to sell innovations. That is why, prior partnership with such universities and research centers could have a relationship with the innovative capacity of startups and microenterprises in patent development. Therefore, we hypothesize that:

Hypothesis 6: Alliances and partnerships are positively associated with higher innovative activities and patent development in e-health startups and microenterprises.

There are many studies in the existing literature that show how a company's performance is influenced by patents. [63] believes that patent development can lower entry barriers and increase competition capacity in the downstream product market [63]. Therefore, patent development can clearly affect enterprise innovation performance. However, whether or not patent development influences innovation performance directly or indirectly depends upon the quality of patents that are developed by enterprises [65]. Patent also contributes to the legitimization of a startup or microenterprise, patent application is one of the symbols of success of technological innovation [64]. Acquiring and maintaining patents, especially high quality patents to protect important technology, can offer competitive advantage to an enterprise and help the acquiring enterprise to defeat direct competitors in the market [65]. Therefore, we hypothesize the positive association of patents and performance

Hypothesis 7: Patents development is positively associated with performance.

20.4 Methodology

20.4.1 Research Context

Prior to this study, an observatory has been created to track more than 200 e-health enterprises in Occitanie region in South of France. This observatory will serve as the fieldwork of this study and our hypotheses are going to be tested on this specific context; however, we choose to study only e-health startups and microenterprises for several reasons. Firstly, startups have a great capacity to be created quickly and use the capabilities of technological innovation to satisfy the needs of a growing market [39, 40]. Startups, thanks to their structure, have a great potential to challenge the traditional healthcare service industry by introducing radical and sustainable innovations of technology [39, 88, 89]. Secondly, by focusing on startups and microenterprises, our data collection sample will be homogeneous to obtain generalizable results and thereby increase the external validity of this research.

E-health startups and microenterprises of our fieldwork are involved in launching different technological innovations in form of e-health products and services such as mobile applications, connected objects, clouds for health data, robots, and online services like telemedicine. The users of the e-health startups and microenterprises are classified as all public (looking for improvement of health status and health indicators) general patients, patients of one specific disease, healthcare professionals, pharmacists, physicians, pharmaceutical laboratories, and hospitals. These e-health startups and microenterprises have different investors at national and regional levels such as Business Angels, the funds of capital risk, banks, venture capital mutual funds (FCPR), etc.

Since the context of this research is Occitanie region in South of France, we refer to the definitions and categorization of enterprises provided by the INSEE (National Institute of Statistics and Economic Studies), the national statistics bureau of France which collects and publishes information about the French economy and carries out the periodic national census.

20.4.2 Enterprise Typology (Microenterprises, SMEs, Intermediate-Sized & Large Enterprises)

According to INSEE, microenterprises or startups are companies employing fewer than ten employees with turnover of less than 2 million euros, while small and medium-sized enterprises (SMEs) have fewer than 250 employees and an annual turnover of less than 50 million euros. Intermediate-sized enterprises are defined as a company that has between 250 and 4999 employees, and a turnover not exceeding 1.5 billion euros. INSEE regards two characteristics in the definition of large enterprises: (1) at least 5000 employees (2) more than 1.5 billion euros in turnover.

The fieldwork of this study only concerns the first category of microenterprises (startups), which supply technological solutions, products, and services in health IT field. Health IT can be defined as an intersection of health services, information technologies, and business in which all the stakeholders and other authorized parties can manage and share their health information interactively and contribute to healthcare and well-being of final users.

Questionnaires are distributed to determine the best format to reduce bias and optimize the scale dynamics; in the last step, after data collection and exploratory factor analysis (EFA), we created preliminary estimates of reliability and validity; and finally, we tested convergent and discriminant validity using confirmatory factor analysis (CFA). After data collection, we tested our hypotheses.

20.4.3 Measurements Development

All the variables and constructs were chosen from the previous studies and the existent literature and were adapted to the context of the current study. Since there were two groups of respondents (founders and employees from e-health startups and healthcare professionals and practitioners who prescribed e-health products and services) two different questionnaires were designed and, in each questionnaire, certain constructs of our proposed model were measured.

The questionnaire, designed for healthcare professionals and practitioners who prescribe e-health products and services, measured three constructs: Information quality, System quality, and Prescriber satisfaction.

Information quality was measured with a seven-scale construct taken from established scale items of [55] on a Likert scale from 1 = strongly agree to 7 = strongly disagree [55]. In the literature of information systems, the construct of information quality has four antecedents. The first three constructs of completeness, accuracy, and currency were taken from established scaled items of [66] which were developed specifically in the context of healthcare [66]. The fourth antecedent of information quality which is transparency was taken from the scale items of Eysenbach et al. on a Likert scale from 1 = strongly agree to 7 = strongly disagree.

System quality was measured with a seven-scale construct taken from established scale items of [55] on a Likert scale from 1 = strongly agree to 7 = strongly disagree. In the literature of information systems, the construct of information quality has five antecedents. The first and fourth antecedents of system quality which are navigation and presentation, respectively, were taken from the scale items of [90] on a Likert scale from 1 = strongly agree to 7 = strongly disagree [90]. The second and third antecedents of system quality which are delay and accessibility were taken from established scale items of [66], which were developed in a research focused on e-health. [66]. The fifth antecedent of system quality which is readability was taken from the scale items of Eysenbach et al. on a Likert scale from 1 = strongly agree to 7 = strongly disagree.

In order to make the constructs adaptable to the context of this research, we renamed customer satisfaction as prescriber satisfaction because in e-health context, patients or final users are not the ones who decide about the adoption of an e-health products and services. Prescribers as health professionals prescribe them to the final users and the prescriber satisfaction construct was taken from [66] which were developed specifically in the context of healthcare [66].

The questionnaire for e-health startup founders and employees was designed to measure four constructs of Human capital, Alliance, Patent, and Performance.

Human capital was measured with a seven-scale construct taken from established scale items of [91] on a Likert scale from 1 = strongly agree to 7 = strongly disagree [91]. This construct deals with the qualification of e-health startup founders and employees and their competence to handle different situations and acquire knowledge and expertise in a continuous manner in order to succeed in innovation.

Alliance was measured with a seven-scale construct taken from established scale items of Landry et al. on a Likert scale from 1 = strongly agree to 7 = strongly disagree which measures the importance of different types of alliances that can contribute to the innovativeness of e-health startups and enterprises. The alliances measured in this construct could vary from big multinational pharmaceutical enterprises to health organizations like hospitals to universities and research centers.

Patent plays a central role in innovation and learning perspective of e-health startups and microenterprises and it was measured with a seven-scale construct taken from established scale items of [65] on a Likert scale from 1 = strongly agree to 7 = strongly disagree which measures the importance of developing or acquiring patents for an e-health startup and also the impact of developing or acquiring (purchasing) patents on their innovative performance [65].

Performance, one of the key perspectives of BSC, included four items identified by Ravichandran and Lertwongsatien and the respondents were supposed to rate the business performance of their e-health startup in terms of productivity, profit rate, return on investment (ROI), and sales revenues.

20.4.4 Sample and Data Collection

The data collection was made from e-health startups and microenterprises enrolled in BIC (Business and Innovation Center) incubator. The data about these enterprises were obtained directly from one branch of BIC called "Montpellier Capital Santé," which is responsible for enterprises in health field. The data were obtained from BIC (Business & Innovation Center) contained an annual directory of companies active in the health IT and a wide range of firm characteristics such as activities, personnel, products, and alliance partners.

Based on the data provided by the BIC, the data collection was made through online questionnaires distributed to startups from November 1 2020 and March 30, 2021. The data collection had two stages: the first stage concerned the founders or employees of e-health startups and microenterprises. This part concerned testing certain constructs of the proposed model such as human capital, alliance patent, and performance. The second stage concerned healthcare professionals in clinics hospitals, rehabilitation or sport centers that prescribed the services and products of e-health startups and microenterprises to patients and final users. This part concerned testing other constructs such as information quality, systems quality, and prescriber satisfaction.

The lead researcher distributed questionnaires directly to the respondents through Qualtrics software. The total of 283 startups out of 318 (including those located in Occitanie region and those located in other French regions but providing healthcare services and products to Occitanie healthcare centers and professionals) enrolled in BIC incubator participated in the survey. 276 respondents finished the questionnaire and out of this number, 11 responses were either incomplete or unengaged. As shown in Table 20.1, we retained 265 responses for further analysis, which was equal to 87 percent of response rate.

Table 20.1 Response rate

Survey participation	Number of responses
Total respondents	305
Refused to participate	22
Survey participated	283
Questionnaires finished	276
Questionnaires incomplete responses	11
Total questionnaires retained	265
Response rate	87%

20.4.5 Data Screening

It is very important to do data screening because it is the only way to make sure that the collected data are accurate, reliable, and correct enough to be able to validate the variable and show relationships among constructs of a theory. We searched in the collected data to find outliers, missing data and unengaged responses and at the end, we found 11 cases with considerable missing data of more than 0.2% and 7 cases of unengaged responses. These 18 cases were removed, and we finally retained 265 cases out of 283. We did not find any outliers and we observed a normal distribution with skewness and kurtosis values between +3 and −3. After these observations, we made sure about the quality of the data to test our proposed model and continued the data analysis.

20.5 Results

20.5.1 Data Analysis and Measurement Evaluation

In this research, we have chosen partial least square (PLS) for testing our proposed model and data analysis [92]. We have several reasons for our choice of PLS. Firstly, PLS is a second-generation component-based structural equation modeling technique and has the capability of evaluating measurement models and structural model simultaneously. The great advantage of PLS is related to its clarity and transparency in explaining complex relationships because it has good features to avoid two important issues: inadmissible solutions and factor indeterminacy (Ringle et al. [93]). Secondly, PLS is totally different from covariance-based structural equation modeling (CB-SEM). It has much fewer statistical identification errors.

There is also evidence from prior research that PLS has the capability to provide similar parameter estimations and that PLS has more power in maximizing variance explained [93]. We have chosen PLS for our data analysis because we intend to explain variance in the different constructs of our proposed model to measure the performance of e-health startups and microenterprises and PLS is an appropriate tool for analyzing data in this research.

Before testing the hypotheses of this study, we made an assessment of the validity measures by using two steps. In the first step, we calculated the factor loading and cross loading of all the constructs as shown in Table 20.2. The loading of each item on its substantive construct was over 0.70, which suggests sufficient convergent validity [55]. In addition, each item's factor loading was much higher than its cross loadings on other constructs, which confirms the sufficiency of discriminant validity as shown in Table 20.3. In the second step, as Table 20.4 shows, the square root of each construct's average variance extracted (AVE) was much greater than the construct's correlations with all other constructs, which suggests sufficient discriminant validity. We examined the reliability of the measurements by computing composite reliability. As Table 20.4 shows, all reliability scores were higher than 0.70. Table 20.4 also shows the calculated each construct's average variance extracted (AVE) and Cronbach's alpha.

20.5.2 *Hypothesis Testing*

The results of our hypothesis testing in Fig. 20.4 show that most of the hypothesized relationships are significant. Three antecedents of information quality were positively associated with this construct. Completeness has the strongest relationship (T test = 4.27, $p < 0.01$), then the relationships of currency (T test = 3.98, $p < 0.01$) and transparency (T test = 4.07, $p < 0.01$) are significant; therefore, H1b, H1c, and H1d are validated. Although accuracy has the lowest significance (T test = 2.08, $p < 0.01$), it exceeds the threshold of 1.96 and H1a is also accepted.

Three antecedents of system quality were positively associated with this construct. Readability has the strongest relationship (T test = 8.86, $p < 0.01$), then the relationships of presentation (T test = 3.53, $p < 0.01$) and delay (T test = 2.81, $p < 0.01$) are significant; therefore, H2e, H2b, and H2d are validated and H2c is rejected because accessibility did not show a meaningful and strong relationship (T test = 0.99, $p > 0.05$). Navigability had the weakest relationship (T test = 0.82, $p > 0.05$) so H2a is rejected.

System quality had a strong positive association with prescriber satisfaction (T test = 19.28, $p < 0.01$); however, the effect of information quality on prescriber satisfaction was not significant (T test = 0.73, $p < 0.01$). H3a is rejected and H3b is accepted. Finally, prescriber satisfaction shows a significant relationship with performance of e-health startups and enterprises (T test = 2.76, $p > 0.05$). Therefore, H4 is accepted.

On the other side of the proposed model, the strongest relationship is shown between human capital and patent and we observe that human capital is positively associated with patent development (T test = 12.99, $p < 0.01$) and H5 is accepted. Alliance also shows a positive association with patent development (T test = 4.79, $p < 0.01$), and H6 is accepted. Finally, patent development showed a positive association with performance (T test = 4.6, $p < 0.01$) and the last hypothesis H7 is accepted.

Table 20.2 Loading and cross loadings

	Accessibility	Accuracy	Alliance	Completeness	Currency	Delay	Human capital	Information quality	Prescriber satisfaction	Presentation	Readability	System quality	Transparency
ACCS1	0.972	−0.016	0.872	−0.064	−0.023	0.780	0.824	−0.039	0.928	0.758	0.831	0.801	0.008
ACCS2	0.969	0.011	0.849	−0.064	0.009	0.806	0.843	−0.034	0.864	0.758	0.841	0.814	0.029
ACCS3	0.959	−0.005	0.842	−0.060	−0.022	0.772	0.813	−0.026	0.829	0.760	0.811	0.788	0.024
ACCU1	−0.029	0.939	0.019	0.654	0.646	0.057	0.032	0.703	−0.045	0.042	−0.003	0.017	0.747
ACCU2	0.001	0.967	0.064	0.660	0.681	0.081	0.066	0.727	−0.025	0.089	0.042	0.033	0.772
ACCU3	0.018	0.958	0.063	0.680	0.669	0.106	0.084	0.716	−0.001	0.082	0.052	0.068	0.757
ALLI1	0.745	0.038	0.904	−0.009	0.011	0.778	0.779	−0.011	0.643	0.915	0.730	0.745	0.041
ALLI2	0.607	0.052	0.831	0.053	0.020	0.669	0.680	0.039	0.529	0.868	0.638	0.693	0.084
ALLI3	0.763	0.108	0.916	0.027	0.068	0.787	0.812	0.044	0.658	0.923	0.780	0.796	0.104
ALLI4	0.972	−0.016	0.872	−0.064	−0.023	0.780	0.824	−0.039	0.928	0.758	0.831	0.801	0.008
COM1	−0.033	0.665	0.013	0.915	0.718	0.042	0.025	0.748	−0.047	0.030	0.002	0.000	0.657
COM2	−0.083	0.563	−0.022	0.860	0.554	0.003	−0.016	0.657	−0.079	0.005	−0.043	−0.053	0.572
COM3	−0.059	0.609	0.007	0.869	0.638	0.027	−0.001	0.610	−0.077	0.032	−0.028	−0.024	0.572
CURR1	−0.003	0.681	0.023	0.703	0.972	0.050	0.038	0.757	−0.004	0.038	0.028	0.014	0.696
CURR2	0.003	0.665	0.028	0.703	0.976	0.060	0.042	0.738	0.006	0.040	0.027	0.027	0.666
CURR3	−0.036	0.687	0.012	0.709	0.969	0.039	0.024	0.730	−0.036	0.032	0.014	0.011	0.681
DLY1	0.794	0.078	0.838	0.041	0.064	0.971	0.949	0.048	0.711	0.806	0.833	0.844	0.075
DLY2	0.767	0.080	0.827	0.001	0.027	0.968	0.935	0.037	0.672	0.807	0.813	0.819	0.082
DLY3	0.786	0.088	0.810	0.038	0.055	0.948	0.921	0.053	0.685	0.776	0.821	0.825	0.075
HUMN1	0.794	0.078	0.838	0.041	0.064	0.971	0.949	0.048	0.711	0.806	0.833	0.844	0.079
HUMN2	0.767	0.080	0.827	0.001	0.027	0.968	0.935	0.037	0.672	0.807	0.813	0.819	0.082
HUMN3	0.786	0.088	0.810	0.038	0.055	0.948	0.921	0.053	0.685	0.776	0.821	0.825	0.075
HUMN4	0.811	0.049	0.831	0.000	0.043	0.843	0.933	0.018	0.731	0.788	0.951	0.878	0.062
HUMN5	0.795	0.000	0.768	−0.061	−0.024	0.768	0.885	−0.043	0.706	0.713	0.935	0.822	−0.005

INFQ1	−0.033	0.729	0.016	0.756	0.716	0.035	0.016	0.968	−0.046	0.039	−0.009	−0.007	0.772
INFQ2	−0.034	0.722	0.000	0.727	0.761	0.058	0.032	0.967	−0.023	0.013	−0.006	−0.008	0.718
NAV1	0.735	0.104	0.761	0.031	0.056	0.790	0.800	0.049	0.655	0.722	0.740	0.748	0.099
NAV22	0.751	0.082	0.807	0.047	0.065	0.818	0.820	0.075	0.635	0.784	0.761	0.768	0.064
NAV3	0.673	0.147	0.719	0.066	0.123	0.766	0.766	0.139	0.550	0.708	0.715	0.716	0.138
PAT1	0.805	0.030	0.806	−0.021	0.028	0.823	0.875	−0.006	0.699	0.760	0.939	0.872	0.068
PAT2	0.775	0.054	0.839	−0.010	0.014	0.825	0.863	−0.001	0.677	0.812	0.849	0.947	0.091
PAT3	0.805	0.030	0.825	−0.037	0.020	0.824	0.869	−0.014	0.703	0.787	0.878	0.965	0.028
PAT4	0.809	0.035	0.826	−0.032	0.018	0.836	0.881	−0.006	0.710	0.787	0.885	0.972	0.048
PERF1	−0.132	−0.013	−0.066	−0.047	−0.039	−0.060	−0.058	0.005	−0.346	−0.023	−0.040	−0.045	−0.013
PERF2	−0.121	−0.023	−0.090	−0.058	−0.032	−0.073	−0.056	0.046	−0.313	−0.062	−0.037	−0.036	−0.014
PERF3	−0.147	−0.010	−0.076	−0.068	−0.047	−0.066	−0.058	−0.002	−0.357	−0.032	−0.036	−0.009	−0.021
PERF4	−0.139	−0.023	−0.096	−0.080	−0.021	−0.098	−0.081	−0.016	−0.346	−0.064	−0.063	−0.058	−0.026
PRES1	0.763	0.108	0.916	0.027	0.068	0.787	0.812	0.044	0.658	0.923	0.780	0.796	0.104
PRES2	0.607	0.052	0.831	0.053	0.020	0.669	0.680	0.039	0.529	0.868	0.638	0.693	0.084
PRES3	0.745	0.038	0.904	−0.009	0.011	0.778	0.779	−0.011	0.643	0.915	0.730	0.745	0.041
READ1	0.811	0.049	0.831	0.000	0.043	0.843	0.933	0.018	0.731	0.788	0.951	0.878	0.062
READ2	0.795	0.000	0.768	−0.061	−0.024	0.768	0.885	−0.043	0.706	0.713	0.935	0.822	−0.005
READ3	0.824	0.040	0.801	−0.013	0.040	0.792	0.867	0.001	0.740	0.744	0.951	0.850	0.058
READ4	0.805	0.030	0.806	−0.021	0.028	0.823	0.875	−0.006	0.699	0.760	0.939	0.872	0.068
SATIS1	0.972	−0.016	0.872	−0.064	−0.023	0.780	0.824	−0.039	0.928	0.758	0.831	0.801	0.008
SATIS2	−0.183	0.028	−0.111	0.054	−0.021	−0.125	−0.135	0.005	−0.537	−0.073	−0.127	−0.098	0.063
SysQ1	0.775	0.054	0.839	−0.010	0.014	0.825	0.863	−0.001	0.677	0.812	0.849	0.947	0.091
SysQ2	0.805	0.030	0.825	−0.037	0.020	0.824	0.869	−0.014	0.703	0.787	0.878	0.965	0.028

(continued)

Table 20.2 (continued)

	Accessibility	Accuracy	Alliance	Completeness	Currency	Delay	Human capital	Information quality	Prescriber satisfaction	Presentation	Readability	System quality	Transparency
SysQ3	0.809	0.035	0.826	−0.032	0.018	0.836	0.881	−0.006	0.710	0.787	0.885	0.972	0.048
TRAN1	0.064	0.569	0.128	0.521	0.521	0.107	0.091	0.574	0.045	0.149	0.082	0.087	0.820
TRAN2	0.052	0.789	0.066	0.594	0.644	0.107	0.096	0.742	0.029	0.070	0.089	0.085	0.897
TRAN3	−0.028	0.720	0.029	0.605	0.587	0.047	0.028	0.681	−0.071	0.057	0.014	0.029	0.911
TRAN4	−0.013	0.647	0.015	0.638	0.659	0.022	0.006	0.649	−0.058	0.029	−0.016	0.000	0.825

Note: accu = accuracy; comp = completeness; curr = currency; tran = transparency; InfQ = information quality; accss = accessibility; dly = delay; navig = navigability; pres = presentation; read = readability; SysQ = system quality; Satis = Prescription satisfaction; Humn = human capital; Alli = alliance; Pat = patent; Perf = performance

Table 20.3 Discriminant validity

	Accessibility	Accuracy	Alliance	Completeness	Currency	Delay	Human capital	Information quality	Navigation	Patent	Performance	Prescriber satisfaction	Presentation	Readability	System quality	Transparency
Accessibility	1.000															
Accuracy	−0.003	1.000														
Alliance	0.884	0.051	1.000													
Completeness	−0.065	0.696	0.000	1.000												
Currency	−0.012	0.697	0.021	0.725	1.000											
Delay	0.813	0.085	0.858	0.028	0.051	1.000										
Human apital	0.855	0.064	0.882	0.004	0.036	0.972	1.000									
Information quality	−0.034	0.749	0.008	0.766	0.763	0.048	0.024	1.000								
Navigation	0.772	0.118	0.818	0.051	0.087	0.849	0.853	0.093	1.000							
Patent	0.844	0.039	0.871	−0.027	0.021	0.874	0.921	−0.007	0.809	1.000						
Performance	−0.145	−0.018	−0.088	−0.068	−0.038	−0.079	−0.068	0.008	−0.063	−0.042	1.000					
Prescriber satisfaction	0.904	−0.024	0.791	−0.075	−0.012	0.717	0.759	−0.036	0.659	0.736	−0.367	1.000				
Presentation	0.785	0.074	0.980	0.026	0.038	0.828	0.842	0.027	0.792	0.831	−0.048	0.679	1.000			
Readability	0.857	0.032	0.850	−0.025	0.024	0.855	0.943	−0.007	0.792	0.938	−0.047	0.762	0.796	1.000		
System quality	0.829	0.041	0.864	−0.028	0.018	0.862	0.906	−0.008	0.798	0.992	−0.039	0.725	0.827	0.906	1.000	
Transparency	0.021	0.795	0.066	0.683	0.700	0.082	0.064	0.770	0.107	0.062	−0.020	−0.017	0.085	0.049	0.058	1.000

Table 20.4 Reliability statistics

	Cronbach's Alpha	rho_A	Composite reliability	Average Variance Extracted (AVE)
Accessibility	0.965	0.965	0.977	0.935
Accuracy	0.951	0.952	0.969	0.911
Completeness	0.857	0.867	0.913	0.777
Currency	0.971	0.971	0.981	0.945
Delay	0.960	0.960	0.974	0.926
Information quality	0.932	0.933	0.967	0.937
Navigation	0.925	0.927	0.953	0.870
Performance _	0.948	0.951	0.963	0.865
Prescriber satisfaction	0.837	0.812	0.802	0.825
Presentation	0.886	0.891	0.929	0.814
Readability	0.959	0.960	0.970	0.891
System quality	0.959	0.959	0.973	0.924
Transparency	0.886	0.895	0.922	0.747
Patent	0.961	0.961	0.972	0.896
Alliance	0.904	0.908	0.933	0.776
Human capital	0.958	0.958	0.967	0.855

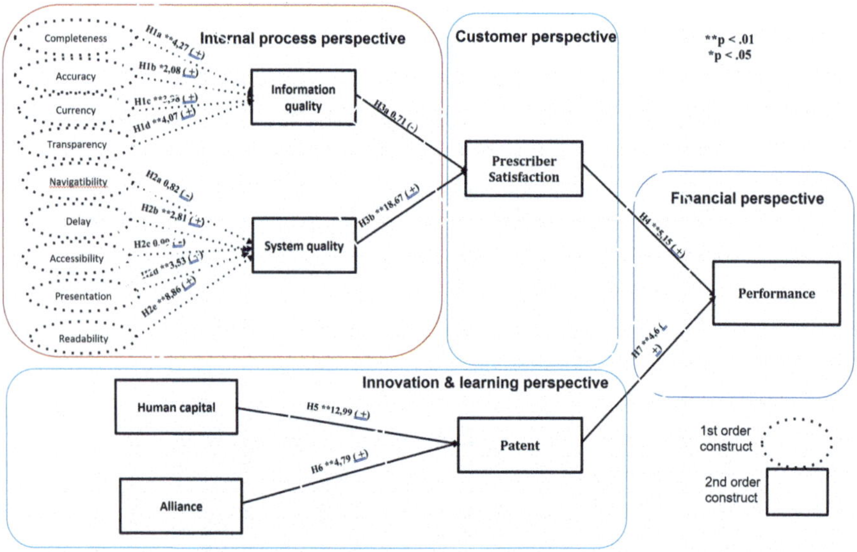

Fig. 20.4 PLS results of model testing

20.6 Discussion

Most hypotheses in this research were validated with meaningful relationships and associations among variables and constructs; however, the details and degree of these associations are very meaningful. All antecedents of information quality were proved to have meaningful relationships, but completeness was the strongest one meaning that healthcare professionals prescribing e-health devices and service expect to receive complete sets of information to cover all details of their patients. This shows that healthcare professionals (prescribers) look for an e-health product or service which can provide them with a complete set of information based on which they make right decisions. In the second position, and slightly lower than completeness, transparency had the highest meaningful relationship (T test = 4.07, $p < 0.01$). Validation of this hypothesis with such a high degree of association shows that new treatment procedures made possible by information technology have more advantages than disadvantages for health professionals because all patient data are simultaneously accessible to the health professionals without any mediation, facilitating their ability to make right decision at the right time. Currency has a position slightly less than the two other antecedents (T test = 3.98, $p < 0.01$). After completeness and transparency, healthcare professionals expect to receive and process patient data instantly and if the circulation of data in e-health products and services is blocked due to a bug or technical problem, it would have negative consequences on their decision making. Accuracy has the least meaningful association with information quality and its value is slightly higher than the threshold (T test = 2.08, $p < 0.01$). This shows that the health professionals do not totally trust and rely on the e-health products and services and they want to have their own verification of accuracy from their own side.

Among antecedents of system quality, navigability has no meaningful association as an antecedent of system quality (T test = 0.82, $p > 0.05$) meaning that the fluidity and comfort of navigating through different pages and interfaces of e-health products and services does not have any impact on the perception of information quality for e-health professionals. Accessibility is also rejected as an antecedent of system quality (T test = 0.99, $p > 0.05$) signifying that it does not account for an important and inevitable criterion for healthcare professionals to regard as a determinant of system quality.

On the other hand, readability had the highest degree of significance (T test = 8.86, $p < 0.01$) on system quality meaning that healthcare professionals want e-health service and product to be understandable for their patients and final users with different levels of literacy so that they can continue adopting the system. Presentation is also accepted as an antecedent of system quality (T test = 3.53, $p < 0.01$), demonstrating that the way e-health services and devices are presented to patients and final users—through different colors and layouts—plays an important role in healthcare professionals' perception of system quality. Last but not the least, delay is an accepted antecedent of system quality (T test = 2.81, $p < 0.01$). This shows that healthcare professionals regard speed and acceleration in flow of information and treatment processes as an important characteristic of a good e-health product or service.

In the next step, the impact of information quality and system quality as two constructs of internal process perspective is tested on prescriber satisfaction. The results show that system quality meaning smooth and easy usage of e-health tools and devices constitutes subscriber satisfaction (T test = 19.28, $p < 0.01$). This strong relationship shows the fact that easy handling of an e-health device or service can be very significative for its continuous use and through this persistent use of patients (final users), e-health professionals will be able to obtain their expected results leading to their satisfaction. On the other hand, the impact of information quality on prescriber satisfaction is rejected (T test = 0.73, $p < 0.01$) showing that healthcare professionals make several verifications of patient information through different ways and information quality by itself does not have that significant impact on prescriber satisfaction.

As expected, subscriber satisfaction had a very important impact on performance of e-health startups and microenterprises (T test = 2.76, $p > 0.05$). This shows that prescribers who are not final users of e-health services and products have a decisive role on the success of their commercialization. The satisfaction of prescribers (healthcare professionals like physicians and nurses) plays an important role in success of e-health services and products on the market and consequently, profitability of e-health startups and microenterprises.

In innovation and learning perspective of the proposed model, human capital is the most important determinant in success and failure in terms of innovation and patent development (T test = 12.99, $p < 0.01$) signifying that e-health startups and microenterprises which care for the training of their competent employees to increase their knowledge and innovative capacities will obtain good results in innovation and development of patents. After human capital, alliances with universities, research centers, and large multinational companies are also important in contributing to the innovation of e-health startups and enterprises (T test = 4.79, $p < 0.01$). This shows that universities, research centers, and big multinational companies have good potentials to provide training and other facilities for e-health startups and their employees to increase their innovative capacities. Finally, the results validate strong association of patent development and performance (T test = 4.6, $p < 0.01$) showing that the innovative capacity of e-health startups and enterprises can lead to higher sales of e-health products and services and therefore higher profitability and performance.

References

1. Agarwal, R., Gao, G., DesRoches, C., Jha, A.K.: Research commentary—the digital transformation of healthcare: current status and the road ahead. Inf. Syst. Res. **21**(4), 796–809 (2010)
2. Kohli, R., Tan, S.S.-L.: Electronic health records: how can IS researchers contribute to transforming healthcare? MIS Q. **40**(3), 553–573 (2016). https://doi.org/10.25300/misq/2016/40.3.02
3. Tian, K., Sautter, P., Fisher, D., Fischbach, S., Luna-Nevarez, C., Boberg, K., et al.: Transforming health care: empowering therapeutic communities through technology-enhanced narratives. J. Consum. Res. **41**(2), 237–260 (2014). https://doi.org/10.1086/676311
4. Reardon, J.L., Davidson, E.: An organizational learning perspective on the assimilation of electronic medical records among small physician practices. Eur. J. Inf. Syst. **16**(6), 681–694 (2007)
5. Leys, M.: Health care policy: qualitative evidence and health technology assessment. Health Policy. **65**(3), 217 (2003). https://doi.org/10.1016/S0168-8510(02)00209-9

6. Pare, G., Meyer, J., Trudel, M.C., Tetu, B.: Impacts of a large decentralized Telepathology network in Canada. Telemed. E-Health. **22**(3), 246–250 (2016). https://doi.org/10.1089/tmj.2015.0083
7. Kelley, H., Chiasson, M., Downey, A., Pacaud, D.: The clinical impact of eHealth on the self-management of diabetes: a double adoption perspective. J. Assoc. Inf. Syst. **12**(3), 208–234 (2011) Retrieved from <Go to ISI>://WOS:000294993500001
8. Hardiker, N.R., Grant, M.J.: Factors that influence public engagement with eHealth: a literature review. Int. J. Med. Inform. **80**(1), 1–12 (2011). https://doi.org/10.1016/j.ijmedinf.2010.10.017
9. Lapointe, L., Mignerat, M., Vedel, I.: The IT productivity paradox in health: a stakeholder's perspective. Int. J. Med. Inform. **80**(2), 102–115 (2011). https://doi.org/10.1016/j.ijmedinf.2010.11.004
10. Lemire, M.: What can be expected of information and communication technologies in terms of patient empowerment in health? J. Health Organ. Manag. **24**(2), 167–181 (2010)
11. De Rosis, S., Barsanti, S.: Patient satisfaction, e-health and the evolution of the patient-general practitioner relationship: evidence from an Italian survey. Health Policy. **120**(11), 1279–1292 (2016). https://doi.org/10.1016/j.healthpol.2016.09.012
12. Connell, N.A.D., Young, T.P.: Evaluating healthcare information systems through an "enterprise" perspective. Inf. Manag. **44**(4), 433–440 (2007). https://doi.org/10.1016/j.im.2007.04.002
13. Lorden, A., Coustasse, A., Singh, K.P.: The balanced scorecard framework-a case study of patient and employee satisfaction: what happens when it does not work as planned? Health Care Manag. Rev. **33**(2), 145–155 (2008)
14. Maiga, A.S., Jacobs, F.A.: Balanced scorecard, activity-based costing and company performance: an empirical analysis. J. Manag. Issues. **15**(3), 283–301 (2003)
15. Lyytinen, K., King, J.L.: Standard making: a critical research frontier for information systems research. MIS Q. **30**, 405–411 (2006)
16. Klecun, E., Cornford, T.: A critical approach to evaluation. Eur. J. Inf. Syst. **14**(3), 229–243 (2005). https://doi.org/10.1057/palgrave.ejis.3000540
17. Klecun, E., Lichtner, V., Cornford, T., Petrakaki, D.: Evaluation as a multi-ontological endeavour: a case from the English National Program for IT in healthcare. J. Assoc. Inf. Syst. **15**(3), 147–176 (2014) Retrieved from <Go to ISI>://WOS:000334580400002
18. Aidemark, L.G.: Cooperation and competition: balanced scorecard and hospital privatization. Int. J. Health Care Qual. Assur. **23**(8), 730–748 (2010)
19. Gurd, B., Gao, T.: Lives in the balance: an analysis of the balanced scorecard (BSC) in healthcare organizations. Int. J. Product. Perform. Manag. **57**(1), 6–21 (2008)
20. Schalm, C.: Implementing a balanced scorecard as a strategic management tool in a long-term care organization. In: SAGE Publications Sage, London (2008)
21. Zelman, W.N., Pink, G.H., Matthias, C.B.: Use of the balanced scorecard in health care. J. Health Care Finance. **29**(4), 1–16 (2003) Retrieved from https://www.ncbi.nlm.nih.gov/pubmed/12908650
22. Sigalas, C.: Empirical investigation of balanced scorecard's theoretical underpinnings. J. Account. Organ. Change. (2015)
23. Kaplan, R.S., Norton, D.P.: Measuring the strategic readiness of intangible assets. Harv. Bus. Rev. **82**(2), 52–63 (2004a)
24. Cleven, A., Mettler, T., Rohner, P., Winter, R.: Healthcare quality innovation and performance through process orientation: evidence from general hospitals in Switzerland. Technol. Forecast. Soc. Chang. **113**, 386–395 (2016)
25. Kaplan, R.S., Norton, D.P.: The strategy map: guide to aligning intangible assets. Strateg. Leadersh. **32**(5), 10–17 (2004b)
26. Messeghem, K., Bakkali, C., Sammut, S., Swalhi, A.: Measuring nonprofit incubator performance: toward an adapted balanced scorecard approach. J. Small Bus. Manag. **56**(4), 658–680 (2018)

27. Chow, C.W., Ganulin, D., Haddad, K., Williamson, J.: The balanced scorecard: a potent tool for energizing and focusing healthcare organization management. J. Healthc. Manag. **43**(3), 263 (1998)
28. Mattke, S., Epstein, A.M., Leatherman, S.: The OECD health care quality indicators project: history and background. Int. J. Qual. Health Care. **18**(suppl_1), 1–4 (2006)
29. Brewer, P.C.: Using the balanced scorecard to measure supply chain performance Peter C Brewer; Thomas WSpeh. J. Bus. Logist. **21**(1), 75 (2000)
30. Inamdar, N., Kaplan, R.S., Reynolds, K.: Applying the balanced scorecard in healthcare provider organizations/Practitioner's response. J. Healthc. Manag. **47**(3), 179 (2002)
31. Aidemark, L.G., Funck, E.K.: Measurement and health care management. Financ. Account. Manag. **25**(2), 253–276 (2009)
32. Fottler, M.D., Erickson, E., Rivers, P.A.: Bringing human resources to the table: utilization of an HR balanced scorecard at Mayo Clinic. Health Care Manag. Rev. **31**(1), 64–72 (2006)
33. Yang, M.-C., Tung, Y.-C.: Using path analysis to examine causal relationships among balanced scorecard performance indicators for general hospitals: the case of a public hospital system in Taiwan. Health Care Manag. Rev. **31**(4), 280–288 (2006)
34. Albertsen, O., Lueg, R.: The balanced Scorecard's missing link to compensation: a literature review and an agenda for future research. J. Account. Organ. Chang. **10**(4), 431–465 (2014)
35. Chang, W.-C., Tung, Y.-C., Huang, C.-H., Yang, M.-C.: Performance improvement after implementing the balanced scorecard: a large hospital's experience in Taiwan. Total Qual. Manag. **19**(11), 1143–1154 (2008)
36. Mutale, W., Stringer, J., Chintu, N., Chilengi, R., Mwanamwenge, M.T., Kasese, N., et al.: Application of balanced scorecard in the evaluation of a complex health system intervention: 12 months post intervention findings from the BHOMA intervention: a cluster randomised trial in Zambia. PLoS One. **9**(4), e93977 (2014)
37. Sasse, C.: A literature review on the balanced scorecard's impact on performance. *LeaderPoint*. Available at: www.leaderpoint.biz (2005). Accessed 24 May 2005
38. Morgan, C.: Structure, speed and salience: performance measurement in the supply chain. Bus. Process. Manag. J. **10**(5), 522–536 (2004)
39. Muhos, M., Saarela, M., Foit, D., Rasochova, L.: Management priorities of digital health service start-ups in California. Int. Entrep. Manag. J. **15**(1), 43–62 (2019)
40. Beaulieu, M., Lehoux, P.: The emergence of health technology firms through their sensegiving activities and competitive actions. Int. J. Innov. Manag. **21**(06), 1750043 (2017)
41. Ohinmaa, A., Hailey, D., Roine, R.: Elements for assessment of telemedicine applications. Int. J. Technol. Assess. Health Care. **17**(2), 190–202 (2001). https://doi.org/10.1017/s0266462300105057
42. Traberg, A., Jacobsen, P.: Rethinking performance evaluation in health care. J. Health Organ. Manag. (2012)
43. Silva, H.P., Viana, A.L.: Health technology diffusion in developing countries: a case study of CT scanners in Brazil. Health Policy Plan. **26**(5), 385–394 (2011)
44. Toften, K., Hammervoll, T.: Niche marketing and strategic capabilities: an exploratory study of specialised firms. Mark. Intell. Plan. (2010)
45. Stanton, J. F., Hawkins, N. M., & Hicks, T. R.: PRESSS Project 1.3: Connection classification and evaluation. PCI journal. **36**(5), 62-71 (1991)
46. Dalgic, T., Leeuw, M.: Niche marketing revisited: concept, applications and some European cases. Eur. J. Mark. (1994)
47. Zhang, M.W., Ward, J., Ying, J.J., Pan, F., Ho, R.C.: The alcohol tracker application: an initial evaluation of user preferences. BMJ Innov. **2**(1), 8–13 (2016). https://doi.org/10.1136/bmjinnov-2015-000087
48. Doyle, P., Stern, P.: Marketing Management and Strategy. Pearson Education (2006)
49. Rogers, A.E., Hwang, W.-T., Scott, L.D., Aiken, L.H., Dinges, D.F.: The working hours of hospital staff nurses and patient safety. Health Aff. **23**(4), 202–212 (2004)

50. Torkzadeh, G., Doll, W.J.: The development of a tool for measuring the perceived impact of information technology on work. Omega. **27**(3), 327–339 (1999)
51. DeLone, W.H.: Determinants of success for computer usage in small business. MIS Q., 51–61 (1988)
52. Carter, N., Day, P., Klein, R.: How Organisations Measure Success: The Use of Performance Indicators in Government. Psychology Press (1995)
53. Flynn, N.: Public Sector Management. Sage Publications (2012)
54. Rouse, J.: Performance management, quality management and contracts. In: Public Management in Britain, pp. 76–93. Springer (1999)
55. Liang, H., Xue, Y., Zhang, Z.: Understanding online health information use: the case of people with physical disabilities. J. Assoc. Inf. Syst. **18**(6), 2 (2017)
56. Halawi, L.A.: Knowledge Management systems' Success in Knowledge-Based Organizations: an Empirical Validation Utilizing the DeLone and McLean IS Success model. Nova Southeastern University (2005)
57. Delone, W.H., McLean, E.R.: The DeLone and McLean model of information systems success: a ten-year update. J. Manag. Inf. Syst. **19**(4), 9–30 (2003)
58. Baum, J.A., Silverman, B.S.: Picking winners or building them? Alliance, intellectual, and human capital as selection criteria in venture financing and performance of biotechnology startups. J. Bus. Ventur. **19**(3), 411–436 (2004)
59. Davidsson, P., Achtenhagen, L., Naldi, L.: *Research on small firm growth: a review.* Paper presented at the 35th EISB Conference: Sustaining the Entrepreneurial Spirit over Time: Implications for Young Companies, Family Businesses, and Established Companies (2005).
60. Bastida, R., Marimon, F., Tanganelli, D.: Alliance success factors and performance in social economy enterprises. Manag. Decis. (2017)
61. Martinez, M.G., Zouaghi, F., Garcia, M.S.: Capturing value from alliance portfolio diversity: the mediating role of R&D human capital in high and low tech industries. Technovation. **59**, 55–67 (2017)
62. Liu, W.L., Zhang, K., Locatis, C., Ackerman, M.: Cloud and traditional videoconferencing technology for telemedicine and distance learning. Telemed. E-Health. **21**(5), 422–426 (2015). https://doi.org/10.1089/tmj.2014.0121
63. Löfsten, H.: Business and innovation resources. Manag. Decis. (2016)
64. Hsu, D.H., Ziedonis, R.H.: Resources as dual sources of advantage: implications for valuing entrepreneurial-firm patents. Strateg. Manag. J. **34**(7), 761–781 (2013)
65. Zhao, L., Xiang, Y., Yi, Q.: Fuzzy front end patent management and innovation performance. Manag. Decis. (2017)
66. Wixom, B.H., Todd, P.A.: A theoretical integration of user satisfaction and technology acceptance. Inf. Syst. Res. **16**(1), 85–102 (2005)
67. Harle, C.A., Golembiewski, E.H., Rahmanian, K.P., Krieger, J.L., Hagmajer, D., Mainous, A.G., Moseley, R.E.: Patient preferences toward an interactive e-consent application for research using electronic health records. J. Am. Med. Inform. Assoc. **25**(3), 360–368 (2018). https://doi.org/10.1093/jamia/ocx145
68. Hatch, B., Tillotson, C., Angier, H., Marino, M., Hoopes, M., Huguet, N., DeVoe, J.: Using the electronic health record for assessment of health insurance in community health centers. J. Am. Med. Inform. Assoc. **23**(5), 984–990 (2016). https://doi.org/10.1093/jamia/ocv179
69. Michel-Verkerke, M.B., Spil, T.A.: The USE IT-adoption-model to predict and evaluate adoption of information and communication technology in healthcare. Methods Inf. Med. **52**(06), 475–483 (2013)
70. Klein, R.: An empirical examination of patient-physician portal acceptance. Eur. J. Inf. Syst. **16**(6), 751–760 (2007). https://doi.org/10.1057/palgrave.ejis.3000719
71. Beamon, B.M.: Measuring supply chain performance. Int. J. Oper. Prod. Manag. (1999)
72. Chan, S., Torous, J., Hinton, L., Yellowlees, P.: Towards a framework for evaluating Mobile mental health apps. Telemed. E-Health. **21**(12), 1038–1041 (2015). https://doi.org/10.1089/tmj.2015.0002

73. Curtright, J.W., Stolp-Smith, S.C., Edell, E.S.: Strategic performance management: development of a performance measurement system at the Mayo Clinic. J. Healthc. Manag. **45**(1), 58–68 (2000)
74. Gunasekaran, A., Patel, C., Tirtiroglu, E.: Performance measures and metrics in a supply chain environment. Int. J. Oper. Prod. Manag. (2001)
75. Lapide, L. (2000). What about measuring supply chain performance. Achieving Supply Chain Excellence Through Technology, 2(2), 287–297
76. Lovaglio, P.G.: Model building and estimation strategies for implementing the balanced scorecard in health sector. Qual. Quant. **45**(1), 199–212 (2011)
77. Pedro, E., Leitão, J., Alves, H.: Back to the future of intellectual capital research: a systematic literature review. Manag. Decis. (2018)
78. Salvi, A., Vitolla, F., Giakoumelou, A., Raimo, N., Rubino, M.: Intellectual capital disclosure in integrated reports: the effect on firm value. Technol. Forecast. Soc. Chang. **160**, 120228 (2020)
79. Abramson, E.L., McGinnis, S., Edwards, A., Maniccia, D.M., Moore, J., Kaushal, R.: Electronic health record adoption and health information exchange among hospitals in New York state. J. Eval. Clin. Pract. **18**(6), 1156–1162 (2012). https://doi.org/10.1111/j.1365-2753.2011.01755.x
80. Condren, M., Studebaker, I.J., John, B.M.: Prescribing errors in a pediatric clinic. Clin. Pediatr. **49**(1), 49–53 (2010)
81. Santell, J.P., Kowiatek, J.G., Weber, R.J., Hicks, R.W., Sirio, C.A.: Medication errors resulting from computer entry by nonprescribers. Am. J. Health Syst. Pharm. **66**(9), 843–853 (2009)
82. Agostini, L., Nosella, A.: A dual knowledge perspective on the determinants of SME patenting. Manag. Decis. (2017)
83. Bewaji, T., Hamilton III, R.D.: The effect of external social resources on firm patent performance: the impact of boards, partnerships and geographic proximity. J. Bus. Entrep. **24**(1) (2012)
84. Cammarano, A., Caputo, M., Lamberti, E., Michelino, F.: Open Innovation and Intellectual Property: a Knowledge-Based Approach. Manag. Decis. **55**(6), 1182–1208 (2017)
85. Goodhue, D.L., Thompson, R.L.: Task-technology fit and individual performance. MIS Q., 213–236 (1995)
86. Igbaria, M., Tan, M.: The consequences of information technology acceptance on subsequent individual performance. Inf. Manag. **32**(3), 113–121 (1997)
87. Seddon, P.B., Staples, S., Patnayakuni, R., Bowtell, M.: Dimensions of information systems success. Commun. Assoc. Inf. Syst. **2**(1), 20 (1999)
88. Filson, D., Oweis, A.: The impacts of the rise of paragraph IV challenges on startup alliance formation and firm value in the pharmaceutical industry. J. Health Econ. **29**(4), 575–584 (2010)
89. Russell, S.: Investor drought and regulatory headwinds slow device innovation. Health Aff. (Millwood). **34**(2), 199–202 (2015)
90. Bart, Y., Shankar, V., Sultan, F., Urban, G.L.: Are the drivers and role of online trust the same for all web sites and consumers? A large-scale exploratory empirical study. J. Mark. **69**(4), 133–152 (2005)
91. Subramaniam, M., Youndt, M.A.: The influence of intellectual capital on the types of innovative capabilities. Acad. Manag. J. **48**(3), 450–463 (2005)
92. Ong, M.H.A., Puteh, F.: Quantitative data analysis: choosing between SPSS, PLS, and AMOS in social science research. Int. Interdiscip. J. Sci. Res. **3**(1), 14–25 (2017)
93. Ringle, C., Da Silva, D., Bido, D.: Structural equation modeling with the SmartPLS. REMark: Revista Brasileira de Marketing. **13**(2) (2015)

Part III
Economic and Human Capital Development

Chapter 21
Business Innovations and Human Capital: Consequences for Economic Growth

Vicente J. Coronel and Carmen Díaz-Roldán

Abstract This chapter aims to investigate the effects of the human capital and the employment in high-tech sectors on the economic growth, given the business environment's level of innovation. The key novelty lies in examining how education and employment variables behave in relation to the extent of innovation dissemination by companies. The study underscores the significance of innovation as a driver of growth and explores methodologies for its measurement. The implications of innovation on education (human capital) and employability (complementarity of productive capital and labor) are examined. The analysis utilizes quantitative methods, employing Dynamic Ordinary Least Squares (DOLS) to estimate cointegrated panel data for the EU 27 during the period 2008–2021. This approach addresses endogeneity issues and mitigates biases associated with small sample sizes. Our results point out that employment in high-tech sectors is the primary driver of economic growth in states leading in innovation.

Keywords Innovation · Human capital · Economic growth

21.1 Introduction

In the literature on economic growth, there are several contributions showing the importance of the innovation processes on productivity (see [1] and the references therein). Griliches and Jones [2, 3] found that growth is generated endogenously through R & D spillovers and that productivity depends on the discovery of new

V. J. Coronel · C. Díaz-Roldán (✉)
Universidad de Castilla-La Mancha, Facultad de Derecho y Ciencias Sociales, Ciudad Real, Spain
e-mail: carmen.diazroldan@uclm.es

F. Schiavone et al. (eds.), *Advanced Perspectives and Trends in Digital Transformation of Firms, Networks, and Society*, Springer Proceedings in Business and Economics, https://doi.org/10.1007/978-3-031-80692-6_21

designs by agents able to use new technologies. In addition, several authors ([4–6], to name a few) address the role of human capital externalities. We find recent contributions emphasizing the role of the strategies of innovation development managed by firms, aimed to improve the technological capabilities of workers ([7–11], for example). As far as we know, the existing literature explores the feedback between firm innovation and the human capital acquisition from a microeconomic perspective. But there is a scarcity of papers considering the macroeconomic effects of a firm's innovation on human capital and their impacts on employability.

This chapter aims to examine the role of innovation as driver of human capital, and its role for enhancing productivity when companies employ high-tech skilled workers and view them as complements to productive capital. Besides that, we consider the function of the degree of innovation performed by companies on the employability of skilled workers.

21.2 Methodology and Data Analysis

After performing a literature review, the analysis utilizes quantitative methods, employing Dynamic Ordinary Least Squares (DOLS) to estimate cointegrated panel data for the EU 27 during the period 2008–2021 using data obtained from Eurostat. The DOLS approach was proposed by [12–14] and addresses endogeneity issues and mitigates biases associated with small sample sizes. DOLS estimates also eliminate the serial correlation present in standard OLS which results in inconsistent estimates for cointegrated panel data.

To conduct the empirical application, we estimate the effects on the GDP growth of those variables related to the expenditure on R&D both in the business sector and in the higher education sector. The labor market variables are likely to indicate the level of digitalization of workers, such as employment in technology and knowledge-intensive sectors and employed persons working from home. We complement the analysis by exploring the role of higher education level of people. Additionally, we follow the European Innovation Scoreboard classification [15] to distinguish among three sets within the EU-27 state members, depending on the degree of innovation disseminated by the firms. In this fashion, we will split the 27 countries into the "Highly innovative" group (North/Central); the "Intermediate innovative" (Southern) group, and the "Scarcely innovative" (or Central and Eastern) group. As far as we know, this way of grouping countries is not common in the empirical literature on human capital and employability, even when innovation is considered a driver of economic growth.

When we look at the descriptive statistics, the values for the mean, the median, the maximum, and the minimum values differ depending on the nature of the series. Regarding the dispersion of the data the maximum value of the standard deviation can be observed for employment rate of women with higher education, while the minimum for the expenditure on R&D. These statistics reveal the heterogeneity of the standards in the EU-27 throughout the period.

21.3 Discussion of Results

In the first step, we estimate the effects on GDP growth for the whole EU. Next, we split the data of the EU-27 into the three groups referred above organized according to the degree of innovation dissemination performed by firms. Summarizing the results, we found that the spending on R&D, both in the business sector and in the higher education sector, shows a negative effect and in the case of education it is significant. Which could reveal that the returns of education cannot be observed in the medium term. And, it also could raise questions such as to what extent we can expect returns of education, whether if the highly educated students emigrate, or even if they do not achieve productive jobs enough.

In addition, our results indicate that employment in high-tech sectors makes the most significant contribution to economic growth in those countries that are leaders in innovation. This could lead us to conclude that there is a feedback relationship between innovation and the quality of the workforce, given that employed persons in high-tech sectors are expected to have a higher degree of human capital. Therefore, these workers act as a complementary factor of production to innovation, enhancing productivity and, consequently, fostering economic growth.

When we pay attention to the gender perspective, we found that the gender gap in employment is observed to have a positive and significant effect in the highly innovative group, which could suggest that, in those countries, men hold more positions in high-tech sectors than women. On the contrary, we find that women who have higher education tend to contribute more than men with similar degrees (except in low-innovative countries), although women's contribution in high-tech sectors is always lower. This result raises questions about the dynamics of the labor market, as it seems that women face difficulties in accessing jobs in high-tech sectors.

21.4 Conclusions

According to our findings, R&D spending does not have a positive effect on growth. Examining the returns to education and analyzing the incorporation of highly qualified graduates into the labor market could help find the explanation. We could examine these issues in the future, as an extension of our work. In addition, our findings regarding the role of women raise questions about the functioning of the labor market, as it seems that women face difficulties in accessing jobs in high-tech sectors. Therefore, when looking at the implications of innovation on education and employability, our results could be a departure point for designing education as well as industrial economic policies. In particular, promoting campaigns encouraging girls to study STEM careers, and implementing policies favoring the employment of women in technological sectors. Finally, and as a wake-up call, it is important to remember that we conducted our analysis for the EU. And, given the expansion of North American, and especially Asian economies, it would be essential to prioritize innovation and worker training in strategic sectors to ensure European leadership at global level, to contextualize our outcomes in a global context.

References

1. Bongers, A., Díaz-Roldán, C., Torres, J.L.: Highly skilled international migration, STEM workers, and innovation. Economics. **16**(1), 73–89 (2022)
2. Griliches, Z.: The search for R&D spill overs. Scand. J. Econ. **94**(1), 29–47 (1992)
3. Jones, C.I.: Time series tests of endogenous growth models. Q. J. Econ. **110**, 420–431 (1995)
4. Acemoglu, D., Angrist, J.: How large are human capital externalities? Evidence from compulsory schooling laws. NBER Macroecon. Annu. **15**, 9–59 (2020)
5. Iranzo, S., Peri, G.: Schooling externalities, technology, and productivity: theory and evidence from US States. Rev. Econ. Stat. **91**(2), 420–431 (2009)
6. Moretti, E.: Workers' education, spillovers, and productivity: evidence from plant-level production functions. Am. Econ. Rev. **94**(3), 656–690 (2004)
7. AlQershi, N., Mohd Mokhtar, S., Abas, Z.: The relationship between strategic innovations, human capital and performance: an empirical investigation. Sustain. Futur. **3**, 100056 (2021). https://doi.org/10.1016/j.sftr.2021.100056
8. Capozza, C., Divella, M.: Human capital and firms' innovation: evidence from emerging economies. Econ. Innov. New Technol. **28**(7), 741–757 (2019)
9. Chabbouh, H., Boujelbene, Y.: Open innovation in SMEs: the mediating role between human capital and firm performance. J. High Technol. Managem. Res. **31**(2), 100391 (2020)
10. Kahn, K., Candi, M.: Investigating the relationship between innovation strategy and performance. J. Bus. Res. **132**, 56–66 (2021)
11. Yue, W.: Human capital and firm innovation: evidence from China's higher education expansion in the late 1990s. Emerg. Mark. Financ. Trade. **60**(3), 500–518 (2024)
12. Kao, C., Chiang, M.H.: On the estimation and inference of a cointegrated regression in panel data. In: Baltagi, B. (ed.) Nonstationary Panels, Panel Cointegration, and Dynamic Panels (Advances in Econometrics), pp. 161–178. JAI Press, Amsterdam (2000)
13. Pedroni, P.: Purchasing power parity tests in cointegrated panels. Rev. Econ. Stat. **83**(4), 727–731 (2001)
14. Stock, J.H., Watson, M.W.: A simple estimator of cointegrating vectors in higher order integrated systems. Econometrica. **61**(4), 783–820 (1993)
15. European Commission: Directorate-General for Research and Innovation. In: Hollanders, H. (ed.) European Innovation Scoreboard 2023. Publications Office of the European Union (2023) https://data.europa.eu/doi/10.2777/119961

Chapter 22
Twin Transformation as an Enabler of Female Entrepreneurship

Anna-Luisa Stöber, Beatrice Zier, Alexander Roth, and Silke Conradi

Abstract In a period characterized by unparalleled technological progress, the terrain of worldwide economies is experiencing a significant transformation through digital upheaval. As nations strive to harness the potential of digital transformation for sustainable development, there exists a critical need to explore and understand the multifaceted dynamics that shape this evolution. One aspect that is receiving increased attention is the intersection between female entrepreneurship and female empowerment as powerful enablers driving the progress of digital transformation. Historically, women have encountered barriers that impede their full participation in economic activities, including entrepreneurship. However, the tide is changing as a growing body of research highlights the pivotal role that women play in fostering innovation, economic growth, and social development. This paradigm shift toward recognizing and harnessing almost untapped potential of female entrepreneurs is integral to realizing the true essence of digital transformation. In this perspective, the process of twin transformation aims to enhance a company through digital transformation coupled with sustainable transformation. Twin transformation in the context of female entrepreneurship pertains to how women can contribute to integrating the aspects of digitization and sustainability in businesses. This chapter follows the stages of small business growth to highlight how twin transformation impacts female entrepreneurship. The authors apply and adapt this framework and conduct a qualitative case study assessing the success story of a platform business startup in care industry. This research extends perspectives on twin transformation, demonstrating the pivotal role of females in transformation processes.

Keywords Female entrepreneurship · Digital transformation · Sustainability · Twin transformation

---

A.-L. Stöber · B. Zier (✉) · A. Roth
Macromedia University of Applied Sciences, Munich, Germany

S. Conradi
heynannyly GmbH, Bayern, Germany

F. Schiavone et al. (eds.), *Advanced Perspectives and Trends in Digital Transformation of Firms, Networks, and Society*, Springer Proceedings in Business and Economics, https://doi.org/10.1007/978-3-031-80692-6_22

22.1 Introduction

Female entrepreneurship has received increasing attention in various economic contexts [1]. Studies showed that companies started by women have a 63% increase in performance compared to companies started by male founders [2]. It is confirmed that women-led businesses can positively impact the economy and innovation [3]. Although more women want to step into self-employment nowadays [4], only 10% of start-ups were founded by female founders [5]. As innovation and international networks drive digital transformation, organizations must keep up to deploy adequate strategies to harness sustainable development and success in competitive industries [6]. Nevertheless, the pivotal role of women in fostering innovation, economic growth, and social development is neglected.

The reasons may be because challenges and societal perceptions still constrain women's entrepreneurial progress today [7]. Take the example of the digital revolution: it can bring immense potential to improve social and economic outcomes for women, for example, by enabling women in all parts of the world to start and grow businesses more effectively than ever before but also poses the risk of perpetuating existing patterns of gender inequality [8]. Compared to men, slow and organic business creation can be cited as a reason in this context [9]. Hence, startups under female leadership are seen as promising and they often integrate social and environmental sustainability firmly into their corporate strategy [10]. Women in entrepreneurship aim for economic success and take on and demonstrably drive social and environmental responsibility—an aspect that has so far received little attention in combination [11]. Facing these complex digitalization and sustainability challenges, the concept of twin transformation as the value-adding reinforcing interplay between digital and sustainability transformation efforts that improve an organization by leveraging digital technologies to enable sustainability and to guide digital progress [12] gains increasing importance.

However, in the face of advancing digitalization, there is little research on the development processes, long-term stability, start-up aspects, and sustainability of women-owned businesses [13]. Ongoing literature only discusses twin transformation and female entrepreneurship separately (e.g., [3, 7, 11, 12]). However, integrating feminist theories with existing theories on entrepreneurship offers a promising way to advance the field [14].

Little is known about specific mechanisms through which both concepts can be integrated, and it is not clear what factors can increase successful business development [15]. Overall, this study aims to understand the female leader's role in twin transformation and its impact on businesses' future success [16]. Accordingly, our research question is: How can the digital and sustainability aspects of twin transformation be integrated into female entrepreneurial activities? To answer the research question, we conduct qualitative research based on an inductive approach where semi-structured interviews are the foundation of empirical findings.

With this case study at hand, we contribute to theory and practice in twofold ways: (1) showcasing a successful example of a female entrepreneurship venture and (2) adapting the model of the stages of entrepreneurial growth of Lewis and Churchill [17] to the digital and sustainability aspects of twin transformation in the context of female entrepreneurial activities. A mixed approach is taken, combining surveys and interviews with female entrepreneurs to investigate the influence on entrepreneurial action and to shed light on the current state of knowledge of female entrepreneurship. This relatively unexplored dimension of entrepreneurship is illuminated, providing insights into the dynamics and impact of Twin Transformation on female entrepreneurship, thereby expanding existing knowledge in this field [18].

The remainder of this paper is organized as follows: Sect. 22.2 provides theoretical background on female entrepreneurship, twin transformation, and the interplay of both. In Sect. 22.3, we introduce the research method and the case study. Section 22.4 demonstrates the applicability of the model of the Stages of entrepreneurial growth of Lewis and Churchill [17]. In Sect. 5, we briefly summarize the key findings and provide topics for future research. Section 6 presents implications to theory, practice, and policy.

22.2 Female Entrepreneurship and Twin Transformation

22.2.1 Defining the Scope and Importance of Female Entrepreneurship

The term entrepreneur encompasses all individuals contemplating starting a business or having already established one [19]. The term entrepreneurship in entrepreneurial research involves, on the one hand, the exploitation of entrepreneurial opportunities and, on the other hand, the creative and innovative entrepreneurial process within a company or during a period of entrepreneurial change [20]. Over recent years, entrepreneurship has been predominantly perceived as a domain associated with males [21]. Thus, there is often a tendency to generalize entrepreneurship research [22] without focusing specifically on women [23]. Consequently, in recent years, the research field of female entrepreneurship has manifested itself as an independent branch within the research. It is comparatively underexplored, leading to increased interest in the exploration of business startups by women [24].

The term female entrepreneurship refers to women undertaking the risks associated with starting their own business and employing others [25]. An analysis of the current literature shows that analyses of female entrepreneurship have not yet been sufficiently investigated [26]. Due to the slow but steady increase in women's entrepreneurial ventures in recent years, the full potential in this area has not exploited yet [3, 27]. The literature also shows that women are more cautious and thorough when starting a business and act more cautiously in the long term [28]. Furthermore, the literature provides evidence that male and female entrepreneurs differ in terms of the industry in which they operate, the products they bring to the market, the goals they pursue, the way they structure their businesses, and the extent of delegation they choose [29]. Various aspects of the topics are well researched, such as emerging and developing countries [30]. But the influence of digitalization has not yet researched thoroughly. The first research deals with digitalization in female entrepreneurship by systematically reviewing the literature of 18 papers published between 2017 and 2022 by discipline, time, methodologies, context, topic, and theoretical emphasis [31]. The results show that the literature on female digital entrepreneurship is inadequate, fragmented, and divergent regarding less practice-based insights.

In this context, the term twin transformation stands for a comprehensive change in the economy and society in which the two key elements of digitalization and sustainability are linked, which has not yet been given sufficient importance in research [32]. As a result, the twin transformation defines the productive interaction between efforts toward digital and sustainable transformation, aiming to enhance the performance of an organization. Digital technologies are actively applied to enhance sustainability, while sustainable practices guide the progress of these digital advancements [12]. Thus, the twin transformation approach aims to harness the benefits and innovations of digital technologies to promote sustainable business models [33].

While the general area of female entrepreneurship is well-researched concerning international comparisons and small and family businesses, there is less research focusing on women's entrepreneurial activities to improve social sustainability regarding equality and equal opportunities. From this perspective, it would be meaningful to incorporate the aspect of twin transformation into research on female entrepreneurship. The fundamental idea is that women can contribute to the economically successful integration of digitization and sustainability in businesses [34]. In this context, the twin transformation emphasizes the importance of integrating digital and sustainability aspects into entrepreneurial activities, with women having the potential to play a crucial role in shaping change. They can develop new approaches to initiate transformation within businesses, connecting economic success with social responsibility [11].

22.2.2 Methods

As this study emphasizes understanding the context within which a phenomenon occurs and how it is influenced by various factors, case study research allows us to thoroughly examine the interplay between context and the subject of study. By gaining a thorough and comprehensive understanding of a specific case's context, processes, and outcomes, we want to identify patterns and lay the groundwork for further investigation. We follow Merriam and Tisdell [35] and pursue the intensive analysis of the individual, for example, a person or community, while stressing developmental factors about the environment or developmental stages. This study adopted the method of Gioia et al. [36] to identify patterns and themes that can generate valuable conclusions on the chosen topic. The Gioia method is a systematic approach to developing new concepts and consists of essential elements [37]. This includes the elaboration of a data structure, the creation of a solid model based on this structure, and the presentation of well-founded results in a convincing manner [38]. The objective of the Gioia approach is to elevate the precision and scientific caliber of inductive investigations [39]. It addresses issues such as sampling, creativity, and reflexivity [40]. In this context, we consider three important quality criteria: objectivity, reliability, and validity [41].

Objectivity means looking at results without being influenced by personal biases. In this context, independent research results were described in detail in this chapter from different perspectives and on the basis of factual arguments. These descriptions were made independently of the researcher or observer, which guarantees objectivity. *Reliability* ensures that the results remain constant and are reproducible. The Gioia method was used to analyze measurement accuracy, stability of results, and replicability. By repeatedly applying this method with the same instruments and methods, it was possible to achieve the same results. *Validity* makes sure that measurements accurately show exactly what was recorded. Applying these standards helps ensure that the research is of high quality and trustworthy. Within the scope of the study, the measurement accuracy of Heynanny's key characteristics was ensured by using fit questions to sample representativeness according to Lewis and Churchill's [17] five stages of small business growth as a valuable and proven tool. This ensures validity.

A qualitative methodology was adopted for this research and semi-structured in-depth interviews were conducted to unravel the complex stages of small business growth by twin transformation. Before data collection, the initial questions were developed from the literature review, and further questions were asked to understand the topic comprehensively. The interview lasted 40 min, and the recordings were transcribed to obtain data for analysis. This case study aims to examine a specific company's female entrepreneurial activities focusing on integrating the digital and sustainability aspects of twin transformation. We are highlighting the challenges they faced and the strategies they employed to overcome them. By delving into this case, we aim to gain insights into the complexities of female entrepreneurial activities in the context of twin transformation and explore the potential impact of such initiatives on corporate performance and stakeholder relationships [42]. Heynanny's journey offers valuable lessons for other companies seeking to embark on similar paths. Through an in-depth analysis of this case, we hope to uncover key factors contributing to the success of female entrepreneurial initiatives in the context of twin transformation and identify best practices for fostering a culture of corporate responsibility.

The conceptualization of organizational development often employs models like life cycles or stages, capturing the organization's evolution from the start to maturity. These models typically delineate the common challenges faced by organizations and their adaptive strategies through growth phases [43, 44, 45]. The scholars adapted or extended the model introduced by Churchill and Lewis [46], which is utilized in this study. The original model is favored for its efficacy in identifying growth-related organizational problems and bottlenecks, although it notably omits the decline phase. It articulates distinct stages of organizational growth with each stage punctuated by a critical crisis point that must be navigated to advance to the subsequent stage.

Heynanny's vision is to help shape a modern working world with equality and equal opportunities for people with care issues. They believe in family-friendly companies that work with them to become an employer of choice and make their employees happy. Heynanny is a corporate benefit for family and career. Through the exclusive all-in-one platform, they provide certified, insured, and trustworthy childcare and elderly care as part of our employee self-service. They are available to all employees of their partner companies regularly and at short notice [47]. A key feature is that the company can finance these services as a corporate benefit. If companies subsidize care hours, the service is free of charge for parents and relatives up to a specific limit. Tax advantages can be claimed on the company side. With heynanny, employees and companies can also take advantage of the monetary benefits for care services, which often go unused due to high administrative barriers [48]. Companies benefit from the regular and short-term relief of their employees and the reduction of downtime. Additionally, this increases employee satisfaction as a return on investment. At the same time, heynanny relieves the companies' HR teams [49]. This concept convinced investors and award juries: As the first investor, allygatr, a Berlin-based venture capital firm specializing in HR tech, joined. Subsequently, several business angels also invested. In July 2023, the team successfully completed the seed financing round with a total volume of 1.6 million euros [48]. Heynanny was awarded Startup of the year by the Founder Summit 2022 and received the HR Startup Award Special Award 2023.

22.2.3 Findings

The case study results will be conceptualized to be able to draw conclusions. We apply the framework outlined in Lewis and Churchill's [17] five stages of small business growth as it serves as a valuable and well-established tool for entrepreneurs, business owners, investors, and other stakeholders involved in the small business ecosystem, helping them navigate the complexities of growth and maximize the chances of success. Table 22.1 compares the interview findings to the theoretical description of the five stages. This procedure enables drawing conclusions and reflecting on supplementary insights.

Table 22.1 Five stages of heynanny's business growth

Stage	Theory	Case study
1	*Existence as a viable business* • The main goal is creation of the business. • The most important issues include obtaining customers, providing the product or service, and finding enough cash to reach viability. • The owner does everything, and any other employees report directly to him or her. • Thus, the critical skills are the owner's operating ability, access to financial resources, and ability to develop the processes and business relationships that will create the company's competitive position in the industry. These are guided by the owner's personal and business goals.	*How was heynanny created? 08/21–03/22* • Business idea created based on mutual interests of founders: influencing equality and enabling women to pursue their careers. • Common values of founders as enabler • Female founder network as platform for business ideas (changes due to post-corona care crisis enabled to rapidly start developing the business idea further). • Network of startup incubator as supporters. • Critical skill in this stage is to be able to "try and error" and be "hands on." • Low costs as priority (e.g., IT/website: self-created webpage à freelancer à agency). • Defining MVP (Minimum Viable Product).

(continued)

Table 22.1 (continued)

Stage	Theory	Case study
2	*Survival* • The business has demonstrated that it is potentially viable and has established a market niche. • Key issues in this stage revolve around the relationship between revenues and expenses. • In the short run, the goal is to break even, with sufficient funds to maintain the capital assets. • In the longer view, the company needs to generate enough cash to grow in. • The critical skills are similar to Stage One.	*Becoming a company 03/22–05/22* • Founding of GmbH/ limited liability company. • Testing of MVP (1000 Nanny, 4 months test in Munich, B2C) afterward roll-out in 10 cities all over Germany. • First early stage investing via venture capital. Investor was necessary to cover IT costs but did not increase strategic potential. • Founders were able to pay themselves their first salary. • Founders are doing everything.
3	*Stabilization* • The business is economically viable and stable enough to support the owners, returning a profit. • It is likely to have evolved into a functional structure, with some professional managers. • The company can remain at this level as long as its environment does not change. • The main issue, then, is what to do next. • The owners may begin to disengage themselves from the business. • The most important skill at this stage is still the owner's operational abilities, with development of business resources and relationships and its financial resources also still important.	*Creating a new normal 10/22–06/23* • Founded quickly and sustainably with three companies. • Further employees with new Skills brought into the team. • Big key skills à Conscious network visibility from minute one. • The most crucial aspect for startups is securing funding "not only for mentoring but also to establish a digital presence." • Next step, target financing round 06/23 approx. 1,6 million EUR raised by business angels. • The German market is open.

(continued)

Table 22.1 (continued)

Stage	Theory	Case study
4	*Growth orientation* • The goal now is to become a big company, and resources and profits are used to fuel growth. • The main issues are acquiring the financial resources and developing the systems to enable that growth. • The owner's personal goals and strategic ability, as well as operational ability, are key skills. • Additional factors that are becoming important include the owner's management skills and ability to delegate and access to personnel resources, especially at the management and staff levels.	*Focusing on growth* • Current market entry in Austria. • Next big step internationalization, e.g., inquiries from UK investors. • To achieve the expansion target, it is important to keep the focus for the personal skills. • Learning from German market entry, good onboarding through a lot of research work, maximum accessibility and correct analysis of the market. • Expand internal processes, developing the owner's management skills, hiring new employees and delegating tasks. • Heynanny is 100% digitally bookable in the end but 100% live in terms of the support you receive—subsidized, high-quality care. • Further digital services: advice can be booked digitally with software, chat function, digital introduction, digital invoices.
5	*Take-off/rapid growth* • The business has developed the resources to grow rapidly. • Problems now result from the focus on how to grow and how to finance that growth and/or manage cash flow. • The organization is decentralizing, and another key issue is delegation and development of control systems. • Due to the demands of rapid growth, all of skills and factors are important, although the owner's operational ability is decreasing in importance and the business resources and relationships have already been established.	*This stage has not yet been reached* • Heynanny didn't reach the requirements to be on the stage yet. • Future outlook of the company.

22.3 Discussion

22.3.1 Enhancing Strategic Development through Multi-layer Analysis

As evident from the findings in the preceding section and the interview results, the startup Heynanny corresponds to the stages enumerated in the five stages of small business growth, but there is more to consider. Investigating additional factors and adjusting strategies aligned with the findings will enhance our comprehension of the company's potential for success. In this context, incorporating three different layers is essential for a more comprehensive analysis and strategic development [50].

The first layer that needs to be added is the digital layer. On this level, the importance of the digital platform is to be emphasized. This pertains to the technological structure and the seamless process a user experiences using the company's platform. This includes all digital services provided through the company's platform to offer the customer an optimal user experience. In this context, as was emphasized in the interview, the company should ideally strive to integrate fast digital processes that can be used 100% digitally in order to enable a decisive expansion or diversification of the product range. The so-called multichannel approach allows every customer to be reached, potentially enhancing the company's success [51]. As a final yet crucial point in this context, the significance of the technical backend and frontend should be highlighted, without which the preceding digital processes would not be possible. An efficient backend refers to the part of a software application not directly seen or used by end-users but responsible for user authentication. It forms the invisible backbone of the application, dealing with data processing, database management, logic, and other non-visible aspects [52]. The frontend represents the interactive user interface used by users (Filipova & Vilão, 53).

The second level that should be considered is the sustainable layer. When it comes to staffing at this level, involving individuals with extensive practical experience is crucial. Long-term and widely accessible orientations should be implemented to promote a sustainable work environment. This includes aspects such as reducing the mental load on employees, promoting retreat options, and considering the health aspects of the workforce. Equality among all employees is a key issue at this level, which aims to ensure that all employees have equal work and career opportunities [54]. In summary, the sustainable layer focuses on holistic and inclusive practices that not only contribute to the well-being of individual employees but also foster a work environment that promotes equality, diversity, and long-term organizational success [55].

The third layer that needs to be added is the female layer. It is crucial to note that statistically, women do not have the same job opportunities as men, especially when they decide to start a family. It is imperative to recognize and eliminate the obstacles that hinder women's career progression and opportunities for professional development. Moreover, the establishment of efficient networks, whether digital or offline, plays a pivotal role in supporting women in their professional journeys. Networking platforms can provide opportunities for mentorship, skill development, and knowledge sharing, contributing to the overall empowerment and advancement of women in the workplace [56].

22.3.2 *Implications*

With the presented approach, we contribute to theory, practice, and policy. By extending the five phases of Lewis and Churchill [17] particularly in the context of female entrepreneurship in the twin transformation we contribute to theory. In our exploration, we initially illustrated how organizations can structure the field of action. Given that the implementation of female entrepreneurship in twin transformation is characterized by ongoing evolution, we have adapted the fundamental concept of Lewis and Churchills' stages to provide a framework for describing the sequential progression of stages that constitute an anticipated path from an initial state to maturity.

Practical implications are created by adapting the methodology and developing a better understanding of the development journey within this transformative process. In this context, examining the successful example of female entrepreneurship using the startup heynanny illustrates the significance and success of female entrepreneurs and the exemplary influence of women in the startup scene. Companies can contribute to advancing gender equality by fostering an environment that recognizes and accommodates the unique challenges faced by female employees. New strategies to promote female entrepreneurship should serve to put in place effective support mechanisms for women entrepreneurs at different stages of their business.

It requires various multi-faceted policy-oriented approaches to enable women to bring their full potential to their professional lives and economic activities [56]. Equal economic participation is not only a matter of social justice but also creates potential for greater prosperity and sustainability in the interest of the socio-ecological market economy [57].

22.4 Conclusion and Limitation

The main goal of this study was to examine how female-led businesses integrate digital and sustainable practices in a process termed twin transformation. The paper presents a case study of a platform business startup in the care industry to illustrate these concepts. The key findings are mapped against the theoretical framework of the stages of small business growth by Lewis and Churchill [17], highlighting how digital and sustainability transformations are integrated at different stages of business development by female entrepreneurs.

While offering valuable insights, this research depicts a single case study approach, primarily serving as an initial exploration into the integration of sustainability and digitalization within organizational structures. Given the singular context of the study of a German care industry company heynanny, the unique insights into the characteristics and challenges may be not necessarily generalizable to other industries. However, from this specific case, future research should aim to broaden the scope, incorporating multiple case studies across diverse industries and regions to validate and potentially expand upon the findings of this initial research.

Despite these limitations, the study opens avenues for further research in female entrepreneurship and the operationalization of sustainability and digitalization within organizations. The insights gained from this exploratory case study can serve as a foundational step toward a more comprehensive integration of sustainable practices in corporate decision-making. Future research could focus on examining the applicability of the proposed stages of growth and sustainability integration in different industries and cultural contexts, thereby enhancing the generalizability of the findings. Researchers could also explore the varying impacts of sustainability practices across different types of organizations, particularly in the context of female-led enterprises.

Taken together, this study contributes significantly to the understanding of entrepreneurial growth stages and the integration of sustainability and digitalization in organizational contexts. By applying a case-based research approach, the study not only affirms existing knowledge about entrepreneurial growth stages but also provides exploratory insights into decision-making processes at each stage. Also, we increase the awareness of female entrepreneurship in the context of digital and sustainable business practices, emphasizing the unique role of women in driving these transformative processes. Overcoming the limitations of this research we exemplify the contribution of case-based research to teaching, research, and the development of both organizations and individuals [58].

References

1. Khan, R.U., Salamzadeh, Y., Shah, S.Z.A., Hussain, M.: Factors affecting women entrepreneurs' success: a study of small- and medium-sized enterprises in emerging market of Pakistan. J. Innov. Entrep. **10**(1), 11 (2021). https://doi.org/10.1186/s13731-021-00145-9
2. Female Founders Fund: Female Founders Fund. https://Femalefoundersfund.com/ (2023)
3. Meyer, D., Meyer, R., Merz, D.: Female Entrepreneurship: Unternehmensgründungen von Frauen im Fokus (2021)
4. McKinsey & Company: Entrepreneurship Zeitgeist 2030. https://www.mckinsey.com/industries/technology-media-and-telecommunications/our-insights/entrepreneurship-zeitgeist-2030 (2021)
5. Boston Consulting Group: Women are under-represented in the European startup ecosystem. https://www.bcg.com/press/13june2023-women-are-under-represented-in-the-european-startup-ecosystem (2023). Accessed 15 Feb 2024
6. Calderon-Monge, E., Ribeiro-Soriano, D.: The role of digitalization in business and management: a systematic literature review. Rev. Manag. Sci. (2023). https://doi.org/10.1007/s11846-023-00647-8

7. Ilie, C., Monfort, A., Fornes, G., Cardoza, G.: Promoting female entrepreneurship: the impact of gender gap beliefs and perceptions. SAGE Open. **11**(2), 215824402110184 (2021). https://doi.org/10.1177/21582440211018468
8. Rosepti, P., Niasari, C.: Female entrepreneurship in leading digitalization as a marketing strategy for prosperous MSMEs and women's empowerment: a case study of Latifa Indonesia. Muslim Bus. Econ. Rev. **1**(2), 305–323 (2022)
9. Ughetto, E., Rossi, M., Audretsch, D., Lehmann, E.E.: Female entrepreneurship in the digital era. Small Bus. Econ. **55**(2), 305–312 (2020). https://doi.org/10.1007/s11187-019-00298-8
10. Gather, C., Biermann, I.: Unternehmen: Warum gründen Frauen seltener? Wirtschaftsdienst (2021) https://www.wirtschaftsdienst.eu/inhalt/jahr/2021/heft/11/beitrag/unternehmen-warum-gruenden-frauen-seltener.html
11. Graf-Drasch, V., Kauffeld, L., Krempf, L., Oberländer, A. M., Teuchert, A.: Driving twin transformation - the interplay of digital transformation and sustainability transformation. In ECIS 2023 Research Papers, 255. https://aisel.aisnet.org/ecis2023_rp/255/ (2023)
12. Christmann, A.-S., Crome, C., Graf-Drasch, V., Oberländer, A.M., Schmidt, L.: The twin transformation butterfly. Bus. Inf. Syst. Eng. (2024). https://doi.org/10.1007/s12599-023-00847-2
13. Genç, M., Öksüz, B.: A fact or an illusion: effective social media usage of female entrepreneurs. Procedia. Soc. Behav. Sci. **195**, 293–300 (2015). https://doi.org/10.1016/j.sbspro.2015.06.345
14. de Bruin, A., Brush, C.G., Welter, F.: Advancing a framework for coherent research on women's entrepreneurship. Entrep. Theory Pract. **31**(3), 323–339 (2007). https://doi.org/10.1111/j.1540-6520.2007.00176.x
15. Civera, A., Meoli, M.: Empowering female entrepreneurs through university affiliation: evidence from Italian academic spinoffs. Small Bus. Econ. **61**(3), 1337–1355 (2023). https://doi.org/10.1007/s11187-022-00729-z
16. Akter, M., Rahman, M., Radicic, D.: Women entrepreneurship in international trade: bridging the gap by bringing feminist theories into entrepreneurship and internationalization theories. Sustainability. **11**(22), 6230 (2019). https://doi.org/10.3390/su11226230
17. Lewis, V., Churchill, N.: The five stages of small business growth. Harv. Bus. Rev. **3**(3) (1987)
18. Müller, S., Kirst, A.L., Bergmann, H., Bird, B.: Entrepreneurs' actions and venture success: a structured literature review and suggestions for future research. Small Bus. Econ. **60**(1), 199–226 (2023). https://doi.org/10.1007/s11187-022-00644-3
19. Iversen, J., Jørgensen, R., Malchow-Møller, N.: Defining and Measuring Entrepreneurship. Now Publishers (2008) https://books.google.de/books?id=Ds8GsBHrdRsC
20. Fueglistaller, U., Müller, C., Müller, S., Volery, T.: Entrepreneurship. Gabler Verlag (2016). https://doi.org/10.1007/978-3-8349-4770-3
21. Basu, A., Casson, M., Wadeson, N., Yeung, B. (eds.): The Oxford Handbook of Entrepreneurship. Oxford University Press (2008). https://doi.org/10.1093/oxfordhb/9780199546992.001.0001
22. Henry, C., Foss, L., Ahl, H.: Gender and entrepreneurship research: a review of methodological approaches. Int. Small Bus. J. **34**(3), 217–241 (2016). https://doi.org/10.1177/0266242614549779
23. Marlow, S.: Gender and entrepreneurship: past achievements and future possibilities. Int. J. Gend. Entrep. **12**(1), 39–52 (2020). https://doi.org/10.1108/IJGE-05-2019-0090
24. Ratten, V., Usmanij, P.: Entrepreneurship education: time for a change in research direction? Int. J. Manag. Educ. **19**(1), 100367 (2021). https://doi.org/10.1016/j.ijme.2020.100367
25. Tomos, F., Kumar, N., Clifton, N., Hyams-Ssekasi, D.: Women Entrepreneurs and Strategic Decision Making in the Global Economy. IGI Global (2019). https://doi.org/10.4018/978-1-5225-7479-8
26. Serafini, G.: Business economics vs political economics: why female entrepreneurship value creation is underestimated at macroeconomic level. Proc. IFKAD, 957–961 (2017)
27. Startupdiversity: Initiative #startupdiversity. https://startup-diversity.de/ (2023)

28. Mütze, J.: Frauen gründen gründlicher—und vorsichtiger. In: Digitalisierung als Chancengeber, pp. 49–54. Springer Fachmedien Wiesbaden (2020). https://doi.org/10.1007/978-3-658-26893-0_5
29. Mukhtar, S.-M.: Differences in male and female management characteristics: a study of owner-manager businesses. Small Bus. Econ. **18**(4), 289–311 (2002) http://www.jstor.org/stable/40229211
30. Corrêa, V.S., Brito, F.R.d.S., Lima, R.M.d., Queiroz, M.M.: Female entrepreneurship in emerging and developing countries: a systematic literature review. Int. J. Gend. Entrep. **14**(3), 300–322 (2022). https://doi.org/10.1108/IJGE-08-2021-0142
31. Alhajri, A., Aloud, M.: Female digital entrepreneurship: a structured literature review. Int. J. Entrep. Behav. Res. (2023). https://doi.org/10.1108/IJEBR-09-2022-0790
32. Ziegler, K.: Die Twin-Transformation von Unternehmen ist ein Turbo-Booster. https://www.ibm.com/blogs/digitale-perspektive/2022/04/nachhaltigkeit-transformation-in-unternehmen/ (2022)
33. Kamble, S.S., Gunasekaran, A., Parekh, H., Mani, V., Belhadi, A., Sharma, R.: Digital twin for sustainable manufacturing supply chains: current trends, future perspectives, and an implementation framework. Technol. Forecast. Soc. Chang. **176**, 121448 (2022). https://doi.org/10.1016/j.techfore.2021.121448
34. Fraunhofer FIT: Fraunhofer-Institut für Angewandte Informationstechnik FIT. https://www.fit.fraunhofer.de/de/publikationen/studien-und-whitepaper.html (2023)
35. Merriam, S.B., Tisdell, E.J.: Qualitative Research: a Guide to Design and Implementation, 4th edn. Jossey-Bass (2015)
36. Gioia, D.A., Corley, K.G., Hamilton, A.L.: Seeking qualitative rigor in inductive research. Organ. Res. Methods. **16**(1), 15–31 (2013). https://doi.org/10.1177/1094428112452151
37. Mees-Buss, J., Welch, C., Piekkari, R.: From templates to heuristics: how and why to move beyond the Gioia Methodology. Organ. Res. Methods. **25**(2), 405–429 (2022). https://doi.org/10.1177/1094428120967716
38. Bouncken, R.B., Qiu, Y., Sinkovics, N., Kürsten, W.: Qualitative research: extending the range with flexible pattern matching. Rev. Manag. Sci. **15**(2), 251–273 (2021). https://doi.org/10.1007/s11846-021-00451-2
39. Magnani, G., Gioia, D.: Using the Gioia Methodology in international business and entrepreneurship research. Int. Bus. Rev. **32**(2), 102097 (2023). https://doi.org/10.1016/j.ibusrev.2022.102097
40. Gehman, J., Glaser, V.L., Eisenhardt, K.M., Gioia, D., Langley, A., Corley, K.G.: Finding theory–method fit: a comparison of three qualitative approaches to theory building. J. Manag. Inq. **27**(3), 284–300 (2018). https://doi.org/10.1177/1056492617706029
41. Heesen, B.: Qualitätskriterien. In: Wissenschaftliches Arbeiten, pp. 11–19. Springer, Berlin Heidelberg (2021). https://doi.org/10.1007/978-3-662-62548-4_3
42. Stam, E., van de Ven, A.: Entrepreneurial ecosystem elements. Small Bus. Econ. **56**(2), 809–832 (2021). https://doi.org/10.1007/s11187-019-00270-6
43. Davidsson, P., Achtenhagen, L., & Naldi, L. (2005). Research on small firm growth: A review. In 35th EISB Conference: Sustaining the Entrepreneurial Spirit over Time: Implications for Young Companies, Family Businesses, and Established Companies (pp. 1–27). IESE Business School
44. Dobbs, M., & Hamilton, R. T. (2007). Small business growth: recent evidence and new directions. Int. J. Entrep. Behav. Res. **13**(5), 296–322.
45. Scott, M., & Bruce, R. (1987). Five stages of growth in small business. Long Range Plan. **20**(3), 45–52.
46. Lewis, V. L., & Churchill, N. C. (1983). The five stages of small business growth. University of Illinois at Urbana-Champaign's Academy for En-trepreneurial Leadership Historical Research Reference in Entrepreneurship
47. heynanny: heynanny. https://www.heynanny.com (2024)

48. Henseleit, L.: heynanny: Die work-life-heroes. https://www.starting-up.de/geschaeftsideen/trends/start-up-report-sextech.html (2023)
49. Nemitz, M.: Heynannyly sichert sich in überzeichneter Seed-Runde 1,6 Mio EUR. Heynannyly sichert sich in überzeichneter Seed-Runde 1,6 Mio EUR (2023)
50. Deloitte: Spans and layers for the modern organization. https://www2.deloitte.com/content/dam/Deloitte/us/Documents/human-capital/us-spans-and-layers-for-the-modern-organization-2020.pdf (2020)
51. Fürst, A., Leimbach, M., Prigge, J.-K.: Organizational multichannel differentiation: an analysis of its impact on channel relationships and company sales success. J. Mark. **81**(1), 59–82 (2017). https://doi.org/10.1509/jm.14.0138
52. Oyeyemi, O.P., Kess-Momoh, A.J., Omotoye, G.B., Bello, B.G., Tula, S.T., Daraojimba, A.I.: Entrepreneurship in the digital age: a comprehensive review of start-up success factors and technological impact. Int. J. Sci. Res. Arch. **11**(1), 182–191 (2024). https://doi.org/10.30574/ijsra.2024.11.1.0030
53. Filipova, O., & Rui V. (2018). "Frontend Development." In Software Development From A to Z, pp. 133–180. Berkeley, CA: Apress. https://doi.org/10.1007/978-1-4842-3945-2_6
54. Grillitsch, M., Schubert, T.: Does the timing of integrating new skills affect start-up growth? Strateg. Entrep. J. **15**(4), 647–684 (2021). https://doi.org/10.1002/sej.1375
55. Salehi, F., Abdollahbeigi, B., Najafabadi, M.K., Sajjady, S.: Role of organizational behavior maturity in the business success of the start-ups. Asian J. Sociol. Res. **4**(1), 241–245 (2021) Retrieved from https://globalpresshub.com/index.php/AJSR/article/view/1255
56. German Federal Ministry for Economic Affairs and Climate Action: Women in business. https://www.bmwk.de/Redaktion/EN/Dossier/women-in-business.html (2024)
57. European Parliament, Directorate-General for Internal Policies of the Union, Bastida, M: Women's Entrepreneurship and Self-Employment, Including Aspects of Gendered Corporate Social Responsibility. European Parliament (2021) https://data.europa.eu/doi/10.2861/284985
58. Naumes, W., Naumes, M.J., Merenda, M.: A case based analysis of the stages of entrepreneurial growth: a preliminary study. Int. J. Case Method Res. Appl. **XIX**(1), 63–73 (2007)

Chapter 23
The Role of Workforce Governance in the Authentic Leadership-Digital Culture Relationship: Toward a Trust-Based Conceptual Model Anchored in the Cultural Context

Kouassi François Ouffouet

Abstract Authentic leadership and digital culture are essentials for successful digital transformation. So is governance, for leadership must use it to transform the workforce otherwise digital transformation will be inconclusive. But, while the mechanism by which leadership influences corporate culture has a growing interest, the relationship between authentic leadership, workforce governance, and digital culture remains unexplored. Moreover, workforce governance and digital culture are still emerging concepts that need clarification. Thus, based on a literature review and the theory of trust, this research proposes a conceptual model of workforce governance and its role in the authentic leadership-digital culture relationship. As SMEs in Côte d'Ivoire face digital transformation challenges, this model was analyzed in context. According to the results, workforce governance concept in a collectivist culture must integrate mutual aid and solidarity mechanisms in order to make ease digital culture building. For this, authentic leadership has to develop, besides dimensions identified in literature, a mutual aid and solidarity relationship with workforce.

Keywords Digital transformation · Collectivist culture · Côte d'Ivoire

23.1 Introduction

Since digitalization uprising, scientific literature has given a growing interest to leadership skills in digital era [1]. Trying to reconcile contemporary and traditional theories, Kane et al. [2] support the need for digital leaders to combine traditional

---

K. F. Ouffouet (✉)
Université Jean Lorougnon Guédé, Daloa, Côte d'Ivoire

F. Schiavone et al. (eds.), *Advanced Perspectives and Trends in Digital Transformation of Firms, Networks, and Society*, Springer Proceedings in Business and Economics, https://doi.org/10.1007/978-3-031-80692-6_23

and recent skills to effectively guide their organization into the future. For this purpose, authentic leadership is emphasized [3, 4] due to the centrality of ethics as common characteristic to authentic leadership [5], "e-leadership" [1] and digital transformation [6].

Digital transformation refers to companies' technology adoption to varying degrees [7]. According to Walumbwa et al. [8], "authentic leadership is a model that through self-awareness, internalized moral perspective, balanced information processing and relational transparency creates a positive work environment, favorable to leader and collaborators development."

Besides authentic leadership, digital culture is another key organizational variable for successful digital transformation [9–11]. We define it as a cultural bank [12] consisting in values and behaviors shared by an organization's members and which promote the organization's flexibility throughout its digital transformation [9]. However, digital culture remains insufficiently explored [13].

According to Fakhri et al. [14], digital leaders leverage digital culture to drive digital transformation, which Williams [15] assimilates to a cultural transformation. Then, we can insight an influence of authentic leadership on digital culture. Yet, except in South Korean context, this influence remains little explored [16].

To involve employees in a successful digital transformation, workforce governance is also a third organizational variable that is needed for leadership [9]. But, whereas, authentic leadership and digital culture relationship is part of a growing body of works calling, through years, scholars to identify mechanisms by which leadership influences corporate culture [9, 17–23], literature have not yet explored workforce governance as a mechanism by which authentic leadership can foster digital culture.

Without workforce transformation, technological change cannot take place [24, 25] or even be sustained [26]. Then, it is useful to restructure workforce management to make ease technology adoption and avoid digitalization failure [15]. However, just as digital governance remains weakly conceptualized [27], workforce governance concept remains diffuse, for digital transformation is not a destination but a journey submitted to various changes and a continual learning [9]. As such, designing a universal model of workforce governance to address challenges of change can bias leadership decision process [28, 29] aiming to meet workforce transformation challenges and reduce digitalization failure caused by a weak digital culture [10].

As building a universal governance model is elusive [30], it is relevant for research, called to analyze workforce governance [9], to build a basic and adaptable conceptualization for leadership. However, as national culture influences technology adoption [31] and leadership [32], this conceptualization can be cultural adjusted to help leaders elaborate their digital strategies. Hence, conceptualizing workforce governance and its interaction with leadership can help to understand mechanisms by which leadership influences digital culture. But this requires a solid theoretical framework grounded in digital context.

For both digital transformation and workforce governance thriving, collaboration is required [27, 33]. Nevertheless, digitalization negatively affects workforce and hinders collaboration by destroying trust relations [34]. So, trust in management becomes critical to influence the way employees feel, think, and behave concerning technological change and technology adoption [34]. Therefore, leaders acting with authenticity [33] and transparency [35] can foster trust in workplace and then collaboration through workforce governance and build a digital culture. But, not only little is known about individuals trusting relationship in technological affected organization [34], a theoretical framework that investigates workforce governance role in the authentic leadership-digital culture relationship is lacking. Therefore, more exploratory studies are required to get a deeper understanding of how trust can be developed in digitally transformed organizations [36].

To address these gaps, we build a conceptual model based on trust to analyze authentic leadership, workforce governance, and digital culture relationship. Doing so, we contribute to fill the lack of studies on corporate culture significance in SMEs [37] whereas they are successful in digital transformation [38]. Moreover, in Africa, we contribute to identify a leadership style to address digital transformation challenges in SMEs [39]. Particularly, we aim to help SME managers in Côte d'Ivoire to define successful digital strategies [40], as well as public authorities to better manage their public actions to promote technologically innovative SMEs [41]. Thus, our research question is: How conceptualize workforce governance and its role in the authentic leadership-digital culture relationship and in particular in the cultural context?

Our aim in this article is to propose a conceptualization of workforce governance and its role in the authentic leadership-digital culture relationship through trust theory. And then adapt it to SMEs' cultural context in Côte d'Ivoire.

23.2 Authentic Leadership, Workforce Governance and Digital Culture Relationship: Toward a Conceptual Model Based on Trust Theory

23.2.1 Digital Transformation: Workforce Governance, a Key but Unconceptualized Concept

As workforce remains essential to any technological change [25], it is convenient to ensure its ongoing involvement in new ways of working adoption [9]. Hence, leadership has to invest in an adequate workforce governance development to succeed digitalization [9]. However, workforce governance is part of the emerging digital phenomena that deserves clarification of their scope and better conceptual distinction [42].

Through a deductive approach, we called literature dealing with digital governance, leadership, and digital transformation. As until 2017 digital transformation was not at the heart of academic debate [43], articles from 2016 to 2023 have been selected. Therefore, we examined 11 articles including 5 literature reviews, two case studies and 4 research articles.

These articles were selected in *European Management Journal*, *Journal of Business Research*, *Sustainable Technology and Entrepreneurship*, *International Journal of Leadership*, *Information System Frontiers*, *Frontiers in Psychology*, *Journal of Knowledge Economy*, *Journal of Decision Systems*, *Contemporary Global Issues in Human Resource Management*, *Asia Pacific Business Review* and research technology management.

Workforce governance in digital era has not been defined in literature, but some studies have focused on techniques and methods used to ensure workforce involvement in digital transformation process [9, 44, 45]. According to Ḩanisch et al. ([27]), "governance consists of establishing inputs and outputs verification rules (control mechanisms), dividing and allocating tasks (coordination mechanisms), aligning competing interests (incentive mechanisms) and mitigating relational vulnerabilities (trust mechanisms)."

Referring to this literature, we identified techniques and methods related to workforce governance involvement in digitalization and associated them with governance dimensions. Tables 23.1 and 23.2 present the results of our deductive analysis.

We also define workforce governance as: "a set of control, coordination, incentive and transparency rules [27] developed in digital workplace by leadership to motivate employees to adopt digital tools, technologies and their new ways of working, so as to maximize the benefits expected from technological investments [30]." From this definition, we conceptualize workforce governance under four dimensions [27]: control, coordination, incentive, and relational transparency mechanisms. This conceptual framework cannot be exhaustive, as digital governance is an emerging field of research [27]. However, it delineates the main dimensions of governance with great precision and can therefore serves as a basis for conceptualizing workforce governance.

Control and coordination are confused in literature [49]. Thus, for an appropriate workforce governance conceptualization, we conclude that control mechanisms consist of distributing decision-making power between employees [43] and monitoring the use of this power [49]. Thus, above the centralization of decision-making process to define roles and tasks of senior managers [37, 50] and to make digital solutions choices or data processing policy [43], leadership has to delegate to lower echelons more decision-making power [43] as employees need more autonomy in digital era [50, 51]. Thus, flattened organizational structures for decentralized decision-making are recommended [35]. However, coordination is needed to converge actions toward the success of organization digitalization objectives [30].

We define coordination mechanisms as leadership efforts to align goals, resources, and activities of interdependent organizational units [49] with those of digitalization [30]. These mechanisms include division [27] and collaboration. The latter is essential to support digital transformation objectives [47] and refers to a group joint effort to achieve common goals. Its success relies on human and technological resource capabilities sharing [30]. Then, collaboration nurtures the need to

Table 23.1 Workforce governance according to a selection of articles on digital leadership and digital governance

Mechanisms authors	Control	Coordination	Incentive	Relational transparency
Du et al. [44]	Building mechanisms on a cultural basis		Hiring on a cultural basis,	On a cultural basis
Sommer [46]	Employee participation and empowerment		Individual to team rewards, Employee digital mindset and agility	Walking the talk (leader)
Guinan et al. [47]	Involvement in decision-making, empowerment	Experienced, diverse and focused teams Daily meetings, extendable objectives, indicators, collaboration	Incentivizing, Risk-taking, Individual rewards based on results and learning, Leadership training for teams, Employee digital mindset and agility	Talking about and learning from failures
Williams [15]	Empowerment		Employee's digital mindset and agility	
Margherita and Braccini [26]			Training teams	
Varshney [45]	Flat and collegial organization pattern	Collaboration	Appropriate skills, learning	
Plekhanov et al. [43]	Centralization and empowerment	Collaboration		

incentive employees to acquire appropriate skills through training [45], equipping themselves with agility and digital mindset [48]. To achieve such an incentive, leadership must establish a trust environment through relational transparency because an open and transparent communication eliminates fear of changing or losing job and enables employees to align themselves with planned decisions and changes [35]. At this stage of analysis, workforce governance, its dimensions, and their interactions are conceptualized in Fig. 23.1.

With this conceptualization, we show that an open and transparent communication will foster a trust environment, within which employees should be granted more decision-making power, to work autonomously under leadership [35]. This empowerment will also motivate employees [35] to learn new skills, agility [52] and get a digital mindset needed for collaboration [30]. The latter will be regulated within the boundaries leadership defines to employee's autonomy [35]. As trust is the cornerstone of workforce governance conceptualization, what leadership style is

Table 23.2 Workforce governance across selected articles on digital leadership and digital governance

Mechanisms authors	Control	Coordination	Incentive	Relational transparency
Tagscherer and Carbon [35]	Reduced hierarchical distance, Autonomy, Agility, decision-making empowerment	Multicultural, multigenerational teamwork and collaboration,	Encourage risk-taking and essay-error practice. Learning from mistakes Open to new ideas, digital mindset	Commitment to partnership, Trust of the leader in his employee, employees trust in their leader ability to succeed in the digitalization, Open, rapid and transparent communication between leader-employee and customer
Ly [48]	Empowerment	Collaboration network	Digital skills, digital mindset and agility	
Trenerry et al. [37]	Top-down and bottom-up strategy	Collaboration	Training and updating digital skills, Communication and problem-solving skills, Technological creativity	Communication
Alrasheedi et al. [9]	Decision-making empowerment		Error tolerance, mindset, Competent workforce, Use of digital tools, Group-oriented rewards	Transparent reward system

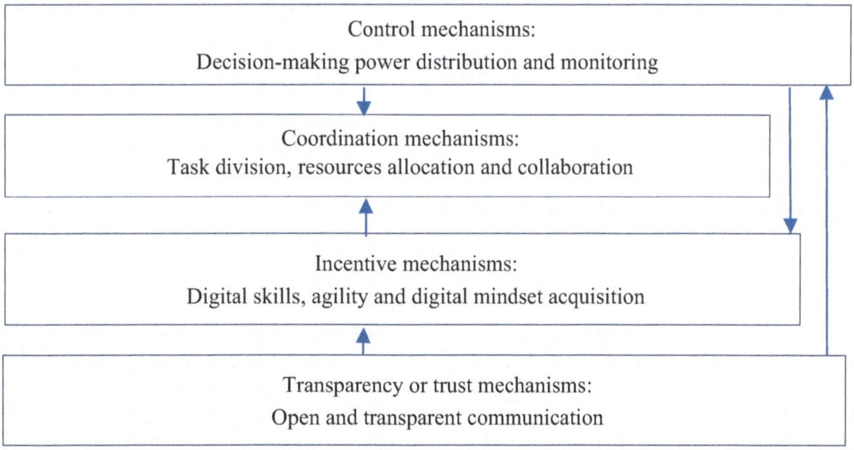

Fig. 23.1 Workforce governance conceptualization and relation between dimensions

appropriate to build trust and maintain it in order to foster effective workforce governance and subsequently digital culture within an organization?

23.2.2 Authentic Leadership Relationship with Workforce Governance and Digital Culture: Trust Theory as a Bonding Factor

Trust in digital era has two perspectives. First, leaders need to trust in their teams and employees. Second, employees have to trust in their leaders' capabilities to successfully implement digital transformation [35]. Then, we focus on interpersonal trust and define it through an adaptation of Fulmer and Gelfand [53] as: "Employee (trustor) willingness, in a digital workplace, to deal with the leadership (trustee) and others collaborators (trustee)."

Authentic leadership in a continuous pursuit of transparency will develop four features [8]. First, self-awareness allows authentic leaders to continually question their actions and activities in order to act in accordance with their values and beliefs. Then, balanced information enables them to objectively analyze all relevant data by sharing collaborators views before deciding. Moreover, through an internalized moral perspective, authentic leaders adopt behaviors consistent with their norms and values in decision-making. Finally, relational transparency helps leaders to build transparent relations with collaborators and also to take care of their development.

Concerning control mechanisms, relational transparency, which requires consensual objectives established on trust basis [54], enables leadership to grant workforce the autonomy required to accomplish their tasks. Because autonomy demands

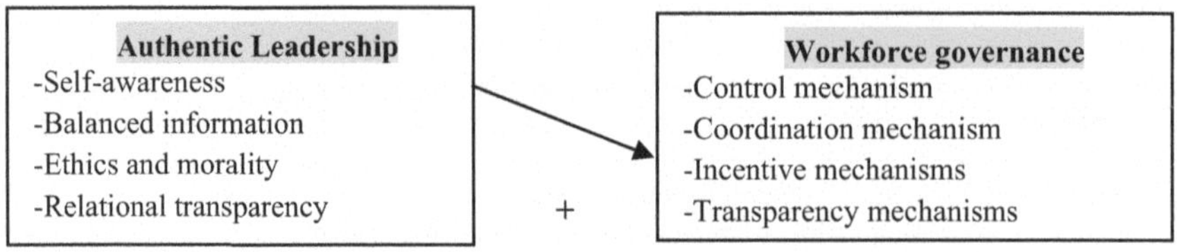

Fig. 23.2 Authentic leadership influence on workforce governance

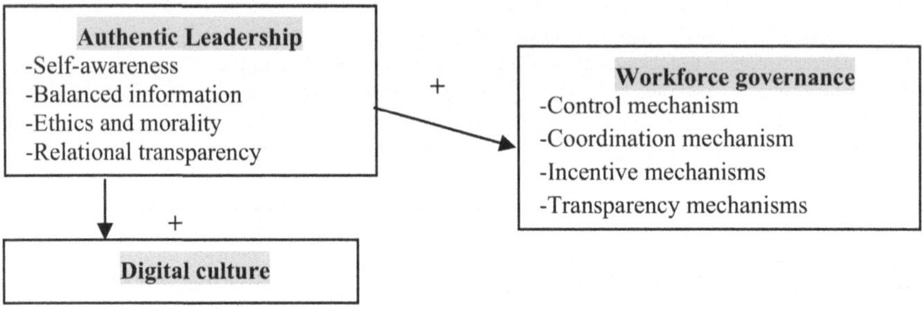

Fig. 23.3 Authentic leadership influence on digital culture and workforce governance

leader to let employees work with a decision-making power that remains monitored, regardless of trusting in their skills [35].

Balanced Information enables authentic leadership to create a workplace that fosters ideas sharing [54], access to information, and unlocking of universes [39]. Self-awareness generalized within the organization through relational transparency or trust will enable improving of employees' digital skills [1].

Authentic leadership is rooted in ethics and morality. The latter are important for the appropriate use of technology [6] and massive data [1] which are crucial to avoid organization being banned or a damaged brand image. These authentic qualities claimed in digitalization enable leadership to foster identification among employees [45, 55]. Thus, we propose that authentic leadership has a positive influence on workforce governance (See Fig. 23.2).

Digital culture requires a strong customer and learning orientation [56]. Indeed, for Sainger [24], leadership must foster a culture of change. For Tagscherer and Carbon [35], it is incumbent on leadership to build a culture of learning and knowledge sharing. As authenticity is frequently mentioned in the context of digital transformation [3, 4, 35], with digital culture requiring trust to be established through marketing relationship with the customer [57], and trust being necessary for learning to develop [52], we propose that authentic leadership has a positive influence on digital culture (see Fig. 23.3).

In the next section, we look at the influence of workforce governance on digital culture.

Table 23.3 Digital culture conceptualization

Creating change	Customer orientation	Organizational learning
– Flexible, easy-to-change work methods. – Positive reaction to competition and change – Acceptance of new working methods. – Cooperation between players to initiate change	– Listening to customers – Addressing customer concerns.	– Failure as an opportunity to learn and improve. – Encouraging and rewarding innovation and risk-taking. – Continuous learning – Information sharing

23.2.3 From Workforce Governance to Digital Culture: Another Trust-Based Relation

Digital transformation aims to shift radically an organizational culture to a digital culture which, remains an adaptive culture [11, 56]. Researchers and practitioners agree on digital culture relevance for digitalization success [38], but literature on digital culture is scant [13] and the concept is insufficiently operationalized [58].

We measure digital culture with the adaptability dimension of the DOCS [59] to avoid risk of no convergent and discriminant validity that could assimilate our measure to an organizational climate [60]. Indeed, CVF [61] is akin to a measure of organizational climate [60]. Moreover, given their recentness, instruments of Costanza et al. [62] and Martínez-Caro et al. [58] need confirmation of their validity and psychometric qualities [62]. Concerning DOCS adaptability dimension [59], it reflects the norms and values specific to an organizational culture [60].

Digital culture incorporates organization's ability to effect change [56]. It also focuses on customer orientation as so far, beyond being mere consumers to be retained, customers become co-creators of digital solutions [63]. Finally, it prioritizes continuous learning, creativity, and digital discovery [43]. These line up with DOCS's three adaptability culture dimensions. Table 23.3 shows these three dimensions [59].

About workforce governance influence on digital culture, employees' autonomy, a key control mechanism, has a positive influence on their adoption of cultural change [24, 35]. Then, digital mindset identified as an incentive mechanism improves adherence to cultural values [15, 64] and facilitates efforts to foster a change culture [11].

Finally, open interpersonal relationship environments convey interpersonal trust [35] and foster needs of development of skills and autonomy [65] required for digital transformation. Thus, coordination mechanisms through collaboration can encourage employees to embrace the cultural change essential for digitalization [11]. Given this, we argue that workforce governance can foster adherence to a digital culture and present our conceptual model (see Fig. 23.4).

The following section presents the Côte d'Ivoire context, methodology, and results in context.

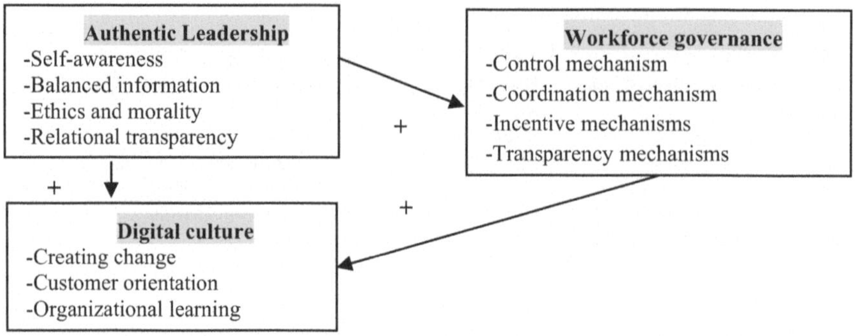

Fig. 23.4 Conceptual model

23.3 Methods

23.3.1 *The Digital Transformation of SMEs in the Cultural Context of Côte d'Ivoire*

Côte d'Ivoire has a young and fast-growing urban population. Thus, digital development according to Word Bank represents an entrepreneurial opportunity [66]. However, SMEs are struggling to take advantage of this opportunity through digital transformation [40].

According to Cangah and Gbédjé [67] of Deloitte, digital transformation has effects on organization and in particular on workforce. Therefore, managers need to conceive and implement adequate governance to support future workforce developments. However, as digitally unskilled, business leaders fail to define a digital strategy [40] to promote SMEs capable to adapt to their markets demands [67].

Then, the role of the conceptual model is to help these leaders. But, as national culture influences technological adoption [31] and leadership [32], we analyze the conceptual model in the context of Côte d'Ivoire. The following section presents our methodological approach and the results.

23.3.2 *Systematic Approach as Methodology*

We adopted Denyer and Tranfield's [68] and Dumez's [69] protocol analysis, structured in four stages: identification of available articles, detection of interesting articles, selection of relevant articles, exploitation of selected articles, and writing.

Criteria used to include or exclude an article are: language, keywords, discipline, document type, and databases. Because, Sub-Saharan African countries including Côte d'Ivoire are not very prolific in the literature on leadership [70], our search for references, started in September 2023, was limited to the following databases: Google Scholar and Google.

Table 23.4 Final articles

Nature	Number	Authors	Theme
Articles	2	Dia [71], Ouffouet [72]	National culture and management
Books	2	Diby [73], Kessy [74]	
Reports	2	Yohou and Picarelli [66], Deloitte [40]	Digital development
	2	Okan [75]; IOS reports [76]	SME and business competitiveness
	1	Global Innovation Index [77]	Innovative countries ranking
	1	IRIS [78]	Human prospective
Interview	1	Cangah and Gbédjé [67]	Digital transformation

This search included articles, academic and professional books dealing with digital business, leadership, and national culture relating to Côte d'Ivoire and its SMEs. These documents were listed in French and English. Given the cross-disciplinary nature of the subject, we also included articles published in management, psychology, and social science disciplines, regardless of publication date. In addition, keywords were used: authentic leadership, leadership and digital transformation, management and culture, digital, and Côte d'Ivoire.

Keywords combined with inclusion or exclusion criteria gave around 30 documents, 11 of which were retained after examining titles and reading abstracts. Table 23.4 details the final articles analyzed.

Based on this literature, we analyzed national culture influence on authentic leadership perception and on workforce governance.

23.4 Findings

23.4.1 Mutual Aid and Solidarity: An Emerging Dimension of Authentic Leadership in a Culture Nearly Resistant to Change

We conducted an analysis on employee's perception of authentic leadership [72] and on the cultural change [71]. The first shows that some perceived dimensions are anchored in cultural context (internalized moral perspective and mutual aid-solidarity), while others are not (self-awareness, balanced information and relational transparency). Conformism, cult of secrecy, hierarchical distance, and collectivism are cultural context's main features.

Overall, with exception of mutual aid and solidarity, employees' perception of authentic leadership [72] is consistent with Walumbwa et al.'s model [8]. Such results are worth considering since in digitalization process employee's perception tend to differ from those of leaders [37]. Table 23.5 posits a synthesis.

The second analysis shows that collectivism, leader myth, cult of secrecy, and conformism are persistent values but keeping one's word is disappearing. Table 23.6 outlines this.

Table 23.5 Authentic leadership perceived in relation with Walumbwa's model and culture

Perceived Authentic leadership dimensions	Dimensions définition by Walumbwa et al. [8]	Employees perception			
		Congruent with Walumbwa et al. [8]		Compliance with cultural context	
		Yes	No	Yes	No
Ethic and morality	Leaders say what they mean and do what they say	x		Keeping one's word	
Self-awareness	Leaders improve themselves by learning from their strengths and weaknesses	x			Leader myth, conformism, secrecy cult
Balanced-information	Leaders hear their employees before a decision	x			Leader myth, hierarchical distance
Relational transparency	Leaders trust employees, set clear objectives, recognize their performance, ensure their personal development and one of their skills	x			Leader myth, hierarchical distance, secrecy cult
Solidarity and mutual aid			Employees need leaders who create a workplace of solidarity and mutual aid	Collectivism	

Table 23.6 Cultural changes in Côte d'Ivoire

Cultural dimensions	Dimensions meaning (Kessy [74], Diby [73])	Value in decline	Persistent values
Keeping one's word	Compliance with employee commitments made by management	Kessy [74] Strategie PME (2013) [75]	
Collectivism	Solidarity and mutual aid		Kessy [74]), Ouffouet [72], Observatoire (2015) [78], Global Innovation Index (2021, 2022) [77]
Leader myth	Leads to avoid contradicting leader in public		
Cult of secrecy	Failure to share knowledge and skills with employees in order to manage power		
Conformism	Counterweight to questioning and innovation		

Below, we present an analysis of cultural features impact on workforce governance in digital context.

23.4.2 Culture in Côte d'Ivoire: Lever or Barrier to Workforce Governance in SMEs?

The cultural context, besides excluding employees from the decision-making process [74], hinders autonomy [75, 76]. Indeed, leadership in SMEs is generally authoritarian [75, 78] and fosters hierarchical structure, centralized decision-making [73, 74], conformism, and cult of secrecy [74]. These values are ineffective in a digital workplace [35]. However, as employee's perception aligns with authentic leadership dimensions, leaders can adopt this style to succeed in their digital transformation.

In addition, incentive mechanisms based on mutual aid and solidarity can be initiated to develop trust [74], a sense of synchronicity between communicators [79] and encourage skills acquisition to facilitate collaboration [30]. Finally, although the culture of keeping one's word, precious in the cultural context [72, 74], is clearly declining in Côte d'Ivoire, a coming back to this cultural trait will foster transparency in workplace. Based on the above, we adapt the model as follows (see Fig. 23.5).

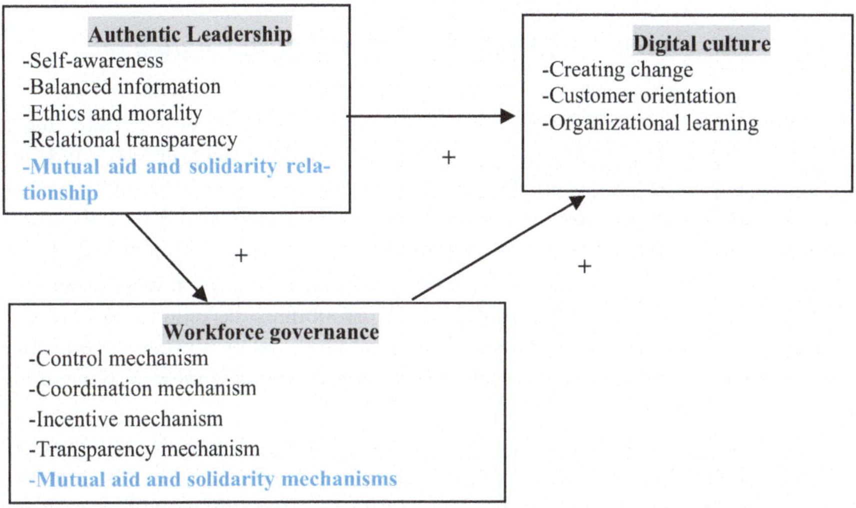

Fig. 23.5 Conceptual model adapted to the cultural context of SMEs in Côte d'Ivoire

23.5 Discussion and Implications

23.5.1 *Authentic Leadership, Workforce Governance, and Digital Culture: Trust, a Decisive Factor*

Our first objective was to propose a conceptualization of workforce governance and to understand its role in the relation between authentic leadership and digital culture. For this, we call for trust theory to demonstrate authentic leadership abilities to foster a trustful digital workplace to incentive employees to learn continually in order to get new skills to collaborate under the coordination mechanisms. Then, employees acquiring digital mindset and autonomy, they easily adopt cultural values of a digital culture.

Our results are supported by Rijswijk et al. [33] who aiming to understand connections between trust and digitalization through a qualitative study, asserts that trust is the prerequisite for digital transformation success which requires an active trust in management by employees. Regarding van den Broek et van Veenstra [80], for trust issues to be overcome, management has to implement governance arrangements to foster more sustainable relations. Lau and Höyng [34] show how employee trust in management regarding digitalization is encouraged to successfully promote technological and implement digital technologies within organizations. So, to control workforce governance in order to develop digital culture, leaders must behave authentic. Moreover, when hiring, leaders have to evaluate employees' perception of authenticity in leadership but also their perception of trusting relation in workplace. Doing so, there will be a fit between hired employees and workplace values and norms.

23.5.2 *The Conceptual Model and the Cultural Challenge*

The second objective of this research was to understand how culture could shape our conceptual model. On this point, we first note that leadership in SMEs in Côte d'Ivoire is influenced negatively by some cultural values. Then, the authoritarian leadership style that prevailed in these organizations cannot foster digitalization. Fortunately, employee's perception of leadership is consistent with authentic leadership, and above all, they appreciate mutual aid and solidarity in leader-member exchange. About this, the Globe project [32] corroborates the influence of culture on leadership. This is also consistent with Zhang et al. [81] who states that differences in perception of authentic leadership can appear among employees depending on the cultural context.

Regarding the workforce governance, Du et al. [44] support the influence of culture in digital transformation process. Indeed, in China, these authors recommend managers of state-owned technology companies, because of employees' hierarchical distance and collectivism profile, to implement mechanisms that foster hierarchical distance and collectivism within the organization if they aim for the success of employees' technology adoption.

Thus, studies show that imitation in collectivist countries fosters technology adoption [82], and collectivism within the group positively influences the adoption and use of ICT [83]. Therefore, mutual aid and solidarity, in the context of Côte d'Ivoire, have to be implemented as mechanism to foster trust and motivation within the workplace [74] to get employees to adhere to digital culture.

23.6 Conclusion and Limits

The aim of our research was to propose a conceptual model of workforce governance and its role in the relationship between authentic leadership and digital culture. Based on an analysis of literature and trust theory, we proposed a workforce governance conceptualization, following both governance protocol suggested by Hanisch et al. [27] and techniques and methods related to workforce governance involvement in digitalization [9, 44, 45]. Thus, the concept of workforce governance is made up of transparency mechanisms, incentive mechanisms, coordination mechanisms, and control mechanisms.

Transparency mechanisms convey trust within the organization to incentive employees to acquire skills and digital mindset in order to engage them in collaborative work with given autonomy regulated by leadership. Making the proposition of this conceptualization, this research contributes to the call of Alrasheedi et al. [9] to uncover which governance styles are preferable to manage employees' empowerment, engagement, and motivation for a successful digital transformation.

We also show that authentic leadership can build trust and maintain it in order to foster effective workforce governance and subsequently digital culture within an organization. This is a response to Arvidson & Melander's call [36] to more exploratory studies to get a deeper understanding of how trust can be developed in digitally transformed organizations.

Moreover, conceptualizing the relation between authentic leadership, workforce governance, and digital culture, this research is a contribution to the persistent call in the twenty-first-century literature to uncover the mechanisms by which leadership influences organizational culture [17–23]. Finally, this study is one of the first dealing with the conceptualization of workforce governance in digital era and the relation between authentic leadership, workforce governance, and digital culture.

In Côte d'Ivoire, this conceptualization is updated because of cultural context which is collectivist. Then, authentic leadership has to develop besides the dimensions of Walumbwa et al. [8] a mutual aid and solidarity relationship with employees, implementing mechanisms of mutual aid and solidarity in workforce governance, so that to engage them in digital culture building.

Despite our search of higher objectivity, our review remains limited. Complete objectivity and comprehensiveness in a grounded approach are illusive because of the engaged protocol and the interpretation. Even if articles are more recent and are literature review articles, we are aware that we can miss out on some valuable contributions from articles not collected. Moreover, workforce governance concept needs to be operationalized in order to test the proposed model.

References

1. Cortellazzo, L., Bruni, E., Zampieri, R.: The role of leadership in a digitalized world: a review. Front. Psychol. **10**, 1938 (2019)
2. Kane, G.C., Phillips, A.N., Copulsky, J., Andrus, G.: How digital leadership is (n't) different. MIT Sloan Manag. Rev. **60**(3), 34–39 (2019)
3. Bennis, W.: Leadership in a digital world: embracing transparency and adaptive capacity. MIS Q. **37**, 635–636 (2013)
4. Myers, D.: Authentic leadership in a digital world. In: Winston, B.E. (ed.) Transparent and Authentic Leadership, Christian Faith Perspectives in Leadership and Business (2021)
5. Hoch, J.E., Bommer, W.H., Dulebohn, J.H., Wu, D.: Do ethical, authentic, and servant leadership explain variance above and beyond transformational leadership? A meta-analysis. J. Manag. **44**(2), 501–529 (2016)
6. Jones, M.: E-leadership case study and the impact of (un)faithful appropriation of technology. In: Proceedings of the 5th International Conference on Management Leadership and Governance, ICMLG 2017 (Johannesburg), pp. 191–199 (2017)
7. Harvard Business Review Analytic Services: Operationalizing Digital Transformation: new Insights into Making Digital Work, 1–12 (2017).
8. Walumbwa, F.O., Avolio, B.J., Gardner, W.L., Wernsing, T., Peterson, S.: Authentic leadership: development and validation of a theory-based measure. J. Manag. **34**, 89–126 (2008)
9. Alrasheedi, N.S., Sammon, D., McCarthy, S.: Understanding the characteristics of workforce transformation in a digital transformation context. J. Decis. Syst. **31**(sup1), 362–383 (2022)
10. Boutetiere, H., Montagner, A., Reich, A.: Unlocking Success in Digital Transformations, pp. 1–14. McKinsey and Company (2018)
11. Pradana, M., Silvianita, A., Syarifuddin, S., Renaldi, R.: The implication of digital organisational culture on firm performance. Front. Psychol. **13**, 840699 (2022)
12. Eden, R., Jones, A.B., Casey, V., Draheim, M.: Digital transformation requires workforce transformation. MIS Q. Exec. **18**(1), 1–17 (2019)
13. Busco, C., González, F., Aránguiz, M.: Factors that favor or hinder the acquisition of a digital culture in large organizations in Chile. Front. Psychol. **14**, 1153031 (2023)
14. Fakhri, M., Pradana, M., Syarifuddin, S., Suhendra, Y.: Leadership style and its impact on employee performance at Indonesian national electricity company. Open Psychol. J. **13**, 321–325 (2020)
15. Williams, M.: How to nurture a transformative digital culture in the workplace. Strateg. HR Rev. **19**(1), 7–9 (2020)
16. Shin, J., Mollah, M.A., Choi, J.: Sustainability and organizational performance in South Korea: the effect of digital leadership on digital culture and employees' digital capabilities. Sustainability. **15**, 2027 (2023)
17. Avolio, B.J., Hannah, S., Reichard, R., Chan, A., Walumbwa, F.: 100 years of leadership intervention research. Leadersh. Q. **20**, 764–784 (2009)
18. Gholamzadeh, D., Khazaneh, A., Nabi, M.: The impact of leadership styles on organizational culture in MAPSA company. Manag. Sci. Lett. **4**(9), 2161–2170 (2014)
19. Giberson, T.R., Resick, C.J., Dickson, M.W., Mitchelson, J.K., Randall, K.R., Clark, M.A.: Leadership and organizational culture: linking CEO characteristics to cultural values. J. Bus. Psychol. **24**(2), 123–137 (2009)

20. Gordana, N.: The influence of organizational culture on leadership—Case study Montenegro. Tech. Gazette. **23**, 191–197 (2016)
21. Kargas, A.D., Varoutas, D.: On the relation between organizational culture and leadership: an empirical analysis. Cogent Bus. Manag. **2**, 1055953 (2015)
22. Schimmoeller, L.J.: Leadership styles in competing organizational cultures. Kravis Leaders. Inst. Leadership. Rev. **10**, 125–141 (2010)
23. Tsui, A.S., Zhang, Z., Wang, H., Xin, R.: Unpacking the relationship between CEO leadership behavior and organizational culture. Leaders. Q. **17**(2), 113–117 (2006)
24. Sainger, G.: Leadership in digital age: a study on the role of leader in this era of digital transformation. Int. J. Leaders. **6**(1), 1 (2018)
25. Schneider, B.: The people make the place. Pers. Psychol. **40**(3), 437–453 (1987)
26. Margherita, E.G., Braccini, A.M.: Industry 4.0 technologies in flexible manufacturing for sustainable organizational value: reflections from a multiple case study of Italian manufacturers. Inf. Syst. Front. **20**, 995 (2020)
27. Hanisch, M., Goldsby, C., Fabian, N., Oehmichen, J.: Digital governance: a conceptual framework and research agenda. J. Bus. Res. **162**, 113777 (2023)
28. Buchanan, D.A., Boddy, D.: The Expertise of the Change Agent. Prentice Hall (1992)
29. Pettigrew, A.M.: Theoretical, methodological, and empirical issues in studying change: a response to Starkey. J. Manag. Stud. **24**, 420–426 (1987)
30. Westerman, G., Bonnet, D., McAfee, A.: Réussir la mutation numérique: comment les technologies numériques transforment les entreprises. In: Nouveaux Horizons. Edition Française 2016. Les Editions Diateino (2014)
31. Srite, M., Karahanna, E.: The role of espoused national cultural values in technology acceptance. MIS Q. **30**, 679–704 (2006). https://doi.org/10.2307/25148745
32. House, R.J., Hanges, P.J., Javidan, M., Dorfman, P.W., Gupta, V.: Culture, leadership, and organizations. In: The GLOBE Study of 62 Societies. Sage, Thousand Oaks (2004)
33. Rijswijk, K., Jasper, R.d.V., Laurens, K., James, A.T.: The enabling and constraining connections between trust and digitalisation in incumbent value chains. Technol. Forecast. Soc. Chang. **186**(Part A), 122175, ISSN 0040-1625 (2023). https://doi.org/10.1016/j.techfore.2022.122175
34. Lau, A., Höyng, M.: Digitalization? A matter of trust: a double-mediation model investigating employee trust in management regarding digitalization. Rev. Manag. Sci. **17**, 2165–2183 (2023)
35. Tagscherer, F., Carbon, C.: Leadership for successful digitalization: a literature review on companies' internal and external aspects of digitalization. Sustain. Technol. Entrep. (2023)
36. Arvidson, A., Melander, L.: The multiple levels of trust when selecting suppliers–insights from an automobile manufacturer. Ind. Mark. Manag. **87**, 138–149 (2020)
37. Trenerry, B., Chng, S., Wang, Y., Suhaila, Z.S., Lim, S.S., Lu, H.Y., Oh, P.H.: Preparing workplaces for digital transformation: an integrative review and framework of multi-level factors. Front. Psychol. **12**, 620766 (2021)
38. Tuukkanen, V., Wolgsjö, E., Rusu, L.: Cultural values in digital transformation in a small company. Procedia Comput. Sci. **196**, 3–12 (2022) https://www.sciencedirect.com/science/article/pii/S1877050921022055
39. Frimousse, S., Peretti, J.-M.: Comment accompagner la transformation digitale des entreprises en Afrique ? EMS Ed. Questi. Manag. **3**(18), 199–225 (2017)
40. Deloitte: Etude sur la maturité numérique des entreprises en Côte d'Ivoire, Rapport Final Mai 2016.
41. Tax appendix to Finance Act no. 2021–899 of December 21, 2021 on the State Budget for 2022.
42. Adner, R., Puranam, P., Zhu, F.: What is different about digital strategy? From quantitative to qualitative change. Strategy Sci. **4**(4), 253–261 (2019)
43. Plekhanov, D., Franke, H., Netland, T.: Digital transformation: a review and research agenda. Eur. Manag. J. **41**, 821–844 (2023)
44. Du, R., Liu, L., Straub, D.W., Knight, M.B.: The impact of espoused national cultural values on innovative behaviour: an empirical study in the Chinese IT-enabled global service industry. Asia Pac. Bus. Rev. **23**(3), 354–372 (2016)

45. Varshney, D.: Digital transformation and creation of an agile workforce: exploring company initiatives and employee attitudes. In: Turkmenoglu, M.A., Cicek, B. (eds.) Contemporary Global Issues in Human Resource Management, pp. 89–105. Emerald Publishing Limited, Leeds (2020)
46. Sommer, A.F.: Agile transformation at LEGO group: implementing agile methods in multiple departments changed not only processes but also employees' behavior and mindset. Res. Technol. Manag. **62**(5), 20–29 (2019)
47. Guinan, P.J., Parise, S., Langowitz, N.: Creating an innovative digital project team: levers to enable digital transformation. Bus. Horiz. **62**(6), 717–727 (2019)
48. Ly, B.: The interplay of digital transformational leadership, organizational agility, and digital transformation. J. Knowl. Econ. (2023)
49. Zeng, R., Grøgaard, B., Björkman, I.: Navigating MNE control and coordination: a critical review and directions for future research. J. Int. Bus. Stud. **54**, 1599–1622 (2023)
50. Chanias, S., Myers, M.D., Hess, T.: Digital transformation strategy making in pre-digital organizations: the case of a financial services provider. J. Strateg. Inf. Syst. **28**, 17–33 (2019)
51. Horvath, D., Szabo, R.Z.: Driving forces and barriers of industry 4.0: do multinational and small and medium-sized companies have equal opportunities? Technol. Forecast. Soc. Chang. **146**, 119–132 (2019)
52. Tsai-Lung, L.: Knowledge transfer: past research and future directions. Bus. Rev. **7**(1), 273–281 (2007)
53. Fulmer, A.C., Gelfand, M.J.: At what level (and in whom) we trust. J. Manag. **38**, 1167–1230 (2012)
54. Contreras, F., Baykal, E., Abid, G.: E-leadership and teleworking in times of COVID-19 and beyond: what we know and where do we go. Front. Psychol. **11**, 590271 (2020)
55. Hess, T., Matt, C., Benlian, A., Wiesböck, F.: Options for formulating a digital transformation strategy. MIS Q. Exec. **15**(2), 103–119 (2016)
56. Autissier, D., Johnson, K., Metais-Wiersch, E.: Du changement à la transformation: stratégie et pilotage de transformation. Dunod (2018)
57. Ndubisi, N.O., Nataraajan, R.: How the young adult consumer segment responds to trusty and committed marketing relationship. Psychol. Mark. **35**, 923–935 (2018)
58. Martínez-Caro, E., Cegarra-Navarro, J.G., Alfonso-Ruiz, F.J.: Digital technologies and firm performance: the role of digital organizational culture. Technol. Forecast. Soc. Change. **154**, 119962 (2020)
59. Denison, D.R.: Organizational culture: can it be a key driver for driving organizational change? In: Cooper, C., Cartwright, S., Earley, P.C. (eds.) International Handbook of Organizational Culture and Climate, pp. 347–372. John Wiley & Sons, London (2001)
60. Chatman, J.A., O'Reilly, C.A.: Paradigm lost: reinvigorating the study of organizational culture. Res. Organ. Behav. **36**, 199–224 (2016)
61. Cameron, K.S., Quinn, R.E.: Diagnosing and Changing Organizational Culture: Based on the Competing Values Framework, Rev. edn. Wiley (Jossey Bass), San Francisco (2006).
62. Costanza, D., Blacksmith, N., Coats, M., Severt, J., DeCostanza, A.: The effect of adaptive organizational culture on long-term survival. J. Bus. Psychol. (2015)
63. Ziaie, A., Shamizanjani, M., Manian, A.: Systematic review of digital value propositions in the retail sector: new approach for digital experience study. Electron. Commer. Res. Appl. **47**, 101053 (2021)
64. Hazarika, I.: Health workforce governance: key to the delivery of people-centred care. Int. J. Healthc. Manag. **14**(2), 358–362 (2019)
65. Deci, E.L., Ryan, R.M.: The "what" and "why" of goal pursuits: human needs and the self-determination of behavior. Psychol. Inq. **11**, 227–268 (2000)
66. Yohou, D.H., Picarelli, N.: The Digital Sector: Central Enabler in Emerging Economies (English). World Bank Group, Washington, D. C (2022)
67. Cangah, J., Gbédjé, L.: Les défis à relever pour réussir la transformation digitale de vos fonctions RH/DELOITTE CI. Tribunes RHmag. **3 mai** (2023)

68. Denyer, D., Tranfield, D.: Producing a systematic review. In: Buchanan, D.A., Bryman, A. (eds.) The Sage Handbook of Organizational Research Methods, pp. 671–689. Sage Publications Ltd, Thousand Oaks (2009)
69. Dumez, H.: Faire une revue de littérature: pourquoi et comment ? Le Libellio d'AEGIS. **7**(2 - Été), 15–27 (2011)
70. Fourie, W., Merwe, F.C., Merwe, B.: Sixty years of research on leadership in Africa: a review of the literature. In: Leadership. SAGE (2015)
71. Dia, A.L.: Le management Africain: mythe ou réalité. Rev. Int. PME. **4**(1), 29–48 (1991)
72. Ouffouet, K.F.: Le leadership authentique dans le contexte culturel de la Côte d'Ivoire: une exploration à partir du terrain des PME. Revue Internationale de Gestion et d'Economie série A - gestion. **5**(10), 118–142 (2021)
73. Diby, K. C.: Management des services publics en Afrique, Retour d'expérience: des raisons d'espérer. NEI-CEDA/Hachette Livre international, deuxième édition (2007)
74. Kessy, M.Z.: Culture Africaine et Gestion de L'entreprise Moderne. Centre d'édition et de diffusions africaines (CEDA), Janvier (1998)
75. Okan, C. (2013). Dalberg: Stratégie de développement des PME ivoiriennes, Financement PARE-PME -Don IDA N°H5280 CI, rapport final.
76. IOS Partners: Etude sur la compétitivité des entreprises ivoiriennes, financée par la banque mondiale, Rapport provisoire, Juillet (2014)
77. WIPO: Classement des pays africains en termes d'innovation. Global Innovation Index (2021; 2022).
78. IRIS: Observatoire de la prospective humanitaire: analyse prospective CÔTE D'IVOIRE 2020, Une période décisive, Novembre (2015)
79. Kelly, S., MacDonald, P.: A look at leadership styles and workplace solidarity communication. Int. J. Bus. Commun. **56**(3), 432–448 (2019)
80. Van den Broek, T., van Veenstra, A.F.: Governance of big data collaborations: how to balance regulatory compliance and disruptive innovation. Technol. Forecast. Soc. Chang. **129**, 330–338 (2018)
81. Zhang, H., Everett, A.M., Elkin, G., Cone, M.H.: Authentic leadership development: theorizing on Chinese philosophy. Asia Pac. Bus. Rev. **18**(4), 587–605 (2012)
82. Lee, S.G., Trimi, S., Kim, C.: The impact of cultural differences on technology adoption. J. World Bus. **48**, 20–29 (2013)
83. Azam, M.S., Quaddus, M.: Examining the influence of national culture on adoption and use of information and communication technology: a study from Bangladesh's SME perspective. Int. Technol. Manag. Rev. **3**(2), 116–126 (2013)

Part IV
Sustainability and Circular Economy

Chapter 24
A Development Model for Circular Economy Between Industrial Parks and Surroundings Areas (CE_IPsa): The Reverse Logistics Strategic Approach

Carlotta D'Alessandro, Giuseppe Ioppolo, Grazia Calabrò, and Giuseppe Caristi

Abstract The CE_IPsa (Circular Economy_Industrial Parks and the surrounding area) project aims to outline a development model to foster interconnection between industrial parks (IPs) and their surrounding area, based on circular economy (CE) and industrial symbiosis (IS) principles. CE offers sustainable solutions, fostering economic growth and environmental health. Despite this, CE has received limited attention in the supply chain context. Encompassing CE, IS, and Industry 4.0 (I. 4.0) digital technologies, (such as Blockchain, Internet of Things, Cyber-Physical Systems, and Cloud Computing), this chapter proposes reverse logistics (RL) as a strategic approach for managing product returns, end-of-life treatment, and recovery to eliminate waste. As part of the CE_IPsa project, the study's purpose is to promote the transition to a net-zero balance, where resource extraction is offset by resource recovery and waste generation is minimized. This chapter shows the preliminary results of the first phase of the CE_IPsa project, building upon a Scooping review, and proposing a framework based on the integration of RL with I. 4.0 technologies as driving force to stimulate the connection between IP and the surrounding context. With a redesigned supply chain that integrates a RL approach empowered by I. 4.0 technologies, businesses within the IPs and the surrounding area can unlock significant potential. This solution optimizes operations and minimizes environmental impact, offering a valuable strategy for policymakers to support the implementation of a circular industrial system. The paper concludes by outlining key challenges, opportunities, and future research directions in harnessing the power of digital transformation for greener businesses.

Keywords Circular economy · Industrial parks · Industry 4.0 technologies

C. D'Alessandro (✉) · G. Ioppolo · G. Calabrò · G. Caristi
Department of Economics, University of Messina, Messina, Italy
e-mail: carlotta.dalessandro@unime.it; giuseppe.ioppolo@unime.it; grazia.calabro@unime.it; giuseppe.caristi@unime.it

F. Schiavone et al. (eds.), *Advanced Perspectives and Trends in Digital Transformation of Firms, Networks, and Society*, Springer Proceedings in Business and Economics, https://doi.org/10.1007/978-3-031-80692-6_24

24.1 Introduction

The CE_IPsa (Circular Economy_Industrial Parks and the surrounding area) project purpose is to propose a development model that enhances interconnection between industrial park (IP) and their surrounding area. This model leverages the fundamentals of circular economy (CE) and industrial symbiosis (IS) to promote resource efficiency, waste reduction, and a more collaborative industrial ecosystem.

Nowadays, the irresponsible practices of certain firms exacerbate current global challenges (increasing population, urbanization, industrial expansion, and climate change), exposing society to sustainability risks [1]. Stakeholder commitment, encouraging collaboration among nations, and inclusion of experts' insights are essential for initiating a shift toward sustainability [2].

In this regard, in today's dynamic landscape, the CE is emerging as a powerful method for various industries, contributing to sustainability goals [3]. CE promotes strategies that create closed-loop material flows, enhancing economic productivity, business effectiveness, and minimizing environmental impact [4].

Indeed, by adopting recovery, redesign, reuse, recycling, sharing, and remanufacturing practices, CE strategies unlock several sustainability advantages, such as resource conservation, and reduced waste generation [5]. While the CE has gained traction, its application within supply chains remains an under-explored area [6].

Supply chains connect suppliers, manufacturers, distributors, etc., through integrated processes to deliver products to customers [7]. Furthermore, they encompass both upstream and downstream activities, including physical goods movement, information flow, and financial transactions [8]. Additionally, a sustainable supply chain could be an important promoter of CE in the context of IP [9].

An IP is a designated area equipped with essential infrastructure, such as transportation systems, to facilitate various industrial operations [10]. In this regard, in the so-called Eco-Industrial Parks (EIP), business collaborate exchanging materials and energy in a closed-loop system.

The design of an efficient supply chain of an EIP is contingent on the realization of an IS network [9]. As per Chertow's definition, IS is: "the part of industrial ecology known as industrial symbiosis engages traditionally separate industries in a collective approach to competitive advantage involving physical exchange of materials, energy, water, and by-products. The keys to industrial symbiosis are collaboration and the synergistic possibilities offered by geographic proximity" [11].

Indeed, IS stands as the primary model for Industrial Parks, safeguarding the environment, fostering CE, and sustainability [12]. Additionally, businesses are increasingly recognizing the need to integrate sustainability principles throughout their supply chains due to their increasing environmental awareness [13].

Among other approaches, reverse logistics (RL) is a key solution to address environmental impacts. Rooted in the principles of CE and IS, RL operations encompass handling product returns, managing end-of-life (EoL) treatment, and engaging in product recovery tasks, all with the objective of eliminating waste. Indeed, in this context, products are designed for cyclical recovery through biological or technical cycles [14].

In fact, RL practices are renowned for their eco-friendliness [15], aiming to create zero waste through "cradle-to-cradle" product design and regenerative actions [16], aligning with CE principles [17].

In addition, the rise of digitalization and Industry 4.0 (I. 4.0) significantly influences supply chain management [18]. Emerging digital technologies, like Big Data analytics, Blockchain, Cyber-Physical Systems (CPS), Cloud Computing, Internet of Things (IoT), and more, are incorporated in I. 4.0 [19]. Such technologies hold great potential in enhancing supply chain efficiency [20, 21]. However, a critical gap exists in research on applying RL integrated with I. 4.0 technologies within the environment of IPs.

Based on these considerations, this research displays the preliminary achievements of the first phase of the CE_IPsa project, proposing a framework based on the integration of RL with I. 4.0 technologies as a driving force to stimulate the connection between IP and the surrounding context, within a broader digital transformation landscape.

Adoption of a redesigned supply chain, including a RL approach empowered by I. 4.0 technologies optimizes operations and minimizes environmental impact [22], offering a valuable strategy for policymakers to bolster the realization of a Circular Industrial System. Therefore, the study addresses the following research question:

RQ: How I. 4.0 technologies can enhance RL to create a circular industrial system by integrating CE and IS principles, in the context of IPs?

By identifying key challenges and opportunities within this approach, the study aims to guide businesses toward more sustainable practices through the adoption of a future-oriented strategy.

After this Introduction, Section 4.2 overviews literature on RL, CE, IS, and I. 4.0 in the context of IPs. Section 4.3 presents a Scoping review on I. 4.0 technologies to enhance RL practices within IPs setting, while Sect. 4.4 presents primary research findings. Finally, Section 4.5 discusses key points, and Sect. 4.6 outlines conclusions and limitations.

24.2 Literature Overview

The growing focus on sustainability has pushed the concept of CE to the forefront. Indeed, in accordance with Kirchherr et al. [23]: "A circular economy is an economic system that is based on business models which replace the 'end-of-life' concept with reducing, alternatively reusing, recycling and recovering materials in production/distribution and consumption processes, thus operating at the micro level (products, companies, consumers), meso level (eco-industrial parks) and macro level (city, region, nation and beyond), to accomplish sustainable development, which implies creating environmental quality, economic prosperity, and social equity, to the benefit of current and future generations."

What is more, aligning with the central tenets of the CE, the principles of IS have gained increased recognition, making it a relevant business model for minimizing waste and maximizing resource recovery [24].

Through collaborative networks, IS addresses resource consumption, waste management, and contamination by transforming waste flows into valuable inputs for other industries [25, 26].

Currently, IS stands as the primary model for IPs, safeguarding the environment, fostering CE, and sustainability [12]. IPs, traditionally centers of production and consumption of resources, are being re-evaluated for their potential role in this new economic model.

RL plays a crucial part in this transformation. Initially focused on maximizing value while minimizing waste, RL broadened to encompass efficient product and information flow beyond traditional supply chains [27].

IPs offer a fertile ground for RL implementation. Indeed, geographical proximity of industries can foster IS [28]. Shared infrastructure and expertise within the park can enhance cost-effectiveness and efficiency of RL operations, but establishing a collaborative culture and overcoming information-sharing barriers remain significant challenges [29].

However, implementing RL presents significant challenges for logistics managers [30], such as lack of commitment, insufficient technological infrastructure, unsupportive government regulations, deficient coordination within the supply chain, and low consumer confidence in remanufactured products [31].

The emergence of I. 4.0 technologies presents a potential solution to these challenges. The transformative potential of I. 4.0 holds significant promise for enhancing sustainability [32], influencing manufacturing, and supply chain management [33].

Furthermore, I. 4.0, characterized by real-time data, and interconnected systems [33], offers significant opportunities to enhance sustainability throughout production and supply chain management [32, 34].

A sustainable circular business model requires a focus on the supply chain, particularly its role in fostering collaboration with partners. Upstream collaboration involves green sourcing, while downstream collaboration focuses on practices like product return, reuse, and recycling [35]. These downstream activities are facilitated by RL.

24.3 Methods

The first step of this study was a Scoping review to investigate the ability of I. 4.0 technologies to strengthen RL practices within IPs. Scoping reviews provide valuable insights into the breadth, range, and nature of existing research on a topic [36].

The research was performed adhering with the PRISMA Extension for Scoping Reviews (PRISMA-ScR) guidelines [37]. The selection process was executed utilizing the Scopus and Web of Science (WoS) databases and applying the keywords "Industrial Park" and "Industry 4.0" OR "Reverse Logistic."

Only articles and reviews in English were selected. The review focused on identifying key I. 4.0 aspects with the potential to improve RL efficiency and effectiveness in an IP setting.

Hence, articles out of this topic were excluded. The initial step yielded a total of 51 items, (26 for Scopus, and 25 for WoS). After duplicate removal, screening and full-text reading, the sample amounted to nine articles.

Building upon this analysis, a framework for integrating these I. 4.0 technologies into RL operations within industrial system is proposed. This framework aims to guide the implementation of a more sustainable and efficient circular economy model within industrial ecosystems.

24.4 Findings

The European Working Group on RL describes RL as "the process of planning, implementing and controlling backward flows of raw materials, in process inventory, packaging, and finished goods, from a manufacturing, distribution or use point, to a point of recovery or point of proper disposal" [38].

Properly managed RL possess the capacity to propel sustainable development and promote advantages by enhancing profits, reducing costs, and elevating customer satisfaction [39].

The RL system involves five activities [40, 41]:

1. Product acquisition: obtaining used products for further use.
2. Gathering of utilized material: collecting materials directly or through distributors, retailers, or third-party logistics providers.
3. Supervision, choice, and sort: determining quantity, quality, and timing of returned products.
4. Recovery process: restoring economic value through reuse, remanufacturing, or recycling.
5. Redistribution: returning items as new, selling at a discount, placing on the secondary market, or donating to charity.

The outputs of the RL system include remanufactured products, recycled materials, and replacement elements [42].

In recent years, there has been a proposal for the notion of Reverse Logistics 4.0, defined as "the sustainable management of all relevant flows and activities for value recovery and/or proper disposal of End-of-Life products by using data-driven and smart technologies enabled individualization and innovative services" [43].

Integrating RL empowered by I. 4.0 technologies, Fig. 24.1 depicts the connection between IP and their surrounding environment built upon the principles of CE and IS to promote the transition to a net-zero balance between resource efficiency and waste reduction.

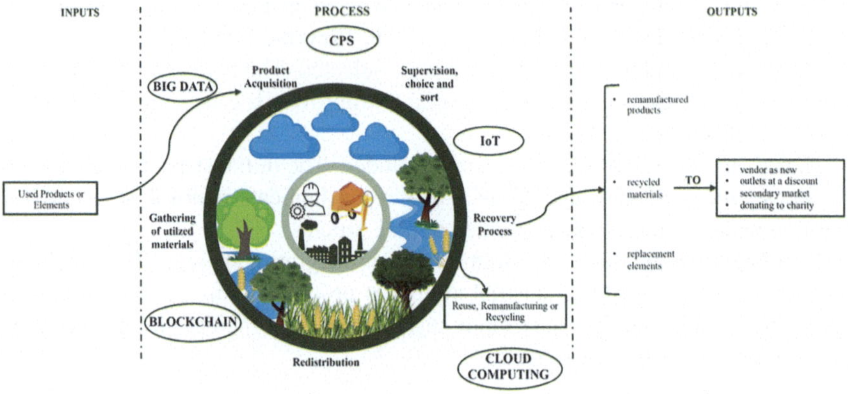

Fig. 24.1 RL empowered by I. 4.0 technologies (Source: Authors)

Table 24.1 Overview of I. 4.0 technologies

I. 4.0 innovations	
IoT	It links sensors and actuators for data exchange in virtual networks [47]
Big data analytics	It supports predictive maintenance, reducing breakdowns by 50% and boosting productivity by 20% [48]
CPS	It merges computational capabilities with real-world operations, allowing computers and networks to oversee and manage physical processes. There's a continuous exchange between physical activities and computational analysis through feedback loops [49]
Cloud computing	It delivers IT services through an extensive, cost-effective network of computing units connected via IP networks. Its foundation lies in the design principles of search engine platforms [50]
Blockchain	It refers to a digital system for carrying out and documenting transactions. It is built upon smart algorithms and aggregated data, all protected through cryptography [51]

Traditional supply chains lacked cooperation among stages, causing information transfer obstacles [44]. In this perspective, with the emergence of I. 4.0, the term Supply Chain 4.0 has been introduced with the aim of augmenting customer satisfaction and overall performances [45].

Supply Chain 4.0 represents the evolution of supply chains toward a smarter model. This approach refers to a network of physical objects combined with digital technologies for sensing, tracking, and information sharing throughout an organization and its value chain members [46]. It embraces the principles of I. 4.0 innovations, incorporating the IoT, Big Data Analytics, CPS, Cloud Computing, and Blockchain (Table 24.1).

Furthermore, with the introduction of I. 4.0 principles in the context of supply chain, the term Digital Supply Chain has been provided. Digital Supply Chain transcend traditional logistics by integrating intelligent technological systems. These systems leverage extensive data management capabilities and foster collaboration and connection across digital hardware, software, and networks. By adopting innovative technologies such as the IoT and Cloud Computing [52], Digital Supply Chain aims to bolster collaboration among organizations.

Moreover, within the CE and IS framework, digitizing the supply chain unlocks significant advantages; real-time data control, as well as optimized resource utilization, reducing costs and aligning with social, economic, and environmental goals [53]. This connection is further supported by research. Study by Zhou and Wang [54] highlights how digitization enhances supply chain flexibility, leading to improved sustainability performance. Similarly, research by Kim and Lee [55] demonstrates that a robust digital infrastructure streamlines operations and positively impacts overall firm performance.

Furthermore, Ye et al. [56] showcase how digital assets improve visibility and agility, enabling better responses to uncertainties. This aligns with Oliveira-Dias et al. [57] who established a positive correlation between digital technology and supply chain agility. Ultimately, the adoption of advanced I. 4.0 technologies like IoT and Big Data is gaining traction, conducting toward increased customer satisfaction and reduced costs, as evidenced by research from Shah et al. [58].

Disruptive technologies, like those encompassed by I. 4.0, hold great potential to enhance the sustainability of RL systems across economic, environmental, and social factors [43].

This paves the way for a novel approach—a redesigned supply chain, encompassing RL empowered by I. 4.0 technologies, as a driving force to fostering interconnection among IP and their surrounding areas, thanks to the application of CE and IS principles.

This approach optimizes operations, leading to increased consumer satisfaction, minimized environmental impact, and economic prosperity. Furthermore, it provides policymakers with a valuable strategy to support the implementation of a circular industrial system (Fig. 24.2).

24.5 Discussions and Implications

The fusion of I. 4.0 technologies with RL is catalyzing a revolution in IPs, driving efficiency through meticulous real-time data analysis and the optimization of resource utilization.

This synergy necessitates that managers not only strategically plan for and invest in cutting-edge technologies but also cultivate an innovative culture to maintain a competitive edge. The financial benefits manifest as cost reductions achieved by minimizing waste and enhancing the recovery of value from returns.

Moreover, collaboration is the linchpin across the supply chain, ensuring a frictionless RL flow. Adherence to environmental regulations is guaranteed through enhanced traceability, which, in tandem with improved return policies and personalized services, elevates customer satisfaction. Furthermore, embracing sustainable practices not only cements the role of managers as environmental stewards but also

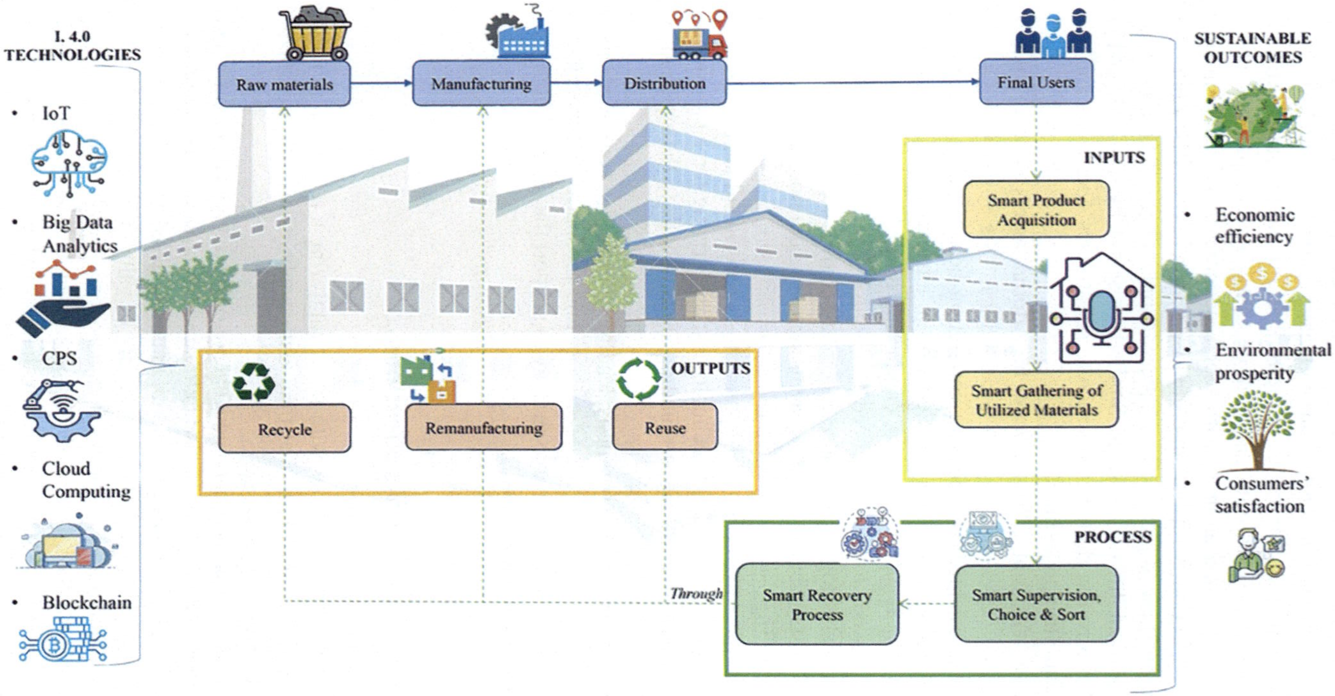

Fig. 24.2 Circular Industrial System integrating RL and I. 4.0 technologies (Source: Authors)

necessitates proactive risk management to navigate emerging challenges like data security.

Additionally, workforce development is imperative, equipping employees with the skills required for advanced RL processes and fortifying IPs to be more resilient, efficient, and sustainable. Concurrently, policymakers grapple with a multifaceted scenario, tasked with formulating regulations that foster the adoption of these technologies and protecting data and upholding ethical standards. Incentives aligned with sustainability objectives propel the CE, while the standardization of data exchange is essential for system interoperability.

Initiatives in education and training are critical to prepare the workforce for impending technological shifts, and prioritizing infrastructure development is key to underpinning sophisticated reverse logistics systems.

Moreover, championing research and development ignites innovation, and cross-sector collaboration disseminates knowledge and best practices, all crucial for policymakers to enable sustainable and proficient supply chains in the epoch of I. 4.0.

24.6 Conclusions and Limitations

As preliminary results of CE_IPsa project (first phase), this study showed RL as a central strategy for managing product returns, EoL processing, and material recovery within a CE framework. I. 4.0 technologies empower RL, creating a resilient, closed-loop production system that minimizes waste.

This integration fosters a sustainable and regenerative model. An important aspect is that RL, enhanced by I. 4.0, offers a solution for overcoming sustainability challenges and establishing a self-sustaining system within IPs.

However, several challenges exist. Further research is necessary to explore this approach in real-world settings, particularly within the context of the IP "Fosso Imperatore" located in Nocera Inferiore, as planned in the CE_IPsa project.

Furthermore, as I. 4.0 is constantly evolving, future research should delve into the continuous exploration of how emerging technologies can further enhance sustainability in the production system.

Implementing I. 4.0 technologies requires substantial investments in infrastructure, training, and data management. Small and medium-sized enterprises within IPs may face resource constraints, hindering widespread adoption.

Addressing these barriers necessitates collaborative efforts among stakeholders, including policymakers, industry associations, and funding bodies. In conclusion, achieving a sustainable production system demands collaboration from various stakeholders.

While this conceptual paper lays the groundwork for integrating RL, and I. 4.0 for a Circular Industrial System, further analysis is needed to broaden the scope of RL and its practical contribution to CE and IS implementation, for enriching interconnection between IPs and surrounding areas.FundingProgetto "CE_IPsa: A development model for Circular Economy between Industrial Parks and surroundings areas" CUP J53D23016630001—codice identificativo P20227NFA4, Department of Economics, University of Study of Messina (Italy).

References

1. Jan, A., Marimuthu, M., Hassan, R., Mehreen.: Sustainable business practices and firm's financial performance in Islamic banking: under the moderating role of Islamic corporate governance. Sustainability. **11**(23), 6606 (2019)
2. D'Adamo, I., Gastaldi, M.: Perspectives and challenges on sustainability: drivers, opportunities and policy implications in universities. Sustainability. **15**(4), 3564 (2023)
3. Zarbakhshnia, N., Govindan, K., Kannan, D., Goh, M.: Outsourcing logistics operations in circular economy towards to sustainable development goals. Bus. Strateg. Environ. **32**(1), 134–162 (2023)
4. Geng, Y., Doberstein, B.: Developing the circular economy in China: challenges and opportunities for achieving 'leapfrog development. Int. J. Sust. Dev. World. **15**(3), 231–239 (2008)
5. Moktadir, M. A., Ahmadi, H. B., Sultana, R., Zohra, F.-T., Liou, J. J. H., & Rezaei, J. (2020). Circular economy practices in the leather industry: A practical step towards sustainable development. Journal of Cleaner Production, **251**, 119737. https://doi.org/10.1016/j.jclepro.2019.119737
6. Hazen, B.T., Russo, I., Confente, I., Pellathy, D.: Supply chain management for circular economy: conceptual framework and research agenda. The Int. J. Logist. Manag. **32**(2), 510–537 (2021)
7. Govindan, K., Soleimani, H., Kannan, D.: Reverse logistics and closed-loop supply chain: a comprehensive review to explore the future. Eur. J. Oper. Res. **240**(3), 603–626 (2015)
8. Chin, T.A., Tat, H.H., Sulaiman, Z.: Green supply chain management, environmental collaboration and sustainability performance. Procedia CIRP. **26**, 695–699 (2015)
9. Zeng, H., Chen, X., Xiao, X., Zhou, Z.: Institutional pressures, sustainable supply chain management, and circular economy capability: empirical evidence from Chinese eco-industrial park firms. J. Clean. Prod. **155**, 54–65 (2017)
10. Barrera, G.A.F., Durany, X.G., Pons, J.R., Erazo, J.G.G.: Trends in global research on industrial parks: a bibliometric analysis from 1996–2019. Heliyon. **7**(8) (2021). https://doi.org/10.1016/j.heliyon.2021.e07778
11. Chertow, M.R.: Industrial symbiosis: literature and taxonomy. Annu. Rev. Energy Environ. **25**(1), 313–337 (2000)
12. Chen, X., Dong, M., Zhang, L., Luan, X., Cui, X., Cui, Z.: Comprehensive evaluation of environmental and economic benefits of industrial symbiosis in industrial parks. J. Clean. Prod. **354**, 131635 (2022)
13. Berberoglu, Y., Kazancoglu, Y., Sagnak, M.: Circularity assessment of logistics activities for green business performance management. Bus. Strateg. Environ. **32**(7), 4734–4749 (2023)
14. Prieto-Sandoval, V., Jaca, C., Ormazabal, M.: Towards a consensus on the circular economy. J. Clean. Prod. **179**, 605–615 (2018)
15. Pushpamali, N., Agdas, D., Rose, T.M., Yigitcanlar, T.: Stakeholder perception of reverse logistics practices on supply chain performance. Bus. Strateg. Environ. **30**, 60–70 (2021)
16. Braungart, M., McDonough, W., Bollinger, A.: Cradle-to-cradle design: creating healthy emissions—A strategy for eco-effective product and system design. J. Clean. Prod. **15**(13–14), 1337–1348 (2007)
17. Julianelli, V., Caiado, R.G.G., Scavarda, L.F., Cruz, S.P.D.M.F.: Interplay between reverse logistics and circular economy: critical success factors-based taxonomy and framework. Resour. Conserv. Recycl. **158**, 104784 (2020)

18. Ivanov, D., Dolgui, A., Sokolov, B.: The impact of digital technology and Industry 4.0 on the ripple effect and supply chain risk analytics. Int. J. Prod. Res. **57**(3), 829–846 (2019)
19. Bai, C., Dallasega, P., Orzes, G., Sarkis, J.: Industry 4.0 technologies assessment: a sustainability perspective. Int. J. Prod. Econ. **229**, 107776 (2020). https://doi.org/10.1016/j.ijpe.2020.107776
20. Jabbour, C.J.C., Fiorini, P.D.C., Ndubisi, N.O., Queiroz, M.M., Piato, É.L.: Digitally-enabled sustainable supply chains in the 21st century: a review and a research agenda. Sci. Tot. Envron. **725**, 138177 (2020)
21. Pourmehdi, M., Paydar, M.M., Ghadimi, P., Azadnia, A.H.: Analysis and evaluation of challenges in the integration of Industry 4.0 and sustainable steel reverse logistics network. Comput. Ind. Eng. **163**, 107808 (2022)
22. Khan, S.A., Laalaoui, W., Hokal, F., Tareq, M., Ahmad, L.: Connecting reverse logistics with circular economy in the context of Industry 4.0. Kybernetes. **52**(12), 6279–6320 (2023)
23. Kirchherr, J., Reike, D., Hekkert, M.: Conceptualizing the circular economy: an analysis of 114 definitions. Resour. Conserv. Recycl. **127**, 221–232 (2017)
24. Baldassarre, B., Schepers, M., Bocken, N., Cuppen, E., Korevaar, G., Calabretta, G.: Industrial Symbiosis: towards a design process for eco-industrial clusters by integrating Circular Economy and Industrial Ecology perspectives. J. Clean. Prod. **216**, 446–460 (2019)
25. Chertow, M.R.: "Uncovering" industrial symbiosis. J. Ind. Ecol. **11**(1), 11–30 (2007)
26. Massard, G., Jacquat, O., Zürcher, D.: International survey on eco-innovation parks. In: Learning from Experiences on the Spatial Dimension of Eco-Innovation (2014)
27. Krstić, M., Agnusdei, G.P., Miglietta, P.P., Tadić, S., Roso, V.: Applicability of Industry 4.0 technologies in the reverse logistics: a circular economy approach based on comprehensive distance based ranking (COBRA) method. Sustainability. **14**(9), 5632 (2022)
28. Ruiz-Puente, C., Jato-Espino, D.: Systemic analysis of the contributions of co-located industrial symbiosis to achieve sustainable development in an industrial park in Northern Spain. Sustainability. **12**(14), 5802 (2020)
29. Steven, M.: Networks in reverse logistics. Supply Chain Manag. Rev. Logist., 163–180 (2004)
30. Phochanikorn, P., Tan, C., Chen, W.: Barriers analysis for reverse logistics in Thailand's palm oil industry using fuzzy multi-criteria decision-making method for prioritizing the solutions. Granul. Comput. **5**(4), 419–436 (2020)
31. Gomes da Silva, F.J., Gouveia, R.M.: Drivers and barriers to cleaner production. Clean. Prod., 375–399 (2020)
32. Garcia-Muiña, F.E., Gonzalez-Sanchez, R., Ferrari, A.M., Volpi, L., Pini, M., Siligardi, C., Settembre-Blundo, D.: Identifying the equilibrium point between sustainability goals and circular economy practices in an Industry 4.0 manufacturing context using eco-design. Soc. Sci. **8**(8), 241 (2019)
33. Sanders, A., Elangeswaran, C., Wulfsberg, J.P.: Industry 4.0 implies lean manufacturing: research activities in Industry 4.0 function as enablers for lean manufacturing. J. Ind. Eng. Manag. **9**(3), 811–833 (2016)
34. Flint, D.J., Larsson, E., Gammelgaard, B., Mentzer, J.T.: Logistics innovation: a customer value-oriented social process. J. Bus. Logist. **26**(1), 113–147 (2005)
35. Zhu, Q., Geng, Y., Lai, K.H.: Circular economy practices among Chinese manufacturers varying in environmental-oriented supply chain cooperation and the performance implications. J. Environ. Manag. **91**, 1324–1331 (2010)
36. McMeekin, N., Wu, O., Germeni, E., Briggs, A.: How methodological frameworks are being developed: evidence from a scoping review. BMC Med. Res. Methodol. **20**, 1–9 (2020)
37. Tricco, A.C., Lillie, E., Zarin, W., O'Brien, K.K., Colquhoun, H., Levac, D., et al.: PRISMA extension for scoping reviews (PRISMA-ScR): checklist and explanation. Ann. Intern. Med. **169**(7), 467–473 (2018)
38. De Brito, M.P., Dekker, R.: A framework for reverse logistics. In: Dekker, R., Fleischmann, M., Inderfurth, K., Van Wassenhove, L.N. (eds.) Reverse Logistics: Quantitative Models for Closed-loop Supply Chains, pp. 3–28. Springer-Verlag, Berlin (2004)

39. Banihashemi, T.A., Fei, J., Chen, P.S.-L.: Exploring the relationship between reverse logistics and sustainability performance: a literature review. Mod. Supply Chain Res. Appl. **1**(1), 2–27 (2019)
40. Ali, A.H., Zalavadia, S., Barakat, M.R., Eid, A.: The role of sustainability in reverse logistics for returns and recycling. Arch. Bus. Res. **6**(7) (2018)
41. Rubio, S., Jiménez-Parra, B.: Reverse logistics: overview and challenges for supply chain management. Int. J. Eng. Bus. Manag. **6**, 12 (2014)
42. Pokharel, S., Mutha, A.: Perspectives in reverse logistics: a review. Resour. Conserv. Recycl. **53**(4), 175–182 (2009)
43. Sun, X., Yu, H., Solvang, W.D.: Towards the smart and sustainable transformation of Reverse Logistics 4.0: a conceptualization and research agenda. Environ. Sci. Pollut. Res. **29**(46), 69275–69293 (2022)
44. da Silva, V.L., Kovaleski, J.L., Pagani, R.N.: Technology transfer in the supply chain oriented to industry 4.0: a literature review. Tech. Anal. Strat. Manag. **31**(5), 546–562 (2019)
45. Okpala, B.: How is supply chain 4.0 tackling the supply chain challenges in the retail sector? (2020)
46. García-Reyes, H., Avilés-González, J., Avilés-Sacoto, S.V.: A model to become a supply chain 4.0 based on a digital maturity perspective. Procedia Comput. Sci. **200**, 1058–1067 (2022)
47. Govindan, K., Kannan, D., Jørgensen, T.B., Nielsen, T.S.: Supply Chain 4.0 performance measurement: a systematic literature review, framework development, and empirical evidence. Transp. Res. E Logist. Transp. Rev. **164**, 102725 (2022)
48. Fernández-Miranda, S.S., Marcos, M., Peralta, M.E., Aguayo, F.: The challenge of integrating Industry 4.0 in the degree of Mechanical Engineering. Procedia Manuf. **13**, 1229–1236 (2017)
49. Dafflon, B., Moalla, N., Ouzrout, Y.: The challenges, approaches, and used techniques of CPS for manufacturing in Industry 4.0: a literature review. Int. J. Adv. Manuf. Technol. **113**(7–8), 2395–2412 (2021)
50. Qian, L., Luo, Z., Du, Y., Guo, L.: Cloud computing: an overview. In: Jaatun, M.G., Zhao, G., Rong, C. (eds.) Cloud Computing. CloudCom 2009 Lecture notes in computer science, vol. 5931. Springer, Berlin (2009)
51. Justinia, T.: Blockchain technologies: opportunities for solving real-world problems in healthcare and biomedical sciences. Acta Inform. Med. **27**(4), 284 (2019)
52. Büyüközkan, G., Göçer, F.: Digital supply chain: literature review and a proposed framework for future research. Comput. Ind. **97**, 157–177 (2018)
53. Arenkov, I., Tsenzharik, M., Vetrova, M.: Digital technologies in supply chain management. In: International Conference on Digital Technologies in Logistics and Infrastructure (ICDTLI 2019), pp. 448–453. Atlantis Press (2019)
54. Zhou, Q., Wang, S.: Study on the relations of supply chain digitization, flexibility and sustainable development—A moderated multiple mediation model. Sustainability. **13**(18), 10043 (2021)
55. Kim, H., Lee, C.: Relationships among healthcare digitalization, social capital, and supply chain performance in the healthcare manufacturing industry. Int. J. Environ. Res. Public Health. **18**, 1417 (2021)
56. Ye, F., Liu, K., Li, L., Lai, K.H., Zhan, Y., Kumar, A.: Digital supply chain management in the COVID-19 crisis: an asset orchestration perspective. Int. J. Prod. Econ. **245**, 108396 (2022)
57. Oliveira-Dias, D., Maqueira-Marín, J.M., Moyano-Fuentes, J.: The link between information and digital technologies of industry 4.0 and agile supply chain: mapping current research and establishing new research avenues. Comput. Ind. Eng. **167**, 108000 (2022)
58. Shah, S., Dikgang, G., Menon, S.: The global perception of industry 4.0 for reverse logistics. Int. J. Econ. Manag. Syst. **4** (2019)

Chapter 25
Unraveling the Role of Artificial Intelligence to Achieve Energy-Related Sustainable Development Goals: A Literature Review and an Empirical Analysis

Martino Bonalumi, Simone Franzò, and Vittorio Chiesa

Abstract Digital technologies have been increasingly debated among scholars and practitioners for their ability to support the evolution of the energy sector, i.e., the so-called energy transition, thus supporting the achievement of Sustainable Development Goals. Nevertheless, there are only a few contributions that aim at shedding light on the enabling roles that can be played by specific digital technologies—such as artificial intelligence—to enable energy transition. This chapter analyzes how artificial intelligence can support the achievement of energy-related Sustainable Development Goals, by shedding light on the different roles it can play in this domain. Such roles are first identified through a literature review and are characterized by two dimensions, i.e., the applications and the domains along the energy value chain. Second, an empirical analysis targeting startups has been carried out to preliminary corroborate findings coming from the literature review.

Keywords Artificial Intelligence · Energy transition · Sustainable Development Goals · Sustainability · Energy · Startups

25.1 Introduction

Digitalization has been increasingly debated among scholars and practitioners as one of the major trends that are currently transforming the society we live in [1, 2]. From the companies' perspective, digitalization can contribute to transforming the way companies create and capture value, thus shaping their business model [3, 4].

M. Bonalumi (✉) · S. Franzò · V. Chiesa
School of Management Politecnico di Milano, Milan, Italy
e-mail: Martino.bonalumi@polimi.it

F. Schiavone et al. (eds.), *Advanced Perspectives and Trends in Digital Transformation of Firms, Networks, and Society*, Springer Proceedings in Business and Economics, https://doi.org/10.1007/978-3-031-80692-6_25

In particular, there has been an increasing interest toward the possibilities enabled by digital technologies (DTs) to help companies achieve sustainability-related goals [5–8].

Sustainable Development Goals (SDGs) have been quite recently introduced by the United Nations (UN) within the so-called "2030 Agenda" with the aim to set the path for a global partnership to foster sustainable development. Among them, SDG 7 ("ensuring access to affordable, reliable, sustainable and modern energy for all") is strictly related to the energy transition, which can be considered as the transformation of the energy sector needed to comply with the objective of limiting the global temperature increase below 2 °C [9]. Interestingly, DTs have been increasingly recognized as energy transition enablers [6]; however, there are only a few contributions that aim at shedding light on the enabling roles that can be played by specific DTs (such as artificial intelligence (AI), blockchain, or big data analytics) to enable energy transition [10].

Among DTs, AI has gone through an unequaled development over the last decades. Accordingly, it is being recognized as one of the most promising DTs and is rapidly spreading into many different sectors, such as healthcare or finance [11–13]. In this vein, scholars have recently addressed the role that AI can play to significantly impact sustainability across the environmental, economic, and social pillars. For example, Zhao and Gómez Fariñas [14] highlight the close relationship between AI and sustainability, due to the importance of sustainable decisions in the context of AI, while Khakurel et al. [15] focus on the impact assessment of AI on sustainable development. Furthermore, Perucica and Andjelkovic [16] raise awareness about the interdependence between AI and environmental sustainability, stressing the need for a more comprehensive approach to this relationship.

Within the existing literature at the intersection between AI and sustainability, a topic under discussion refers to the relationship between AI and SDGs. Indeed, existing studies acknowledge the potential for AI to act as both an enabler and inhibitor for sustainable development [17]. On the one hand, AI has the potential to act as an enabler for numerous targets across the SDGs. Yigitcanlar et al. [18] state that AI may act as an enabler on 134 targets across all SDGs, usually through technological improvement that might help overcome limitations of current technologies. Wakunuma et al. [19] demonstrated a link between the development of AI and the attainment of SDGs, further emphasizing the potential for AI to contribute positively to sustainable development. On the other hand, AI also might have a potential inhibitor effect on the achievement of SDGs. Vinuesa et al. [20] highlight that AI may inhibit 58 targets across the SDGs, indicating the potential for AI to impede progress in certain areas of sustainable development.

To the best knowledge of the authors, there are only a few studies analyzing the roles that can be played by AI to support the achievement of energy-related SDGs. Ababneh [21] highlights the increasing number of studies in the policy literature focusing on governing AI to benefit climate change mitigation and sustainability under the SDGs, emphasizing the need for policy regulations specific to climate informatics. Cowls et al. [22] analyze the role of AI in combating global climate change, emphasizing the opportunities, challenges, and recommendations for leveraging AI to address climate change. These studies underscore the potential of AI to contribute to climate change adaptation and mitigation, aligning with the goals of sustainable development. However, a comprehensive picture of the different roles that AI can play to achieve energy-related SDGs is still lacking [23], thus calling for further research on the topic.

Accordingly, the paper helps filling this gap by shedding light on the roles that can be played by AI to achieve energy-related SDGs. In particular, the paper aims at answering the following research question: Which roles can be played by AI to foster the achievement of energy-related SDGs?

To this aim, a bibliometric literature review was first carried out to identify such roles. 35 roles that AI can play to achieve energy-related SDGs were identified, being characterized by different applications (i.e., the activity/task that AI algorithms can carry out) and domains (i.e., the stages along the energy value chain where AI can be exploited to support the achievement of energy-related SDGs). Furthermore, an empirical analysis that involved 281 startups dealing with AI-based applications in the energy sector was carried out, to preliminary validate evidence coming from the literature review.

The remainder of the chapter is structured as follows. The second section shows the methodological process that has been followed to carry out the study. The third section presents the findings that emerged from the literature review and the following empirical analysis. Finally, the last section provides concluding remarks, along with limitations and avenues for future research.

25.2 Methodology

The literature review is aimed at identifying the roles that AI can play to achieve energy-related SDGs. To these aims, a bibliometric review was conducted, based on publications on Scopus. The search process lasted around 6 months (from August 2023 to January 2024). The query used included two keywords: "Artificial intelligence" and "energy." The query allowed to identify around 7500 contributions, which have been filtered through their title, abstract, and keywords according to their relevance against the objective of the present study. The overall process led to a final sample of more than 4000 publications. The main reason for exclusion was the unrelatedness to the energy sector. Each contribution has been classified to have only one domain and one application, originating a single "intersection" of the two dimensions. A few papers with a more general scope, which could not fit in a single domain and/or application, have been labeled as "general" and are not included in the analysis (less than 3% of the sample).

The following empirical research has been conducted by targeting start-ups that leverage AI for energy-related purposes. The analysis of startups is well-suited to examine the impact of emerging phenomena such as the spread of AI in the energy sector, due to several reasons. First, startups play a crucial role in driving innovation and technological advancements in various industries, as they are often at the forefront of adopting new technologies and business models [24], making them an ideal subject for studying the impact of AI on business operations and models. Moreover, startups are known for their agility and adaptability, making them ideal subjects for observing the integration of AI into business operations and the subsequent impact on their business models [25]. Additionally, the use of AI and machine learning in business operations has been recognized as a significant area of research, emphasizing the need to systematically analyze the scientific literature relating to the application of AI and machine learning in industry [26]. This systematic analysis can be effectively conducted through the study of startups, which are often early adopters of AI technologies and can provide valuable insights into the practical implications of AI on business models.

The analysis mainly relied on the Crunchbase database, which has been exploited to identify the relevant startups by using "artificial intelligence" as a keyword and limiting the analysis to the energy sector and only to companies founded from 2013 onwards. The research allowed to identify around 600 startups, which then were checked on an individual basis (through their website). The methodology of selection relied on a desk analysis of the business model of the startups—mainly in terms of the product/service offered, target client (e.g., B2C, B2B, B2B2C), and revenue model (e.g., spot payment and service fee)—to check: (1) whether their activities involved the development of AI applications and (2) toward which sector they were applied to. Only firms that clearly implemented AI applications in the energy sector were included in the analysis. Furthermore, startups that are no longer active were excluded from the sample. The final sample included 281 start-ups, which have been classified along the two dimensions that emerged through the literature analysis and that characterize the different roles played by AI to achieve energy-related SDGs (i.e., AI applications and domains covered). While contributions have been classified univocally along the two dimensions (applications and domains), startups have been labeled into one or more, due to their usually larger set of activities carried out.

25.3 Results

25.3.1 Bibliometric Review

Figure 25.1 shows the distribution of the identified contributions per year. A clear growing pattern in the last few years emerges as the number of publications in 2023 is almost ten times higher compared to the one recorded in 2014. This shows the very huge recent interest in AI applications in the energy sector.

The analysis of the contributions led to the identification of 13 different AI applications to support the achievement of energy-related SDGs, which refer to 7 different domains [27, 28, 29], as detailed in the following tables.

Table 25.1 provides a definition for each identified AI application, while Fig. 25.2 shows the relative spread of such applications within the sampled contributions.

The most diffused AI applications refer to energy assessment and asset optimization (around 15% of the sampled contributions), followed by predictive maintenance, individual and collective consumption forecast, grid stability, and generation forecast.

Table 25.2 provides a definition of the seven identified domains along with its relationship with the energy-related SDGs.

The two most diffused domains by far are the ones related to transmission and distribution and renewable energy generation, as shown in Fig. 25.3. Energy consumption also captures a lot of attention, for both industrial and residential applications.

Finally, Table 25.3 shows the 35 roles (each one being identified as an application-domain pair) that AI can play to support the achievement of energy-related SDGs, as well as the related frequency among the analyzed contributions. It is worth highlighting that the two most diffused are both related to renewables, i.e., generation forecast and asset optimization. This is not surprising, as the non-predictability of generation assets has been an issue as their percentage in electricity generation mix increased. To this aim, more and more sophisticated algorithms have been developed to have the most accurate forecast possible; right now, this is achieved through machine learning algorithms. Moreover, in the development of renewable power plant, it is extremely important to optimize the layout, which AI can help with [30].

The other most diffused AI roles refer to the transportation and distribution domain and tackle grid stability and collective energy consumption forecast. This is consistent with the fact that electricity grid management is facing a higher level of complexity, due to the dramatic diffusion of distributed generation assets. This transformation requires a different approach in the management of the grid, which is the reason behind the diffusion of algorithms to optimize energy flows and to better forecast the load on close to real-time [28].

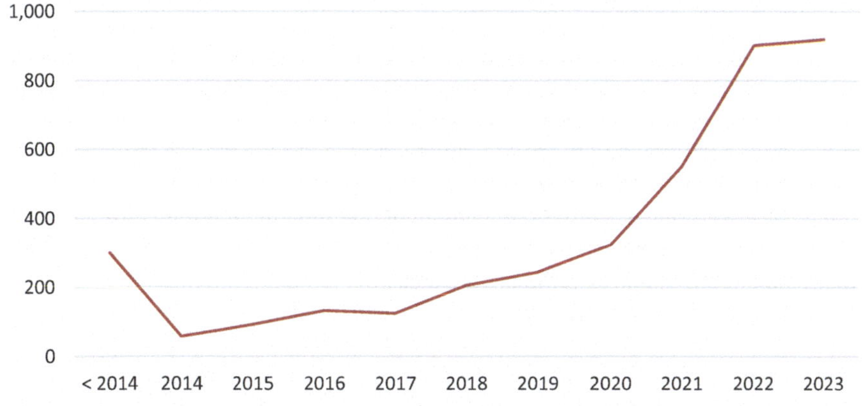

Fig. 25.1 Number of contributions included in the analysis per year of publication

Table 25.1 AI applications in the energy sector

AI applications	Definitions
Generation forecast	Algorithms that forecast the amount of energy produced by generation plants (usually from non-programmable renewable assets, such as PV and wind)
Generation asset design optimization	Algorithms that help to optimize the layout of a generation asset during the design phase (e.g., the position of wind towers in a wind farm)
Energy assessment	Algorithms that evaluate energy usage of buildings and or processes to identify possible inefficiencies (e.g., energy audit of a building characteristics to implement energy efficiency measures)
Predictive maintenance	Algorithms that implement predictive analytics to better forecast when an asset would need maintenance
Grid stability	Algorithms that allow to optimize the management of electricity grids, especially when they include Distributed Generation (DG) assets
Asset optimization	Algorithms that modify certain operating parameters of an asset to optimize its operating conditions (e.g., with reference to an industrial process)
Collective consumption forecast	Algorithms that forecast the electricity demand of a set of energy users (e.g., an energy community)
Individual consumption forecast	Algorithms that forecast the electricity demand of a single energy user (e.g., a household)
Price forecast	Algorithms that forecast energy commodities prices (e.g.: Electricity, gas)
Bidding strategy	Algorithms that create an optimal bidding strategy for players operating in energy commodities markets (e.g., Traders)
Demand response	Algorithms that allow the implementation of demand response programs, which are a modulation of final consumers' electricity demand to help balancing the load on electricity grids
Vehicle-grid-integration (VGI)	Algorithms that optimize the electric vehicle charging process and the possible interactions with the grid, either mono-directional (V1G) or bi-directional (V2G).
Storage optimization	Algorithms that allow to optimize the charge and discharge of energy storage systems

Finally, among the most diffused AI roles, it is worth mentioning the energy assessment for residential energy consumption (closely followed by the energy assessment for industrial energy consumption). This evidence confirms the potential role of AI to help selecting the most powerful measures to reduce energy consumption or identifying inefficiencies [31].

The literature analysis also brings into light that some AI roles are characterized by an AI application that is mainly related to one or a few domains, while other roles refer to applications that cover many domains.

Regarding the former, the most evident example refers to "generation forecast," which is associated to the renewable energy generation domain. Indeed, one of the most important areas of application of AI and ML algorithms is the forecast for non-programmable generation assets, such as PV and wind plants. These algorithms

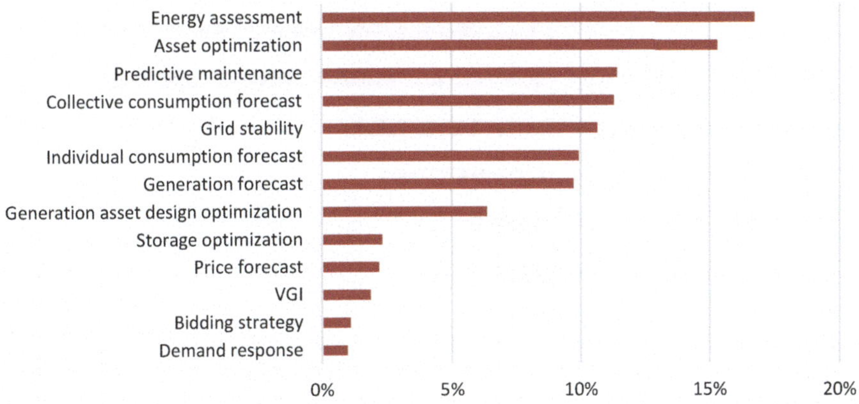

Fig. 25.2 Distribution of AI applications

Table 25.2 Description of domains and related SDGs. The SDGs indicated in the table refer to: 7.1—affordable, reliable, and modern energy services; 7.2—increase renewable energy; 7.3—increase energy efficiency; 7.a—promote investment in energy infrastructure and clean energy technology; 11—sustainable cities

Domains	Definition	Associated SDGs
Renewable energy generation (RES)	Stage of the energy value chain related to the generation of electricity from renewable energy sources (e.g., PV, wind)	7.2
Non-renewable energy generation (NRES)	Stage of the energy value chain related to the generation of electricity from traditional energy sources (e.g., coal, natural gas)	7.1
Transportation and distribution (T&D)	Stage of the value chain related to the transmission and distribution of energy carriers (mainly electricity, but also oil and gas). It includes all the infrastructures "behind the meter" and micro-grids	7.a
Residential energy consumption (REC)	Stage of the value chain related to the final consumption of energy by residential users/consumers	7.3
Industrial energy consumption (IEC)	Stage of the value chain related to the final consumption of energy by non-residential users/consumers (e.g., offices and industry)	7.3
Electric vehicles (EVs)	Stage of the value chain related to the consumption of energy by electric vehicles	Sustainable transport (related to SDG 11)
Commodity market (CM)	Energy markets (e.g., electricity and gas)	7.1

allow to increase the accuracy of the forecast compared to existing methods, both in the long and the short term. This has a direct consequence on the management of the electricity grids (by reducing the amount of reserves needed to balance the system), but also on the remuneration of generators (since they might incur in lower imbalance costs, having a generation closer to the bid they submitted, which is based on

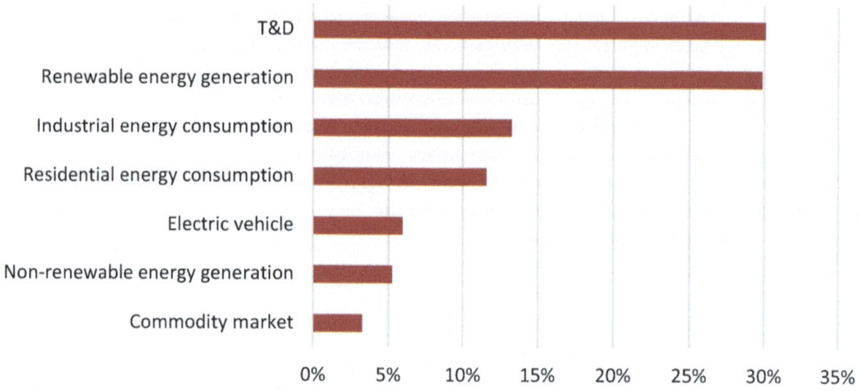

Fig. 25.3 Distribution of domains

Table 25.3 Frequency of application domain

	CM	RES	EV	T&D	NRES	IEC	REC	Total
Asset optimization	–	9.3%	0.1%	1.8%	1.6%	1.9%	0.6%	*15.3%*
Bidding strategy	1.1%	–	–	–	–	–	–	*1.1%*
Collective consumption forecast	–	–	–	7.2%	–	2.4%	1.7%	*11.3%*
Generation asset design optimization	–	4.7%	–		1.6%	–	–	*6.3%*
Demand response	–	–	–	1.0%	–	–	–	*1.0%*
Energy assessment	–	–	2.1%	3.9%	–	5.0%	5.5%	*16.6%*
Grid stability	–	1.4%	–	8.0%	–	1.3%	–	*10.6%*
Generation forecast	–	9.7%	–	–	–	–	–	*9.7%*
Individual consumption forecast	–	–	1.2%	2.8%	–	2.2%	3.7%	*9.9%*
Predictive maintenance	–	4.3%	0.4%	4.0%	2.1%	0.6%	–	*11.4%*
Price forecast	2.2%	–	–	–	–	–	–	*2.2%*
Storage optimization	–	0.5%	0.3%	1.4%	–	–	–	*2.3%*
VGI	–	–	1.9%	–	–	–	–	*1.9%*
Total	*3.4%*	*29.9%*	*6.0%*	*30.1%*	*5.3%*	*13.3%*	*11.6%*	*100%*

the forecast itself). Regarding the latter, energy assessment is one of the most "transversal" applications, as it covers four different domains. The same applies for asset optimization: indeed, a lot of publications focus on the possibility to optimize operations using AI, which may provide higher yields and consequently higher returns of investments, crucial for the "marketability" of these power plants.

Regarding the latter, one of the most diffused is "Predictive maintenance": predictive maintenance is growing rapidly, overtaking the "traditional" on-fault or preventive maintenance. AI algorithms allow to better estimate when certain assets or components are close to the end-of-life, enabling cost savings compared to other solutions. It is a quite transversal application due to the possibility to implement it in every asset and because of the relevance, O&M has in the cost along the life of the asset itself. Finally, among domains it is worth mentioning one which heavily relies on AI, which is transportation and distribution, whose main applications are related with grid stability and generation and consumption forecast (either collective or individual). Indeed, as the electricity systems become more and more complex, algorithms are increasingly needed to manage the large variety of inputs of the different assets involved.

25.4 Empirical Analysis

Figure 25.4 shows the year of foundation of the 281 startups included in the sample. It emerges a reduction in the number of startups that have been founded in the last 4 years. Even though this evidence appears to be in contrast with the growing interest that emerged from the literature analysis, there are some possible explanations for this trend. A possible explanation for this phenomenon is that it may take some time for startups to be recognized and registered into databases (such as Crunchbase, from which startups were selected), which might explain the quite limited number of startups founded in the last few years.

When looking at the country of origin of the startups analyzed, Europe and North America are the most represented areas (Fig. 25.5), with 41% and 36% of the total sample, respectively.

The analysis of startups confirmed the applications and domains characterizing AI roles that have been identified through the literature review. A few notable differences emerged, as detailed hereafter.

About AI applications, Fig. 25.6 shows that energy assessment is still the most diffused one (around 20% of the total sample), thus confirming the interest toward energy efficiency measures, which are strictly related to SDG 7.3. Predictive maintenance closely follows, being implemented by almost 20% of the sample: this

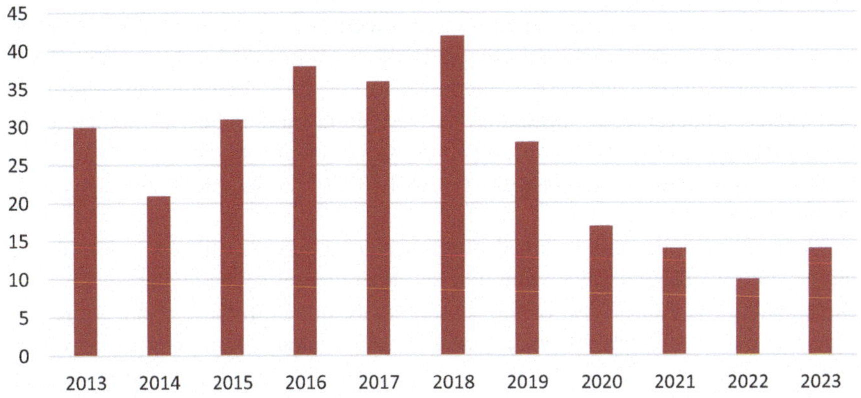

Fig. 25.4 Startup per year of foundation

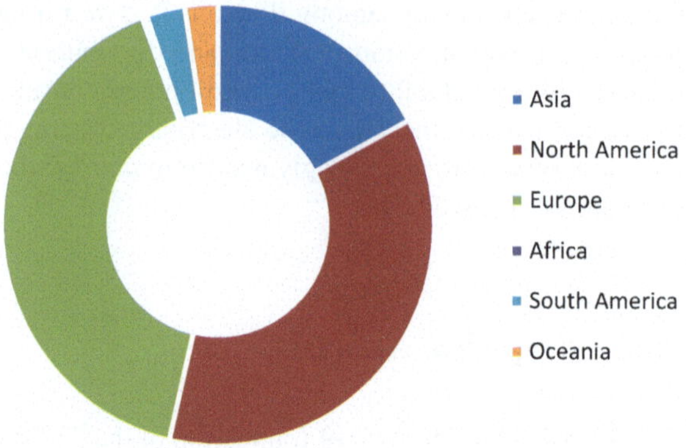

Fig. 25.5 Startup per continent

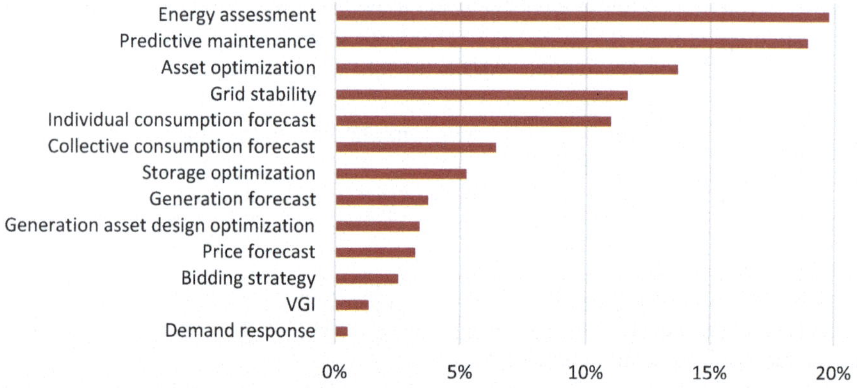

Fig. 25.6 Distribution of AI applications among startups

application is more "transversal," due to its impact of the different assets along the value chain. However, it is especially important to help develop renewable power plants (SDG 7.2) and a reliable electricity grid (SDG 7.a).

Furthermore, the startups analysis confirmed the seven domains identified through the literature review and especially the two most diffused ones, i.e., T&D and renewable energy generation, as shown in Fig. 25.7. Energy consumption is well represented as well, especially for residential and industrial applications. Among the least diffused domains, it is interesting to notice that commodity market is more present in the startup analysis (around 10% of the sample) than in literature (around 3%).

Asset optimization, often linked to renewable energy, is still commonly found among startups (around 13%), although less than predictive maintenance, which is implemented by almost 20% of the sample analyzed. Both these applications are typically related to renewable energy plants, which is one of the main areas where startups focus. Also, around the 10% mark, there are different applications related to energy flow prediction and management, such as grid stability and individual

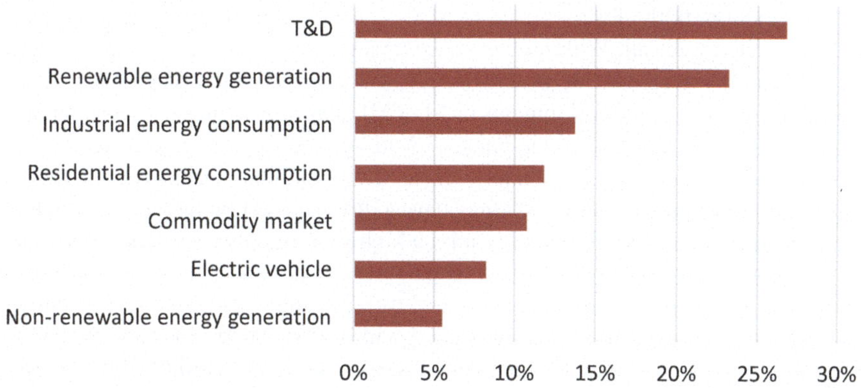

Fig. 25.7 Distribution of domains among startups

consumption forecast. All of these are related to transmission and distribution domain, whose development attracted a variety of firms with innovative services and business models in the last few years.

The most notable difference compared to the literature analysis is about generation forecast, as startups analysis shows a limited coverage by analyzed companies. This is likely due to the "in-house" development (or acquisition) of this kind of application by large players, such as electric utilities or independent power producers (IPPs), due to the high relevance for their business model, which leaves less space for startups to develop their own business model around it.

25.5 Conclusion

The 17 interconnected SDGs and 169 targets outlined in the 2030 Agenda for Sustainable Development have been a focal point for researchers and policymakers. The potential of AI to address issues associated with SDGs has led to increased discussions about its applications [14, 20].

The present study contributes to the existing literature by showing how AI can spread as a powerful tool to significantly impact the achievement of the energy-related SDGs, i.e., by shedding light on the different roles it can play in such a context. In particular, results show that AI has the potential to optimize energy markets and power systems, transforming the way energy is produced, distributed, and consumed. Furthermore, AI is effectively used to improve the efficiency and performance of processes, including those related to renewable energy sources [28, 30]. This paper confirmed the strong and growing interest toward AI applications in the energy sector and allowed to identify the most interesting AI roles that may support the transformation of the industry. Indeed, the first part focused on a bibliometric review of activities carried out by AI, which are transforming the energy sector, starting from two of its "pillars," renewable energy generation and transportation and distribution, both closely linked to SDG 7, which aims at ensuring affordable and clean energy for all. Within these two domains, there is room for improvement for applications related to the optimization of operations, which AI can help with: indeed, the most recent developments of AI or ML algorithms, allow for a better and more precise forecast in different "areas." A more precise generation forecast, for example, allows to reduce imbalances on electricity markets, increasing the profitability for the actors involved. This may have a direct impact on target 7.2, which is about the diffusion of RES in energy mix: a higher profitability for renewable generation assets may help with their "marketability," hence fostering their diffusion. Moreover, predictive maintenance is quickly replacing "on-fault" or preventive maintenance, reducing the costs associated with O&M; apart from an impact on profitability, predictive maintenance may also increase the reliability of the energy infrastructure (targets 7.1 and 7.a). Another AI application revealed by the analysis is the energy assessment to support energy efficiency measures, which is directly correlated with target 7.3.

The second part of the analysis involved a broad analysis of startups, which were chosen as the "empirical" setting upon which verify the results coming from the literature review. The startup analysis confirmed most of the findings from the literature review for both the domains (around one out of two startups focuses either on transportation and distribution or on renewable energy) and applications. This shows the great potential that AI has to transform the energy sector and thus help to achieve decarbonization objectives necessary to comply with the SDGs.

Apart from academic implications, this study can be useful for practitioners and policymakers. First of all, it helps identify the most promising AI roles to address the different energy-related SDGs, thus facilitating professionals working within the energy sector to understand and implement AI technologies with the overarching goal of expediting advancements toward the SDGs. Moreover, the identification of AI applications may help policymakers in identifying critical aspects of their implementation and their consequences on business models of the actors involved, in a crucial sector such as the energy one. This is especially true in areas, like the European Union, where a strong attention is being paid to AI regulation.

Finally, the authors acknowledge the presence of some limitations characterizing the present study. First, the identification and selection of startups is based on a database (Crunchbase), where startups are classified according to their sector of activity. There is a first "bias" related to this classification: startups whose core activity is not related with the energy sector may not be included in the search. Furthermore, the database suffers from a "survivability bias," which means that: (1) startups may not be included in the database in the first few months/years of their foundation because it is difficult to detect and identify them; (2) the ones that fail in the first stages of development might be underrepresented in the total sample [32]. Moreover, even though startups are well suited to represent the changes in the industry, limiting the analysis to them is also a limitation for the paper. Indeed, a possible improvement on the investigation could be related to the inclusion of incumbents, to understand whether their approach and positioning (in terms of AI application exploited and domains) differ from startups. Moreover, the information collected comes from a desk analysis based on secondary sources. A complementary assessment might be carried out by direct interviews to relevant firms, to better explore their approach toward AI applications. Finally, as a possible avenue for future research, each of the most interesting and diffused applications that emerged from the analysis might be investigated singularly. This might allow to explore the peculiarities and the pros and cons in depth, identifying the possible development of a single application, especially in relation to startups. Finally, it would be crucial to address ethical, diversity, and sustainability considerations to ensure that AI contributes positively to the achievement of these goals. The potential of AI to optimize energy systems and processes, coupled with the need for ethical and sustainable AI development, presents a complex landscape that requires further research and policy attention to harness the full potential of AI for energy-related SDGs.

References

1. Coreynen, W., Matthyssens, P., Van Bockhaven, W.: Boosting servitization through digitization: pathways and dynamic resource configurations for manufacturers. Ind. Mark. Manag. **60**, 42–53 (2017). https://doi.org/10.1016/j.indmarman.2016.04.012
2. Vial, G.: Understanding digital transformation: a review and a research agenda. J. Strateg. Inf. Syst. **28**(2), 118–144 (2019). https://doi.org/10.1016/j.jsis.2019.01.003
3. Kamalaldin, A., Linde, L., Sjödin, D., Parida, V.: Transforming provider-customer relationships in digital servitization: a relational view on digitalization. Ind. Mark. Manag. **89**, 306–325 (2020). https://doi.org/10.1016/j.indmarman.2020.02.004
4. Kohtamäki, M., Parida, V., Oghazi, P., Gebauer, H., Baines, T.: Digital servitization business models in ecosystems: a theory of the firm. J. Bus. Res. **104**, 380–392 (2019). https://doi.org/10.1016/j.jbusres.2019.06.027
5. Cardinali, P.G., De Giovanni, P.: Responsible digitalization through digital technologies and green practices. Corp. Soc. Responsib. Environ. Manag. **29**(4), 984–995 (2022). https://doi.org/10.1002/csr.2249
6. Mondejar, M.E., Avtar, R., Diaz, H.L.B., Dubey, R.K., Esteban, J., Gómez-Morales, A., Hallam, B., Mbungu, N.T., Okolo, C.C., Prasad, K.A., She, Q., Garcia-Segura, S.: Digitalization to achieve sustainable development goals: steps towards a Smart Green Planet. In: Science of the Total Environment, vol. 794. Elsevier B.V (2021). https://doi.org/10.1016/j.scitotenv.2021.148539
7. Parida, V., Sjödin, D., Reim, W.: Reviewing literature on digitalization, business model innovation, and sustainable industry: past achievements and future promises. In: Sustainability (Switzerland), vol. 11(2). MDPI (2019). https://doi.org/10.3390/su11020391
8. Pigola, A., da Costa, P.R., Carvalho, L.C., da Silva, L.F., Kniess, C.T., Maccari, E.A.: Artificial intelligence-driven digital technologies to the implementation of the sustainable development goals: a perspective from Brazil and Portugal. Sustainability. **13**(24) (2021). https://doi.org/10.3390/su132413669
9. Gielen, D., Boshell, F., Saygin, D., Bazilian, M.D., Wagner, N., Gorini, R.: The role of renewable energy in the global energy transformation. Energ. Strat. Rev. **24**, 38–50 (2019). https://doi.org/10.1016/j.esr.2019.01.006
10. Barinova, V., Devyatova, A., Lomov, D.: The role of digitalization in the global energy transition1,2. Int. Organ. Res. J. **16**(4), 1–18 (2021). https://doi.org/10.17323/1996-7845-2021-04-06
11. Goralski, M.A., Tan, T.K.: Artificial intelligence and sustainable development. International. J. Manag. Educ. **18**(1) (2020). https://doi.org/10.1016/j.ijme.2019.100330

12. Met, İ., Kabukçu, D., Uzunoğulları, G., Soyalp, Ü., Dakdevir, T.: Transformation of business model in finance sector with artificial intelligence and robotic process automation. In: Contributions to Management Science (2020). https://doi.org/10.1007/978-3-030-29739-8_1
13. Zahlan, A., Ranjan, R.P., Hayes, D.: Artificial intelligence innovation in healthcare: literature review, exploratory analysis, and future research. Technol. Soc. **74** (2023). https://doi.org/10.1016/j.techsoc.2023.102321
14. Zhao, J., Gómez Fariñas, B.: Artificial intelligence and sustainable decisions. Eur. Bus. Organ. Law Rev. **24**(1), 1–39 (2023). https://doi.org/10.1007/s40804-022-00262-2
15. Khakurel, J., Penzenstadler, B., Porras, J., Knutas, A., Zhang, W.: The rise of artificial intelligence under the lens of sustainability. Technologies. **6**(4), 100 (2018). https://doi.org/10.3390/technologies6040100
16. Perucica, N., Andjelkovic, K.: Is the future of AI sustainable? A case study of the European Union. Transf. Govern. People Process Policy. **16**(3), 347–358 (2022). https://doi.org/10.1108/TG-06-2021-0106
17. Palomares, I., Martínez-Cámara, E., Montes, R., García-Moral, P., Chiachio, M., Chiachio, J., Alonso, S., Melero, F.J., Molina, D., Fernández, B., Moral, C., Marchena, R., de Vargas, J.P., Herrera, F.: A panoramic view and swot analysis of artificial intelligence for achieving the sustainable development goals by 2030: progress and prospects. Appl. Intell. **51**(9), 6497–6527 (2021). https://doi.org/10.1007/s10489-021-02264-y
18. Yigitcanlar, T., Desouza, K.C., Butler, L., Roozkhosh, F.: Contributions and risks of artificial intelligence (AI) in building smarter cities: insights from a systematic review of the literature. In: Energies, vol. 13(6). MDPI AG (2020). https://doi.org/10.3390/en13061473
19. Responsible AI, SDGs, and AI Governance in Africa
20. Vinuesa, R., Azizpour, H., Leite, I., Balaam, M., Dignum, V., Domisch, S., Felländer, A., Langhans, S.D., Tegmark, M., Fuso Nerini, F.: The role of artificial intelligence in achieving the Sustainable Development Goals. Nat. Commun. **11**(1) (2020). https://doi.org/10.1038/s41467-019-14108-y
21. Ababneh, L.N.: Climate change, climate informatics, and AI: information analysis. In: Handbook of Climate Change Management: Research, Leadership, Transformation, vol. 5, pp. 3933–3943. Springer International Publishing (2021). https://doi.org/10.1007/978-3-030-57281-5_287
22. Cowls, J., Tsamados, A., Taddeo, M., Floridi, L.: The AI gambit: leveraging artificial intelligence to combat climate change—opportunities, challenges, and recommendations. AI Soc. **38**(1), 283–307 (2023). https://doi.org/10.1007/s00146-021-01294-x
23. Kindylidi, I., Cabral, T.S.: Sustainability of AI: the case of provision of information to consumers. Sustainability. **13**(21) (2021). https://doi.org/10.3390/su132112064
24. Weber, M., Beutter, M., Weking, J., Böhm, M., Krcmar, H.: AI startup business models: key characteristics and directions for entrepreneurship research. Bus. Inf. Syst. Eng. **64**(1), 91–109 (2022). https://doi.org/10.1007/s12599-021-00732-w
25. Saura, J.R., Reyes-Menéndez, A., Dematos, N., Correia, M.B.: Identifying startups business opportunities from UGC on twitter chatting: an exploratory analysis. J. Theor. Appl. Electron. Commer. Res. **16**(6), 1929–1944 (2021). https://doi.org/10.3390/jtaer16060108
26. Cioffi, R., Travaglioni, M., Piscitelli, G., Petrillo, A., De Felice, F.: Artificial intelligence and machine learning applications in smart production: progress, trends, and directions. In: Sustainability (Switzerland), vol. 12(2). MDPI (2020). https://doi.org/10.3390/su12020492
27. Bhattacharyya, S.C.: Energy economics: concepts, issues, markets and governance. In: Energy Economics: Concepts, Issues, Markets and Governance. Springer, London (2011). https://doi.org/10.1007/978-0-85729-268-1
28. Motlagh, N.H., Mohammadrezaei, M., Hunt, J., Zakeri, B.: Internet of things (IoT) and the energy sector. In: Energies, vol. 13(2). MDPI AG (2020). https://doi.org/10.3390/en13020494
29. International Renewable Energy Agency: Enabling technologies: innovation landscape (n.d.)

30. Sayed-Mouchaweh, M.: Artificial intelligence techniques for a scalable energy transition: advanced methods, digital technologies, decision support tools, and applications. In: Artificial Intelligence Techniques for a Scalable Energy Transition: Advanced Methods, Digital Technologies, Decision Support Tools, and Applications. Springer International Publishing (2020). https://doi.org/10.1007/978-3-030-42726-9
31. Mehmood, M.U., Chun, D., Zeeshan, H., Jeon, G., Chen, K.: A review of the applications of artificial intelligence and big data to buildings for energy-efficiency and a comfortable indoor living environment. In: Energy and Buildings, vol. 202. Elsevier Ltd. (2019). https://doi.org/10.1016/j.enbuild.2019.109383
32. International Energy Agency. (2022). https://www.iea.org/data-and-statistics/data-tools/energy-start-up-data-explorer

Chapter 26
BPM and BPR for Green Innovation in the Shipping Industry

Bianca Vottero, Giovanni Satta, and Massimo Caruso

Abstract As the maritime business faces increasing challenges concerning sustainability, maritime operators are urged to adopt new technological solutions for improving the environmental and social impact deriving from their operations.

In this context, new evaluation tools are needed, allowing operators to carefully evaluate and select the most viable technological solutions for pursuing their sustainability goals, also considering the specificities of the business which makes decision-making particularly complex. For this aim, the study proposes an ad hoc methodological procedure for investigating the economic, social, and environmental implications that affect the operational processes of shipbuilding companies and the related financial performances when introducing a new green technological solution in the ship manufacturing process. More in detail, the analysis assesses the main implications originating from the introduction of an electric propulsion system in the ship manufacturing process as an alternative to the traditional propulsion systems. To this aim, the main theoretical constructs and practical tools of Business Process Management and Business Process Reengineering are employed. The study allows the identification of the main types of implications arising from the implementation of the technology, and the allocation of the implications within the shipbuilding process, allowing a balanced and holistic evaluation.

Keywords Business process management and reengineering · Sustainability · Shipping

B. Vottero (✉) · G. Satta
Università degli Studi di Genova, Genoa, Italy

M. Caruso
Università degli studi di Palermo, Palermo, Italy

F. Schiavone et al. (eds.), *Advanced Perspectives and Trends in Digital Transformation of Firms, Networks, and Society*, Springer Proceedings in Business and Economics, https://doi.org/10.1007/978-3-031-80692-6_26

26.1 Introduction

The maritime industry and shipbuilding sector are currently facing significant challenges regarding sustainability aspects [1]. The growing awareness among industry operators and stakeholders concerning the impact of their activities on environmental and social sustainability is driving maritime operators to improve their sustainability performance and to adapt to the constantly evolving international regulatory framework [2]. Given the complex nature of the industry, maritime operators must adopt a cautious approach when evaluating performance implications before making technical investments: the capital-intensive nature of the business, the long financial cycles, and related payback periods in the maritime and shipbuilding sectors make investment decisions particularly complex [3]. As a result, new and multifaceted evaluation criteria capable of assessing the impacts on both operational and financial long-term sustainability of the companies have to be identified and, consequently, inserted within the decision-making process of both maritime and shipbuilding companies.

In this context, there is a need to identify priority tools for reinterpreting decision-making processes from a sustainable perspective. Business Process Management (BPM) and Business Process Reengineering (BPR) practices, along with related managerial tools such as flowcharts, blueprints, and scripts, emerge as strategic managerial perspectives and tools for supporting the decision-making process [4] and for developing new sound decision support systems (DSS).

26.2 Objective

The manuscript aims at applying both theoretically and empirically BPM and BPR perspectives and related managerial tools within the maritime and shipbuilding sectors to identify new methodologies for supporting decision-making process, given the specificities and challenges that nowadays shape the industry. More in detail, the study proposes an ad hoc methodological procedure for investigating the economic, social, and environmental implications that affect the operational processes of a shipbuilding company and the related financial performances when introducing a new green technological solution in the ship manufacturing process. The analysis, indeed, assesses the main implications originating from the introduction of an electric propulsion system in the manufacturing process as an alternative to the traditional propulsion systems, which are currently mostly adopted in the industry.

For the aim of the study, the following research objective has been set:

RO: *"to explore and analyse how the principles of Business Process Management (BPM) and Business Process Reengineering (BPR) can be applied to the decision-making process in order to improve technical and financial performances of a shipbuilding company when moving from conventional to electric propulsion engines."*

The prospect of utilizing tools such as flowcharts and blueprints becomes crucial for modeling and optimizing the decision-making process related to the introduction of new green investment solutions in the industry, by promoting an approach that considers the mostly affected manufacturing sub-process, and, as a consequence, key operational, economic, financial, social and environmental Key Performance Indicators (KPIs).

For the purposes of the study, the manuscript first proposes an ad hoc methodological framework for the analysis of functional processes to support researchers, analysts, and managers in identifying the costs and environmental externalities associated with the introduction of green technologies and innovations in the maritime and shipbuilding sectors. Subsequently, the methodological approach is tested through an empirical application to the shipbuilding industry, focusing on the introduction of new electric marine engines.

26.3 Method

For the purpose of the study, an ad hoc methodological procedure, grounded on the Business Process Management (BPM) and the Business Process Reengineering (BPR) managerial perspective is developed. The methodological procedure consists of eight main steps as reported below (Fig. 26.1).

The first step of the methodological procedure aims at identifying the new green and sustainable technologies technically feasible, which can be introduced within a specific operational/manufacturing process in the maritime and shipbuilding sector.

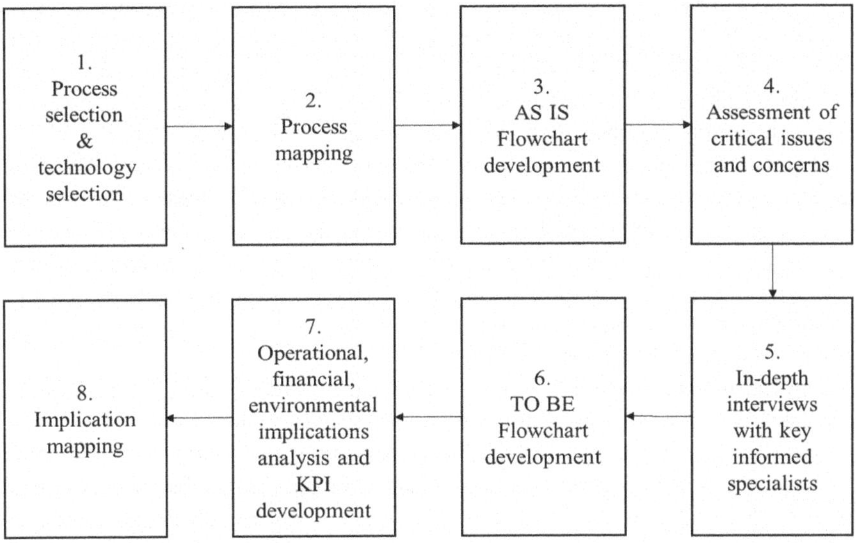

Fig. 26.1 Methodological steps

The analysis aims at scrutinizing those technological options which are considered to have higher influence on internal processes, considering both informational/organizational and productive/operational facets, within shipyards and maritime enterprises. Specifically, electric propulsion emerged as a solution of paramount significance for the sector's green transition and has therefore been selected as case study for the purpose of the analysis.

Step 2 and Step 3 are directed to map the overall operational/manufacturing process target in the analysis and provide the "AS-IS" flowchart, respectively. These methodological steps, indeed, allow to better understand technical and operational activities included in the process, to elucidate the sequences of actions and the potential interdependencies among process activities, thus providing useful insights for identifying critical issues and concerns, as well as potential improvement opportunities embedded in each subprocess.

Subsequently, an in-depth analysis of each key operational activity of the operational/manufacturing process is undertaken to discern subcomponents, subprocesses, and the intricate interrelationships that define the shipbuilding process in the "as is" scenario (i.e., the scenario involving the implementation of electrical propulsion solutions). When it comes to the empirical case selected for testing the proposed methodological framework, the integration of a traditional propulsion system in the shipbuilding manufacturing overall process is the target process. Upon the completion of the process flowcharts, the discernment of the most critical phases followed (step 4), whether pertaining to environmental, operational, economic-financial, or other considerations, which stood to gain the most from the incorporation of the technological solution under investigation. These pivotal phases were annotated using a distinctive graphic element for emphasis.

Following the identification of the phases or process components with high criticality concerning the implementation of electric propulsion solutions, a more detailed analysis of these phases was conducted to achieve a more comprehensive understanding of the shipbuilding processes. This activity was made possible following in-depth interviews organized with industry experts, who provided valuable insights regarding the structure of the process under examination (step 5).

Step 6 of the methodological process entails the development of a new flowchart delineating the constituent elements, phases, and activities inherent to the process when integrating a new technological solution. For the aim of the study, leveraging the insights obtained from interviews with experts, a "to be" flowchart has been developed showing the different articulation of phases and components involved in the shipbuilding processes when influenced by the implementation of the electric propulsion technologies under scrutiny.

Step 7 is aimed at providing a more comprehensive understanding of the impacts associated with the implementation of the investigated solution within the overall process. To this purpose, a set of Key Performance Indicators (KPIs) has to be identified as being best able to represent and quantify the impact that the introduction of the technology can have on the process. The selected KPIs have been classified into the following categories: "Economic & Financial," "Environment & Sustainability," "Shipowner Efficiency," "Shipyard Efficiency," and "Compliance." For each KPI, a unit of measurement has to be defined to qualitatively or quantitatively assess the impact that the introduction of the technology under consideration can have in the context considered by each KPI. Referring to the empirical analysis conducted within this research, the analysis included an evaluation of the shipbuilding operations mostly affected by the introduction of an electric propulsion system through an ad-hoc KPI system, and an assessment of the relevance of each indicator within the identified process, indicated by a color coding.

Once the main implications of the introduction of the innovation or the new green technology (implementing the electric propulsion solution in our empirical case) in the studied processes are evaluated, the methodological procedure concludes by placing the identified KPIs in the process stages mostly affected by them (Step 8). To facilitate understanding, each KPI was associated with an icon, which was then placed in the flowchart in correspondence with the process stages most affected by the indicator. The different categories of KPIs were marked with an identifying color for ease of interpretation.

26.4 Findings

The unique methodological framework for applying BPR and BPM managerial perspective and tool within the decision-making process of the companies operating in the maritime and shipbuilding sectors previously reported and discussed is then empirically applied to a relevant shipbuilding case, i.e., the introduction of an electric ship engine aimed at improving the environmental performance of the ship.

The outcome of the empirical analysis is reported and detailed in Fig. 26.2, which illustrates the shipbuilding process with a graphical indication of the process stages most affected by various types of implications resulting from the implementation of electric propulsion systems. The flowchart graphically identifies the stages affected by the introduction of the new technical solution, namely, the procurement, design, outfitting, testing and commissioning, and delivery phases. Most of the impacts generated pertain, positively or negatively, to the categories of KPIs named "Shipyard Efficiency," "Economic & Financial," and "Environment & Sustainability."

The application of the proposed methodological procedure to a real-life case from the industry allows to demonstrate how the industry should benefit from the application of BPR and BPM perspective and managerial tools when evaluating the introduction of available green and sustainable technologies, taking into account operational, economic, financial, social and environmental implications. In particular, the empirical analysis performed in line with the proposed methodological procedure provides a number of valuable insights.

First, it is demonstrated that the implementation of electric propulsion solutions can imply a series of critical issues at the production process level, mostly related to procurement, new technical skills requested, and timing for project completion. Second, the innovation characterizing the technology in question and the complexity in terms of materials required and its implementation can have significant

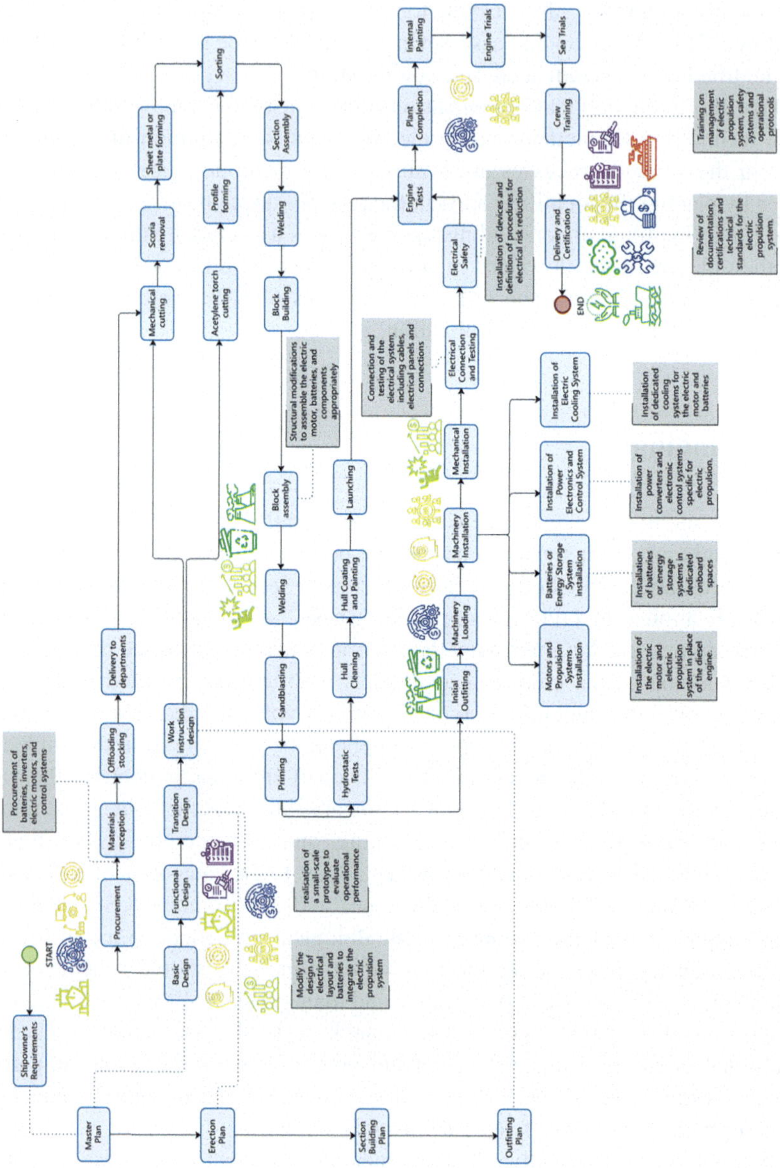

Fig. 26.2 Process steps and KPIs for shipbuilding activities: electric motor

implications in terms of overall production costs, which may be considerably higher than those associated with the implementation of traditional propulsion solutions. However, electric propulsion seems to require lower long-term maintenance costs, offsetting the higher expenditure on construction and installation. Finally, it is necessary to consider the significant repercussions that the implementation of electric propulsion technologies can have in terms of environmental performance: the solution in question allows for the elimination of greenhouse gas and particulate emissions, potentially even from a life cycle perspective if the electricity powering the engine is generated from renewable sources. However, it is necessary to bear in mind that this implication does not concern the shipbuilding process strictly speaking, but the result of such process in terms of the environmental performance achieved by the ship under construction.

26.5 Implications and Conclusions

The investigation enriches the academic debate concerning decision-making processes within the maritime and shipbuilding domains. By synthesizing theoretical constructs from business process management with a pragmatic examination of green technological innovations, this research introduces a novel methodological paradigm. Specifically, it presents an innovative framework for scrutinizing the multifaceted ramifications—financial, social, and environmental—of incorporating sustainable technological solutions into companies' production processes within the sector. This contribution also lays a foundation for further scholarly inquiry into the emergent dynamics of sustainability within the maritime and shipbuilding sectors.

On a managerial front, this study offers insights for industry practitioners tasked with navigating the complexities of technological integration. By delineating a structured model for assessing the introduction of green technologies into production workflows, it provides a systematic approach to delineate the main costs and benefits related to the investment decision. By adopting this integrated approach, industry players can position themselves as vanguards of sustainable enterprise in an ever-changing landscape.

References

1. Lee, J., Lee, J., Lee, C., Kim, Y.: Identifying ESG trends of international container shipping companies using semantic network analysis and multiple case theory. Sustainability. **15**(12), 9441 (2023)
2. Nõmmela, K., Kõrbe Kaare, K.: Maritime policy design framework with ESG performance approach: case of Estonia. Economies. **10**(4), 88 (2022)
3. Akan, E., Bayar, S.: An evaluation of ship investment in interval type-2 fuzzy environment. J. Oper. Res. Soc. **73**(8), 1768–1786 (2022)
4. Ghose, A., Hoesch-Klohe, K., Hinsche, L., Le, L.S.: Green business process management: a research agenda. Aust. J. Inf. Syst. **16**(2) (2010)

Chapter 27
Metaverse into Events: Implications for Sustainable Development Goals

Francesca Igini, Arturo Cafaro, and Mario Calabrese

Abstract This chapter explores the multifaceted implications of integrating digital tools and metaverse into events, with a specific focus on the Phygital Sustainability Expo. The metaverse, representing an innovative evolution of the Internet, has significant potential to advance Sustainable Development Goals (SDGs), across diverse sectors. Through a conceptual-based observation, this research analyzes the impact of using the metaverse in events on Sustainable Development Goals (SDGs), including partnerships, inclusive education, and climate action. In addition, a predictive approach, using linear interpolation, is utilized to measure how much the metaverse may affect the magnitude of the event in the future. The findings contribute to bridging gaps in the current academic discourse surrounding the metaverse's intersection with sustainability offering actionable insights for entities navigating this evolving landscape. Limitations include the case study focus and the relatively short dataset time frame, suggesting avenues for future research expansion.

Keywords Metaverse · SDGs · Events · Fashion

27.1 Introduction

In contemporary research, there is an escalating focus on the metaverse [1]. The metaverse, formed by the fusion of the prefix "meta" (indicating transcendence) and the suffix "verse" (short for universe), constitutes a computer-generated realm

F. Igini (✉) · A. Cafaro · M. Calabrese
Faculty of Economics, Department of Management, Sapienza University of Rome, Rome, Italy
e-mail: francesca.igini@uniroma1.it

F. Schiavone et al. (eds.), *Advanced Perspectives and Trends in Digital Transformation of Firms, Networks, and Society*, Springer Proceedings in Business and Economics, https://doi.org/10.1007/978-3-031-80692-6_27

characterized by a cohesive value system and a distinct economic system interlinked with the physical world [2]. This technology is characterized as a novel iteration of the Internet that incorporates virtual reality (VR) headsets, blockchain technology, and avatars to integrate the physical and virtual realms [3]. Notably, Stata's analysis indicates a substantial growth trajectory for the metaverse market, projecting a value of US$74.4 billion in 2024 and an anticipated annual growth rate (CAGR 2024–2030) of 37.73%. This trajectory is expected to culminate in a forecasted market volume of US$507.8 billion by 2030 [4, 5]. The versatile features of the metaverse make it a valuable tool for advancing sustainable development reducing carbon emissions across various sectors and contributing to the achievement of the SDGs [6, 7]. The fashion industry is increasingly embracing this technology. In 2022, the global fashion landscape began and ended with a metaverse-based virtual fashion week, organized with Decentraland [4, 5]. This shift to virtual realms is not just a pandemic response but represents a sustained trend. Post-epidemic, traditional fashion weeks are transforming into digital media fashion weeks, significantly impacting audience reach, operational costs, consumption formats, and consumer engagement [8]. However, the scholarly comprehension of the metaverse in the context of fashion and beauty remains ambiguous, indicating a gap in academic understanding [3].

This study investigates the impact of hosting events in both online and metaverse environments, using the Phygital Sustainability Expo as a case study in which, for the first time in 2023, the metaverse was used. This event exemplifies how innovation fosters collaborations and enhances sustainability awareness in the fashion sector.

Advances in the metaverse emphasize the importance of integrating sustainability and responsibility but the academic discussion in this area remains incomplete [1]. The current academic conversation lacks comprehensive insights into the environmental consequences of virtual consumption both positively and negatively [9]. Furthermore, the academic understanding of the metaverse in the context of fashion remains ambiguous, indicating a gap in academic understanding [3]. This denotes the lack of a defined academic perspective on the environmental implications of using virtual resources within the metaverse. This knowledge gap underscores the importance of more in-depth investigation to obtain a comprehensive understanding of the possible positive implications on the SDGs and the environmental impact generated using the metaverse especially in the fashion industry. Our research aims to fill this gap. Through an examination of the online aspect of the Phygital Sustainability Expo, our study aims to uncover insights related to the effects and significance of using the metaverse as a tool for digital events, with a specific focus on sustainability. By adopting this approach, our study aims to contribute not only to the specific discourse on the metaverse, but also to the broader academic understanding of the multiple implications of digital events in different contexts.

Therefore, two research questions were developed to achieve the goal:

RQ1. How does the use of metaverse in events impact the achievement of the SDGs?

This question explores how integrating the metaverse into events impacts Sustainable Development Goals (SDGs), aiming to reveal transformative potential. Through a literature review and conceptual observation, specific SDGs influenced by events were identified and the implications of using the metaverse on these SDGs were assessed, providing insights into the interconnected dynamics.

RQ2. How much does the use of digital tools and metaverse impact the number of participants and how much will it impact the future?

To comprehend the influence exerted by the utilization of digital tools and metaverse on the current and prospective number of participants online of the event under scrutiny, a predictive methodology employing linear interpolation was employed.

The chapter is structured as follows: Sect. 27.2 explores the theoretical foundations, providing a comprehensive background. Section 27.3 details the event data and methodology, outlining the research approach. Moving to Sect. 27.4, the findings from the analysis are presented. Section 27.5 delves into a Discussion, exploring the broader implications of the results. Section 27.6 "Conclusion and Limitations" summarizes key discoveries, their relevance, theoretical contribution to the academic landscape, and study limitations, suggesting avenues for future research.

27.2 Literature Review

The metaverse is an interactive, extensive network of three-dimensional virtualized worlds in real time. It enables concurrent and enduring engagement with an unlimited number of users, marked by a personal sense of presence and continuous data, including identities, histories, rights, items, communications, and payouts [10].

Numerous scholarly inquiries have delved into the opportunities presented by the metaverse in mitigating environmental impacts and fostering sustainability. For instance, blockchain technology is identified as pivotal in addressing trading-related challenges, albeit with a caveat of substantial energy consumption [11]. Despite this, the transformative potential of metaverse technology is accentuated, suggesting its significant role in expediting the realization of Sustainable Development Goals [12]. Moreover, to achieve the Sustainable Development Goals (SDGs), an effective approach is precisely to harness and repurpose the innovation capabilities of companies to create new products, procedures, and business models aligned with the SDGs [13]; thus, the metaverse could serve as a tool for achieving these goals due to its high innovation. The metaverse's distinctive capability to establish a seamless interconnection between real and virtual life emerges as a catalyst for reducing resource consumption in the production of goods and products [14]. Moreover, it holds promise in curtailing daily activities requiring transportation, such as commuting to work [15], and in minimizing energy consumption within office buildings [16, 17], notwithstanding its own energy-intensive nature. Therefore, the metaverse, as an increasing digital space, presents a distinctive opportunity to align with Sustainable Development Goals, particularly evident in the field of events. Event planning research has increasingly recognized the pivotal role it plays in addressing key Sustainable Development Goals (SDGs). Events plays a crucial role in advancing Sustainable Development Goals (SDGs) by serving as a platform for fostering partnerships (SDG 17), facilitating educational opportunities (SDG 4), and addressing climate change (SDG 13) [18–21]. Through the collaborative engagement of diverse stakeholders, the dissemination of knowledge and skills, and the adoption of sustainable practices, event planners could contribute to the collective efforts aimed at achieving sustainable development objectives. Metaverse helps event planners follow corporate, social, and environmental guidelines by cutting carbon emissions from travel and catering, promoting environmental education, awareness, and responsible behavior [22]. Also, metaverse events go beyond typical virtual gatherings, evolving into curated occasions on dedicated platforms. This transformative shift departs from traditional virtual events, showcasing meticulous curation for immersive experiences in the metaverse [23]. Even fashion and retail technologies advanced, connecting consumers to the metaverse through the integration of physical and digital spaces. This integration has given rise to an omnichannel commerce environment [24]. Fashion designers are leveraging these developments to create virtual fashion shows and catwalks within the metaverse, thereby presenting their designs to a global audience [25]. The application of metaverse technology proves particularly relevant in the fashion industry, where achieving sustainability necessitates a profound digital transformation [26]. In fact, this sector significantly contributes to environmental challenges as the second-largest global polluter after the oil industry, leading to various adverse environmental consequences [27, 28]. The industry's influence extends across multiple sectors, notably impacting human rights, carbon emissions, water consumption, pollution, and gender inequality [29]. Dewalska-Opitek [30] demonstrates that metaverse fashion shows boost global accessibility, eliminate costs related to physical events, and contribute to sustainability by reducing carbon footprints. This underscores the metaverse's growing significance as a transformative force, particularly in the fashion sector, aligning with broader sustainability goals.

27.3 Methods

In the first phase of the methodology, an investigation is carried out, focusing on the relevant literature, to elucidate the multiple implications that the use of the metaverse in events can have on the Sustainable Development Goals (SDGs), which is considered to be a conceptually based observation [31]. Sustainable Development imperatives have gained heightened prominence globally. The United Nations responded to diverse environmental and social challenges with the proactive formulation of the 2030 Agenda, comprising 17 goals and 169 targeted outcomes.

The study analyzes relevant literature to discern the impact of events on specific SDGs and explores the metaverse's potential contributions to their achievement.

Event planning research focuses on key Sustainable Development Goals: SDG 17 (Partnerships), SDG 4 (Education), and SDG 13 (Climate Action) [18–21], emphasizing collaboration, education, and innovation; it reflects commitment to sustainability. Subsequently, the inquiry expanded to encompass scholarly works that delve into the specified Sustainable Development Goals (SDGs) in connection with the metaverse. The exploration was conducted utilizing the Scopus dataset. The search query employed for this literature review was TITLE-ABS-KEY (metaverse) AND (impact) AND (sustainable AND development AND goals) OR (sdg*). The outcome of this search yielded a total of 102 documents. The applied filters for this search were Document type "Article"; Publication stage "Final"; Language "English," and it was restricted to all Open Access. Given that this is, as explicitly stated, an emerging topic, no temporal constraints were imposed during the search process. With the inclusion of the filters, the number of documents found was reduced to 57. Subsequently, a reading of all papers was carried out, and a screening was performed. There were 12 papers found to be related to the effect of the metaverse on the SDGs 4, 17, and 13. The papers mainly focus on the relation between metaverse and education and metaverse and Sustainability.

Furthermore, this work adopts a focused approach, centering its analysis on a case study to delve into the intricacies of a specific issue within distinct contexts, aligning with the insights [32]. The fashion industry is a significant contributor to environmental issues, particularly evident in the environmental impact of traditional fashion weeks. For instance, a study conducted by the Zero To Market project in 2020 by fashion technology company ORDRE in collaboration with the Carbon Trust revealed that travel between fashion capitals like Milan, London, Paris, and New York alone emitted 241,000 tons of CO_2 [33]. Beyond these visible metrics, events like fashion weeks have a broader environmental impact, including energy consumption in venue operations, material selection for event infrastructure, catering services' resource consumption and food waste management, and waste production throughout each event. So, the focus of this work is on the "Phygital Sustainability Expo," a key sustainability exposition in fashion and design, currently in its fourth edition in 2023 at the Imperial Forums in Rome. The dataset is generously provided by the Sustainable Fashion Innovation Society, the event's organizing entity (Fig. 27.1).

The data on total reach represent the complexity of the users reached by the Sustainable Fashion Innovation Society through the communication strategies implemented using different channels, actions, and tools both online and offline (Figs. 27.2 and 27.3).

ONLINE TOTAL REACH RECAP - PHIGITAL SUSTAINABILITY EXPO 2023				
TOTAL	**70.000.000**	DIGITAL	43.000.000	**80%**
		OFFLINE	27.000.000	**20%**

Fig. 27.1 Total reach of Phygital Sustainability Expo 2023 (author's own elaboration based on data provided by the Sustainable Fashion Innovation Society)

Digital channels	Actions	Tool	Total reach
Paid media champaign	Impressions Reach Likes Shares and new followers	Meta Instagram Linkedin	2.000.000
Influencer marketing	90 international cross-sectors influencer/opinion leaders	Kaby Lame Josè Mourinho Agil Mamyyev	18.000.000
Online Articles by main institutional partners	Main partners dedicated spaces on thei official websites	Fao Poste Italiane European Commission European Parmiament Italian Trade Agency Italian Ministry of Foreign affairs Rome Capitol City Ferrovie dello Stato Lega di calcio serie B Enel CNH industrial	10.000.000
Media Partners	Prime time tv shows on the event Full streaming of the event Streaming of the press conference of the event	Rai Today.it Forbes	13.000.000
Social media Marketing	Great visibility for all partners on official channels	Instagram Linkedin Facebook	30.000 users

Fig. 27.2 Communication strategies online of Phygital Sustainability Expo 2023 (author's own elaboration based on data provided by the Sustainable Fashion Innovation Society)

Offline channels	Actions	Tool	Total reach
Print media	Newspaper articles	Corriere della sera Milano Finanza Repubblica Libero La Stampa Il Messaggero Il Giornale Forbes Fortune	15.000.000
Out of home ADS	1 Ledwall in New York Times Square Event's Poster on roman busses 10 Led walls in Rome and Milan Event's billboards on all national trains	Nasdaq ATAC Urban Vision Ferrovie dello stato	12.000.000

Fig. 27.3 Communication strategies online of Phygital Sustainability Expo 2023 (author's own elaboration based on data provided by the Sustainable Fashion Innovation Society)

To better understand the scale of the event, an overview of the countries of origin of attendees in presence could be generated from the data provided (Fig. 27.4).

The analyzed event, with a global reach, highlights the need to understand the impact of adopting digital tools and the metaverse in its current year and future implications. This scrutiny is crucial, given its dual impact on reducing travel-related factors and reaching a broader audience.

To understand the impact that the use of the digital and metaverse has had and will have in the future on the number of participants of the event representing the case study, a linear interpolation prediction was used. The predictive mechanism of linear interpolation operates on the principle of utilizing the values associated with known data points to extrapolate and estimate the value corresponding to an unknown data point [34]. The procedural sequence entails ascertaining the linear

Division by country of participants			
15.000 Partecipants			
		Italy	**85%**
Europe	**76%**	Netherlands	5%
		Rest of Europe	10%
USA	**12%**		
Latin America	9%		
Rest of World	3%		

Fig. 27.4 Overview of offline participants of Phygital Sustainability Expo 2023 (author's own elaboration based on data provided by the Sustainable Fashion Innovation Society)

Year	**Participants in streaming**
2020	1.000
2021	5.000
2022	10.000
2023	42.000

Fig. 27.5 Overview of online participants of Phygital Sustainability Expo through years (author's own elaboration based on data provided by the Sustainable Fashion Innovation Society)

equation that encapsulates the relational dynamics between the established data points, subsequently employing this derived equation to extrapolate and compute the value assigned to the unknown point [35]. Given the recent inception of this event, the dataset at our disposal spans a relatively condensed time frame, encapsulating a four-year period from 2020 to 2023, the results then represent an approximation that could be implemented through a larger time frame.

To predict the trend of event participants in streaming for the year 2024, manual linear interpolation was conducted using data from the event's first to fourth editions (2020, 2021, 2022, and 2023). This approach established a linear relationship between the year and associated values, which was then used to predict the participant count for 2024 (Fig. 27.5).

27.4 Findings

RQ1 explores how metaverse integration in events impacts the achievement of Sustainable Development Goals (SDGs), revealing transformative potential in sustainable development. With the purpose of responding to this question a literature review based on the state of the art of the impact of the metaverse on the various SDGs identified was carried out; a conceptual observation was then conducted in order to provide a comprehensive understanding of the topic in the field of events. The study concludes by exploring the interconnected dynamics between events, metaverse integration, and sustainable development goals, highlighting the benefits derived from this approach (Fig. 27.6).

The exploration of SDG 4—"Ensure inclusive and equitable quality education and promote lifelong learning opportunities for all"—The metaverse is anticipated to be pivotal in integrating new technologies into education [36]. Çelik and Baturay [37] found positive impacts on vocabulary learning outcomes in a study involving 86 high school students, emphasizing the metaverse's immersive and interactive features. Park and Kim [38] highlight its potential to overcome temporal and spatial constraints in education, providing equal opportunities and addressing physical limitations. Hassani et al. [39] emphasize the positive impact on learning experiences and equitable access to resources, eliminating barriers related to location, language, and disability. Qadir and Fatah [40] underscore the potential of the metaverse to enhance distance learning through effective collaboration and shared virtual environments.

Building upon the insights derived from the studies, the metaverse emerges not only as a transformative force in education but also as a versatile platform capable of hosting diverse events, ranging from promotional activities to cultural, artistic, and sports events and festivals. The potential of the metaverse, demonstrated through empirical studies, signifies its multifaceted impact on both education and event organization, marking it as a groundbreaking innovation with far-reaching implications. In education, the metaverse provides immersive and interactive learning experiences, which, when used within an event, can ensure the dissemination of knowledge to a wide audience. This comprehensive approach aims to improve accessibility and engagement by addressing the diverse needs of the audience.

SDG 4	Çelik, F., & Baturay, M. H. (2024). The effect of metaverse on L2 vocabulary learning, retention, student engagement, presence, and community feeling. BMC psychology, 12(1), 58.
	Jamshidi, M., Dehghaniyan Serej, A., Jamshidi, A., & Moztarzadeh, O. (2023). The meta-metaverse: ideation and future directions. Future Internet, 15(8), 252.
	Qadir, A. M. A., & Fatah, A. O. (2023). Platformization and the metaverse: Opportunities and challenges for urban sustainability and economic development. EAI Endorsed Transactions on Energy Web, 10(1).
	Hassani, H., Huang, X., & MacFeely, S. (2022). Enabling Digital Twins to Support the UN SDGs. Big Data and Cognitive Computing, 6(4), 115.
	Park, S., & Kim, S. (2022). Identifying world types to deliver gameful experiences for sustainable learning in the metaverse. Sustainability, 14(3), 1361.
SDG 17	Chen, P. K., & Huang, X. (2023). Enhancing supply chain resilience and realizing green sustainable development through the virtual environment of the metaverse. Sustainable Development.
	De Giovanni, P. (2023). Sustainability of the Metaverse: A transition to Industry 5.0. Sustainability, 15(7), 6079.
SDG 13	Cao, W., Cai, Z., Yao, X., & Chen, L. (2023). Digital Transformation to Help Carbon Neutrality and Green Sustainable Development Based on the Metaverse. Sustainability, 15(9), 7132.
	Pachouri, V., Singh, R., Gehlot, A., Pandey, S., Akram, S. V., & Abbas, M. (2024). Empowering sustainability in the built environment: A technological Lens on industry 4.0 Enablers. Technology in Society, 76, 102427.
	Dwivedi, Y. K., Hughes, L., Baabdullah, A. M., Ribeiro-Navarrete, S., Giannakis, M., Al-Debei, M. M., ... & Wamba, S. F. (2022). Metaverse beyond the hype: Multidisciplinary perspectives on emerging challenges, opportunities, and agenda for research, practice and policy. International Journal of Information Management, 66, 102542.
	Jauhiainen, J. S., Krohn, C., & Junnila, J. (2022). Metaverse and Sustainability: Systematic Review of Scientific Publications until 2022 and Beyond. Sustainability, 15(1), 346.
	Joseph, A. (2023). An Exploratory Study on Metaverse and SDGs. In How the Metaverse Will Reshape Business and Sustainability (pp. 83-93). Singapore: Springer Nature Singapore.

Fig. 27.6 Literature review on metaverse and SDGs (author's own elaboration)

Specifically, within SDG 17—"Strengthen the means of implementation and revitalize the Global Partnership for Sustainable Development"—According to De Giovanni [11], the metaverse functions as an innovative platform for hosting virtual events, providing immersive and interactive experiences that are conducive to attracting a wider audience and fostering collaborative engagements among diverse stakeholders. On the other hand, the investigation conducted by Chen and Huang revealed that the virtual environment within the metaverse encourages close spatial proximity and provides multisensory feedback, eliciting emotional responses during interactions and knowledge sharing. These characteristics enhance mutual influence and contribute to mitigating perceived cooperation risks.

Based on these findings, interactions within the metaverse are identified as a catalyst for the creation of new partnerships and collaborations. This dynamic has the potential to contribute significantly to the advancement of the Sustainable Development Goals, aligning closely with the articulated aspirations of SDG 17 in events as well. The metaverse, through its immersive and collaborative features, through events, emerges as a powerful tool for fostering synergies among diverse stakeholders, thereby promoting collective efforts toward sustainable development.

In the context of SDG 13 "Take urgent action to combat climate change and its impacts," digital technology, as emphasized by Cao et al. [12], significantly contributes to optimizing enterprise production and operations for carbon-neutral objectives and green, sustainable development goals. Pachouri et al. [41] underline the role of digitalized methods from Industry 4.0 in enhancing the built environment for sustainability. Jauhiainen et al. [1] focus on the metaverse, highlighting its potential to reduce carbon emissions by substituting physical goods with digital equivalents and optimizing the physical world through simulations and increased awareness. Dwivedi et al. [42] and Joseph [31] both stress the metaverse's role in reducing carbon footprint through decreased travel, promoting work efficiency, and facilitating the replication of buildings and cities for a positive ecological impact.

In conclusion, the collective evidence suggests that the integration of digital technology and the metaverse into various facets of human activities holds promise for achieving sustainable development objectives. The transformative influence of digital technology and the metaverse extends beyond the optimization of industrial processes, encompassing a revolution in how events and interactions are conducted. This paradigm shift, notably applicable to travel-related activities, holds the potential to significantly reduce environmental footprints. The metaverse, with its capacity to replicate physical spaces and enable virtual interactions, emerges as a key player in mitigating the environmental impact traditionally associated with travel. By offering alternatives to physical meetings, events, and sightseeing, these technologies pave the way for a more sustainable approach to human interactions, aligning with the imperative of reducing carbon emissions and fostering ecological responsibility.

In addressing RQ2 a linear interpolation was performed with the aim of understanding the trend and predicting the number of future online participants of the Phygital Sustainability Expo with the purpose of understanding the impact (current and future) of using digital tools and the metaverse (Fig. 27.7).

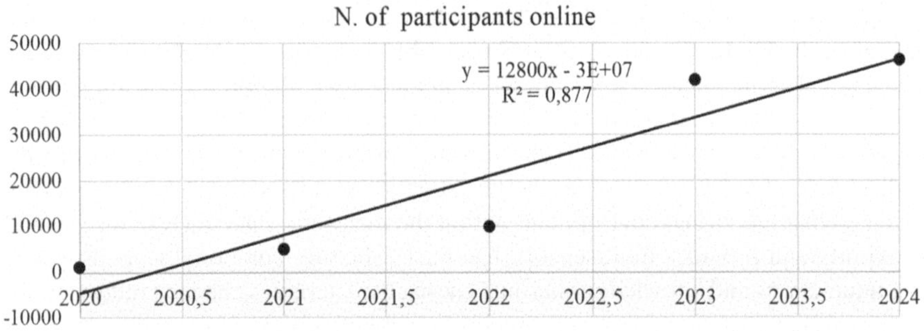

Fig. 27.7 Linear interpolation graphic on inline participants (author's own elaboration based on data provided by the Sustainable Fashion Innovation Society)

Projected through linear interpolation techniques, the estimate of online participants for the year 2024 amounts to approximately 46,500 individuals, reflecting a careful analytical prediction of the anticipated audience engagement in the virtual realm of the event.

The coefficient of determination, R2, measures the goodness of fit of the linear model to the data. It ranges from 0 to 1, where 1 indicates a perfect fit. In this case, 2 = 0.877R2 = 0.877, which means that approximately 87.7% of the variability in the predicted values (based on the linear model) can be explained by the linear relationship with the year. It's crucial to note that this is an approximation, and a more extensive dataset would provide a better perception of the situation, considering the approximation of the data with a limited set.

In addition, we report a significant increase from the year 2023, when, for the first time, the metaverse was used to display the "narrated fashion show" online.

27.5 Discussion and Implications

RQ1 delves into the impact of utilizing the metaverse in events on the achievement of Sustainable Development Goals (SDGs). The examination of various studies reveals a profound and multifaceted influence that the metaverse can have on event organization, contributing to the broader agenda of sustainable development.

The metaverse, seen as a transformative force, goes beyond education, extending to various events. Its immersive features demonstrate a commitment to improving accessibility and engagement and make it a cutting-edge innovation for training and event management. In fact, the wide reach that such technology enables plays a crucial role in knowledge dissemination, and its use within sustainability-related events, such as the Phygital Sustainability Expo, would promote sensitization among many people.

Also, interactions facilitated within the metaverse are recognized as catalysts for forging novel partnerships and collaborations. The metaverse, used in events, through its immersive and collaborative features, emerges as a powerful tool for fostering synergies among diverse stakeholders, thereby propelling collective efforts toward sustainable development. In sustainability-related events such as the Phygital Sustainability Expo, creating partnerships is critical. Using new technologies, as demonstrated, collaborations can be generated between different companies and between companies and institutions, facilitating the initiation of changes in an increasingly sustainable direction.

The blend of digital technology and the metaverse offers potential for achieving-Sustainable Development Goals, transforming how events and interactions occur, particularly in travel-related activities. The metaverse, by replicating physical spaces and enabling virtual interactions, plays a crucial role in reducing the environmental footprint associated with travel. Providing alternatives to in-person meetings and events, these technologies lay the groundwork for a more sustainable approach to human interactions, aligning with the imperative of reducing carbon emissions and fostering ecological responsibility.

In the context of RQ1, the multifaceted impact of the metaverse, as highlighted, underscores its role as a transformative force with far-reaching implications for sustainable development also in the event fields.

In response to RQ2, the use of both digital tools and metaverse in events has been shown to have a significant impact on the number of participants, both current and future. According to the analysis conducted, the number of online participants in the Phygital Sustainability Expo event was predicted to reach approximately 46,500 people in 2024. This value was obtained using the predictive methodology of linear interpolation, which revealed a strong correlation and explained about 87.7% of the variability in the predicted values based on the linear model and time progression.

These results indicate a significant impact of the platforms on the number of online participants, suggesting an important growth in public involvement. As mentioned before, the Phygital Sustainability Expo is an event dedicated precisely to raising awareness of sustainable issues in the fashion sector, and with different platforms and the metaverse more people are being reached, which will grow over the years, also fostering collaborations between the various brands in an increasingly sustainable perspective.

The results suggest that integrating the metaverse into events can advance sustainability practices and broaden audience engagement, with implications for various sectors. Adoption of the metaverse can benefit industries, organizations, and policymakers by fostering inclusive engagement and integrating sustainability principles into virtual environments.

27.6 Conclusion and Limitations

In conclusion, the chapter explores the impact of using digital tools and metaverse in events, focusing particularly on their influence on the Sustainable Development Goals (SDGs). The metaverse, recognized as a transformative force, extends beyond education to various events, exhibiting immersive features that enhance accessibility and engagement. The study highlights the role of the metaverse in creating partnerships and collaborations, in line with SDGs, especially in sustainability-related events such as the Phygital Sustainability Expo. In addition, the integration of digital technology and the metaverse in events has the potential to achieve Sustainable Development Goals, particularly in reducing the environmental footprint associated with travel. The study predicts a significant increase in online participants at events such as the Phygital Sustainability Expo, indicating a growing impact on audience engagement. The findings underscore the transformative potential of the metaverse across industries, including fashion, to advance sustainability practices and foster inclusive audience engagement in this field. The identified research gap concerns the limited academic discussion of the implications of the metaverse for sustainable development. This work, especially in the context of fashion events, addresses and fills this gap. Through our research, we have provided insights into how the emergence and integration of the metaverse in fashion events can influence sustainability efforts. By reviewing the relevant literature and the case study analyzed, this work leads to a more comprehensive understanding of the potential implications for sustainability within this evolving scenario. The study also suggests broader implications for industries, organizations, and policymakers, highlighting the role of the metaverse in integrating sustainability principles into virtual environments and achieving sustainability goals. Overall, the chapter highlights how the metaverse is a dynamic tool that can reshape events, engage diverse audiences, and contribute to a more sustainable future.

Limitations of this work include the focus on a specific case study, which suggests the possibility of future expansion to include a wider range of events in different sectors. In addition, limitations stem from the relatively short time spans of the dataset, as the event being analyzed has been ongoing for only a few years. It is recognized that there is a need to extend the analysis to events with data covering a wider time frame to provide further insights. In addition, the current state of academic discourse remains incomplete, and further investigation is needed to understand the implications of the metaverse on sustainable development. This gap provides an opportune space for further research that provides valuable insights specifically designed for brands and associations venturing into the burgeoning realm of the metaverse.

References

1. Jauhiainen, J.S., Krohn, C., Junnila, J.: Metaverse and sustainability: systematic review of scientific publications until 2022 and beyond. Sustainability. **15**(1), 346 (2022)
2. Maden, A., Yücenur, G.N.: Evaluation of sustainable metaverse characteristics using scenario-based fuzzy cognitive map. Comput. Hum. Behav. **152**, 108090 (2024)
3. Lee, L.H., Braud, T., Zhou, P., Wang, L., Xu, D., Lin, Z., et al.: All one needs to know about metaverse: A complete survey on technological singularity, virtual ecosystem, and research agenda. *arXiv preprint arXiv:2110.05352* (2021)

4. STATA INSIGHTS, Metaverse Worldwide, (2023a)
5. STATA INSIGHTS, Fashion in the metaverse, (2023b)
6. Bibri, S.E., Allam, Z.: The Metaverse as a virtual form of data-driven smart cities: the ethics of the hyper-connectivity, datafication, algorithmization, and platformization of urban society. Comput. Urban Sci. **2**(1), 22 (2022)
7. Johri, A., Joshi, P., Kumar, S., Joshi, G.: Metaverse for sustainable development in a bibliometric analysis and systematic literature review. J. Clean. Prod. **435**, 140610 (2024)
8. Dai, R.: The impact of digital fashion week on fashion trends in China. J. Educ. Humanit. Soc. Sci. **17**, 76–83 (2023)
9. Pellegrino, A., Stasi, A., Wang, R.: Exploring the intersection of sustainable consumption and the Metaverse: a review of current literature and future research directions. Heliyon. **9**(9), e19190 (2023)
10. Ball, M.: The Metaverse: and How it Will Revolutionize Everything. Liveright Publishing (2022)
11. De Giovanni, P.: Sustainability of the Metaverse: a transition to industry 5.0. Sustainability. **15**(7), 6079 (2023)
12. Cao, W., Cai, Z., Yao, X., Chen, L.: Digital transformation to help carbon neutrality and green sustainable development based on the metaverse. Sustainability. **15**(9), 7132 (2023)
13. Montera, R., Nevi, G., Cucari, N., De Falco, S.E.: How firms adjust their SDG adoption in response to COVID-19 outbreak: a regional perspective. Corp. Gov. Int. J. Bus. Soc. **24**(3), 559–575 (2023)
14. Allam, Z., Sharifi, A., Bibri, S.E., Jones, D.S., Krogstie, J.: The metaverse as a virtual form of smart cities: opportunities and challenges for environmental, economic, and social sustainability in urban futures. Smart Cities. **5**(3), 771–801 (2022)
15. Kshetri, N., Dwivedi, Y.K.: Pollution-reducing and pollution-generating effects of the metaverse. Int. J. Inf. Manag. **69**, 102620 (2023)
16. Al-Emran, M.: Beyond technology acceptance: development and evaluation of technology-environmental, economic, and social sustainability theory. Technol. Soc. **75**, 102383 (2023)
17. Vlăduțescu, Ş., Stănescu, G.C.: Environmental sustainability of Metaverse: perspectives from Romanian developers. Sustainability. **15**(15), 11704 (2023)
18. Di Gerio, C., Fiorani, G., Giuseppe, P.: Sustainability and large-scale events: the case of the extraordinary Jubilee of mercy and the application of sustainable development goals. J. Sustain. Dev. **13**(3), 71 (2020)
19. Fiorani, G., Di Gerio, C.: The Italian public administration and the 2030 agenda: a best practice event. In: STRATEGICA 2018-Challenging the Status Quo in Management and Economics. SNSPA AND TRITONIC (2018)
20. Liu, Y.Q., Lei, W.S.C.: Towards sustainable practice in the event industry: insights from practitioners. Event Management. **25**(3), 213–226 (2021)
21. Mackay, C., Julien, A., Hirst, J.: Sustainability and sustainable event management. In: *Cases for Event Management and Event Tourism*, p. 105 (2023)
22. Kark, S., Singh, S.: Metaverse experiences, events: and sustainability. In: Virtual Events Management: Theory and Methods for Event Management and Tourism, p. 175 (2023)
23. Piccioni, N.: From physical to metaversal events: an exploratory study. Ital. J. Mark. **2023**, 1–16 (2023)
24. Weiss, C.: Fashion retailing in the metaverse. Fash. Style Pop. Cult. **9**(4), 523–538 (2022)
25. Mu, X., Zhang, H., Shi, J., Hou, J., Ma, J., Yang, Y.: Fashion Intelligence in the Metaverse: Promise and Future Prospects. Artif. Intell. Rev. **57**(3), 67 (2023)
26. Periyasamy, A.P., Periyasami, S.: Rise of digital fashion and metaverse: influence on sustainability. Digital Econ. Sustain. Dev. **1**(1), 16 (2023)
27. Gupta, R., Kushwaha, A., Dave, D., Mahanta, N.R.: Waste management in fashion and textile industry: recent advances and trends, life-cycle assessment, and circular economy. In: Emerging Trends to Approaching Zero Waste, pp. 215–242 (2022)
28. Raj, M.M., Bajpai, P.: Mathematical analysis of the environmental impact of throwaway fashion. Eur. J. Sustain. Dev. **11**(4), 211–211 (2022)

29. Ali, A., Riyadh, I.A., Mohammed, A.: Environmental sustainability in fashion industry through sharing economy. Int. J. Res. Publ. Rev. **3**(11), 1657–1660 (2022). https://doi.org/10.55248/gengpi.2022.3.11.20
30. Dewalska-Opitek, A.: Digital revolution in fashion industry–to what extent can Metaverse become a predominant business-to-customer model in the digital era. Humanitas Univ. Res. Pap. Manag. **24**(2), 31–47 (2023)
31. Joseph, A.: An exploratory study on Metaverse and SDGs. In: How the Metaverse Will Reshape Business and Sustainability, pp. 83–93. Springer Nature Singapore, Singapore (2023)
32. Sandars, J.: Case study research. In: How to Do Primary Care Educational Research, pp. 121–124. CRC Press (2021)
33. Casciani, D., Chkanikova, O., Pal, R.: Exploring the nature of digital transformation in the fashion industry: opportunities for supply chains, business models, and sustainability-oriented innovations. Sustain. Sci. Pract. Policy. **18**(1), 773–795 (2022)
34. Pownuk, A., Kreinovich, V.: Why linear interpolation? Математические структуры и моделирование. **3**(43), 43–49 (2017)
35. Dyba, T., Hakulinen, T.: Comparison of different approaches to incidence prediction based on simple interpolation techniques. Stat. Med. **19**(13), 1741–1752 (2000)
36. Jamshidi, M., Dehghaniyan Serej, A., Jamshidi, A., Moztarzadeh, O.: The meta-metaverse: ideation and future directions. Future Internet. **15**(8), 252 (2023)
37. Çelik, F., Baturay, M.H.: The effect of metaverse on L2 vocabulary learning, retention, student engagement, presence, and community feeling. BMC Psychol. **12**(1), 58 (2024)
38. Park, S., Kim, S.: Identifying world types to deliver gameful experiences for sustainable learning in the metaverse. Sustainability. **14**(3), 1361 (2022)
39. Hassani, H., Huang, X., MacFeely, S.: Enabling digital twins to support the UN SDGs. Big Data Cogn. Comput. **6**(4), 115 (2022)
40. Qadir, A.M.A., Fatah, A.O.: Platformization and the metaverse: opportunities and challenges for urban sustainability and economic development. EAI Endorsed Trans. Energy Web. **10**(1), 3842 (2023)
41. Pachouri, V., Singh, R., Gehlot, A., Pandey, S., Akram, S.V., Abbas, M.: Empowering sustainability in the built environment: a technological lens on industry 4.0 enablers. Technol. Soc. **76**, 102427 (2024)
42. Dwivedi, Y.K., Hughes, L., Baabdullah, A.M., Ribeiro-Navarrete, S., Giannakis, M., Al-Debei, M.M., et al.: Metaverse beyond the hype: multidisciplinary perspectives on emerging challenges, opportunities, and agenda for research, practice and policy. Int. J. Inf. Manag. **66**, 102542 (2022)

Part V
Territorial and Maritime Innovations

Chapter 28
"Blue" Transformations: Toward an Analytical Model of Maritime Digitalization

Anastasia Christodoulou and Martin Jes Iversen

Abstract The digitalization of the shipping industry represents a significant technological transformation that is comparable in impact to previous technological shifts such as the move from sail to steam and from steam to motorization. In addition to the direct alteration of fundamental shipping processes, digitalization also affects customer relationships, internal operations, and value propositions. This study focuses on the transformative potential of digital technology in the shipping industry, particularly in terms of its impact on shipping strategy and business models. To fully understand the digital transformation in the maritime industry, we suggest in the conclusion an analytical model based on the fundamental assumption that the driver in digitalization concerns Shifts in Costs (SiCs). The analytical model includes insights from maritime business models, shipping technologies, and digital innovation. The concept of SiCs served as a starting point in order to examine the critical economic consequences from the transition from analog to digital representation that is the driving force behind the shift toward digitalization. Along with the development of novel, redefined business models and streamlined organizational processes that generate or capture extra value for the shipping companies and stakeholders, the ability to capture value and invest in new digital services and products also depends heavily on decision-makers' mindset and external knowledge of the innovation itself. Expanding one's mindset is essential for the innovation journey to generate improvements from data-driven innovation by accessing new digital platforms and learning about innovations that might otherwise remain unknown.

Keywords Shipping · Digitalization · Business models · AI · Strategy · Innovation

---

A. Christodoulou (✉)
Department of Maritime Studies, University of Piraeus, Piraeus, Greece
e-mail: a.christodoulou@unipi.gr

M. J. Iversen
Department of Strategy and Innovation, Copenhagen Business School, Frederiksberg, Denmark
e-mail: mji.si@cbs.dk

F. Schiavone et al. (eds.), *Advanced Perspectives and Trends in Digital Transformation of Firms, Networks, and Society*, Springer Proceedings in Business and Economics, https://doi.org/10.1007/978-3-031-80692-6_28

28.1 Introduction

Digitalization could be considered as a form of technological transformation that is similar in impact—but different in nature—to previous technological changes in the shipping industry, such as the shift from sail to steam engines and from steam engines to motorization. Throughout history, technological advancements have played a significant role in shaping strategies in the shipping industry [1]. The maritime industry relies heavily on information flow because of its global nature, which requires the exchange of information between different regions and stakeholders, such as shipowners, ship operators, ports, shippers, and freight forwarders.

When examining the development of information flow and communication throughout history, we can identify an important technological breakthrough that revolutionized data exchange in the maritime industry: the Morse code and telegraphy, which were created in 1837 and established the first international digital standard. As time progressed, starting from the 1950s, there was a gradual shift toward digitalization, primarily driven by its ability to significantly decrease costs, enhance the quality of services, and generate or capture additional value [2].

Digitalization in the shipping industry revolves around the optimization and simplification of various aspects of the industry through the use of digital technologies and data, with the core being an improved information flow [3]. The adoption of automation and data analytics has played a crucial role in optimizing ship and cargo movement, including the use of digital planning and scheduling systems, as well as real-time tracking and monitoring of ships and cargo [4]. Additionally, digital platforms and tools have improved communication and collaboration between different stakeholders in the industry, enabling the sharing of information and data, and coordination between shipping companies, ports, and logistics providers [5, 6]. The use of digital tools can also enhance safety and security through digital navigation and weather forecasting systems and monitoring compliance with safety and environmental regulations [7].

In addition to the direct advantages of improving various aspects of the shipping process, the digital transformation of shipping also involves changing customer relationships, internal processes, and value propositions. This study aims to explore the profound impact of digitalization on shipping strategy and maritime business models, which goes beyond merely optimizing existing procedures and processes. This impact leads to the actual redefinition of shipping strategies in the digital age, which is the focus of this work rather than the technical features of new technologies. The chapter analyzes the strategic consequences and opportunities of shipping digitalization rather than solely focusing on the advanced technological features of emerging technologies.

The structure of this chapter is as follows. Section 5.2 presents the methodology used and the analytical review that has been conducted in this work. Sections 5.3 and 5.4 describe how digitalization can affect shipping technologies and business models, respectively, while Sect. 5.5 highlights the importance of digital innovation for the "blue" digital transformation. The new "logic" of Shifts in Costs (SiCs) and the impact of digitalization on maritime economics are introduced in Sect. 5.6. Section 5.7 presents the four dimensions of digital integration in shipping proposed for the analysis of maritime digitalization. Finally, the conclusions drawn from this work and the relevant policy implications are presented in Sect. 5.8.

28.2 Methodology

In order to understand and analyze the strategic implications and opportunities of the digital transformation of shipping and, especially the profound impact of digitalization on shipping strategy and maritime business models, which goes beyond just optimizing existing procedures and processes, and leads to the actual redefinition of shipping strategies in the digital age, we proceeded with a comprehensive literature review of the academic and "grey" articles that dealt with the shipping digitalization. The existing literature is, at the moment, limited but emerging, with the vast majority of articles published during the last decade, and most of them focusing on the technological dimension and features of digital transformation rather than its potential to drastically redefine maritime business models [2, 4, 7–10].

As a result, we have used concepts and theories already developed in the management and economics literature and narrowed these into a maritime context taking into account the particular operational and organizational features of the shipping industry. Recognizing the need of exploring the "blue" digital transformation from a "holistic" perspective, an analytical review was conducted that considered the impact of digitalization on three distinct areas; namely, (a) shipping digital technologies, (b) maritime business models, and (c) digital innovation (Fig. 28.1). The analytical review was based on the existing academic literature in the respective areas.

For digital technologies and advancements to be successfully implemented and accepted, appropriate business models must be developed; business models that highlight the value-added benefits of these technologies to existing products and services, and their potential to meet customers' specific needs [11]. Another parameter incorporated in our analytical review is the importance of expanding one's mindset to facilitate data-driven innovation and access new digital platforms emphasized by Andersen and Pedersen [12]. Decision-makers must step out of their comfort zones and linear ways of thinking to broaden their mindsets and incorporate digital innovations into their company's processes.

The use of this review facilitates the comprehension and evaluation of the strategic significance and possibilities brought about by the digital revolution in the shipping industry considering the extensive influence of digitalization on shipping strategies and maritime business models; an influence that involves more than just improving current practices and workflows, but also entails a complete redefinition of shipping strategies in the digital era.

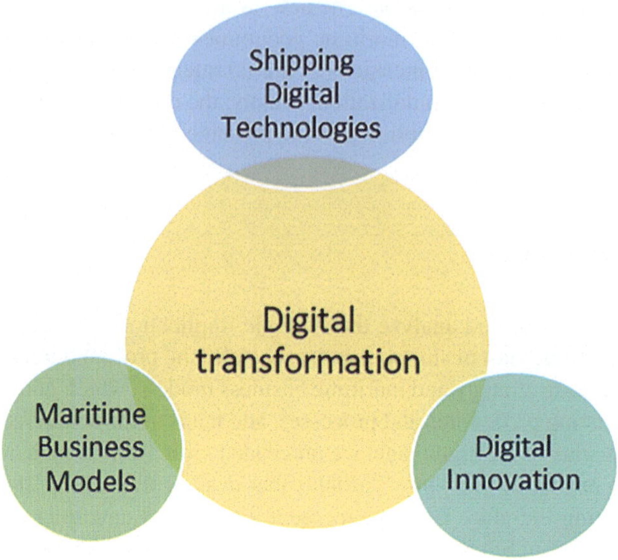

Fig. 28.1 Analytical review process. Source: Own elaboration based on Tijan et al. [7], Verdin and Tackx [11], and Andersen and Pedersen [12]

28.3 "Blue" Digital Transformation and Technologies

Maritime informatics could be defined as the utilization of information systems to enhance the efficiency, safety, and ecological sustainability of the global shipping industry. In essence, the fundamental focus of maritime informatics is the development and implementation of novel analytics that can be leveraged to optimize operational performance and support decision-making within the shipping sector by utilizing digital data sharing. In this respect, five distinct digital technologies with direct benefits for the shipping operations have been identified and analyzed (Table 28.1).

1. *Artificial intelligence (AI)* can be combined with *Big Data Analytics (BDA)* to process large amounts of data related to weather and sea conditions, ship location, speed, and electronic sea charts in order to enhance safe navigation. AI can improve safety and optimize commercial and operational activities, such as fleet planning and compliance monitoring for environmental regulations [3, 6, 9, 13].
2. The utilization of *Internet of Things (IoT)* and *cloud computing (CC)* in shipping networks and operations can revolutionize the industry by enabling remote monitoring and tracking of vessels and cargoes. This can enhance the reliability, safety, and transparency of the transport system, according to studies by Chun et al. [14], Han et al. [10], Sanchez-Gonzalez et al. [2], and Xue and Xin [15]. By incorporating digital platforms and tools such as IoT and CC, supply chain logis-

Table 28.1 "Blue" digital technologies

"Blue" digital technologies
Artificial intelligence (AI)/Big data analytics (BDA)
Internet of Things (IoT)/Cloud computing (CC)
Blockchain
Virtual reality (VR)
Digital Twins/Advanced simulation

tics can be optimized and streamlined, reducing the need for manual coordination and communication between different stakeholders. The integration of digital sensors and tracking systems can improve visibility and transparency throughout the shipping process, minimizing the risk of lost or damaged cargo and identifying potential supply chain bottlenecks.

3. The implementation of *blockchain* technology in the shipping industry can have a significant impact on verification costs by providing a secure and transparent way for digital operational processes and transactions, such as documentation, bill of lading, and customer clearance. Additionally, the use of blockchain technology can improve cybersecurity by creating a chain of secure transactions that make container tracking more accurate and automating commercial transactions of shipping companies [2, 3, 16–19].
4. To improve crew training and prepare for emergency situations, *Virtual Reality (VR)* can be used to create 3D computer graphics of ship environments that enable practice of ship operations through overlaid computer-generated visualizations. This can reduce costs and lead to less human errors, mistakes, and delays during the shipping process [20, 21].
5. To minimize the risks and expenses of introducing new technologies, *Digital Twins* can offer real-time data and create virtual models via *Advanced Simulation* [22, 23]. Additionally, Digital Twins can aid in predictive analytics for optimizing vessel energy performance, reducing fuel consumption and related emissions, and ultimately lowering fuel and operational expenses.

28.4 "Digital" Business Models

The digital transformation of the industry involves more than just incorporating digital technologies. Instead, it necessitates and also results in the development of novel, redefined business models and streamlined organizational processes that generate or capture extra value for the company and its clients. Consequently, it is valuable to attempt to provide a definition of business models. Although there is no clear theoretical definition of the term "business model," according to Teece [24], "a business model embodies nothing less than the organizational and financial 'architecture' of a business. (…) To put it differently, a business model is a conceptual representation of the company's business strategy that describes how it interacts with customers, manages costs and revenue, meets customer needs, and responds to competitive pressures."

The effective implementation and acceptance of digital technologies and technological advancements rely greatly on the development of appropriate business models that outline the value-added benefits of these technologies to existing products and services and their potential to meet particular customer needs. There are two broad distinct definitions associated with the added value of products and services: value creation and value capturing [11]. Value creation is "in line with the microeconomic concept of the utility of a company's offering for its customers, whether it enhances the quality of life for a final consumer (B2C) or increases the profitability of a company (B2B)." It is clear that creating value is a fundamental requirement for introducing a product or service to the market.

Going beyond value creation, value capturing refers to the ability to generate profits sustainably from the production and delivery of a product or service, measured by the net profit obtained after accounting for the resources used. To ensure long-term profitability and a strong market position, it is crucial to develop a "suitable" business model that defines the added value to customers, expected or potential profits for the company, and sustainable production and delivery of the product or service.

In the context of digital transformation in the shipping industry, it is essential to develop successful business models to highlight the value created by new digital products and services. However, the sustainable introduction and value capturing of these products/services in the shipping market require optimizing or redefining existing business models, defining added value to customers, expected/potential profits, and conditions for long-term profitability.

To maintain their advantageous position, shipping companies must continuously improve their processes to react to various external factors and stay competitive in the market. The advantages of incorporating digitalization into maritime business models become evident when considering the current energy efficiency improvements required for all vessels above 5000 gross tonnage (GT) under the Initial IMO Strategy for reducing GHG emissions from shipping. Starting in 2023 with the introduction of the Carbon Intensity Indicator (CII), vessels will be rated based on their energy efficiency performance in terms of CO2 emissions per cargo-carrying capacity and nautical mile. To improve or maintain their rating, vessels can invest in digital innovations to optimize voyage planning by considering weather routing and just-in-time (JIT) arrival of vessels at ports. This streamlines supply chain logistics and reduces the risk of potential bottlenecks in the supply chain.

28.5 Digital Innovation in Shipping

The benefits of digital innovation in the shipping industry are clear and straightforward. However, the ability to capture value and invest in new digital services and products relies heavily on the decision-makers mindset and their external knowledge of the innovation itself [12]. According to Andersen and Pedersen [12], expanding one's mindset is essential for the innovation journey in order to "generate improvements from data-driven innovation by accessing new digital platforms and learning about innovations that might otherwise remain unknown." Put differently, decision-makers must step out of their comfort zone and linear way of thinking and broaden their mindset to incorporate digital innovations into their company's processes [25].

The shipping industry is traditionally conservative, with ships having long lifetimes of up to 25 or 30 years and operating in a highly competitive and volatile market. This makes it challenging to invest in new digital technologies and tools and to explore uncharted territory. However, it is essential for stakeholders in the maritime supply chain to take an active role in driving digitalization in shipping. Customers, such as shippers and freight forwarders, can demand improved services that utilize digital tools and platforms to ensure safe, transparent, timely, and energy-efficient transport of their cargoes. Shipping companies must also adapt their business models to capture added value through digital products and processes, meeting specific customer needs and remaining competitive.

Regulatory bodies and ports can play a role in promoting and facilitating the integration of digital technology into maritime supply chains. Given their strategic location, ports are essential hubs within these chains. By sharing real-time information with shipping companies and vessels, ports can improve scheduling, reduce waiting times, and optimize cargo loading and unloading. This leads to a more efficient maritime supply chain, with lower costs, emissions, and improved transport services overall. Additionally, digital monitoring and navigation systems can help enhance compliance with safety and environmental regulations.

28.6 "Blue" Digital Transformation and Shifts in Costs: A New 'Logic'

While analyzing the different parameters of our analytical review, it is important to keep in mind what has already been stated in the introduction; the fact that the primary driver of digitalization lies not in its technological features but in its economic consequences. To understand whether and how digitalization affects maritime economic activity, the concept of SiCs—noted by Goldfarb and Tucker [26]—is used as a starting point recognizing the fact that the transition from analog to digital representation results in significant cost shifts.

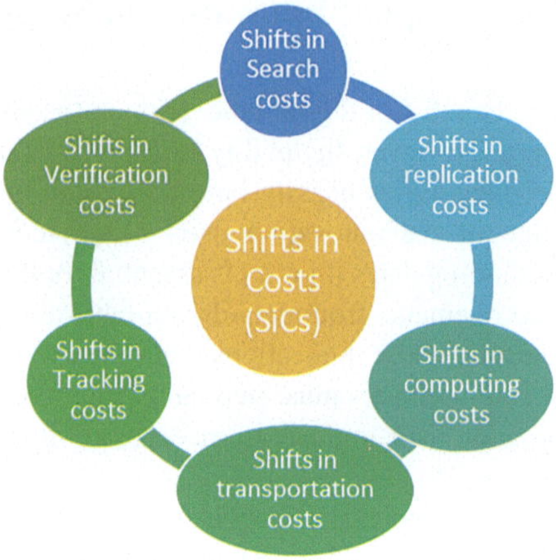

Fig. 28.2 Types of SiCs from digital transformation. Sources: Own elaboration based on Goldfarb and Tucker [26]

To simplify the complexity of the digital transformation, Goldfarb & Tucker have proposed five types of SiCs: Shifts in search, replication, transportation, tracking, and certification costs (Fig. 28.2). In addition to the SiCs identified by Goldfarb and Tucker [26], we also spotted another SiC strongly related to the maritime industry, the Shift in computing costs, coming from the use of AI and BDA to convert data and information into knowledge, intelligence and, finally, competence and significantly decrease—in this way—the costs associated with decision-making processes.

We will now provide examples from the maritime industry to illustrate each of these six types of SiCs.

1. *Shifts in Search costs.* The use of digital platforms in the shipping industry is an example of how shifts in search costs can be observed. By providing real-time information on shipping routes, schedules, and prices, these platforms can significantly reduce the time and resources spent on searching for the best chartering options. This also facilitates direct communication between shippers and carriers, which in turn reduces the need for intermediaries and lowers search costs even further [27].
2. *Shifts in replication costs.* In the shipping industry, digitalization can result in a decrease in replication costs by removing the need for manual documentation and data entry. This can be achieved by using digital systems for tasks like inventory management, invoicing, and documentation. Electronic documents like electronic bills of lading that enable the digital exchange of information and eliminate the need for multiple paper copies are a good example of how digitalization can reduce replication costs in the shipping industry [8].
3. *Shifts in transportation costs.* Digitalization has significantly reduced transportation costs in the shipping industry by enabling the transfer of information with low or almost no additional costs across long distances [28]. The ability to transfer large amounts of information in real-time has made it possible to connect vessels and other assets through smart solutions such as the Internet of Things (IoT) linked to engines and other critical parts, as well as virtual reality (VR) meetings and training for seafarers. Previously, it was crucial to transfer infor-

mation from headquarters to vessels, but with digitalization, the costs have been reduced dramatically.

4. *Shifts in Tracking Costs.* Digital tracking technologies can significantly reduce tracking costs in shipping by providing real-time information and visibility into the location and status of shipments. This helps shipping companies achieve greater accuracy and reliability in tracking operations, which in turn reduces the need for manual data entry and updates. Shippers can benefit from real-time tracking of their shipments along the transport chain, which can significantly increase information transparency and improve proactivity on potential delays through real-time information sharing [29].
5. *Shifts in Verification Costs.* Digitalization can bring significant changes in verification costs in shipping by reducing the reliance on manual inspections and increasing the precision of the verification process. Adoption of digital tools can streamline the verification of compliance with regulatory standards, thereby reducing the need for manual compliance checks and associated costs. Additionally, the use of blockchain technology can provide enhanced accuracy and reliability in verification, particularly in tracking shipments along the supply chain [30, 31].
6. *Shifts in Computing Costs.* By collecting and analyzing knowledge from past and current situations, AI algorithms can assess future situations and increase competence through correct decisions and successful reactions that are based on situational awareness and experience [32]. In this way, the cost of decision-making processes can significantly decrease along with substantial improvements of the safety and energy efficiency in the vessels.

While maritime information systems and digital technologies are necessary for the digitalization of shipping, the driving force behind this shift is the significant reduction in costs that it offers. The concept of SiCs is incorporated into our analytical model in order to identify whether and to what extent digitalization affects maritime economics (Fig. 28.3).

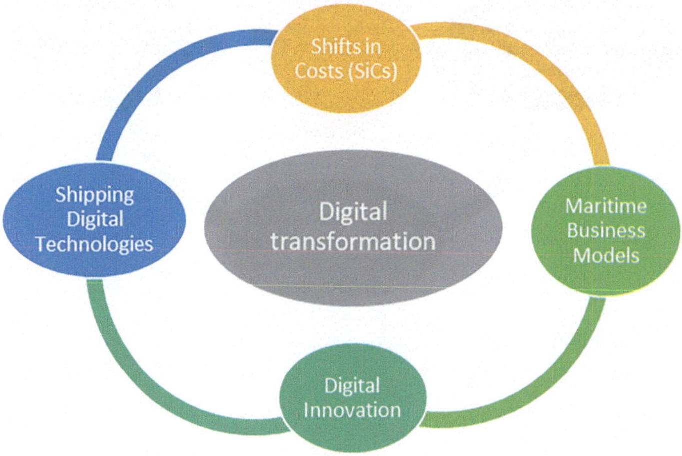

Fig. 28.3 Analytical model

28.7 Dimensions of Digital Integration in Shipping

The introduction highlights the complexity of analyzing digitalization as a transformative force in international shipping and emphasizes the need for a holistic approach to understanding its various aspects. The primary reason for focusing on digitalization is its economic impact, specifically the six SiCs. However, the challenge lies in translating these SiCs into a coherent and comprehensive shipping strategy. The authors propose that SiCs can serve as engines of integration, enabling the integration of maritime headquarters with remote vessels at an unprecedented level of density, as well as facilitating profound collaboration between public authorities and private companies, particularly in relation to customers and data-driven decision-making.

In order to simplify the analysis of global shipping in the digital age, we have identified four specific dimensions of integration (Fig. 28.4) that can serve to define the overall strategy for international shipping and could be investigated in future work:

1. The impact of digitalization on the *operational integration between sea and shore* should initially examine the traditional relation between land (HQs, technical and commercial operations) and sea (the bridge and the functions) and then also consider some of the most advanced digital solutions (e.g., SmartVessel). Such a comparison could lead to significant findings with regards to the changes in relations and information exchange between land and sea, and, most impor-

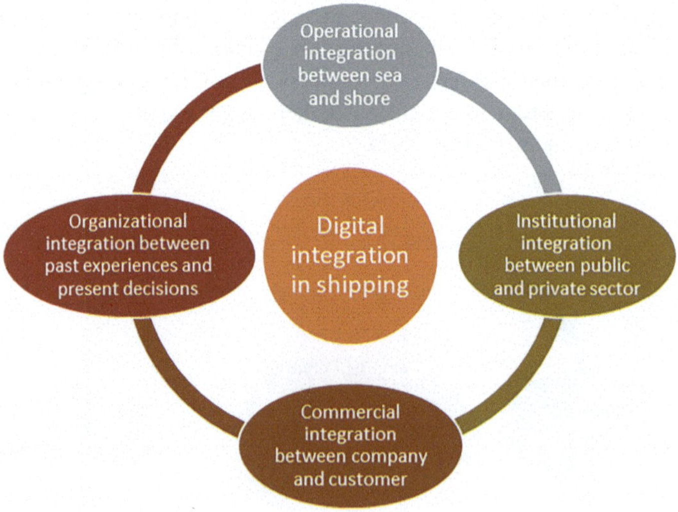

Fig. 28.4 Dimensions of digital integration in shipping

tantly, the impact of such changes on the financial, environmental, and safety performance of vessels, but also the welfare of seafarers.

2. Starting from the changing relation between global, regional, national, and local authorities and private companies (shipowners/operators) over the years, digitalization can significantly influence the *institutional integration between public and private shipping organizations*. Nowadays several authorities (IMO, EU, national maritime administrations, port authorities) have become increasingly active in regulating safety and environmental issues related to maritime activities and digitalization can play a key role in regulation compliance through accurate and digital monitoring and reporting processes.
3. Besides the operational and institutional integration that can be deeply influenced from digitalization in shipping, the *commercial integration between company and customer* should also be analyzed. Digital tools and platforms—and, most importantly, advanced information flows and data exchange—can create new relations with customers through customer-based value propositions. A redefinition of existing business models to capture the created value of digital solutions and better meet customers' needs will, at the same time, increase the competitive advantage of the company under vibrant and dynamic market conditions.
4. *Organizational integration between past experiences and present decisions*. The impact of the use of AI and big data analytics in decision-making processes within the shipping company is probably an obvious dimension that is worth further exploring. The digital decision-making processes create new opportunities for optimization of organizational and strategic processes such as route planning, HR processes, chartering, and bunkering. We can simply compare a typical day in the chartering department of a dry bulk company around the year 2000—flow of information via telephone calls, faxes—and how digitalization has changed these processes at the time being and even go one step forward and consider how they will look like around the year 2030.

28.8 Conclusions and Policy Implications

In this article, we attempted to approach the digital transformation of the shipping industry from a different perspective than the technological one; we focused on the organizational and structural changes that such a transformation could bring in customer relationships, internal procedures, and value propositions. This article focused on how digital technology can revolutionize organizations, their operational processes, and business models.

To fully understand the digital transformation in the maritime industry, an analytical model was developed that considers the impact of digitalization on four distinct areas: cost shifts, maritime business models, shipping technologies, and digital innovation. Successful business models are crucial for the sustainable introduction and value capturing of new digital products and services in the shipping market. However, this requires optimizing or redefining existing business models, defining added value to customers, expected/potential profits, and conditions for long-term profitability. The ability to capture value and invest in new digital services and products also relies heavily on decision-makers' mindset and external knowledge of the innovation itself.

The concept of SiCs served as a starting point in order to examine the critical economic consequences from the transition from analog to digital representation. The reduction in costs is the driving force behind the shift toward digitalization, which necessitates the development of novel, redefined business models and streamlined organizational processes that generate or capture extra value for shipping companies and stakeholders. The authors propose that SiCs can serve as engines of integration, enabling the integration of maritime headquarters with remote vessels at an unprecedented level of density, as well as facilitating profound collaboration between public authorities and private companies, particularly in relation to customers and data-driven decision-making.

The four dimensions of integration identified to analyze global shipping in the digital era are abstract and may work together in an ecosystem integration. Future work can look into each dimension separately with the aim to understand how information flows were organized in each dimension before digitalization, what is the current situation with digitalization, and then examine the strategic directions the sector is heading in with regard to each dimension of integration.

References

1. Stopford, M.: Maritime Economics 3e. Taylor and Francis (2009)
2. Sanchez-Gonzalez, P.-L., Díaz-Gutiérrez, D., Leo, T.J., Núñez-Rivas, L.R.: Toward digitalization of maritime transport? Sensors. **19**(4), 4 (2019)
3. de la Peña Zarzuelo, I., Freire Soeane, M.J., López Bermúdez, B.: Industry 4.0 in the port and maritime industry: a literature review. J. Ind. Inf. Integr. **20**, 100173 (2020)
4. Ichimura, Y., Dalaklis, D., Kitada, M., Christodoulou, A.: Shipping in the era of digitalization: mapping the future strategic plans of major maritime commercial actors. Digit.Bus. **2**(1), 100022 (2022)
5. Dalaklis, D., Christodoulou, A., Ölcer, A.I., Ballini, F., Dalaklis, A., Lagdami, K.: The port of Gothenburg under the influence of the fourth stage of the industrial revolution: implementing a wide portfolio of digital tools to optimize the conduct of operations. Marit. Technol. Res. **4**(3), 253844 (2022)
6. Lam, J.S.L., Zhang, X.: Innovative solutions for enhancing customer value in liner shipping. Transp. Policy. **82**, 88–95 (2019)
7. Tijan, E., Jović, M., Aksentijević, S., Pucihar, A.: Digital transformation in the maritime transport sector. Technol. Forecast. Soc. Chang. **170**, 120879 (2021)
8. Aiello, G., Giallanza, A., Mascarella, G.: Towards Shipping 4.0. A preliminary gap analysis. Procedia Manuf. **42**, 24–29 (2020)
9. Ellingsen, O., Aasland, K.E.: Digitalizing the maritime industry: a case study of technology acquisition and enabling advanced manufacturing technology. J. Eng. Technol. Manag. **54**, 12–27 (2019)

10. Han, H., Liu, Z., Wang, X., Li, S.: Research of the relations among cloud computing, internet of things, big data, artificial intelligence, block chain and their application in maritime field. J. Phys. Conf. Ser. **1927**(1), 012026 (2021)
11. Verdin, P., Tackx, K.: Are You Creating or Capturing Value? Harvard Kennedy School (2015)
12. Andersen, M.M., Pedersen, T.: Data-Driven Innovation: why the Data-Driven Model Will Be Key to Future Success. Routledge & CRC Press (2021)
13. Lambrou, M., Watanabe, D., Iida, J.: Shipping digitalization management: conceptualization, typology and antecedents. J. Shipp. Trade. **4**(1), 11 (2019)
14. Chun, D.-H., Roh, M.-I., Lee, H.-W., Ha, J., Yu, D.: Deep reinforcement learning-based collision avoidance for an autonomous ship. Ocean Eng. **234**, 109216 (2021)
15. Xue, C.T.S., Xin, F.T.W.: Benefits and challenges of the adoption of cloud computing in business. Int. J. Cloud Comput. Serv. Architect. **6**(6), 01–15 (2016)
16. Ahn, Y.-G., Kim, T., Kim, B.-R., Lee, M.-K.: A study on the development priority of smart shipping items—Focusing on the expert survey. Sustainability. **14**(11), 11 (2022)
17. Balci, G.: Digitalization in container shipping: do perception and satisfaction regarding digital products in a non-technology industry affect overall customer loyalty? Technol. Forecast. Soc. Chang. **172**, 121016 (2021)
18. Kuhn, K., Bicakci, S., Shaikh, S.A.: COVID-19 digitization in maritime: understanding cyber risks. WMU J. Marit. Aff. **20**(2), 193–214 (2021)
19. Lafourcade, P., Lombard-Platet, M.: About blockchain interoperability. Inf. Process. Lett. **161**, 105976 (2020)
20. Mallam, S.C., Nazir, S., Renganayagalu, S.K.: Rethinking maritime education, training, and operations in the digital era: applications for emerging immersive technologies. J. Mar. Sci. Eng. **7**(12), 12 (2019)
21. Sharma, A., Kim, T.-E., Nazir, S.: Implications of automation and digitalization for maritime education and training. In: Carpenter, A., Johansson, T.M., Skinner, J.A. (eds.) Sustainability in the Maritime Domain: Towards Ocean Governance and Beyond, pp. 223–233. Springer International Publishing (2021)
22. Fonseca, Í.A., Gaspar, H.M.: Challenges when creating a cohesive digital twin ship: a data modelling perspective. Ship Technol. Res. **68**(2), 70–83 (2021)
23. Sullivan, B.P., Desai, S., Sole, J., Rossi, M., Ramundo, L., Terzi, S.: Maritime 4.0—Opportunities in digitalization and advanced manufacturing for vessel development. Procedia Manuf. **42**, 246–253 (2020)
24. Teece, D.J.: Business models, business strategy and innovation. Long Range Plan. **43**(2), 172–194 (2010)
25. Bergmann, M., Primor, O., Chrysostomou, A. (2021). Digital data sharing for enhanced decision-making. Progress in IS, 155–165 (2012).
26. Goldfarb, A., Tucker, C.: Digital economics. J. Econ. Lit. **57**(1), 3–43 (2019)
27. Wang, Y., Sarkis, J.: Emerging digitalisation technologies in freight transport and logistics: current trends and future directions. Transp. Res. E Logist. Transp. Rev. **148**, 102291 (2021)
28. Iversen, M.J.: The Norden Voyage. Lindhardt & Ringhof (2021)
29. Christodoulou, A., Kappelin, H.: Determinant factors for the development of maritime supply chains: the case of the Swedish forest industry. Case Stud. Transp. Policy. **8**(3), 711–720 (2020)
30. Savrul, M., Incekara, A.: The effect of R&D intensity on innovation performance: a country level evaluation. Procedia. Soc. Behav. Sci. **210**, 388–396 (2015)
31. Sikorski, J.J., Haughton, J., Kraft, M.: Blockchain technology in the chemical industry: machine-to-machine electricity market. Appl. Energy. **195**, 234–246 (2017)
32. Bergmann, M.: Public-private-cooperation models. In: e-Maritime Conference, Brussels (2012)

Chapter 29
How the Startup Studios Pay Attention to the Territory: An Exploratory Analysis

Matteo Di Liberto

Abstract Advances in Digital Technologies are pushing toward the digital transformation of every sphere of human activity and knowledge. AI has been identified as one of the main factors of transformation of the company with a transversal scope to the different sectors of economic life and work that seems to have no analogy with what happened during the previous technological revolutions (Dagnino, CNEL 3, 2024; Gudergan and Mugge, Hawaiian international conference of system science, Whitepaper, 2017). In this era of the digital transformation of society, this work intends to answer the question about the way in which business incubators, and in particular the startup studios, pay attention to the territory. We know that a startup studio is a kind of company creators. They solve real problems by combining great business ideas with the best entrepreneurial talent to execute those ideas. The studios then test these ideas and support them with funding and resources to launch and grow powerful and scalable startups. The research will investigate the presence of the issue of social sustainability, environmental and governance in the values pursued and processes employed by the Italian startup studios to build new companies, therefore in general in the definition of their business models, also taking into account the disturbing and falsificatory phenomenon of greenwashing. Avoiding greenwashing is a key element in building a resilient and sustainable business model over time (Digital360, https://www.esg360.it/environmental/esg-tutto-quello-che-ce-da-sapere-per-orientarsi-su-environmental-social-governance, 2023).

The ambition would be to check whether the statements of principle follow concrete ESG-oriented actions in the creation of new companies launched by the startup studios.

Keywords Startup studios · ESG sustainability · Greenwashing · Qualitative analysis

M. Di Liberto (✉)
University of Salento, Lecce, Italy
e-mail: matteo.diliberto@unisalento.it

F. Schiavone et al. (eds.), *Advanced Perspectives and Trends in Digital Transformation of Firms, Networks, and Society*, Springer Proceedings in Business and Economics, https://doi.org/10.1007/978-3-031-80692-6_29

29.1 Introduction

Advances in Digital Technologies from Artificial Intelligence (AI), metaverse, Big Data Analysis, cloud computing, and the Internet of Things (IoT) along with their use are pushing toward the digital transformation of every sphere of human activity and knowledge.

Precisely AI has been identified as one of the main factors of transformation of the company with a transversal scope to the different sectors of economic life and work that seems to have no analogy with what happened during the previous technological revolutions [1, 2].

In a context of such sudden changes, this work intends to answer the question of research about the way in which business incubators, and in particular the startup studios, pay attention to the Territory.

If it is true that more and more Startup studios are spreading in Italy, Little is yet known about how these economic actors look to the Territory in terms of social sustainability, environmental, and governance [3].

We know that a startup studio is a kind of company creators. They find solutions to real problems by matching promising business ideas with top talent. Studios then test these ideas and support them professionally and financially to launch and grow successful startups.

Startup studios are best thought of as institutional co-founders who provide initial investments. Their approach is called "parallel entrepreneurship." It starts with market research and continues with the generation of ideas and tests. Those ideas that make through the process turn into start-ups and enter the market. Each company is born in direct response to a validated and specific market demand.

The aim is to make the new company operationally and financially viable, then spin it out well prepared for follow-on investment targeting an eventual M&A, IPO, or other exit. Since startups are founded by Startup Studios, these latter normally hold significant stakes in the companies they launch. Startup studios are not "earlystage" investors; they are real founders [4].

A Startup studio is therefore a "factory" of startups, a generator of new business, which proceeds from a methodological and accurate analysis of the market and its most promising trends in terms of needs and demand for goods or services, and then set up a business model focused on the fast scalability of a selected number of projects, which allows the creation from scratch of a portfolio of high-quality and low-risk startups.

According to reports made by the two main international communities in the sector, Global Startup Studio Network (GSSN) in 2020—better known today as Marrow—Global Accelerator Network—and Studiohub in 2021, the success rate of startups produced by a venture builder moves in a range between 35% and 70% of the total, while so-called "traditional" startups, according to Forbes, fail in 90% of cases [5].

Along this line, the research will investigate the presence of the issue of social sustainability, environmental, and governance in the values pursued and processes employed by the startup studios of our country to build new companies, therefore in general in the definition of their business models, also taking into account the disturbing and falsificatory phenomenon of greenwashing. In this regard, it would be interesting to check whether the statements of principle follow concrete ESG-oriented actions in the creation of new companies launched by the startup studios. Avoiding greenwashing is not only an ethical duty but also a key element in building a resilient and sustainable business model over time. In this context, B2B companies have the opportunity not only to drive change through their internal policies but also to positively influence the entire value chain. It is true that every business choice has an impact that goes beyond the confines of the company itself [6].

To investigate how the startup studios pay attention to the territory, therefore, the research work will make a qualitative exploratory analysis conducted on the main Italian startup studios and will help to advance knowledge of the little-investigated phenomenon of startup studios in our country with implications for academics and public managers in terms of addressing the destructive policies of local development impact on employment and growth.

29.2 Objective

The main objective of the present work is to investigate how the new business models of the start-up studio in the digital age pay attention to the territory, here considered not simply as a place that houses the production facilities of companies, but the set of subjects—institutional, associative, informal and liquid—participating in the same value creation chain (SYMBOLA Report, [7]).

We wonder what impact the business models of start-up studios have on the territory, its needs, its resources, and its environmental assets to protect and enhance, and this does not seem negligible or less important to establish that a business incubator, and in particular a Startup studio can be said to be successful in this new era dominated by AI.

Since economic activity brings real prosperity, the definition of its true purpose cannot be separated from social and environmental considerations, including responsible innovation which is not only or primarily aimed at increasing cost efficiency or maximizing profits but also at increasing the prosperity of all stakeholders: investors, workers, consumers, society and the environment [8].

29.3 Methods

The methodology adopted to answer the research question consists in a qualitative analysis of some successful cases of the Italian startup studios to explore the presence of the thematic of social, environmental, and governance sustainability, in the definition of their business models.

The investigation will follow the following steps: (a) mapping of the most successful Italian startup studios; (b) analysis of the relevant websites to identify the values pursued declared by each startup studio, their business strategies, and company documents confirming that the study Avoid greenwashing; (c) Analysis of the Process of creation and launch of New startups; (d) verify that New startups Are ESG oriented based on ESG indicators; (e) targeted interview with each startup study selected conducted online or offline; (f) data collection and interpretation; (g) final analytical report.

29.4 Findings

The results expected at the end of the survey process will allow: (1) identifying, first of all, the most successful startup studios in Italy; (2) identifying their values; (3) analyzing the process of creating and launching new ESG-oriented startups; (4) to know in depth, through semi-structured interviews, their success stories, the challenges on their development path, the results achieved and the impact of their activities on social and environmental welfare.

In this regard, the most significant and original result will be to know just how the activity of the startup studios impacts on social and environmental well-being.

29.5 Implications

The research does not purport to examine the differences between business incubators and startup studios, nor does it dwell on the reasons for the entrepreneurial success of the startup studio. These could be the subject of subsequent research.

The main implications for researchers concern the advancement of knowledge in the field of startup study still little explored, with repercussions on policymakers who could draw from these knowledge investment opportunities for the sustainable development of their territories by acting as facilitators of new business initiatives.

The ambition of the research is to provide political actors with a solid knowledge base to adopt disruptive policies in the field of local development, with effects on the progress and improvement of the quality of life of citizens of areas and contexts at risk of marginalization of our country as new job opportunities, employment, self-employment could be promoted, contributing in turn to greater widespread wellbeing.

References

1. Dagnino, E.: Intelligenza Artificale e mercati del lavoro. Casi e materiali di discussione: mercato del lavoro e contrattazione collettiva, Prima rassegna ragionata della letteratura economica e giuridica. CNEL. **3** (2024)
2. Gudergan, G., Mugge, P.: The gap between the practice and the theory of transformation. In: Hawaiian International Conference Of System Science, Whitepaper (2017)
3. UN SDG Report.: www.un.org/sustainabledevelopment/progress-report (2019). Accessed 10 Feb 2024.
4. Mamazen, IH. (2023, July). Savvy Tinkers Studio, Venture Studio Associates: Redesigning Entrepreneurship - A Guide to Investing in Startup Studios, Whitepaper.
5. Economyup.: https://www.economyup.it/innovazione/startup-studio-che-cosa-sono-come-funzionano-le-fabbriche-di-startup-e-quali-sono-in-italia (2022). Accessed 10 Feb 2024.
6. Digital360. https://www.esg360.it/environmental/esg-tutto-quello-che-ce-da-sapere-per-orientarsi-su-environmental-social-governance (2023). Accessed 10 Feb 2024.
7. SYMBOLA, UnionCamere, Consorzio Aaster: Cohesion is competition, New geographies of value production in Italy I Quaderni di Symbola, report 2014 (2014).
8. European Commission Research and Innovation DG (ed.) (2021, January). Industry 5.0. Towards a sustainable, human-centric and resilient European industry, Policy paper.

Chapter 30
Territorial Development Domain Versus Digital Transformation and Society 5.0: The Tirreno-Eolie LAG Experience

Carlotta D'Alessandro, Katarzyna Szopik-Depczyńska, Alberto Bongiorno, Giuseppe Caristi, and Giuseppe Ioppolo

Abstract Digital Transformation is a necessary component of territorial development processes. However, the Italian country system, characterized by small villages with a high level of protection of environmental and cultural assets, is not uniformly prepared for this transformation. While there is a recognized need for the introduction of digital innovation in terms of technology and services, a society that has not yet embraced 5.0 poses a barrier to this transformation. This document analyzes the Management Plan of a system of Municipalities in the Province of Messina, Italy, organized in a Local Action Group (LAG), (also known as GAL, Gruppo di Azione Locale in Italian), trying to understand whether the demand for territorial development is combined with a model of digital transformation. In particular, the tool created by the authors pinpointed areas to promote growth and emphasized the need to include digital technologies in the LAG's development strategy.

The findings aim to spark discussion, acknowledging that the scientific understanding of digital transformation's impacts is still evolving. They underscore the necessity for new decision-makers actors, namely, digitally oriented governance, to lead an essential transformation that cannot be postponed.

Keywords Digital transformation · Territorial development · Digital governance

C. D'Alessandro (✉) · A. Bongiorno · G. Caristi · G. Ioppolo
Department of Economics, University of Messina, Messina, Italy
e-mail: carlotta.dalessandro@unime.it; alberto.bongiorno@studenti.unime.it; giuseppe.caristi@unime.it; giuseppe.ioppolo@unime.it

K. Szopik-Depczyńska
Department of Corporate Management, Institute of Management, University of Szczecin, Szczecin, Poland
e-mail: katarzyna.szopik-depczynska@usz.edu.pl

F. Schiavone et al. (eds.), *Advanced Perspectives and Trends in Digital Transformation of Firms, Networks, and Society*, Springer Proceedings in Business and Economics, https://doi.org/10.1007/978-3-031-80692-6_30

30.1 Introduction

This document explores the topic of digital transformation and the related digital divide through a real-life local development project. Digital transformation encompasses the habits and changes associated with planning and implementation, relying on emerging business paradigms and state-of-the-art technological solutions [1]. It encompasses the use of digital technology across various aspect of society. The term "digital divide" originated during the Clinton administration in the United States to denote the disparity in access to telematic services among the American population. Today, it is a globally used term indicating the disparity between those with adequate Internet access and those without, whether by choice or circumstance, leading to social, economic, and cultural disadvantages. It is evident that digital transition and the digital divide are distinct yet interconnected phenomena with a shared objective: to achieve universal inclusion.

The National Village Plan operates within this framework, serving as instrument to support the regeneration of villages and small municipalities within the National Recovery and Resilience Plan (PNRR). Its goal is to foster economic and social development in areas distant from major urban centers and initiate the cultural revival of small towns. Specifically, Investment 2.1, focused on enhancing the attractiveness of villages under the M1C3—Tourism and Culture 4.0 initiative, allocates a total funding of 1020 million euros. This funding is divided into two intervention lines to facilitate the submission and approval of projects aimed at revitalizing 250 Italian villages.

"Villages of the Future" stands as a strategic pillar within the National Innovation Plan 2025—strategy for technological innovation and digitalization presented in December 2019 by the Minister for Technological Innovation and Digitalization. Its aim is to incentivize population settlement in these regions, stimulate local economies and tourism, and showcase the cultural, historical, and artistic richness of isolated territories to a wider audience through digitalization.

This initiative ensures that, through digitalization and advanced technologies, remote areas are bridged with the broader territory, promoting sustainability and boosting the quality of life for villages and businesses.

Numerous investments are driving digital transformation, including PNRR's investment 1.4, labeled "Digital services and digital citizenship." Both the public and private sectors are actively promoting and reinforcing the adoption of organizational models and smart working solutions in what is now a post-pandemic landscape. However, despite initiatives like sample projects showcasing meta-villages and the digitalization of cultural heritage, a significant gap remains.

Building digital infrastructure guarantees the progression of the digital economy. The establishment of digital infrastructure is crucial for advancing the digital economy. Expanding broadband networks, a cornerstone of the digital framework, has seen rapid expansion alongside the digital economy rise. Considering its economic implications and benefits, as dissemination of information and data exchange regardless of geographical constraints, have accelerated the integration of information technology across various sectors [2]. Despite the acceleration of the last 5 years, there are areas of the country that remain in serious difficulty due to the use of the Internet by citizens (ranking 26 out of 28); the integration of digital technologies within companies (22 out of 28) and the Human Capital/Digital skills necessary for digital transformation (28 out of 28), as recorded by the Digital Economy and Society Index (Desi), indicator, according to the EY Digital Infrastructure Index report. The valorization of small towns encompasses more than just digital transformation, simply as digitalization alone does not generate a positive process of repopulation or economic relaunch. Conserving and enhancing cultural significance of quaint town, distinguished by its distinctiveness and individuality, whether as an independent entity or as part of the broader regional context, presents several challenges [3].

However, despite these initiatives, digital divide persists in many rural areas, hindering their full participation in the benefits of digital transformation. This research aims to bridge this gap by exploring how a real-life local development project leverages digital tools while considering the specific needs and challenges of these communities.

Therefore, while increasing complexity and interconnections can enhance a system's potential for innovation and societal advancement, prioritizing "smartness" alone may not suffice. This raises the question: "smartness" is truly a short-term goal for the small city model characterized by a richness and concentration of heritage and natural values? This research is based on the practical experiences of collaboration and local development (Management Plan) within a territorial system (LAG, also known as GAL, Gruppo di Azione Locale in Italian) centered on sustainable economic-cultural valorization, highlighting the local perspective on digital transformation.

After this Introduction section, a synthetic literature review on digital transformation and local development, and a key research question is designed. Using a case study at the local scale, method section, and case study introduce the development plan and the connection with the research question. Results, via SWOT analysis, carry out the absence of a context digital transformation oriented. In the Discussion, motivation and implications highlight the necessary role of a Society 5.0. The Conclusion represents a starting point for academia, practitioners, and society, to discuss, introduce, and measure digital innovation at the local scale.

30.2 Literature Review

As a preliminary introduction to identifying the effect of digital transformation on local development strategy, a comprehensive literature review was conducted, according to Reis et al. [4], focused on the Research Profile as a quantitative characterization of the selected publications (observing Date of publication, Academic source and citations, Articles methodology, and Keywords of the articles), together with a content analysis, in order to carry out a qualitative characterization, identifying a category system and revising the papers texts within a classification framework, to interpret and summarize the emerging findings Mayring's [5].

Besides, this section highlights the literature review findings. To resume the results, it emerges the lack of academic guidance on the topic (digital transformation matched with local development strategies). In particular, there is a bias on the direct implications on regulators and practitioners in terms of policy development and effective strategy building [6]. From another side, Artificial Intelligence, Blockchain, Cloud Computing, and Data Analytics, also noted as "ABCD" technologies, are the main drivers for digital transformation as innovation in the decision-making process of microenterprises [7].

The literature review tries to carry out the research question "How do local governments enhance their organizational capacity to implement digital transformation?". Adding the keyword "smart city," emerges that the local government along with those leading digital transformation initiatives, can be instrumental in forging partnerships with diverse stakeholders. This cooperative approach can then bridge the gap between the goals, resources, and needs of the various actors involved [8]. The complexity of the transformation as innovation in urban structure and behaviors needs funds (public and private) and reinforcing the city government's organizational capacity to implement the smart city strategy, engaging all the actors (citizens as well) in a new governance model.

The final starting point, here just summarized, is that digital technology will change business processes and promote the transformation of internal structure and value creation approaches of enterprises, but also impact the life of the community, changing behavior and relationships between public system and private one, generating an expansion of the "space" (the real added by digital). Innovative territorial pilot policies, focused on local development, can intercept the main influencing factors and identify the barriers to promoting a digital transformation, re-thinking society, and understanding the spillover effect of the digital revolution.

30.3 Methods and Case Study

Starting with the statement that innovative territorial pilot policies can boost digital transformation at the local scale, an analysis was conducted by the Management Plan of the Tirreno-Eolie Local Action Group (LAG) in Messina, Sicily, to gauge the role of digital transformation in fostering growth and development within small municipalities focused on tourism, environmental protection, and cultural heritage. Local development can be considered as a complex of qualitative-qualitative processes to co-create value for the community and the business system respecting an equilibrium between priorities and internal and external factors/resources. In summary, the Management Plan introduces medium-term planning, starting from the guiding values, provides for an integrated analysis of the state of the places by identifying the forces of change underway, then identifies the future objectives that can be achieved through intervention options and possible strategies (land-use delineation), evaluates the probable impacts on the local system, chooses the action plans to achieve the set goals, defines the coordination and implementation methods and verifies their achievement through a series of indicators that implement the systematic monitoring of the results over time [9, 10].

In particular, following the approval of the Sicily Rural Development Plan by the European Commission (EC) decision C(2015) 8403 on November 24, 2015, LAGs were established; EC promotes policies that encourage sustainable economic performance, such as initiatives to boost occupation and innovation [11].

LAGs serve as the means for implementing and realizing the objectives outlined by PNRR within the Sicilian region. LAGs are tasked with managing funds and operations in areas with populations ranging from 60,000 to 150,000 inhabitants, fostering active public partnerships, and possessing technical and legal frameworks capable of achieving priority objectives. These goals include empowering local communities, promoting new management methods for economic growth, advocating for responsible practices in agriculture, and enhancing the efficiency and transparency of the food supply chain.

More specifically, LAGs, driving regional development across rural areas of the European Union, comply with the LEADER method [12]. This is a method issued by the European Commission to enhance rural growth. As such, the LEADER method should promote innovative design for the development paths grounded in local insights and an understanding of regional complexities.

The research project is overseen by LAG Tirreno-Eolie named after its geographical area covering the eastern part of the Messina region. It encompasses coastal territories, including the Milazzo Plain and the Aeolian Islands.

Due to the morphological peculiarities of the territory, the Wind Islands (Aeolian Islands) deserve a separate description. The archipelago which is part of the district is made up of seven islands that represent a very rich and diversified artistic, natural, environmental, archaeological, and historical heritage. The Aeolian Islands, therefore, represent an extraordinary museum of history: from remote prehistory, all eras have left homes, monuments, and memories that increase the interest in these small Mediterranean lands with a rich past to the point that the area is described as one of the most attractions of the entire Mediterranean region.

Formed by volcanoes, these islands have unparalleled beauty of the coast and seabed, diverse and exciting landscapes, unique rock formations, and lush vegetation. In this region, green and fertile areas coexist, such as the islands of Salina or Alicudi, but also steep and arid cliffs, such as Stromboli (home to an important permanently active volcano). The cliffs plunge straight into the sea. Such a rich landscape (of land and sea) has given rise to a highly dispersed housing structure, with 20 settlement centers (but only 4 municipalities, of which 3 on the island of Salina alone), which host more than 14,000 residents. The significant aspect of the LAG district, in addition to its natural and uncontaminated environmental beauty, lies within the individual municipalities, which contribute to the vast heritage.

To facilitate a cohesive development and governance strategy, the LAG adopts a programmatic tool known as the Management Plan. This tool analyzes strategic aspects to outline a sustainable development scenario on a territorial scale, incorporating tools for evaluating and monitoring outcomes. Rather than prescribing detailed instructions for activities, the Management Plan provides coordinated and shared guidance, harmonizing interventions within a programmatic framework aligned with local needs.

The plan's effectiveness lies in the ability to correctly interpret the cultural, economic, and social strengths of the provincial territory into distinct competitive advantages. It should serve as a tool for innovative processes by evaluating both currently available resources and those potentially available but not yet used, fostering cohesion in a territorially diverse environment for competitive gains. Resources should be directed toward a strategically coherent project, avoiding a disjointed collection of actions, and instead fostering synergistic development effects across the entire territory. It is important, therefore, that resources are channeled toward a "project" that has its own "strategic coherence." It, therefore, cannot be a generic sum of actions to be carried out, but rather a concerted and unitary system capable of causing synergistic effects of development of the territory.

The strategic development of a territory cannot rely solely on isolated interventions or occasional incentives for economic growth. It requires fostering self-organization among the various stakeholders to leverage the area's resources and capabilities. The territory of the LAG must therefore be treated as a "complex strategic subject" capable of promoting a development policy and adopting the necessary tools to achieve the objectives that are predefined.

A comprehensive planning approach must address the ongoing challenges of land management systematically rather than episodically. This entails explicitly defining priorities and guidelines through participatory co-planning across institutional levels.

To formulate the "Development Plan of the Municipalities of the Partnership—Report on the State of the Territorial System" a document overseen by the LAG Tirreno-Eolie, comprehensive data collection and organization were imperative. This process aimed to assess the current state of the territory and delineate a vision, objectives, and innovative actions to foster its sustainable development (also considering digital transformation as the driver of local competitiveness). The approach adopted took into consideration both the internal variables of the territory (through the mapping of natural, semi-natural, and anthropogenic ecosystems, floristic, faunal, and avifaunal species as well as the consideration of hydrobiological and hydrogeological parameters) and external ones (such as the management of tourism system), which can influence the position of the Tirreno-Eolie district compared to other similar areas in terms of characteristics and market aspirations of the competence of other regional LAGs. The analysis focuses on key areas, to identify the digital transformation inclination:

- The territorial context.
- The biodiversity context.
- The social context.
- The economic context.
- The cultural context.

- The involvement of interested parties in the territory.

The authors' tool underscored growth themes and the integration of digital technologies into the LAG's development model.

30.4 Findings

The territorial analysis, focusing on the environment due to the area's natural significance, culminated in a SWOT analysis (Table 30.1). The evaluation of the area's positioning is considered a critical and potential aspect of the local tourism sector. Key macro-points included:

- The current state of the territorial context.
- The characterizing elements of each municipality under LAG Tirreno-Eolie jurisdiction.
- The urban planning component (from a road, railway, and nautical point of view).
- The main ecosystems.

The key idea that emerges from the preliminary study of the context is also through the programmatic reports provided by the Administration.

Specifically, the preliminary study of the context, supplemented by programmatic reports from the Administration, highlights a central idea: to guide the LAG administration in developing integrated management approaches for the territorial system, leveraging local resources as a foundation for a new economy. The aim is to craft a Plan fostering a fresh territorial consciousness focused on protecting, enhancing, and sustainably developing territorial resources. This involves identifying cohesive actions and effective partnerships to transform the entire Province into an integrated innovative system.

Emphasizing emerging needs, environmental preservation emerges as a key driver for optimizing the tourism sector and other emerging industries. However, digital transformation doesn't represent a significant driver for the local development plan of Tirreno-Eolie LAG.

Differently, safeguarding the territorial heritage is crucial for sustainable development and the well-being of future generations, necessitating a responsible approach to development. This entails prudent resource management, especially at the territorial level, to avoid irreversible interventions that could jeopardize biodiversity—a priceless asset for Sicily and humanity. This strategic choice enables the enhancement of the area's rich environmental heritage, fostering prosperity and garnering national and international acclaim for excellence.

Besides, implementing such a decision poses several challenges. There's a risk that a policy for quality tourism may remain theoretical without concrete actions, undermining its effectiveness and public perception. Timely and coherent interventions are therefore necessary to translate policies into tangible outcomes.

Table 30.1 SWOT analysis of LAG district (Source: Authors elaboration based on [14])

Strengths	Weaknesses
High environmental and natural value of the area Low anthropization Territorial connotation (coastal area) Large agricultural surfaces and quality agricultural products Strong cultural identity and characterization of the territory Sustainable development policies of the metropolitan area	Weak production system Population declines in municipalities far from the coast Difficult accessibility to parks and natural areas of the area Poor connection between tourism and agriculture Low technological intensity Low propensity for digital innovation Low digital culture
Threats	*Opportunities*
Lack of attention to the natural resource with consequent loss of attractiveness of the area Risk of pollution and deterioration of the biodiverse component Conflict between the needs of mass tourism and those of nature tourism Risk of uncontrolled and ungoverned exploitation of the areas and speculation	Attractiveness of protected areas toward high-income tourist segments Evolution of tourism as a dynamic and driving productive reality Potential of a quality supply chain between tourism and agriculture Regenerative capacity of the coastal strip in the area Territorial marketing for well-characterized and certified area products Possibility to develop a new digital governance New digital trends in next generation New business within the digital environment/market

Moreover, embracing sustainable development may spark conflicts with local landowners and proponents of projects that fail to meet quality standards. Despite promises of employment and economic growth, resistance from such parties could impede the tourism policy's effectiveness. Municipal administrations must respond with negotiation and transformation efforts to ensure sustainable outcomes.

Additional challenges stem from the necessity of prioritizing investments and allocating resources to initiatives aligned with the vision of quality tourism. Evaluating these interventions involves not only assessing their environmental impact but also their effectiveness in meeting established objectives. This demands clarity, political resolve, and adept negotiation skills.

In contrast to the traditional development model focused on preserving physical values, the emergence of digital transformation raises questions about its impact on societies with limited innovation readiness and availability. These societies often face barriers-both physical and ideological—that prioritize conservation over progress.

Nevertheless, the Plan introduces enticing elements that pave the way for innovative possibilities, including investments directed at fortifying and amplifying the unique features of the production system, services, and local culture by integrating new digital services along the coastal area. This integration fosters collaboration among operators across various sectors, such as food and wine itineraries, cultural routes, and more.

Furthermore, the diversification of services and functions is a characteristic of areas transitioning toward a knowledge-based society of services. This diversification promotes synergy among activities catering to tourists, tourism services, and local populations. Such robust exchange and interaction can initiate a positive cycle, stimulating innovation through informal cross-sectoral collaboration and local economic empowerment.

A practical example arises from the Municipalities' initiative to develop an app and website providing information on separate waste collection. Additionally, local authorities will establish a toll-free number for citizen assistance and clarifications. The app will grant access to the collection calendar, guide, waste dictionary, collection center information, and other useful features. This digital innovation, which aligns with implementing sustainable development policies, aims to increase citizen engagement and offers opportunities for learning about and appreciating the local landscape. The authors advocate for the creation of a "Territorial Action Plan" across the 9 ATOs, employing interdisciplinary and participatory methods involving all social categories to share knowledge and skills. The tourist potential in Sicily and the province of Messina is significant, yet the lack of organized services poses a major hindrance. Specifically, gaps include the need for a tailored tourist accessibility/mobility offer focusing on proposals and information services to enhance the entire provincial and regional tourism system and well-known centers. In this context, the demand for digital services is particularly critical.

30.5 Discussions and Implications

A territory can be seen as a blend of skills, and experiences, with variable percentages, of skills, i.e., those distinctive specificities that determine its complexity, an updated synthesis of past experiences, and a starting point toward all transformations/evolutions. In a globalized economy, this unique identity allows a territory to stand out. Furthermore, territorial development can represent an identity mechanism that allows the emergence of a place on a globalized and unified system.

However, focusing too heavily on uniqueness can obscure a territory's core strengths. It is crucial to understand what sets a place apart and leverage its natural advantages. The digital revolution plays a vital role here. By acknowledging the inherent differences within a region, technology can become a tool for balanced development across economic, social, and environmental aspects. The three axes of sustainable development fully represent the guidelines that must be analyzed and measured to have a coherent framework in all digital land management processes.

Sustainable development principles provide a framework for managing digital transformation within a territory. This approach requires an evolution of social capital, embracing the collaborative and information-rich environment of "Society 5.0" [13]. Local development thrives on community involvement and a sense of shared responsibility.

The Local Territorial System embodies this collective spirit. It brings together different stakeholders to act as a unified force. Complex scenarios and the multilayered relationships between these actors necessitate the creation of collaborative platforms that facilitate communication and foster new forms of governance. Territorial governance, acting as an internal network, plays a critical role in shaping local development strategies. This involves identifying a territory's inherent potential and translating it into actionable steps that enhance its value and promote its unique qualities, maintaining and increasing the competitive advantage. Governance moves beyond a singular body, becoming the central driver of local development policies. To effectively manage and oversee decision-making, governance needs tools for measuring performance and evaluating outcomes. These could include management control measures, service quality assessments, and the impact on the community. Achieving this vision requires a dual focus: increasing local knowledge within the community and integrating a digital model alongside traditional development tools. This allows territories to leverage their unique strengths and thrive in a globalized world.

Indeed, policymaking involves a series of actions through which an entity formulates its territorial policies, considering the context in which it operates and the potential social, economic, and environmental impacts of its decisions on the administered territory. In fact, sustainability issues have become paramount for the global world [15].

Within this framework, it is essential to differentiate between three key decision-making levels: strategic, operational, and organizational.

Policymaking involves dialogue between political actors who interpret and challenge community expectations. Strategic decisions are quantifiable to measure impact. Implementation, on the other hand, relies on internal public administration capacities.

Social responsibility and proactive management are crucial for translating policies into action. Digital governance can help foster territorial competitiveness while pre-serving organic digital innovation.

30.6 Conclusions and Limitations

Land management, especially when linked to regeneration efforts, can significantly impact the environment, economy, and society. Starting from the literature analysis', through the case study of the Tirreno-Eolie LAG, the Management Plan highlights that to ensure that digital transformation acts as a catalyst and not as a barrier, the support of a new social model is necessary. The limit lies precisely in the ability of a territory to generate a digitally oriented governance. Theoretical underpinnings suggest that the process of digitalization within governance and public management revolves around a complex interaction between individuals, technology, and institutional structures. Only from this collaboration, it is possible to make sustainable technological choices adapted to the specific territory.

However, overcoming some obstacles is crucial. These include limited local awareness of digital tools, data accessibility issues, and resistance to change. In this sense, the effective management model, functionally directing territorial development on a local scale, must be able to correlate techno-digital variables with economic-environmental ones to all territorial aspects and resolve the barriers.

Integrated service management systems, even in the absence of a single platform, can represent an aid to governance, offering valid decision-making support. Existing research and experiences can also serve as useful guides for integration processes.

It is important to remember that models and tools are constantly evolving and cannot replace human decision-making, especially in public management. However, advances in digital solutions and the "on-life" approach of younger generations represent powerful forces for change. Ultimately, change is inevitable, and successful territorial management requires embracing it, starting from the implications of the literature review, the need for the academy to study the relationship between digital transformation and territorial system, moving from theory to practice, is clear.

Funding Co-financed by the Minister of Science under the "Regional Excellence Initiative" Program. Co-financed by the Project "Tirreno-Eolie 2030—Ripensare il Territorio, valorizzare la biodiversità, promuovere i prodotti tipici"—linee di intervento della ricerca e le risorse umane Prot.n. 0166315 del 21/12/2022.

References

1. Anthony Jnr, B.: Distributed ledger and decentralised technology adoption for smart digital transition in collaborative enterprise. Enterp. Inf. Syst. **17**(4), 1989494 (2023)
2. Guo, B., Hu, P., Lin, J.: The effect of digital infrastructure development on enterprise green transformation. Int. Rev. Financ. Anal. **92**, 103085 (2024)
3. Tajani, F., Sica, F., De Paola, P., Morano, P.: A networking-economic model to enhance the cultural value in small towns. Smart Sustain. Built Environ. (2024)
4. Reis, J., Amorim, M., Melão, N., Matos, P.: Digital transformation: a literature review and guidelines for future research. Trends Adv. Inf. Syst. Technol. Vol. 1. **6**, 411–421 (2018)
5. Mayring, P.: Qualitative content analysis. Companion Qual. Res. **1**(2), 159–176 (2004)
6. Guandalini, I.: Sustainability through digital transformation: a systematic literature review for research guidance. J. Bus. Res. **148**, 456–471 (2022)
7. Long, Y., Liu, L., Yang, B.: The effects of enterprise digital transformation on low-carbon urban development: empirical evidence from China. Technol. Forecast. Soc. Chang. **201**, 123259 (2024)
8. Gasco-Hernandez, M., Nasi, G., Cucciniello, M., Hiedemann, A.M.: The role of organizational capacity to foster digital transformation in local governments: the case of three European smart cities. Urban Govern. **2**(2), 236–246 (2022)
9. Ioppolo, G., Saija, G., Salomone, R.: From coastal management to environmental management: the sustainable eco-tourism program for the mid-western coast of Sardinia (Italy). Land Use Policy. **31**, 460–471 (2013)

10. Nowak, M., Śleszyński, P., Cheba, K., Blaszke, M., Szopik-Depczyńska, K., Ioppolo, G.: Flexibility of land use plans: between supporting development and opportunism. Evidence from Poland. Land Use Policy. **135**, 106939 (2023)
11. De Pascale, A., Di Vita, G., Giannetto, C., Ioppolo, G., Lanfranchi, M., Limosani, M., Szopik-Depczyńska, K.: The circular economy implementation at the European Union level. Past, present and future. J. Clean. Prod. **138658** (2023)
12. Food and Agriculture Organization of the United Nations Homepage, Council Regulation (EC) No 1698/2005 of 20 September 2005 on Support for Rural Development by the European Agricultural Fund for Rural Development (EAFRD). https://eur-lex.europa.eu/legal-content/EN/TXT/PDF/?uri=CELEX:32005R1698&from=CS. Accessed 11 February 2024.
13. Liashenko, O., Dluhopolskyi, O.: Uncovering the interplay between social welfare preferences and society 5.0 achievement: implications for sustainable development. Econ. Innov. Econ. Res. J. **12**(1), 175–197 (2024)
14. Assessorato Dell'agricoltura, Dello Sviluppo Rurale E Della Pesca Mediterranea Dipartimento Dell'agricoltura (2024). Allegato 2- Attuazione delle Strategie di Sviluppo Locale (ASSL)
15. Szopik-Depczyńska, K., Cheba, K., Vikhasta, M., Depczyński, R.: New form of innovations related to the environment-a systematic review. Procedia Comput. Sci. **192**, 5039–5049 (2021)

Chapter 31
Legitimacy and Innovation. Looking into the Maritime Industry

Michele Acciaro, Maria Cristina Pietronudo, Marco Ferretti, and Marcello Risitano

Abstract This chapter focuses on the concept of legitimacy within the maritime industry, highlighting its pivotal role in shaping firm behaviors and innovation outcomes. Legitimacy is understood as the societal acceptance of a firm's adherence to norms and values, which significantly influences strategic decisions and overall performance. Specifically, the research centers on external legitimacy, where stakeholders outside the firm validate its credibility within the community, and investigates how this legitimacy—encompassing regulative, normative, and cognitive dimensions—interacts with network embeddedness. This includes both relational and structural ties with stakeholders, to analyze its impact on innovation outcomes. Drawing on insights from institutional theory and the family firm literature, the study aims to elucidate the complex dynamics that link legitimacy and network relationships to varying levels of innovation performance in maritime family firms. Anticipated findings suggest that high normative and cognitive legitimacy, combined with strong network ties, may paradoxically lead to lower innovation performance. The chapter contributes to both family firm and maritime management literature by elucidating underexplored dynamics of legitimacy and network effects on innovation.

Keywords Legitimacy · Innovation · Maritime firms

M. Acciaro
Copenhagen Business School, Copenhagen, Denmark

M. C. Pietronudo (✉) · M. Ferretti · M. Risitano
Parthenope University of Naples, Naples, Italy
e-mail: mariacristina.pietronudo@uniparthenope.it

F. Schiavone et al. (eds.), *Advanced Perspectives and Trends in Digital Transformation of Firms, Networks, and Society*, Springer Proceedings in Business and Economics, https://doi.org/10.1007/978-3-031-80692-6_31

31.1 Introduction

The chapter explores the concept of legitimacy in the maritime industry. Legitimacy signifies the perception of societal acceptance regarding a firm's adherence to a set of social values that exist in both its external and internal environments. More precisely, legitimacy is a generalized perception or assumption that actions are desirable, proper, or appropriate within some socially constructed systems of norms, values, beliefs, and definitions [1] (p. 574). It may stem from the adherence to formal norms and regulations formulated by the government and the industry (regulatory legitimacy), or it may arise from moral or ethical considerations and cognitive alignment with a group of stakeholders (normative and cognitive legitimacy, respectively) [2]. The relevance of this multidimensional construct is due to it being a key factor for the growth and the survival of many enterprises [3]. Legitimacy depends in fact on firms' strategic actions but also on the validation of firms' behavior from internal stakeholders (employees) or external stakeholders (such as clients, suppliers, investors) [1]. According to [4], legitimacy explains the social embeddedness of the organization. Scholars typically attribute positive repercussions to embeddedness and legitimacy. Legitimate companies often experience an increase in the accumulation of resources, the absorption of knowledge, and consequently, an increase in profitability and the entrepreneurial spirit essential for innovation [5]. The concept, widely explored in the field of new firms, family firms, or small and medium enterprises analyzing a plethora of industries (wine industry, tourism industry), was completely ignored by management scholars interested in understanding maritime sectors.

Few preliminary studies have mentioned legitimacy under the umbrella of the institutionalization theory [6, 7], considering legitimacy as a variable to measure the level of institutionalization of maritime firms. Nevertheless, mentioned studies do not propose and do not refer to consequences of the institutionalization at firms' levels, neglecting whatever repercussions on the performance of institutionalized firms. As the management literature shows, the legitimacy is a relevant concept which would help explain firms' behaviors with respect to their own internal and external environment and to clarify drivers of performance [8]. To complete and enrich a too immature literature on maritime companies, this research draws from the literature on the legitimacy of family firms [8]. Numerous companies in the maritime industry have been founded and managed by families for generations, such as Grimaldi and Aponte in Italy, Maersk in Denmark, and Hapag-Lloyd in Germany. These companies range from small, local enterprises to large multinational corporations. As family firms, they have developed strong relationships within the industry or the broader contexts in which they operate. These relationships have contributed to the creation of legitimacy, more precisely external legitimacy, causing sometimes indirect and other times direct consequences on their business.

31.2 Objective

This chapter focuses on external legitimacy, defined as the endorsement by external stakeholders who recognize the worthiness of a firm, perceiving it as a valuable member of a community. Specifically, this study explores the relationship between legitimacy and innovation. Existing literature on family firms discusses the positive effects of legitimacy on firm financial performance and, to a lesser extent, on innovation outcomes. When analyzing the relationship between innovation and legitimacy, scholars often refer to the introduction of new products or services, particularly describing instances where innovation lacks legitimacy [4]. Some studies [9] examine the relationship between organizational legitimacy and firm innovation, yielding contrasting results: constraints vs benefits for innovation. Moreover, recent research in family firm management, drawing from institutional theory, highlights the negative consequences of excessive embeddedness in the business ecosystem [10]. This chapters, specifically, intends to follow the recent research stream to explain the less positive link between legitimacy and innovation. We draw upon the literature on network embeddedness [11] to explore how other network characteristics, in addition to legitimacy, may produce contradictory effects on innovation performance. These characteristics include the frequency of relationships and the structure of connections with core stakeholders within the firm's social capital. In exploring the relationship between external legitimacy and low level of innovation performance we contribute twofold. In the field of family firms, we delve into when and how family firm networks may lead to low innovation outcomes. In the field of maritime management literature, we examine legitimacy to elucidate overlooked aspects of innovation. The chapter aims to identify causal relationships that may affect various levels of innovation, investigating relationships related to different forms of legitimacy (normative, regulative, and cognitive) while also considering additional conditions related to network embeddedness (such as relational and structural embeddedness). The chapter intends to address the following research question: "Which combinations of legitimacy and network embeddedness are likely to produce different levels of firm innovation performance?"

31.3 Methods

To explore how legitimacy and network embeddedness conditions contribute to different levels of innovation outcome, the study will intend to propose a fuzzy-set qualitative comparative analysis (fsQCA) [12]. The selected method should elucidate which combinations of conditions among different forms of legitimacy and various aspects of network embeddedness are associated with specific innovation output results. We will use a sample of 30 maritime family firms' companies located in South Italy, a geographical context characterized by maritime vocation and by family corporations. Maritime family firms will select applying the industry criterion and the family firms' definition by [13]. The data will be collected using a survey to send to business owners and managers. The outcome measure will be the innovation performance, measured through a Likert scale and using items following [14]. Concerning casual conditions: legitimacy (regulative, normative, and cognitive) will be assessed using 12 items as [2, 15]; the network embeddedness will be measured considering the relational and the structural embeddedness inspired by [16, 17].

31.4 Findings

In our preliminary exploration of the factors contributing to different levels of innovation outcomes, we anticipate that normative and cognitive legitimacy, coupled with strong relational and structural embeddedness, will be associated with low levels of innovation performance. Family firms, in general, and maritime firms, specifically, may be influenced by the social capital and the closeness of the link with core stakeholders, particularly when they do not play a key role in the network. Moreover, we expect that the more legitimacy is associated with informal values and culture, the more firms will tend to not destroy the legitimacy gained since it is difficult to rebuild. Unlike research of the network embeddedness [4] this research explains the risk of embeddedness not only recurring to the redundancy of resources, but exploring the concept of legitimacy that pushes companies to conformism in the short term and institutionalism in the long term.

31.5 Implications

These insights are critical for understanding the behavior of maritime companies, as they may face dilemmas between their activities' results and maintaining legitimacy, potentially impacting their outcomes. This research will yield practical and theoretical implications that will be further explored in a comprehensive version.

References

1. Suchman, M.C.: Managing legitimacy: strategic and institutional approaches. Acad. Manag. Rev. **20**(3), 571–610 (1995)
2. Scott, W.R.: Institutions and Organizations, Foundations for Organizational Science. SAGE, Thousand Oaks (1995)
3. Díez-Martín, F., Prado-Roman, C., Blanco-González, A.: Beyond legitimacy: legitimacy types and organizational success. Manag. Decis. **51**(10), 1954–1969 (2013)

4. Huang, J., Li, M., Mao, L., Zhang, T., Li, J.: The impact of network embeddedness on radical innovation performance-intermediators of innovation legitimacy and resource acquisition. Int. J. Technol. Policy Manag. **17**(3), 220–239 (2017)
5. Zimmerman, M.A., Zeitz, G.J.: Beyond survival: achieving new venture growth by building legitimacy. Acad. Manag. Rev. **27**(3), 414–431 (2002)
6. Demir, B., Tavacıoğlu, L.: Developing a scale to measure the institutionalization level of maritime family businesses. J. ETA Marit. Sci. **9**(4), 243–255 (2021)
7. Turhaner, K., Nas, S.: Determining the level of institutionalization in family-owned shipping businesses. J. ETA Marit. Sci. **9**(2), 138–153 (2021)
8. Berrone, P., Duran, P., Gómez-Mejía, L., Heugens, P.P., Kostova, T., van Essen, M.: Impact of informal institutions on the prevalence, strategy, and performance of family firms: a meta-analysis. J. Int. Bus. Stud. **53**(6), 1–25 (2020)
9. Van Dijk, S., Berends, H., Jelinek, M., Romme, A.G.L., Weggeman, M.: Micro-institutional affordances and strategies of radical innovation. Organ. Stud. **32**(11), 1485–1513 (2011)
10. Heider, A., Hülsbeck, M., von Schlenk-Barnsdorf, L.: The role of family firm specific resources in innovation: an integrative literature review and framework. Manag. Rev. Q. **72**(2), 483–530 (2022)
11. Morrish, S.C., Earl, A.: Networks, institutional environment and firm internationalization. J. Bus. Ind. Mark. **36**(11), 2037–2048 (2021)
12. Ragin, C. C.: Measurement Versus Calibration: A Set-Theoretic Approach (2008)
13. Anderson, R.C., Duru, A., Reeb, D.M.: Investment policy in family controlled firms. J. Bank. Financ. **36**(6), 1744–1758 (2012)
14. Higón, D.A.: The impact of ICT on innovation activities: evidence for UK SMEs. Int. Small Bus. J. **30**(6), 684–699 (2012)
15. Wang, Z., Zhou, Y.: Business model innovation, legitimacy and performance: social enterprises in China. Manag. Decis. **59**(11), 2693–2712 (2021)
16. Gulati, R., Sytch, M.: Dependence asymmetry and joint dependence in interorganizational relationships: effects of embeddedness on a manufacturer's performance in procurement relationships. Adm. Sci. Q. **52**(1), 32–69 (2007)
17. Granovetter, M.S.: The strength of weak ties. Am. J. Sociol. **78**, 1360–1380 (1973)

Chapter 32
Ecosystemic Challenges on the Path of Digital Transformation: Exhibition Organizer's Perspective

Romanova Anastasia

Abstract The main goal of this study is to propose a conceptual framework for studying event organizers' business model through the ecosystem lens. The study explores the impact of digital maturity determinants on event organizers' business model innovation as well as changes in event organizers' ecosystem on three levels: organizer-visitor, organizer-partners, and organizer-exhibitor. Based on the mixed-method approach, the study systematically analyzed the innovation of the exhibition organizer's business model and the process of reshaping relationships within its business ecosystem. Quantitative research is prefaced by qualitative research to reveal and propose digital maturity determinants to assess the effect of digitalization on the exhibition organizer's business model. Then it is followed up by semi-structured interviews. The results indicated that digital maturity determinants can be applied to measure the effect of digitalization on the exhibition organizer's business model. Digital literacy, usage of digital platforms, and implementing digital strategy can define the innovative business model of the exhibition organizer, reshaping relationships, and network configuration within the organizer's ecosystem. The study proposes a conceptual framework for studying event organizer's business model through the ecosystem lens and fills the gap as the research has paid insufficient attention to the digital transformation phenomenon in relation to business model innovation in the event industry.

Keywords Digital maturity · Business model · Event industry · Exhibitions · Digital technology

---

R. Anastasia (✉)
Graduate School of Economics and Management, Ural Federal University, UrFU, Ekaterinburg, Russia

F. Schiavone et al. (eds.), *Advanced Perspectives and Trends in Digital Transformation of Firms, Networks, and Society*, Springer Proceedings in Business and Economics, https://doi.org/10.1007/978-3-031-80692-6_32

32.1 Introduction

The event industry covers a wide range of services for organizing business and non-business events. The exhibition sector, special events sector, and meetings sector are considered as three main sectors in the event industry. The global market for exhibition organizing estimated at US$19.7 billion in the year 2022 is projected to reach a revised size of US$60.7 Billion by 2030, growing at a CAGR of 15.1% over the analysis period 2022–2030. Exhibitions are among the significant enablers of global trade and economic growth [1]. The event industry landscape has been undergoing changes stipulated by digitalization and the development of digital technologies that create a highly competitive business environment for event companies. Companies that fail to keep up with digital trends risk failing their competitors [2]. From this perspective, competitive advantages can be achieved through innovating business models [3, 4, 5–7] as well as reaching digital maturity [8–11]. In the report of the Global Association of the Exhibition Industry, the impact of digitalization and internal management challenges were named as the most important business issues in 2022 with new digital products, business model adjustments, and responding to customer digitalization needs specified [12].

A business model is to be interpreted as made up of different components aiming at creating, delivering, and capturing value [13–16]. Network or partnership is a common feature in business model literature. It is proposed that firms recognize that multiple stakeholders (internal and external) need to interact and cooperate in order to face challenges [17]. In relation to business models of event organizers, the network component is represented by a large pool of venues, advertising or communications agencies, and services suppliers that are involved in planning, organizing, and delivering trade shows or exhibitions. All these actors are co-creating a value proposition and building an ecosystem where changes in one component cause changes in the other ones resulting in reshaping and reconfiguration of the whole partner network.

There is much research on different aspects of the digitalization phenomenon in the event industry [18, 19, 20–22]. But there is a lack of studies about digital transformation or implementing digital technologies in relation to business model innovation and even less is written about it in the exhibition context. Business models can be transformed under the influence of digital technologies. At the same time implementing digital technologies contributes to the level of a company's digital maturity (DM). The study finds the network or partners as a key component of the exhibition organizer's business model and explores the relationships between the organizer's DM and business model addressing the following research questions:

RQ1: Which determinants of DM can be applied to measure an impact of digital technologies on the network component in the exhibition organizer's business model?

RQ2: How can the innovative network component change the relationships on three levels: organizer-visitor, organizer-partners, and organizer-exhibitor.

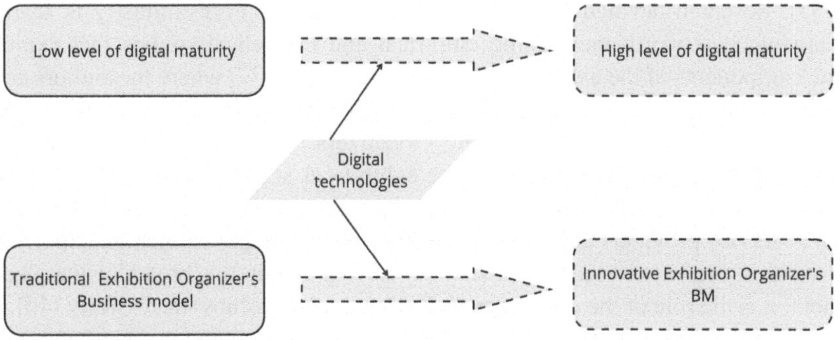

Fig. 32.1 The scheme of relationships between the organizer's DM and business model under the influence of digital technologies

The schematic view of relationships between the organizer's DM and business model through digital technologies is presented in Fig. 32.1.

The structure of the chapter is as follows. In the second section, an overview of the existing literature devoted to DM and business model innovation is given. Section 32.3 describes the research methodology applied to study an impact of DM determinants on the event organizer's business model innovation. The results of hypothesis testing and followed-up semi-structured interviews with the Russian exhibition organizers are presented in Sect. 32.4. Lastly, the chapter closes with a discussion of findings, implications, and a conclusion.

32.2 Digital Maturity and Business Model Innovation

Business model innovation can be considered as the main instrument for companies to adapt to environmental changes and continuously create value [23–25]. Depending on the approach, different scholars propose different sets of components to describe a company's business model. Wirtz et al. [26] give an overview of selected business model components ranked by spectrum of the components and the intensity of use in the scientific literature. With this in mind it is reasonable to consider the impact of digital technologies on a company's business model through its components. The network business model component is widely discussed in business model literature. Collaboration within ecosystem stakeholders increases their capacity to transform organizational structures by influencing each other [27]. Ecosystemic challenges that change the way of how an organization creates and delivers value by orchestrating new interactive and interorganizational processes are considered in studies describing digital platforms as one of the ways to reinvent traditional business models [9, 28, 29, 30]. Among scholars there are attempts to propose conceptual frameworks to study connections between digital transformation and innovations in business models through digital strategy [31, 32] or dynamic capabilities [7,

33–35]. Research devoted to the business model in the event industry is scarce; meanwhile it is worth mentioning empirical and theoretical studies that explore some components of the exhibitors' business model [36, 37] where the authors consider the innovation of exhibitors' business model and how to reshape their interaction with users through innovation. Organizers, exhibitors, and visitors are mentioned as the key stakeholders in the exhibitions sector [38]. However, a large pool of stakeholders are involved in organizing, managing, and delivering events such as service providers and suppliers as well as venues, governmental authorities, universities, media, and industry associations [39]. Within this broad business ecosystem, it is the role of the organizers to facilitate relationships successfully [40]. In terms of business model these stakeholders relate to such a component as "network" or "key partners."

Due to the complexity of such phenomena as digital transformation [41, 42], this study refers to the DM as one of the key elements and antecedents within the digital transformation setting [43]. The higher a company's maturity level, the lower the risks in adoption of new technologies [44]. Thus, the DM is generally assessed to get a deeper understanding of how companies can handle digital technologies and get competitive advantages. The DM can be presented and assessed through a set of dimensions combined in a DM model. Based on the results of literature reviews the most DM models are developed in the manufacturing and industry 4.0 domain [10, 45, 46, 47]. There are only so many studies providing directions on how to adapt these models to specific domains and no model developed to evaluate the DM in the event industry has been published at the time the literature research was performed. For the purposes of the study, we adapted dimensions that can define the DM of the exhibition organizer factoring in insights obtained in the result of interviews with subject matter experts and literature review.

To answer the first research question about the determinants of DM that can be applied to measure an impact on the network component in the exhibition organizer's business model, literature review followed up by expert interviews using Delphi technique was conducted. Delphi technique was chosen as reasonable due to the lack of empirical evidence in scientific literature in the domain under study [48]. Relying on the assumption that dynamic capabilities are relevant to DM assessment and companies need to develop dynamic capabilities specifically for digital transformation [49] three elements—innovation, strategy, and culture—found in DM models were summarized in one dimension "dynamic capabilities." The following dimensions were determined: technology, dynamic capabilities, customer experience, processes, and digital ecosystem. The study mainly focuses on interorganizational relationships and the network component of the exhibition organizer's business model. In light of this only two dimensions—dynamic capabilities and ecosystem—and their six determinants were chosen to test hypotheses and get an understanding of how determinants chosen affect a company's innovative opportunities and the network component enabling event organizers to innovate their business model.

32.2.1 Digital Ecosystem

The most experts pointed out the importance of networking and cooperation for the event organizer. With this in mind such a dimension as "ecosystem" was defined. As one of the determinants of DM it implies that a company works as part of a digital ecosystem and uses technology platforms to enable customizable solutions and seamless interactions with stakeholders including governmental authorities and clients.

Based on the findings of the aforementioned empirical studies and experts' understanding of the ecosystem dimension of DM, this study proposes the three hypotheses regarding the exhibition organizer's innovative opportunities in relation to the network component of its business model as follows:

H1: Using digital platforms positively affects the exhibition organizer's business model innovation.
H2: Collaboration with academicians in the area of event industry and digital technologies positively affect the exhibition organizer's business model innovation.
H3: Digital literacy among ecosystem's actors positively impacts on the exhibition organizer's business model innovation.

32.2.2 Dynamic Capabilities

Dynamic capabilities are defined as the firm's ability to integrate, build, and reconfigure internal and external competences to address rapidly changing environment [7]. Current scientific research considering dynamic capabilities often examines the concept in close relation to digital transformation [49–51] and business model, innovation, or strategy [16, 35, 52]. The extent to which dynamic capabilities are developed can determine companies' ability to become mature in terms of implementing digital technologies. The dimension includes basic processes such as innovation, learning, and agile management and processes specific to digital transformation such as employees' digital skills improvement and developing digital strategy. This study considers well-developed dynamic capabilities to have a positive effect on the exhibition organizer's business model innovation and proposes the remaining three hypotheses as follows:

H4: Using agile management technologies positively affects the exhibition organizer's business model innovation.
H5: Lack of employees' digital literacy negatively affects the exhibition organizer's business model innovation.
H6: Digital strategy positively affects the exhibition organizer's business model innovation.

32.3 Methodological Background

This study mainly investigated the relationships between determinants of digital maturity of the exhibition organizer and innovative opportunities and as a consequence its capabilities to transform and innovate its business model. Due to the lack of studies proposing digital maturity model or devoted to digital maturity in the event industry and in order to answer the research questions, the study employs the mixed-method approach. The mixed use of quantitative and qualitative methods should provide deeper insights, as well as broadening the research perspective to provide valuable contributions in addressing the same research problem [53]. The results obtained through the quantitative research provide general descriptions of the relationships among variables but the more detailed understanding of what the statistical tests or effect sizes actually mean is lacking. Qualitative data and results can help build that understanding [54]. In the study, quantitative research is prefaced by qualitative research to reveal required digital maturity determinants and followed up by semi-structured interviews to get a detailed understanding of how the innovative network component can change the relationships on three levels: organizer-visitor, organizer-partners, and organizer-exhibitor. The research framework is depicted in Fig. 32.2.

For this study, a panel of 12 experts was formed. We defined an expert as a practitioner in the event industry who had considerable and personal experience in exhibition and trade fair organizing and worked in the event company that actively uses digital technologies for the last 5 years. Experts were chosen from the leading exhibition organizers who have a membership in professional event organizations: Ural Congress Bureau (UCB), Russian Union of Exhibitions and Fairs (RUEF), National Association of Event Organizers (NAOM), and International and International Congress and Convention Association—ICCA (see Table 32.1). Based on the feedback from the group of subject matter, hypotheses were developed and a set of

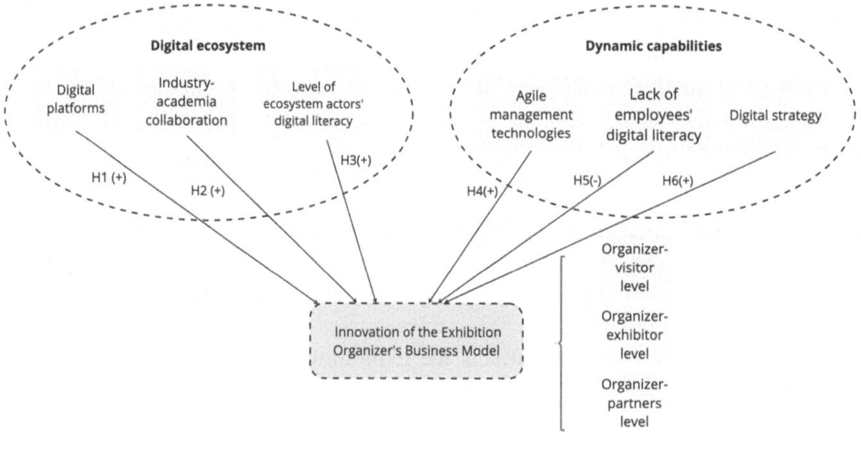

Fig. 32.2 Research framework

Table 32.1 Panel expert's profile

Expert	Job position	Experience	Membership in prof. association
Expert 1	Concept manager	15	ICCA
Expert 2	Site operations manager	9	ICCA
Expert 3	Managing director	12	ICCA
Expert 4	Head of bureau	6	UCB
Expert 5	Chief technical officer	9	ICCA
Expert 6	Project manager	16	RUEF
Expert 7	Project manager	12	NAOM
Expert 8	Business development director	10	RUEF
Expert 9	Head of event organizing department	13	RUEF
Expert 10	Senior exhibition manager	9	NAOM
Expert 11	Project manager	15	ICCA
Expert 12	General director	20	RUEF

Table 32.2 Measurement and coding of research variables

Code	Description
DP	Digital platforms
DLP	Digital literacy of partners
COL	Collaboration with academician in the area of digital technologies and event industry
AMT	Agile management technologies
DLE	Digital literacy of employee
DS	Digital strategy
BMI	Business model innovation

determinants of DM was coded and finalized. All the question items in our survey were measured using a 5-point Likert scale. Table 32.2 gives a list of determinants by two DM dimensions adapted for the exhibition sector of the event industry. The exemplary description of one DM's determinant as a variable is given in Table 32.3.

To validate the research framework a quantitative survey was conducted. The survey base comprised exhibition organizers meeting the following requirements: registered as companies involved in conference or exhibition organizing activity according to the Russian Classifies of Economic Activities, registered no later than 2019, and have membership in the National Congress Bureau or the National Association of Event Organizers (NAOM). These organizations are founded to develop the Russian market of business and event organizers, represent and protect their interests, and set common standards of doing business. In total, we obtained the list of 217 exhibition organizers. Exhibition centers were excluded from the list of respondents as they have their own exhibition facilities and their business model differs from the traditional business model of the exhibition organizer especially in terms of partner networks. Online questionnaire was elaborated, created, and distributed through 179 exhibition organizers and 75 questionnaires were returned resulting in a response rate of 41.8%. Electronic delivery aids in reducing the costs

Table 32.3 Exemplary description of DM determinant as a variable for the study

Code	Variable/Item
DS	Digital strategy DS 1. Well-developed digital strategy significantly enhanced innovative opportunities of our company in relation to business model. DS 2. Our company needs a digital strategy to innovate its business model. DS 3. Our company is willing to develop digital strategy to innovate its business model.

Table 32.4 Interviewees' profile

Exhibition organizer	Y. of foundation	Job position	Venue	Exhibition theme
A	2005	Events sales manager	Russia	Construction and building
B	2015	Senior exhibition manager	UAE, Dubai	Business, exhibition management
C	1998	General director	Russia	Tourism and hospitality
D	2002	Business development director	Russia	Agricultural

associated with survey distribution and increases the reliability of responses by removing the need for rekeying [55]. Data were collected between the end of October 2023 and mid-December 2023.

The questionnaire content included two parts. The first part aimed at getting respondents' profile information; the second part was intended for surveying the impact of DM determinants on the company's business model innovation. The collected data were analyzed using SPSS 27 and the methods involved included descriptive statistics, reliability test, correlation, and multiple regression analysis. Then qualitative research in the form of semi-structured interview was performed to get a deeper understanding of how the higher level of DM can help the organizer to innovate the network component of its business model and how relationships within its network are reshaped on three levels: organizer-partners, organizer–exhibitor, and organizer-visitor under the impact of digital technologies. Semi-structured online interviews offering the opportunity to integrate open-ended and theoretically driven questions [56] were used. Four respondents who participated in the quantitative survey were interviewed with the main goal to get a deeper understanding of how the exhibition organizer perceives the prospects of its business model innovation under the impact of digital technologies. Two event managers and two top managers from SMEs who organize more than ten events per year including exhibitions abroad were interviewed (Table 32.4). Particularly, we asked the interviewees to discuss their experience of using digital platforms during the exhibition and the results of the quantitative study. Using digital platforms was one of the main selection criteria. Each interview lasted between 30 min and one and a half hours.

32.4 Findings

The description of companies' profile in terms of distribution by the number of events organized per year, company's size, and annual revenue is given in Table 32.5. Small and medium-sized companies (SMEs) prevail in the sample including micro-sized businesses (37%) with the number of employees less than 15 and annual revenue up to 120 million roubles (49%).

The results of quantitative research were processed via SPSS 27 software. Descriptive analysis provides an opportunity to reveal the level of agreeableness among respondents within each determinant. As shown in Table 32.6, the respondents' current agreeableness scored on average between 1.85 and 4.2. The top six

Table 32.5 Sample's features

Number of events per year	N	%	Employees number	N	%
Less than 5	13	17	Less than 15	37	49.4
From 5 to 10	35	46	From 16 to 100	38	50.6
From 10 to 25	24	32	From 101 to 250	0	0
More than 25	3	4			
Position of respondent	**N**	**%**	**Annual revenue, roubles**	**N**	**%**
CEO, general director	19	25	Up to 120 mln	49	65.4
Department head, event manager	39	52	From 120 to 800 mln	26	34.6
CTO, business development director	17	23	More than 800 mln	0	0

Source: authors own study

Table 32.6 Descriptive statistics results by items

Item ID	Mean	S.D.	Item ID	Mean	S.D.
BMI1	4.27	1.041	BMI2	3.00	1.166
AMT2	3.69	970	AMT3	2.85	1.287
AMT4	3.62	1.169	COL3	2.81	1.767
DLE2	3.62	852	BMI3	2.81	1.297
DLE3	3.50	1.208	DS1	2.73	1.343
DLE1	3.50	949	DLP1	2.69	1.490
AMT1	3.46	1.104	DS2	2.38	1.444
DP2	3.38	1.235	COL1	2.23	1.243
DP3	3.27	1.485	DS3	1.96	1.216
COL2	3.04	1.113	DLP2	1.96	1.248
DP1	3.04	1.248	DLP3	1.85	1.377
			COL3	2.81	1.767

Table 32.7 Cronbach's α coefficient for DM determinants

Determinants	ID	Cronbach's α
Business model innovation	BMI	0.739
Employees' digital literacy	DLE	0.895
Agile management technologies	AMT	0.791
Digital strategy	DS	0.855
Partners' digital literacy	DLP	0.796
Collaboration	COL	0.834
Using digital platforms	DP	0.913

items with highest agreeableness exceeding 3.5 belong to innovation opportunities—"our company needs to reinvent its business model," agile management technologies—"our company is willing or has already used agile management technologies to encourage business model innovation," and the level of employees' digital literacy—"the level of employees' digital literacy is high enough or no barrier on the way to business model innovation."

At this stage we also tested the reliability of all Likert-scale questions meant to evaluate each determinant (item) of DM. Cronbach's α coefficient was calculated to ensure the internal consistency of items within each dimension. All determinants show good values (>0.73) depicted in Table 32.7 which is above the minimum acceptable level of reliability (0.7) [57, 58].

Correlation analysis performed (Kendall rank correlation coefficient) provides us with information regarding the existence of correlation between dependent and independent variables. According to the analysis results each independent variable achieved significant highly positive correlations with the dependent variable when the significance level was 0.05, as shown in Table 32.8. High positive correlations were obtained between independent variables as well, which can be a predictor of multicollinearity effect.

To assess the multicollinearity a variance inflation factor (VIF) test was run while building the multiple regression model in SPSS. If a VIF value is greater than 5 it means the presence of collinearity between the independent variables [58]. During our study we obtained the VIF values of all the independent variables between 1.85 and 4.03. The study revealed no collinearity between the independent variables, and the data were suitable for multiple regression analysis. Table 32.9 describes the result of the regression analysis for DM determinants and their impact on the exhibition organizer's innovative opportunities in relation to its business model. The value of the adjusted R-square implies that the six independent variables (employees' and partners' level of digital literacy, usage of agile techniques and digital platforms, industry-academia collaboration, and developed digital strategy) explain 74.3% of the variance in the dependent variable—exhibition organizer's business model innovation—while the remaining 19.7% can be explained by other variables.

In relation to three determinants of DM in dynamic capabilities dimension only the relation between the level of employees' digital literacy and digital strategy and

Table 32.8 Correlation analysis of variables

	BMI	DS	AMT	DP	DLP	DLE	COL
BMI	1	.697**	.661**	.703**	.556**	.617**	.655*
DS	.697**	1	.646*	.699	.673**	.450*	.746**
AMT	.661*	.646*	1	.784**	.640*	.622**	.508**
DP	.703**	.699	.784**	1	.693**	.565**	.694**
DLP	.556**	.673**	.640*	.693**	1	.441	.761**
DLE	.617**	.450*	.622**	.565**	.441	1	.509**
COL	.655*	.746**	.508**	.694**	.761**	.509**	1

**Correlation is significant at 0.01 level (two-tailed)
*Correlation is significant at 0.05 level (two-tailed)

Table 32.9 Multiple regression analysis results

	Dynamic capabilities dimension			*Digital ecosystem dimension*		
Variables	DS	AMT	DLE	DLP	DP	COL
B	0.253	0.180	0.174	0.208	0.293	−0.122
SE B	0.205	0.302	0.236	0.270	0.235	0.210
Beta	0.290*	0.159	0.195*	0.211*	0.231*	−0.139*
t-value	1.239	0.597	0.736	0.770	1.246	−0.580
Sig.	0.009	0.458	0.002	0.001	0.003	0.011
VIF	2.961	3.822	3.758	4.031	1.858	3.081
Model summary						
R-square	0.792					
Adjusted R-square	0.743					
F	5.827					
Significance	0.001					

the organizer's innovative opportunities was found to be significant. It means the higher level of employees' digital literacy and the higher quality of digital strategy developed the higher level of the organizer's innovative opportunity by $b = 0.290$, $p = 0.009$, and $b = 0.195$, $p = 0.002$, respectively. Within the research framework of the study it means as well that digital literacy and digital strategy positively impact on the network component in the organizer's business model resulting in changing the organizer's relationships with an exhibitor, a visitor, and partners. This supports H5 and H6, where the well-developed digital strategy and adequate level of employees' digital literacy positively affect the exhibition organizer's innovative opportunities. The level of significance for the agile management determinant is too high indicating that H4 was not supported. The fact can be explained by the project type of exhibition organizers' activities and the implementation of agile technologies supposed to be a tone-of-voice in the industry.

From the digital ecosystem dimension there were three proposed variables and all of them were statistically significant. This implies that the increasing level of partners' digital literacy, intensification of collaboration, and usage of digital platforms will increase the probability of the organizer's business model innovation. However, the connection between collaboration and business model innovation is weak

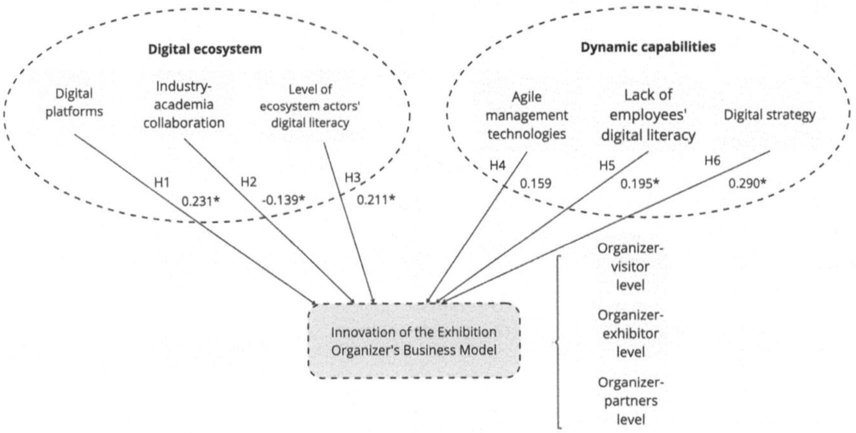

Fig. 32.3 The results of the study. *significance at $p < 0.01$; standardized coefficients are reported

(b = − 0.122, p = 0.011). The negative value suggests that increasing the level of collaboration will lead to deterioration of the organizer's innovative opportunities in relation to its business model. Therefore, H2 is not supported. The results of hypothesis testing are shown in Fig. 32.3. The strongest predictor of the exhibition organizer's business model innovation is digital strategy with b = 0.290 followed by the usage of digital platforms and digital literacy of both partners within the organizer's business ecosystem and employees. The results of the study are depicted in Fig. 32.3.

In the result of qualitative research we obtained insights regarding the linkage between the exhibition organizer's business model innovation and such DM determinants as digital platforms (Table 32.10). Despite the fact that unified digital infrastructure for the event organizer is a far-reaching goal, in most cases various digital platforms where the exhibition organizer and other stakeholders including its business partners could interact have the potential to enhance visitors' and exhibitors' experience resulting in reconfiguration of the whole business network of the event organizer through the inclusion of digital platform developers as a new ecosystem actor.

We also asked the interviewees to comment on the hypothesis regarding the positive effect of collaboration with academicians in the area of event industry and digital technologies with business model innovation as the hypothesis was not supported and the linkage turned out to be weak and negative. One of the possible explanations was given by interviewee D, representing the organizer of the largest agricultural exhibitions in Russia since 2002:

> […] I would say that one of our key activities as the organizer is a constant interaction everywhere with everyone. Of course we see a great potential in collaboration. We don't have our own exhibition facilities and have to rely on external resources when we deliver our events. Our company tries to be on the top of the industry trends but I would prefer business collaboration over collaboration with academicians. Maybe it's not so wise but that's our reality.

In terms of digital ecosystem dimensions and its impact on business model innovation, it makes sense to reconsider the notion of collaboration with an emphasis on

Table 32.10 Exhibition organizer's interview example quotations

Interviewee	Example quotations
A	Mobile apps can be a real salvation for the organizer but meanwhile, it's a double-edged sword. Yes, we could attract and reach three times more online participants than previous year and a number of views exceeded 200,000, we reduced the expenses on printed products but at the same time we paid for the developer's services, technical support and it also took additional efforts and skills from our employees. To assess the advantages of digital platforms usage further calculations are needed
B	We've tried a new mobile app from the external provider customized for our purpose. A set of application opportunities was enormous but we started with a few ones. Nevertheless, even with limited functionality we've got extremely positive feedback from our participants and customers. Invest projects competition took place among the exhibition booths, with the help of the digital platform we as the organizer could enhance activities and ease the navigation for visitors
C	Today, event apps are essential for top-notch exhibitions, serving not just for exposition but as a hallmark of the organizer's business acumen in the industry. The app facilitates diverse event activities through features like meeting scheduling, push notifications, gamification, and feedback gathering. Although primarily for visitors and exhibitors, we envision the app becoming a complete digital platform for all partners
D	Our company embraces advantages we can get from various digital technologies and additional value we can create for our participants but I don't get my hopes up. We've been working in the exhibition sector for more than 20 years and work with different venue providers. Must say that even the usual internet connection and other technical aspects of the venue is always a challenge for the organizer. You can be on the edge of all technological trends but when your partners can't support your efforts…it frustrates you a lot

partnership with software developers and providers. In the last section of the interview, the respondents were asked to discuss the matter of changes in the organizer's relationships with visitor, exhibitor, and partners in the result of innovation of the network component caused by digitalization.

32.5 Discussion and Implication

In assessing the influence of digital technologies on the event organizer's business model innovation, two main DM dimensions—digital ecosystem and dynamic capabilities—divided into six hypotheses were analyzed. Previous studies confirmed that dynamic capabilities are related to business model innovation and stressed the importance of dynamic capability as a source of competitive advantage [7, 33, 52]. The findings of our study have indicated that developed digital strategy as one of the aspects of dynamic capabilities is the strongest predictor of the organizer's business model innovation with the significance level equal to 0.015. The digital strategy encourages exhibition organizers to find new ways to renew their business model. Digital strategy of the exhibition organizer focuses on using various digital channels and technologies to promote their events. This includes leveraging online platforms, technologies, and tools to reach a broader audience, increase event visibility, increase ticket sales, and engage with potential attendees. H5 regarding digital literacy of employees was supported as well and demonstrated the high level of significance. Most respondents pointed out an adequate level of their employees' digital skills and literacy that support their innovative efforts. It implies exhibition organizers will create new competitive advantages with technology.

The findings proved that determinants from another DM dimension, digital ecosystem, were suitable to consider the organizer's business model innovation through their lens but one predictor of business model innovation demonstrated the weak linkage. The hypothesis regarding the positive effect of collaboration with academicians on the organizer's business model innovation was not supported. Respondents explained that due to the rapid changes in the business environment they did not see the point of developing industry–academia collaboration and preferred to invest in developed tools instead of theoretical research (Interview/Exhibition organizer D).

The hypothesis that the usage of digital platforms positively affects business model innovation was supported. But the findings of semi-structured interviews revealed that the notion of digital platforms was misinterpreted or narrowed down by respondents. Most respondents described their experience of usage of special-purpose mobile applications but not the full-on digital platforms. It is noteworthy to discuss the results in light of previous studies. Interviewees expressed the concern regarding ambiguity of mobile apps implementation during the event, which is in line with studies [18, 22, 59, 60]. Multiple case studies conducted revealed that through digital platforms and tools, firms can establish clear channels for communication and collaboration, ensuring that innovation initiatives are in line with the strategic direction of the organization [61]. But assessing an impact of special-purpose mobile apps on customer experience is important but the exhibition organizer is operating in a business ecosystem with multiple stakeholders where introducing any digital technology or innovation leads to reshaping the whole network's configuration and requires not only financial investment but a new mindset from all stakeholders [39]. It implies that efforts of the event organizer are not always enough to obtain competitive advantages.

The study has several theoretical and managerial implications. In terms of methodology the study proposes a conceptual framework for studying the event organizer's business model through the ecosystem lens which factors in specific features of the exhibition sector and fills the gap as research has paid insufficient attention to digital transformation phenomenon in relation to business model innovation in the event industry and even less from the exhibition organizer's perspective. The study reveals two major determinants of digital maturity—dynamic capabilities and digital ecosystem—that can be used to measure the impact of digital technologies on the network component in the exhibition organizer's business model.

The study considers the connection between DM determinants and business model innovation as well. For each determinant the study describes and discusses relevant variables that could be used by managers and exhibition organizers to get in-depth understanding how they can innovate their business model through the reshaping and partner network's reconfiguration. The first one is the usage of digital platforms. Digital platforms help the exhibition organizer to enhance activities and ease the navigation for visitors, collect useful data about visitors' preferences and behavior, and reduce expenses on printed products. The second ones are digital strategy and digital literacy of ecosystem actors. When innovating a business model, especially the network component where a lot of actors are involved, strategists and managers need to understand the importance of digital literacy among not only their own employees but all ecosystem actors as well. The level of digital literacy contributes a lot to the innovative opportunities of the exhibition organizer.

32.6 Conclusion and Limitation

Based on the study results it can be concluded that determinants of DM as an antecedent of digital transformation are associated with the exhibition organizer's business model innovation. The most relevant determinants are digital literacy of employees and partners and digital strategy, while the importance of collaboration with academics and implementing agile management technologies was not confirmed. The findings of semi-structured interviews revealed that the notion of digital platforms was often narrowed down by respondents. Meanwhile, even the usage of special-purpose applications as an example of digital platforms significantly changes the network component of the organizer's business model as it leads to the appearance of a new actor in the event ecosystem. Along with the other DM determinants the usage of digital platforms gives the exhibition organizer new competitive advantages and reshapes its relationships with all stakeholders. On the organizer-visitor level DM of the exhibition organizer implying an active usage of digital technologies can significantly enhance visitors' offline and online experience, allow the organizer to reach a broader audience, and collect more data about this group of stakeholders. In terms of organizer-exhibitor relationships digital platforms demonstrate a great potential. On this level digital technologies and additional digital channels enable the organizer to deliver exhibitors more valuable services, for instance, meeting appointment system, contract conclusion, booths' activities, and others. Any changes and innovations in the network component of the exhibition organizer's business model on the organizer-partners level are the most difficult issue. The organizer interacts with multiple suppliers, providers, and temporary services in order to organize and manage the exhibition. Thus, one new actor in the network component, for instance, a mobile application developer providing electronic navigation tools, can lead to deintensification ties with partners from the traditional printing industry. The structure of the chapter is as follows. The second section provides an overview of the existing literature.

This study has a few limitations that can drive future research. First, the analysis in this study was based on 75 valid samples. A larger sample size is suggested to further research for better representation of findings. Second, we collected samples only from exhibition organizers who do not have their own exhibition facilities though exhibition centers often conduct exhibitions acting both as the organizer and as the provider adding complexity to the research of the exhibition organizer's business ecosystem [39]. A set of DM determinants is suggested to be corrected in light of the results. Particularly, such determinants as "collaboration" should be reconsidered. In terms of semi-structured interviews, there are some perspectives as well.

References

1. Research and Market Report: Exhibition Organizing – Global Strategic Business Report [online]. https://www.researchandmarkets.com/reports/5139493/exhibition-organizing-global-strategic#cat-pos-3 (2023). Accessed 12 Feb 2024
2. Henfridsson, O., Mathiassen, L., Svahn, F.: Managing technological change in the digital age: the role of architectural frames. J. Inf. Technol. **29**(1), 27–43 (2014). https://doi.org/10.1057/jit.2013.30
3. Chesbrough, H.: Business model innovation: opportunities and barriers. Long Range Plan. **43**(2/3), 354–363 (2010)
4. Kaplan, S.: The Business Model Innovation Factory: How to Stay Relevant When the World Is Changing. John Wiley & Sons (2012)
5. Fitzgerald, M., Kruschwitz, N., Bonnet, D., Welch, M.: Embracing digital technology: a new strategic imperative. MITSloan Manag. Rev. **55**, 1–12 (2013). https://doi.org/10.1057/palgrave.ejis.3000650
6. Gassmann, O., Frankenberger, K., Csik, M.: The St. Gallen business model navigator TM. Int. J. Prod. Dev. **18**(3), 249–273 (2013)
7. Teece, D.: Business models and dynamic capabilities. Long Range Plan. **51**(1), 40–49 (2017). https://doi.org/10.1016/j.lrp.2017.06.007
8. Colli, M., Berger, U., Bockholt, M., Madsen, O., Møller, C., Wæhrens, B.V.: A maturity assessment approach for conceiving context-specific roadmaps in the Industry 4.0 era. Annu. Rev. Control. **48**, 165–177 (2019)
9. Remane, G., Hanelt, A., Nickerson, R.C., Kolbe, L.M.: Discovering digital business models in traditional industries. J. Bus. Strateg. **38**(2), 41–51 (2017). https://doi.org/10.1108/JBS-10-2016-0127
10. Teichert, R.: Digital transformation maturity: a systematic review of literature. Acta Univ. Agric. Silvic. Mendel. Brun., Mendel University Press. **67**(6), 1673–1687 (2019)
11. Tutak, M., Brodny, J.: Business digital maturity in Europe and its implication for open innovation. J. Open Innov. Technol. Mark. Complex. **8**(1), 27 (2022). https://doi.org/10.3390/joitmc8010027
12. UFI Report: The Global Exhibition Barometer [pdf]. https://www.ufi.org/wp-content/uploads/2023/01/30th_UFI_Global_Barometer_january_2023.pdf (2023). Accessed 23 Jan 2024
13. Bocken, N.M.P., Short, S.W., Rana, P., Evans, S.: A literature and practice review to develop sustainable business model archetypes. J. Clean. Prod. **65**, 42–56 (2014). https://doi.org/10.1016/j.jclepro.2013.11.039
14. Chesbrough, H.: Business model innovation: it's not just about technology anymore. Strateg. Leadersh. **35**(6), 12–17 (2007). https://doi.org/10.1108/10878570710833714
15. Foss, N.J., Saebi, T.: Fifteen years of research on business model innovation: how far have we come, and where should we go? J. Manag. **43**(1), 200–227 (2017)
16. Teece, D.J.: Business models, business strategy and innovation. Long Range Plan. **43**, 172–194 (2010). https://doi.org/10.1016/j.lrp.2009.07.003
17. Zott, C., Amit, R.: Business model design: an activity system perspective. Long Range Plan. **43**(2–3), 216–226 (2010). https://doi.org/10.1016/j.lrp.2009.07.004

18. Hradecky, D., Kennell, J.: Organizational readiness to adopt artificial intelligence in the exhibition sector in Western Europe. Int. J. Inf. Manag. **65**(C), 102497 (2022). https://doi.org/10.1016/j.ijinfomgt.2022.102497
19. Davidson, R.: Business Events. Routledge (2018)
20. Getz, D., Page, S.J.: Event Studies: Theory, Research and Policy for Planned Events. Routledge (2019)
21. Laing, J.: Festival and event tourism research: Current and future perspective. Tourism Management Perspectives. **25**, 165–168 (2018)
22. Neuhofer, B., Magnus, B., Celuch, K.: The impact of artificial intelligence on event experiences: a scenario technique approach. Electron. Mark. **31**(3), 601–617 (2020)
23. Caputo, A., Marzi, G., Pellegrini, M.M.: The internet of things in manufacturing innovation processes: development and application of a conceptual framework. Bus. Process. Manag. J. **22**(2), 383–402 (2016). https://doi.org/10.1108/BPMJ-05-2015-0072
24. Karimi, J., Walter, Z.: Corporate entrepreneurship, disruptive business model innovation adoption, and its performance: the case of the newspaper industry. Long Range Plan. **49**(3), 342–360 (2016)
25. Spieth, P., Schneckenberg, D., Matzler, K.: Exploring the linkage between business model (&) innovation and the strategy of the firm. R&D Manag. **46**(3), 403–413 (2016)
26. Wirtz, B., Pistoia, A., Ullrich, S., Göttel, V.: Business models: origin, development and future research perspectives. Long Range Plan. **49**(1), 36–54 (2016). https://doi.org/10.1016/j.lrp.2015.04.001
27. Sklyar, A., Kowalkowski, C., Tronvoll, B., Sörhammar, D.: Organizing for digital servitization: a service ecosystem perspective. J. Bus. Res. **104**, 450–460 (2019)
28. Baber, W.W., Ojala, A., Martinez, R.: Effectuation logic in digital business model transformation: insights from Japanese high-tech innovators. J. Small Bus. Enterp. Dev. **26**(6/7), 811–830 (2019). https://doi.org/10.1108/JSBED-04-2019-0139
29. Russo-Spena, T., Tregua, M., D'Auria, A., Bifulco, F.: A digital business model: an illustrated framework from the cultural heritage business. Int. J. Entrep. Behav Res. **28**(8), 2000–2023 (2022). https://doi.org/10.1108/IJEBR-01-2021-0088
30. Pizzi, S., Leopizzi, R., Caputo, A.: The enablers in the relationship between entrepreneurial ecosystems and the circular economy: the case of circularity.com. Manag. Environ. Qual. **33**(1), 26–43 (2022). https://doi.org/10.1108/MEQ-01-2021-0011
31. Mahboub, H., Sadok, H.: Implementing enterprise digital transformation: a contribution to conceptual framework design. Nankai Bus. Rev. Int. **14**(1), 35–50 (2023). https://doi.org/10.1108/NBRI-06-2022-0067
32. Wang, H., Feng, J., Zhang, H., Li, X.: The effect of digital transformation strategy on performance: the moderating role of cognitive conflict. Int. J. Confl. Manag. **31**(3), 441–462 (2020). https://doi.org/10.1108/IJCMA-09-2019-0166
33. De Aro, E.R., Perez, G.: Identification of dynamic capabilities in open innovation. Innov. Manag. Rev. **18**(2), 118–128 (2021). https://doi.org/10.1108/INMR-10-2019-0120
34. Matarazzo, M., Penco, L., Profumo, G., Quaglia, R.: Digital transformation and customer value creation in made in Italy SMEs: a dynamic capabilities perspective. J. Bus. Res. **123**, 642–656 (2021). https://doi.org/10.1016/j.jbusres.2020.10.033
35. Oliveira-Dias, D., Kneipp, J.M., Bichueti, R.S., Gomes, C.M.: Fostering business model innovation for sustainability: a dynamic capabilities perspective. Manag. Decis. **60**(13), 105–129 (2022)
36. Sarmento, M., Simões, C.: Trade fairs as engagement platforms: the interplay between physical and virtual touch points. Eur. J. Mark. **53**(9), 1782–1807 (2019). https://doi.org/10.1108/EJM-10-2017-0791
37. Simonov, K.V.: Innovative business model of exhibition company: modeling methodology testing. J. Mod. Compet. **14**(3), 106–123 (2020)
38. Tafesse, W., Skallerud, K.: Towards an exchange view of trade fairs. J. Bus. Ind. Mark. **30**(7), 795–804 (2015)

39. Bauer, T., Borodako, K.: Trade show innovations — organizers implementation of the new service development process. J. Hosp. Tour. Manag. **41**, 197–207 (2019)
40. Jin, X., Weber, K.: Developing and testing a model of exhibition brand pre- ference: the exhibitors' perspective. Tour. Manag. **38**, 94–104 (2013). https://doi.org/10.1016/j.tourman.2013.02.018
41. Hess, T., Matt, C., Benlian, A., Wiesböck, F.: Options for formulating a digital transformation strategy. MIS Q. Exec. **15**(2), 123–139 (2016)
42. Sousa-Zomer, T.T., Neely, A., Martinez, V.: Digital transforming capability and performance: a microfoundational perspective. Int. J. Oper. Prod. Manag. **40**(7/8), 1095–1128 (2020)
43. Nasiri, M., Saunila, M., Ukko, J.: Digital orientation, digital maturity, and digital intensity: determinants of financial success in digital transformation settings. Int. J. Oper. Prod. Manag. **42**(13), 274–298 (2022). https://doi.org/10.1108/IJOPM-09-2021-0616
44. Da Costa, L.S., Munhoz, I.P., Pereira, L., Akkari, A.: C: assessing the DM of micro and small enterprises: a focus on an emerging market. Procedia Comput. Sci. **200**, 175–184 (2022)
45. Schumacher, A., Erol, S., Sihn, W.A.: Maturity model for assessing Industry 4.0 readiness and maturity of manufacturing enterprises. Procedia CIRP. **52**, 161–166 (2016)
46. Pirola, F., Cimini, C., Pinto, R.: Digital readiness assessment of Italian SMEs: case-study research. J. Manuf. Technol. Manag. **31**(5), 1045–1083 (2020)
47. Schwer, K., Hitz, C., Wyss, R., Wirz, D., Mironne, C.: Digital maturity variables and their impact on the enterprise architecture layers. Probl. Perspect. Manag. **16**(4), 141–154 (2018)
48. Murphy, M.K., Black, N.A., Lamping, D.L., McKee, C.M., Sanderson, C.F.B., Askham, J., Marteau, T.: Consensus development methods, and their use in clinical guideline development. Health Technol. Assess. **2**(3), i-88 (1998)
49. Ellström, D., Holtström, J., Berg, E., Josefsson, C.: Dynamic capabilities for digital transformation. J. Strateg. Manag. **15**(2), 272–286 (2022)
50. Marx, C., de Danielly, P., Falk, U.: Dynamic capabilities & digital transformation: a quantitative study on how to gain a competitive advantage in the digital age. In ECIS 2021 Research Papers, 58 (2021)
51. Weritz, P., Braojos, J., Matute, J.: Exploring the antecedents of digital transformation: dynamic capabilities and digital culture aspects to achieve digital maturity. In AMCIS 2020 Proceedings, 22, (2020)
52. Sahebalzamani, S., Jørgensen, E.J.B., Bertella, G., Nilsen, E.R.: A dynamic capabilities approach to business model innovation in times of crisis. Tour. Plan. Dev. **31**, 1308–1333 (2022)
53. Bartunek, J.M., Bobko, P., Venkatraman, N.: Toward innovation and diversity in management research methods. Acad. Manag. J. **36**(6), 1362–1373 (1993)
54. Creswell, J.W., Plano Clark, V.L.: Designing and Conducting Mixed Methods Research, 3rd edn. Sage, Thousand Oaks, CA (2018)
55. De Bruin, T., Freeze, R., Kulkarni, U., Rosemann, M.: Understanding the main phases of developing a maturity assessment model. In Australasian Conference on Information Systems (2005)
56. Galletta, A.: The semi-structured interview as a repertoire of possibilities. In: Mastering the Semi-Structured Interview and Beyond: From Research Design to Analysis and Publication. New York University Press, New York (2013). https://doi.org/10.18574/nyu/9780814732939.003.0002
57. Nunnally, J.C.: Psychometric theory' 25 years ago and now. Educ. Res. **4**(10), 7–21 (1975)
58. Hair, J., Black, W., Babin, B., Anderson, R., Tatham, R.: Multivariate Data Analysis, 6th edn. Pearson Prentice Hall, Upper Saddle River, NJ (2006)

59. Talantis, S., Shin, Y.H., Severt, K.: Conference mobile application: participant acceptance and the correlation with overall event satisfaction utilizing the technology acceptance model (TAM). J. Conv. Event Tour. **21**(2), 100–122 (2020)
60. Lau, C.K.H., Chui, C.F.R., Au, N.: Examination of the adoption of augmented reality: a VAM approach. Asia Pac. J. Tour. Res. **24**(10), 1005–1020 (2019)
61. Priyono, A., Hidayat, A.: Fostering innovation through learning from digital business ecosystem: a dynamic capability perspective. J. Open Innov.: Technol. Mark. Complex. **10**(1), 100196 (2024). https://doi.org/10.1016/j.joitmc.2023.100196

Chapter 33
Navigating Efficiency: A Comprehensive Review and Strategies for Advancing Truck Appointment Systems in Seaports

Daniela Tocchi, Fabio Danilo Cristiano, Valentin Carlan, and Vittorio Marzano

Abstract The increasing speed to respond to trade demand and the complexity of supply chains operations often lead to inefficiencies in seaports. One of the most critical issues is truck congestion. This has been addressed in the literature from different perspectives. A common proposed approach to reduce congestion in ports is implementing a Truck Appointment System (TAS), a tool aiming at coordinating the truck arrival scheduling and operations with the cargo handling in port terminals. This paper aims at analysing potential implementation strategies to develop an efficient TAS model, based on a twofold perspective, combining literature review findings and logistic stakeholders' needs. Insights drawn from practical implementations of TAS have been gathered through interviews to logistic operators involved in cargo handling in two European ports (i.e. Antwerp-Brugge and Salerno). Also, the approach proposed in this study enables to define a tailored TAS solution for the port of Salerno, offering a generalized yet highly adaptable stepwise implementation strategy.

Keywords Truck appointment system · Maritime container terminal · Port efficiency · Port optimization

D. Tocchi (✉) · F. D. Cristiano · V. Marzano
Department of Civil, Architectural and Environmental Engineering, University of Naples Federico II, Naples, Italy
e-mail: daniela.tocchi@unina.it

V. Carlan
Department of Transport and Regional Economics, University of Antwerp, Antwerp, Belgium

F. Schiavone et al. (eds.), *Advanced Perspectives and Trends in Digital Transformation of Firms, Networks, and Society*, Springer Proceedings in Business and Economics, https://doi.org/10.1007/978-3-031-80692-6_33

33.1 Introduction and Objective

In the ever-evolving landscape of global trade, seaports stand as pivotal nodes connecting economies and facilitating the movement of goods across continents. The burgeoning complexity of global supply chains and the increasing trade demand placed on port operations often lead to underused capacity and bottlenecks. Regular causes such as truck congestion at terminals and in the port result in unforeseen costs for logistics operators. In response to these issues, the implementation of new technologies in applications such as Truck Appointment Systems (TAS) has garnered attention as a promising solution to streamline truck movements, enhance operational efficiency (better use of available capacities), and mitigate congestion within port facilities.

The TAS is a digital platform managed by terminals or port authorities that allows road transport operators to book an appointment for pick-up and drop-off operations of cargo. The adoption of TAS presents a paradigm shift in port logistics, revolutionizing the traditional model of truck handling and scheduling. By allowing for the systematic allocation of arrival slots, TAS not only optimizes resource utilization but also empowers stakeholders with greater predictability and transparency in terminal operations. This type of platform opens the possibility to serve as a single booking point for terminals active in the area of the same port's ecosystem. Hence, the successful deployment of TAS is not devoid of hurdles.

This paper endeavours to analyse these challenges by critically examining barriers and implementation strategies that can overcome obstacles and maximize the efficiency of Truck Appointment Systems in two different port environments. To this purpose, this study adopts a twofold perspective analysis to include both technological advancement and implementation strategies presented in the literature so far and the stakeholders' needs. Thus, first, an exhaustive review of existing literature, combined with insights drawn from practical implementations is presented. Also, results from interviews with representatives of logistic stakeholders involved in cargo handling in two European ports (i.e. Antwerp-Brugge and Salerno) are analysed.

Furthermore, this study seeks to clarify the multiple dimensions of TAS, offering a holistic understanding of its impact, challenges, and potential, paving the way for the development of robust implementation strategies that can overcome obstacles and maximize the efficacy of terminal operations.

Finally, the proposed approach enables to define a tailored TAS solution for the port of Salerno, through the amalgamation of stakeholder demands with the emergent capabilities of ICT, offering a generalized yet highly adaptable stepwise implementation strategy.

33.2 Literature Review

The first step of the proposed analysis is based on an in-depth review of multifaceted dimensions of the Truck Appoint System, synthesizing existing research to provide insights into its evolution, impact, challenges, and future directions within the realm of terminal and/or port management.

During the last years, a growing number of studies paid attention to solving gate congestion problems at seaport terminals by managing truck arrivals (and departures). To manage this issue, one of the possible solutions is the adoption of a Truck Appointment System (TAS). As briefly recalled in the introduction, the Truck Appointment System (also called Truck Guidance System or Vehicle Booking System) is an information system—usually managed by terminals—which allows truckers (or road transport operating companies) to book their arrival at the terminal in a specific time slot of the gate opening time windows [1]. In some cases, this system can be mandatory or optional. In the optional case, the terminal allows truck drivers to arrive at the terminal without a booking but does not guarantee to handle the cargo within a predetermined service level, which is usually guaranteed for appointments booked in advance.

The genesis of TAS can be traced back to the late twentieth century, with initial implementations primarily focused on optimizing container terminal operations. Indeed, the literature review shows that only one study dealing with the implementation of truck appointment systems for bulk cargo has been published recently (e.g. [2].).

Early studies on Truck Appointment System highlight the potential benefits of TAS in terms of reducing truck turn times, minimizing queuing delays, and improving overall terminal productivity [3–7]. Also, research proposed by authors like [8–10] further emphasizes the significance of adopting a truck appointment system to tackle the issues of truck arrival unpredictability and traffic congestion affecting seaport container terminal. Moreover, other studies dealt with the impact of adopting the truck appointment system on the scheduling of trucks providing container pickup and delivery service (drayage) to a port ([11, 12]). Furthermore, some authors investigated the potential benefit of adopting the TAS in mitigating port-related environmental impacts through reducing emissions ([13–15]).

Usually, in conventional truck appointment systems, the terminal operator sets in advance the maximum number of trucks that can be served during each time window and road transport operators book appointments for their trucks in a defined time slot. However, in this framework, when the maximum value of trucks that can be served in a certain time window is reached, additional appointments for truck arrivals are rejected. Thus, this strategy solves the problem only from the terminal operator's perspective, causing several issues for truck companies in satisfying the delivery requirements of containers. To deal with this twofold perspective, [16] proposed a new truck appointment system based on a negotiation process involving both container terminals and trucking companies.

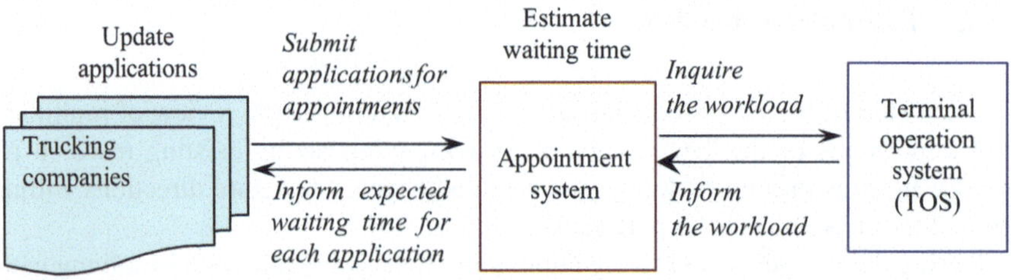

Fig. 33.1 Proposed truck appointment system by Phan and Kim, K.H. (2015)

In detail, as illustrated in Fig. 33.1, when a trucking company submits an appointment request for a preferred delivery time to the system, the TAS estimates the queue length for that time window, accounting for terminal workload (through a TOS) and informs the trucking company on expected waiting time for the required appointment. Then, based on waiting costs, truck availability, and rescheduling costs, the company may decide to adjust its proposal and submits a new appointment application. This process repeats until all appointments are confirmed.

The idea that a collaborative approach involving the different stakeholders operating in the port activities and, more in general, in the maritime supply chains has been identified as a key innovation to reach competitive advantages. [17] investigated a new approach in managing seaport-hinterland access infrastructure. The authors proposed an innovative architectural design for an advanced Truck Guidance System (TGS), crafted through the integration of stakeholder needs with the emergent capabilities of ICT, offering a generalized yet highly adaptable stepwise implementation strategy. To this purpose interviews with logistic stakeholders in the port of Antwerp have been carried out, providing an overview of current data managing solutions related to trucks' movements and challenges faced by port users working with this data. Main findings of this research suggest potential savings and performance improvements for road transport operators, along with enhanced planning, infrastructure management, and client services for other supply chain actors. Also, port administrations can drive the adoption of a truck guidance (appointment) system fostering efficiency across logistics.

33.3 Research Directions

The review of the literature paints a compelling picture of the potential of the implementation of the TAS within port ecosystems and a need for research in this area. The potential findings have profound implications for port authorities, terminal operators, trucking companies, and policymakers since the adoption of TAS enables facilitating data-driven decision-making and enhancing operational resilience. Also, the review emphasizes the need for research exploring emerging technologies like artificial intelligence (AI) and the Internet of Things (IoT) to enhance TAS effectiveness (to anticipate the need of making a new booking and derive the exact date and time from historical data, and then book an appointment). Comparative analyses spanning multiple regions could offer insights into TAS adoption's contextual factors and impact on port efficiency and sustainability.

In this respect, this research adopted the approach proposed by [17] to the port of Salerno, involving the stakeholder in the study of ad hoc solutions to manage trucks' movements within the terminal and the port area. Thus, first, an As-Is analysis will be conducted in the port of Salerno in order to identify the technological improvements needed for the proper implementation of the TAS. To account also for stakeholders' needs, the analysis will be extended to logistics operators involved in the activities of the Salerno port (e.g. port authority, trucking companies, and terminal operators). In detail, firstly the market analysis will be performed via face-to-face interviews to the operators involved in the container shipping sector. The surveys will address several features, namely general informal about the company (e.g. main activities, annual turnover, business units, no. of employees), the ICT tools currently adopted, information about data used in daily operations (e.g. type of data, sources, data prices), willingness to adopt TAS solution. Finally, a comparative analysis of the outcomes derived for both ports (port of Antwerp-Brugge and port of Salerno) will be performed to investigate potential implementation strategies of the TAS and the associated stakeholder benefits. Evidence-based insights derived from both theoretical literature and practical empirical data will be provided, bridging the gap between theory and practice to facilitate informed decision-making in port operations.

Also, this approach enables the definition of TAS implementation strategies tailored to the specific needs and challenges of the port of Salerno. In this respect, the impact of the development of the TAS on the road city network will be taken into account in this analysis. Indeed, port freight flows can generate several issues in terminal surroundings (e.g. congestion and pollutant emissions). The implementation of a TAS can help in better managing the flow of trucks to the port (e.g. lowering the number of empty-truck trips), thereby reducing congestion on the external roads, improving safety, and allowing truck drivers and logistics companies to plan their routes and schedules more effectively.

This paper recognizes the dynamic nature of port operations and the need for adaptable solutions that can evolve in tandem with changing demands and technological advancements. Through a forward-looking lens, it outlines a strategic roadmap for the future advancement of TAS, emphasizing the integration of emerging technologies, collaborative partnerships, and data-driven insights to propel port efficiency to new heights.

In conclusion, from a wider perspective, this paper intends to serve as a beacon guiding seaport authorities, policymakers, and industry stakeholders towards a future where Truck Appointment Systems stand as cornerstones of operational excellence, fostering sustainable growth and resilience in maritime logistics.

Acknowledgements This research has been supported by the European Union under the PNRR of NextGenerationEU, National Sustainable Mobility Center (CN00000023, MUR Decree n. 1033—17/06/2022, Spoke 10).

References

1. Caballini, C., et al.: A combined data mining–optimization approach to manage trucks operations in container terminals with the use of a TAS: application to an Italian and a Mexican port. Transp. Res. Part E Logist. Transp. Rev. **142**, 102054 (2020)
2. Neagoe, M., et al.: Exploring congestion impact beyond the bulk cargo terminal gate. In: Logistics 4.0 and Sustainable Supply Chain Management: Innovative Solutions for Logistics and Sustainable Supply Chain Management in the Context of Industry 4.0. Proceedings of the Hamburg International Conference of Logistics (HICL), vol. 26, pp. 61–80. epubli GmbH, Berlin (2018)
3. Guan, C., Liu, R.: Container terminal gate appointment system optimization. Marit. Econ. Logist. **11**, 378–398 (2009)
4. Huynh, N.: Reducing truck turn times at marine terminals with appointment scheduling. Transp. Res. Rec. **2100**(1), 47–57 (2009)
5. Zhang, R., Yun, W.Y., Kopfer, H.: Heuristic-based truck scheduling for inland container transportation. OR Spectr. **32**, 787–808 (2010)
6. Chen, G., Yang, Z.: Optimizing time windows for managing export container arrivals at Chinese container terminals. Marit. Econ. Logist. **12**, 111–126 (2010)
7. Zhao, W., Goodchild, A.V.: Using the truck appointment system to improve yard efficiency in container terminals. Marit. Econ. Logist. **15**, 101–119 (2013)
8. Giuliano, G., et al.: Evaluation of the terminal gate appointment system at the Los Angeles/ Long Beach ports. METRANS Project, 04–06 (2008)
9. Huynh, N., Walton, C.M.: Robust scheduling of truck arrivals at marine container terminals. J. Transp. Eng. **134**(8), 347–353 (2008)
10. Maguire, A., et al.: Relieving congestion at intermodal marine container terminals: review of tactical/operational strategies (2010)
11. Namboothiri, R., Erera, A.L.: Planning local container drayage operations given a port access appointment system. Transp. Res. Part E Logist. Transp. Rev. **44**(2), 185–202 (2008)
12. Shiri, S., Huynh, N.: Optimization of drayage operations with time- window constraints. Int. J. Prod. Econ. **176**, 7–20 (2016)
13. Schulte, F., Gonzalez, R.G., VOß, S.: Reducing port-related truck emissions: coordinated truck appointments to reduce empty truck trips. In: Computational Logistics: 6th International Conference, ICCL 2015, Delft, The Netherlands, September 23–25, 2015, Proceedings 6, pp. 495–509. Springer International Publishing (2015)
14. Schulte, F., et al.: Reducing port-related empty truck emissions: a mathematical approach for truck appointments with collaboration. Transp. Res. Part E Logist. Transp. Rev. **105**, 195–212 (2017)
15. Wasesa, M., et al.: Impact of overbooking reservation mechanism on container terminal's operational performance and greenhouse gas emissions. Asian J. Shipp. Logist. **37**(2), 140–148 (2021)
16. Phan, M.-H., Kim, K.H.: Negotiating truck arrival times among trucking companies and a container terminal. Transp. Res. Part E Logist. Transp. Rev. **75**, 132–144 (2015)
17. Carlan, V., et al.: Toward implementing a fully automated truck guidance system at a seaport: identifying the roles, costs and benefits of logistics stakeholders. J. Ship. Trade. **4**(1), 12 (2019)

Chapter 34
Strategic Monitoring in Port Authority Management: Implementing an Original Digital Platform

Marcello Risitano, Mariarosalba Angrisani, Giuseppe La Ragione, and Marco Ferretti

Abstract Strategic monitoring is essential for enhancing the efficiency, security, and sustainability of port authority operations. Traditional monitoring approaches often lack integration and real-time analysis, limiting their effectiveness in decision-making. This paper presents an innovative, data-driven Decision Support System (DSS) designed to optimise strategic monitoring in port authority management. The study builds upon existing literature and insights from the EU-funded PortForward project, emphasizing the increasing role of digitalization in port governance. Through a systematic literature review, key challenges and opportunities in port performance monitoring are identified, leading to the development of a multidimensional dashboard. The proposed DSS framework incorporates four performance domains: port authorities and concessionaires, logistics and operations, market trends, and green challenges, each assessed through tailored Key Performance Indicators (KPIs). By integrating digital technologies such as IoT and cloud computing, the dashboard enables port authorities to monitor concessionaire activities effectively while aligning with sustainability objectives and European policy directives. The findings highlight the value of multidimensional performance assessment tools in fostering port competitiveness and stakeholder engagement. This study provides managerial and policy implications for port authorities transitioning toward efficiency-driven and environmentally responsible operational models.

Keywords Port authority · Performance · Decision support system

---

M. Risitano · M. Angrisani (✉) · G. La Ragione · M. Ferretti
Department of Management and Quantitative Studies (DISAQ), Parthenope University of Naples, Naples, Italy
e-mail: mariarosalba.angrisani@uniparthenope.it

F. Schiavone et al. (eds.), *Advanced Perspectives and Trends in Digital Transformation of Firms, Networks, and Society*, Springer Proceedings in Business and Economics, https://doi.org/10.1007/978-3-031-80692-6_34

34.1 Introduction

In the context of sea-land transportation and logistics, strategic monitoring represents a managerial function able to support operations within port authorities, thus ensuring efficiency, security, and environmental sustainability. However, traditional approaches to strategic monitoring often lack integration and real-time analysis capabilities, hindering effective decision-making. Against this backdrop, research and projects developed in the last decade have aimed at conceiving and implementing systemic dashboards to evaluate port performance. In particular, the EU project Port Forward (https://www.portforward-project.eu/) embodies an international consortium which seeks to establish an integrated knowledge base of port performance at a meso-level perspective, through a holistic approach to a smarter, greener, and more sustainable port ecosystem of innovative EU ports. Given the listed premises, our study proposes an innovative data-driven Decision Support System (DSS) tailored for strategic monitoring in port authority management. We first present an overview of port authority management and its challenges [1, 2], underlying the relevance of digitalisation processes in implementing viable solutions to face the requirements of an increasingly technological advanced sector such as the maritime transport and its interactions with port authorities [3]. Starting from the definition of strategic monitoring in optimising port operations and decision-making [4–6], our study aims to apply the main elements of a Decision Support System (DSS) to approach a novel and data-driven monitoring framework able to support the endeavours of the port management in monitoring the performance of the business actors operating in ports in a strategic and integrated way.

34.2 Objective

The research described in the present paper primarily investigates the main features and patterns of the strategic monitoring processes performed by Port Authorities (PAs) in order to assess the performance and effectiveness of the activities of the maritime concessionaires operating within them. Subsequently, the study sets grounds for the adoption of an innovative Decision Support System (DSS) built upon a multidimensional performance assessment dashboard [2]. More specifically, the tool is designed to comply with and enhance the managerialisation trend experienced by Port Authorities [5]. Such a tendency underscores the significance of strategic monitoring in port operations as a pivotal factor for achieving sustainable competitive advantage in the logistics and maritime industries [7]. Empirical evidence suggests that many ports traditionally utilise business metrics to assess critical performance aspects such as port efficiency (e.g., traffic volume throughput, container throughput per hectare) and port effectiveness (e.g., port choice, service delivery effectiveness). Additionally, financial indices such as return on investment and capital structure are commonly considered.

Notwithstanding the intensely competitive maritime environment, only a scarce number of ports are beginning to adopt integrated measurement models to monitor port performance comprehensively. Literature over the past few decades has extensively shown that these models identify key performance indicators (KPIs) as essential criteria for achieving strategic goals in maritime competitiveness while incorporating various perspectives of port stakeholders. Indeed, this approach aligns with Kaplan and Norton's Balanced Scorecard (BSC) Model [8], i.e. a multidimensional dashboard that encompasses economic and financial indicators, along with those related to customers, internal business processes, and learning and growth perspectives, in order to monitor business strategy and develop strategic maps [9]. Moreover, modern ports employing the governance model of "cluster managers" acknowledge the importance of engaging salient stakeholders—both business and social—in monitoring port competitiveness to create superior value propositions within the port community [6].

Additionally, several EU projects champion the development of innovative sustainable solutions in ports, focusing on improving energy efficiency, emissions reduction, enhanced recycling performance, and the utilisation of renewable energy sources. These initiatives align with the European memorandum Connecting Europe Facility (CEF), which advocates for the repositioning of ports as hubs for maritime transport, energy, and innovation, promoting smart port digitalisation and decarbonisation.

A valuable example is provided by the above-mentioned PortForward (https://www.portforward-project.eu/), which combines smart technologies through ICT solutions and enhances the exchange of information allowing the dashboard to benefit from the use of a combination of different modes of transport and the integration of different technologies. Indeed, its orientation towards sustainability is supported by the adoption of green technologies. Thus, PortForward contributes to implement innovative solutions to reduce emissions and energy consumption in ports, leveraging renewable sources and digital technologies like the Internet of Things and cloud computing. Moreover, the project proposes the implementation of an innovative Port Authority Dashboard (PAD) in selected ports to enhance strategic management and port efficiency from a smart and green perspectives. Initiatives like the mentioned one help port managers and policymakers in comparing different seaports and aligning government strategies with the European Commission's policy objectives.

Since the managerialisation of Port Authorities demands a consistent approach to monitoring performance, the scope of the observation should encompass both the level of individual actors and the entire port system. Establishing an effective strategic control system, along with its integration with other management and operational control systems, represents an essential phase for the organisation to effectively pursue its strategic objectives. This necessitates the establishment of clear performance targets and measures that not only drive long-term value but also address short-term demands [5]. Accordingly, the attempt to identify core activities in port management acquires a noteworthy importance to assess and preserve competitive advantage. Also, the identification of said critical activities is functional for the selection of KPIs to monitor port strategies effectively, considering unity of purpose alongside differential measurement perspectives.

Hence, the resort to performance monitoring reveals to be a key instrument in supporting the implementation of a successful port strategy, requiring the development of a comprehensive set of indicators able to grasp the diverse dimensions contributing to port performance. Literature on port performance management over the past few decades has mainly focused on port performance measurement, encompassing individual performance metrics, performance measurement frameworks, and the correlation between performance systems and the port environment [5]. In particular, research approaching port performance analysis has traditionally focused on three separated aspects, namely a micro-level concerning specific actors' performance, a meso-level which falls beyond individual organisations, implying the engagement of numerous stakeholders across organisational boundaries; and a macro-level regarding the port's impact on macroeconomic systems. Our present study mainly concentrates on the micro- and meso-level analysis.

34.3 Methods

For the data collection process, a systematic literature review was conducted and a search was performed across the Scopus academic database [10]. The exclusion of further databases such as IEEE or Web of Science resides in the fact that the Scopus database includes a wider number of indexed articles. Additionally, its science focus is more directed towards humanities and social sciences. A systematic literature review protocol [10] was chosen to provide robustness to the analysis process. In performing the results query on Scopus, keywords such as "strategic monitoring", "strategic management", "decision support system", and "dashboard" were selected in combination with the "port authority" to identify relevant studies. Articles were screened based on predefined inclusion and exclusion criteria, focusing on those discussing strategic monitoring techniques, DSS applications in port management, and innovative approaches to decision support, such as dashboards.

34.4 Findings

Results emerging from the systematic review identify a range of studies highlighting the importance of strategic monitoring in port authority management and the potential benefits of DSS implementation, underscoring the need for an innovative DSS tailored to address these challenges and enhance strategic monitoring capabilities within port authorities. Thus, an updated and enlarged framework for the Decision Support System (DSS) in port authority management can be presented, describing the key components of the framework, as well as the activities addressed to the implementation of decision-making processes in PA. According to the provided findings, our study proposes an innovative multidimensional data-driven performance measurement dashboard as a tool for monitoring and proactively managing the port system. More specifically, the novel multidimensional performance assessment dashboard presented hereto serves as a proactive management tool specifically tailored for the PAs and considers several port characteristics, including the nature of the port authority's objectives, the functions of the port, and the legal entity of the PAs.

Additionally, the study outlines a model describing the way in which the evolution of PAs functions and strategic objectives can be linked to a more extensive utilisation of Key Performance Indicators (KPIs) and a multifaceted dashboard tool. Specifically, the proposed conceptualisation enhances the adoption of digital technologies to ease the development of a broader set of KPIs and the management of robust relationships with stakeholders [3]. The proposed framework is meant to serve as a proactive management tool specifically tailored for the Port Authority, embedding a wide set of port characteristics, including the nature of the port authority's objectives, the functions of the port, and the legal entity of the PAs. The analysis of port competitiveness necessitates a contextual evaluation encompassing both the attainment of pertinent economic performance indicators in business and the environmental sustainability of operations and logistics processes. Moreover, findings from Castellano et al. [11] enrich the discourse on the role of green policies in critical activities undertaken by some PAs in recent years, including energy-saving programmes, pollution reduction measures, and innovative waste management initiatives. These efforts are supported by stakeholder engagement and environmental clauses in concession contracts. Also, they point out the pivotal role of a multidimensional dashboard as a tool for monitoring green port efficiency and supporting environmental initiatives aligned with EU port policies and IMO regulations.

Since the analysis embraces both a micro and a meso perspective, the framework of the upgraded dashboard comprises four performance domains aimed at measuring port competitiveness, namely:

1. Port authorities and concessionaries (actors related KPIs);
2. Logistics and operations (efficiency related KPIs);
3. Market trends (industry related KPIs);
4. Green Challenges (sustainability related KPIs).

The KPIs related to each of the listed categories can be drawn according to the perspective of the port authority as the subject in charge of monitoring concessionaries.

In particular, KPIs falling within the first domain, Port authority and concessionaries, assess the pivotal role of the mentioned actors in fostering economic growth within the port community (meso-level). This evaluation encompasses factors such as the number of concessionaires, port-related firms, employment figures, foreign direct investment, and gross added value.

Conversely, performance assessment within the Logistics and operations domain facilitates the monitoring of port competitiveness by employing indicators pertaining to internal port efficiency, such as dwell time, ship turnaround time, modal split, and customs clearance duration, along with the strategic management and planning.

The Marketing trends domain includes KPIs related to demand conditions and customer perceptions. This encompasses factors like maritime traffic, market share, user satisfaction (e.g., terminal, shipper, logistics provider), and customer experience (e.g., passengers, cruisers).

Finally, the domain addressing Green Challenges is functional for PAs to evaluate port performance from an environmental perspective. This involves scrutinising aspects such as Corporate Social Responsibility (CSR) reporting, the presence of environmental monitoring programmes, daily truck traffic, waste production volumes by type, and CO2 emissions.

34.5 Implications

This study contributes to enrich knowledge and expertise on the utilisation of strategic monitoring activities in PAs, addressing a significant gap in academic literature. Previous studies predominantly adopted a micro-level and focused solely on specific aspects of port performance, such as operational efficiency and financial performance. On the other hand, the present research explores a wide array of functional areas within PAs, advocating for the proactive monitoring of emerging analytical dimensions such as market trends and green challenges.

Evidence provided by the present study highlights the importance of integrated port area management in enhancing port efficiency from both an economic and an environmental perspective, underscoring the innovative role of multidimensional models to support port monitoring activities. Such an approach validates the significance of fostering economic, social, and environmental values in collaboration with key stakeholders amidst the backdrop of port competitiveness. Since Port Authorities at the EU level are urged to develop innovative infrastructures to facilitate the supply of vessels with sustainable energy sources like hydrogen or LNG and promote decarbonisation in the transport sector and efficient utilisation of maritime resources, a multidimensional monitoring dashboard is able to digitally integrate all green indicators.

The findings of this study yield practical managerial implications for public managers and policymakers within Pas which are encouraged to acknowledge the benefits of implementing profit-oriented managerial practices in PAs. Particularly, the introduction of strategic monitoring dashboards could be highly beneficial for those PAs transitioning to efficiency-driven management routines.

On a final note, the development of a data-driven digital dashboard should not be perceived solely as an internal management exercise aimed at evaluating corporate performance. Rather, port governance should recognise that a suitable set of KPIs addressing key functional areas serves as a powerful tool for enhancing stakeholder satisfaction and customer loyalty. For, a paradigm shift from an internal to an external perspective in KPI development represents a significant strategic advancement in monitoring activities.

Acknowledgments This research has been supported by the European Union under the PNRR of NextGenerationEU, National Sustainable Mobility Center (CN00000023, MUR Decree n. 1033—17/06/2022, Spoke 10).

References

1. Notteboom, T., Pallis, A., Rodrigue, J.-P.: Port Economics, Management and Policy. Routledge, New York (2022) 690 pages/218 illustrations. ISBN 9780367331559
2. Parola, F., Ferrari, C., Tei, A., Satta, G., Musso, E.: Dealing with multi-scalar embeddedness and institutional divergence: evidence from the renovation of Italian port governance. Res. Transp. Bus. Manag. **22**, 89–99. ISSN 2210-5395 (2017). https://doi.org/10.1016/j.rtbm.2016.12.005
3. Heilig, L., Voß, S.: Status quo and innovative approaches for maritime logistics in the age of digitalization: a guest editors' introduction. Inf. Technol. Manag. **18**, 175–177 (2017). https://doi.org/10.1007/s10799-017-0282-z
4. Dooms, M., Van Der Lugt, L., De Langen, P.W.: International strategies of port authorities: the case of the port of Rotterdam authority. Res. Transp. Bus. Manag. **8**, 148–157 (2013)
5. Ferretti, M., Parmentola, A., Parola, F., Risitano, M.: Strategic monitoring of port authorities activities: proposal of a multi-dimensional digital dashboard. Prod. Plan. Control. **28**(16), 1354–1364 (2017). https://doi.org/10.1080/09537287.2017.1375146
6. Pallis, T., Parola, F., Risitano, M., Ferretti, M.: Marketing strategies of port authorities: a multi-dimensional framework. In: Paper Submitted at 16th EURAM Annual Conference "Manageable Cooperation?" June 1–4, 2016. Université Paris-Est Créteil (UPEC), Paris (2016)
7. Langenus, M., Dooms, M.: Port industry performance management: a Meso-level gap in literature and practice? Int. J. Log. Res. Appl. **18**(3), 251–275 (2015). https://doi.org/10.1080/13675567.2015.1026885
8. Kaplan, R.S., Norton, D.P.: Using the balanced scorecard as a strategic management system. Harv. Bus. Rev. **85**(7–8), 150–161+194 (1996)
9. Kaplan, R.S., Norton, D.P.: Strategy Maps: Converting Intangible Assets into Tangible Outcomes. Harvard Business Press, Boston, MA (2004)
10. Moher, D., Liberati, A., Tetzlaff, J., Altman, D.G.: Preferred reporting items for systematic reviews and meta-analyses: the PRISMA statement. BMJ. **339**, b2535 (2009). https://doi.org/10.1136/bmj.b2535
11. Castellano, R., Ferretti, M., Musella, G., Risitano, M.: Evaluating the economic and environmental efficiency of ports: evidence from Italy. J. Clean. Prod. **271**, 122560. ISSN 0959-6526 (2020). https://doi.org/10.1016/j.jclepro.2020.122560

Part VI
Marketing and Consumer Engagement

Chapter 35
Marketing in the Metaverse: A Comprehensive Bibliometric Literature Review

Luca Venturini and Annarita Sorrentino

Abstract The Metaverse is an emerging digital paradigm that has the potential to transform marketing dynamics and create unprecedented opportunities for brand-consumer interactions. This study undertakes a thorough bibliometric literature review on marketing in the Metaverse, meticulously analyzing the evolution of the field, its pivotal research domains, and the directions it is poised to take in the future. The study utilizes bibliometric analysis to identify dominant trends, author and institutional collaborations, and unexplored research gaps. The findings emphasize the growing importance of immersive technologies, engagement strategies, and user experience personalization as critical factors contributing to the success of marketing initiatives in the Metaverse. Furthermore, the review illuminates the integration of virtual and augmented reality, gamification, and digital fashion as pivotal elements that shape consumer behavior and brand positioning within this emerging digital ecosystem. The study also discusses the implications of these advancements for businesses and researchers, emphasizing the need for interdisciplinary approaches to address the challenges and opportunities of Metaverse marketing. This review offers a valuable framework for scholars and practitioners seeking to understand the development and potential of this evolving digital landscape.

Keywords Metaverse · Marketing · Bibliometric literature review

L. Venturini (✉) · A. Sorrentino
University of Naples Parthenope, Naples, Italy
e-mail: luca.venturini@studenti.uniparthenope.it

F. Schiavone et al. (eds.), *Advanced Perspectives and Trends in Digital Transformation of Firms, Networks, and Society*, Springer Proceedings in Business and Economics, https://doi.org/10.1007/978-3-031-80692-6_35

35.1 Introduction

The Metaverse postulation was first proposed in Stephenson's science fiction book "Snow Crash" [1]. Stephenson's Metaverse was envisioned as a virtual reality-based substitute for the Internet, providing an immersive online world in which users could escape reality using their avatars. In recent years, virtual reality (VR), augmented reality (AR), and mixed reality (MR) have introduced many innovations in devices and experiences for both businesses and consumers. Academics have analyzed virtual worlds [2, 3] and immersive technologies [4–6] to predict their potential evolution and understand their application domains. Since the announcement of the Metaverse created by Mark Zuckerberg and the subsequent rebranding of Facebook to Meta, interest in understanding the opportunities related to it has rapidly increased among scholars and practitioners [7–11]. The definition of the Metaverse is not yet unanimous [12]. However, it appears to be moving away from a conceptualization rooted in the mere concepts of virtual worlds [13, 14] and the evolution of VR [15, 16]. Instead, it is turning toward combining different technologies that can mix reality and virtuality [17, 18]. The increasing diffusion of new Head-Mounted Display VR devices has led to a consequential increase in possible applications in different domains, opening an era of the Metaverse. Indeed, like any other digital revolution, its application is expected to proliferate, with a projected market size of 2961.55 billion dollars by 2032 [19] and an expected user base of more than 700 million by 2030 [20]. Therefore, it is important to analyze this evolving topic's theoretical and managerial implications, with particular attention to possible marketing applications.

35.2 Objective

The research objective is to conduct a thorough bibliometric literature review to systematically map and comprehend the research trends of Metaverse marketing and the emergent research themes about this topic. Additionally, the research seeks to pinpoint influential authors and publications that have made noteworthy contributions to this discourse, highlighting intellectual influences and the evolving trajectory of thought in the field. The aim is to yield an overview of the marketing Metaverse landscape, offering critical insights for scholars, marketing practitioners, and business decision-makers. To achieve this goal, the research answers the following research questions:

1. RQ1: What countries, journals, and authors have exerted the biggest impact on current Metaverse-related marketing research?
2. RQ2: In what primary thematic areas is the Metaverse applied to marketing?
3. RQ3: What implications can a Metaverse-Marketing literature offer to marketers, firms, and academics?

35.3 Methods

A bibliometric analysis was conducted using the PRISMA protocol [21] to search and select scientific contributions for systematic analysis. Bibliometrics is a reliable approach for searching and analyzing a high volume of scientific data [22], providing insight into the current state of an area or research subject [23]. The bibliometrics for this study were performed with the help of the bibliometrics R package [23] through Biblioshiny, an online application with an intuitive and well-organized interface [24]. The search was carried out using the Web of Science (WoS) databank, a comprehensive resource covering various scientific fields and containing numerous citations related to specific areas [23]. The search query used was TS = Marketing AND TS = Metavers*. Initially, 237 items were retrieved from the WoS search engine using this query. Inclusion/exclusion criteria were then applied [25], starting with a time filter to limit the study to scientific papers published after 2021, coinciding with the Meta announcement by Mark Zuckerberg on October 28, 2021. Contributions published in 2024 were excluded due to the ongoing nature of the year at the time of drafting [24]. Further criteria were applied to select only journal articles, review articles, and proceedings papers; the decision to include proceedings papers is entirely due to the novelty of the research topic. "Business" and "Management" research categories were selected based on criteria established by Aria and Cuccurullo [23]. Non-English publications were excluded to ensure replicability and an international audience [26]. However, no reduction was applied as all the papers in the dataset were in English. Following these steps, abstracts were independently assessed for relevance and bias risk [27]. Only contributions unanimously deemed suitable for the research purpose were included, resulting in a reduced dataset of 52 items. In the final screening phase, the remaining articles were read in full text based on predefined inclusion/exclusion criteria [28], and four additional papers were excluded. Data analysis included descriptive analysis and network extraction. Bibliographic coupling and thematic analysis were used for the mapping of the current research front, as they are recommended approaches in bibliometrics.

35.4 Findings

The data collection yielded 48 scientific papers published between 2022 and 2023, distributed across 35 sources. Within this sample, 43 papers were published as journal articles, 3 as review articles, and 2 as proceedings articles. In total, 155 scientists contributed to these publications, of which 7 scientists were identified as single authors. The authors involved in writing on the topic of the marketing Metaverse came from 32 different countries, with a predominant contribution from the USA, India, the UK, China, South Korea, and Italy. This global distribution indicates a broad international participation in the scholarly discourse on Metaverse Marketing

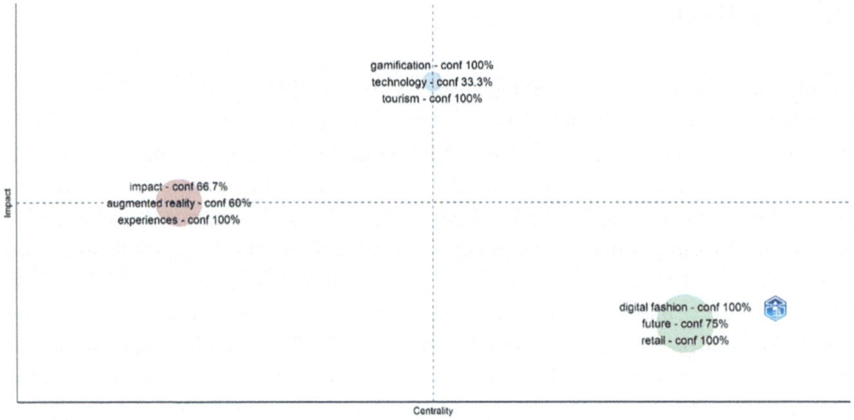

Fig. 35.1 Clustering by coupling. Source: Biblioshiny

during the last 2 years. Clustering by coupling analysis (Fig. 35.1) indicates the existence of three main clusters related to:

1. The impact of AR on the experience
2. The relationship between gamification and technology topics applied to tourism
3. Metaverse is the future of retail in the digital fashion industry

A comparison of these clusters with the thematic analysis conducted on the same sample (Fig. 35.2) reveals any trending themes during this period.

35.5 Implications

The theoretical and managerial implications of the bibliometric analysis of marketing in the Metaverse are significant and promising. The international distribution of scholars and the presence of distinct thematic clusters indicate the diversity and complexity of research perspectives. The multidisciplinary nature of this emerging field is highlighted by the number of authors involved, requiring expertise in marketing, technology, and consumer behavior to converge. On the managerial side, identifying thematic clusters can guide companies in making strategic decisions when considering adopting the Metaverse in their marketing approach. For instance, understanding the influence of AR on the customer experience could influence the design of innovative advertising campaigns and reshape all the touchpoints along the customer journey. Similarly, exploring the role of gamification in the tourism context could suggest new ways of consumer engagement. Recognizing the Metaverse as the future of digital retail in fashion could direct companies toward targeted investments in immersive technologies. These theoretical and managerial

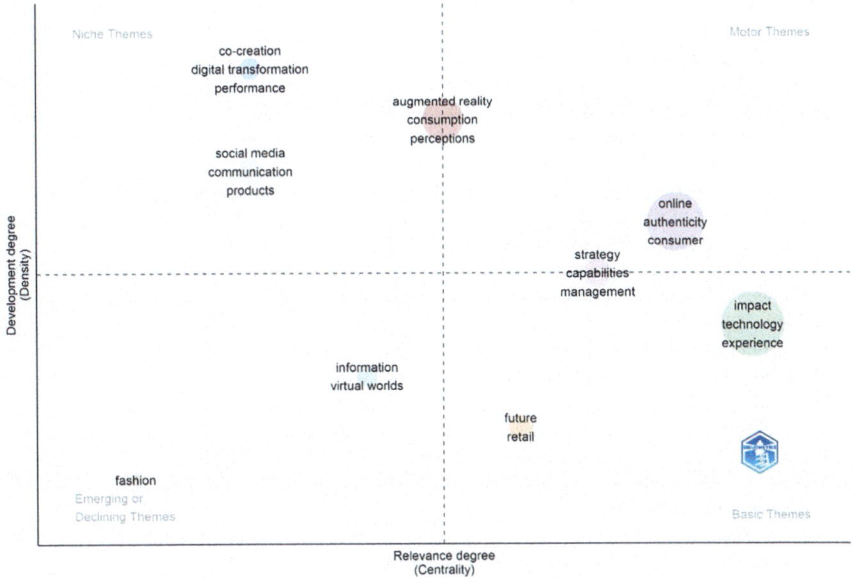

Fig. 35.2 Thematic analysis. Source: Biblioshiny

implications offer a critical framework for academics, marketers, and business leaders seeking to understand and fully exploit emerging opportunities in the Metaverse context.

References

1. Stephenson, N.: Snow Crash: a Novel, 2003rd edn. Spectra (1992)
2. Bainbridge, W.S.: The scientific research potential of virtual worlds. Science. **317**(5837), 472–476 (2007)
3. Papagiannidis, S., Bourlakis, M., Li, F.: Making real money in virtual worlds: MMORPGs and emerging business opportunities, challenges and ethical implications in metaverses. Technol. Forecast. Soc. Chang. **75**(5), 610–622 (2008)
4. Wedel, M., Bigné, E., Zhang, J.: Virtual and augmented reality: advancing research in consumer marketing. Int. J. Res. Mark. **37**(3), 443–465 (2020)
5. Flavián, C., Ibáñez-Sánchez, S., Orús, C.: The impact of virtual, augmented and mixed reality technologies on the customer experience. J. Bus. Res. **100**, 547–560 (2019)
6. Simoni, M., et al.: Boosting the pre-purchase experience through virtual reality. Insights from the cruise industry. J. Hosp. Tour. Technol. **13**(1), 140–156 (2022)
7. Koohang, A., et al.: Shaping the metaverse into reality: a holistic multidisciplinary understanding of opportunities, challenges, and avenues for future investigation. J. Comput. Inf. Syst., 1–31 (2023)

8. Buhalis, D., Lin, M.S., Leung, D.: Metaverse as a driver for customer experience and value co-creation: implications for hospitality and tourism management and marketing. Int. J. Contemp. Hosp. Manag. **35**(2), 701–716 (2022)
9. Dwivedi, Y.K., et al.: Metaverse marketing: how the metaverse will shape the future of consumer research and practice. Psychol. Mark. **40**(4), 750–776 (2023)
10. Gursoy, D., et al.: Metaverse in services marketing: an overview and future research directions. Serv. Ind. J. **43**(15–16), 1140–1172 (2023)
11. Simoni, M., Sorrentino, A., Venturini, L.: Understanding the potentiality of the metaverse in the redesign of the tourist experience: an explorative analysis in Italy. In: Conference on Managing Tourism Across Continents 2023 (2023)
12. Dwivedi, Y.K., et al.: Metaverse beyond the hype: multidisciplinary perspectives on emerging challenges, opportunities, and agenda for research, practice and policy. Int. J. Inf. Manag. **66**, 102542 (2022)
13. Barry, D.M., et al.: Evaluation for students' learning manner using eye blinking system in Metaverse. Procedia Comput. Sci. **60**, 1195–1204 (2015)
14. Owens, D., et al.: An empirical investigation of virtual world projects and metaverse technology capabilities. ACM SIGMIS Database. **42**(1), 74–101 (2011)
15. Prisco, G.: A virtual world space agency. Futures. **41**(8), 569–571 (2009)
16. Müller, F.: Remembering in the Metaverse: Preservation, Evaluation, and Perception. University of Basel (2012)
17. Barrera, K.G., Shah, D.: Marketing in the metaverse: conceptual understanding, framework, and research agenda. J. Bus. Res. **155**, 113420 (2023)
18. Buhalis, D., Leung, D., Lin, M.: Metaverse as a disruptive technology revolutionising tourism management and marketing. Tour. Manag. **97**, 104724 (2023)
19. Research, E.: Metaverse market by component (hardware, software, services), by application (social events & trade shows, health surgery), by platform (desktop, mobile), by offering, by technology, by end-use, and by region forecast to 2032. Available from: https://www.emergen-research.com/industry-report/metaverse-market (2023)
20. World Economic Forum: This chart shows how big the metaverse market could become. Available from: www.weforum.org/agenda/2023/02/chart-metaverse-market-growthdigital-economy/ (2023)
21. Moher, D., et al.: Preferred reporting items for systematic review and meta-analysis protocols (PRISMA-P) 2015 statement. Syst. Rev. **4**, 1–9 (2015)
22. Donthu, N., et al.: How to conduct a bibliometric analysis: an overview and guidelines. J. Bus. Res. **133**, 285–296 (2021)
23. Aria, M., Cuccurullo, C.: bibliometrix: an R-tool for comprehensive science mapping analysis. J. Inform. **11**(4), 959–975 (2017)
24. Maione, G., Cuccurullo, C., Tommasetti, A.: Biodiversity accounting: a bibliometric analysis for comprehensive literature mapping. Sustain. Acc. Manag. Policy J. (2023) (ahead-of-print)
25. Parmentola, A., et al.: Is blockchain able to enhance environmental sustainability? A systematic review and research agenda from the perspective of Sustainable Development Goals (SDGs). Bus. Strateg. Environ. **31**(1), 194–217 (2022)
26. Polese, F., et al.: Performance evaluation and measurement in public organizations: a systematic literature review. Int. J. Bus. Adm. **8**(1), 106–117 (2017)
27. Adams, C.A., Larrinaga, C.: Progress: engaging with organisations in pursuit of improved sustainability accounting and performance. Account. Audit. Account. J. **32**(8), 2367–2394 (2019)
28. Manetti, G., Bellucci, M., Oliva, S.: Unpacking dialogic accounting: a systematic literature review and research agenda. Account. Audit. Account. J. **34**(9), 250–283 (2021)

Chapter 36
Identifying the Value Proposition of Modular Manufacturing Information Systems for Digital Transformation

Timoleon Farmakis, Stavros Lounis, Efstathios Plitsos, Giulio Vivo, Julien Mascolo, Olivier Jotz, and Tobias Hohmann

Abstract This study explores the potential value proposition(s) of reconfigurable manufacturing. Within the scope of an EU-funded project, a system is to be developed that enables modularity in production at a product and factory level. Utilising the Value Proposition Canvas as an analytical framework, the research amalgamates insights from a workshop involving key stakeholders. The goal is to identify the different trajectories of value proposition brought forth by such a system in relation to the observed product-market fit. Furthermore, it presents an outlook for introducing the identified value proposition(s) into the shop floor through a process of digital transformation. The findings support that such a system can have several potential value propositions that, in turn, enable process, customer and business model transformation, providing valuable insights to practitioners and researchers. Digital

T. Farmakis (✉) · S. Lounis
Eltrun The E-Business Research Center, Athens University of Economics and Business, Athens, Greece
e-mail: timoleonfarmakis@aueb.gr; slounis@aueb.gr

E. Plitsos
University of Piraeus, Piraeus, Greece
e-mail: stathisp@unipi.gr

G. Vivo
CRF S.C.p.A, Orbassano, Italy
e-mail: giulio.vivo@crf.it

J. Mascolo
ILTAR ITALBOX, Bairo, Italy
e-mail: julien.mascolo@iltar-italbox.it

O. Jotz
SEW USOCOME SAS, Haguenau, France
e-mail: olivier.jotz@usocome.com

T. Hohmann
FFT Produktionssysteme GmbH & Co. KG, Fulda, Germany
e-mail: tobias.hohmann@fft.de

F. Schiavone et al. (eds.), *Advanced Perspectives and Trends in Digital Transformation of Firms, Networks, and Society*, Springer Proceedings in Business and Economics, https://doi.org/10.1007/978-3-031-80692-6_36

Twins, simulation, optimisation, predictive maintenance and production design, and reconfiguration are vital technologies such a system must have, to alleviate the pains of the end users and provide gains to manufacturers.

Keywords Digital transformation · Reconfigurable manufacturing systems · Modularity · Value proposition

36.1 Introduction

With modern manufacturing landscapes constantly changing, the role information systems (IS) and digital technologies play in facilitating the transformation is constantly and consistently growing [1]. Subsequently, the required digital transformation of manufacturing firms to cope with the changes has also become an increasingly critical area of interest in academia and industry [2]. Industries are evolving to account for agility and efficiency, and reconfigurable manufacturing systems (RMS) are at the frontline of this change [3–5]. According to Pansare et al. [6], modern RMS offers low-cost manufacturing, rapid responsiveness and high-quality products that are required in today's dynamic market environment. As Wang et al. [7] note, this environment is becoming increasingly demanding, necessitating mass customisation and personalisation. In this context, RMSs can play a critical role in helping businesses meet their customers' unique needs and stay competitive. Consumers demand products according to their specific needs, which makes manufacturers seek flexible production supporting structures [8].

Information systems for traditional manufacturing (which commonly has a rigid structure) cannot cope with the individualised nature of the evolved production needs [9]. To this end, information systems that support reconfigurable manufacturing can facilitate swift adjustments to shifting specifications in real-world settings [10]. An effective design solution for achieving system reconfigurability and adaptability is modularity [11, 12], giving rise to modular manufacturing IS, which also constitutes the focal point of this research. However, there are some challenges, including technical complexity, organisational restructuring, cultural shifts in an engineering mindset, demanding project management, long-term strategic alignment and the need to incorporate the business perspective [13] leading to the need to examine the change under the lens of Digital Transformation (DX).

While extensive literature exists concerning the technical dimensions of RMS, more information needs to be documented about what strategic value such an approach can add to manufacturing businesses [3, 4]. Additionally, there is a need for further research, as RMS needs substantial investments from companies [6]. This lack of knowledge is a limiting factor concerning adopting modular manufacturing IS within the manufacturing environment [14]. Identifying and communicating their value proposition can be essential for manufacturers, as it significantly influences their decision to adopt these technologically advanced solutions [15].

Furthermore, identifying the value proposition of such systems can lead to improvements and ensure that they meet customer needs effectively. To this end, this study aims to identify a value proposition for modular manufacturing IS that fits the market needs through a structured workshop with technology partners and pilots (customers) in the context of the MODAPTO project. Additionally, the study provides insights into how the identified value propositions can be integrated into the shop floor through a digital transformation (DX) process, addressing the research gap highlighted by Kraus et al. [16]. The motivation for this study also arises from the four participating industrial companies from various sectors implementing these technologies in their operations. On top of that, understanding RMS's enabling technologies has been a prominent topic for researchers for years [17].

The chapter is structured as follows: initially, a literature review on the digital transformation of manufacturing firms, RMS and modularity is conducted; this is followed by presenting the methodology; and finally, the results, discussion, conclusion and limitations are presented.

36.2 Literature Review

36.2.1 Digital Transformation

The diffusion of advanced technologies such as AI, optimisation and IoT has disrupted business operations, making DX a topic of great interest for academia and industry [18], particularly in manufacturing [19]. Transformation in manufacturing includes implementing digital technologies to enhance operational effectiveness, adjust to evolving customer needs and develop novel goods and services [20]. In particular, RMS is crucial for DX in manufacturing firms [21] as these systems can transform manufacturing companies' production mode, allowing mass customisation [22], an essential strategic consideration in modern manufacturing [23]. DX implies the adoption of an IS capable of supporting the transition and the maintenance of an RMS environment. Assuming a company has a monolithic production line and decides to make it modular. Once it does it, then it has an RMS production. An IS could support either making this transition (through simulation to assess the investment) or monitoring and managing this production setup.

A seminal Dx research by Kraus et al. [16] emphasises the critical importance of examining how emerging technologies can effectively support and drive DX, leading to the need to examine the introduction of emerging technologies and processes (i.e. modular manufacturing). This can be accomplished stepwise, as exemplified in the Doukidis et al. [24] research, which categorised companies' digital transformation efforts into four pillars, each with its levels of stepwise change of introducing systems and services (Fig. 36.1). This framework can thus serve as a lens to assess the transformation outcome of advanced technologies in manufacturing companies [25].

CE: Customer Experience Transformation	BP: Business Processes Transformation	BM: Business Model Transformation	OT: Organizational Transformation
1) Comprehension of Customer's Needs and Behavior 1.1) CRM and Loyalty Scheme 1.2) Business Analytics 1.3) Personalization	1) Business Processes Digitalization and Redesign 1.1) Automation and Redesign 1.2) Agile Development, Collaboration, and Integration 1.3) New Supporting Services	1) Digital Organizational Modifications 1.1) Supplementary Digital Channel 1.2) Digital Modifications 1.3) Digital Innovations	1) Integrated Business Strategy 1.1) Compliance 1.2) High Investments 1.3) Integrated Strategy
2) Upgrading Customer's Experience 2.1) Customer Journey 2.2) Multichannel 2.3) Omnichannel	2) Support and Improvement of Employees' Daily Tasks 2.1) Info Reporting and Support 2.2) Personalized Services 2.3) DSS and Artificial Intelligence	2) Establish New Digital Activity/ Service and Business Model 2.1) Customer-Centric Targeting 2.2) New Digital Services 2.3) New Business Models	2) New Organizational Structures/Functions/Alliances 2.1) New Structure 2.2) Supervision Process 2.3) Alliances
3) Automation and Digitization of Sales Processes 3.1) Information Reporting and Support 3.2) Upgrade Salesperson's Role 3.3) Customer Self-Service	3) Business Performance Management 3.1) KPIs and Balanced Scorecard 3.2) Real-Time Reporting and Analytics; Continuous Integration 3.3) Precision Management	3) Digital International Expansion 3.1) Consolidation of Subsidiaries 3.2) International Service/Support 3.3) Think Global and Act Local	3) Human Resources Development, Talent Acquisition, and Exploitation 3.1) Training and Talent Acquisition 3.2) New Methods 3.3) Innovation Structures

Fig. 36.1 Digital transformation pillars. Adopted from Doukidis et al. [24]

36.2.2 RMS

RMS has become a viable set of supporting systems for keeping manufacturing companies competitive in a global setting where demand for their products and services changes due to shifting market conditions and economic pressures [26]. RMS can quickly adapt to shifts in the market or the demands of production. A manufacturing system must be reconfigurable to address several vital factors efficiently: (1) swiftly adapting to customer demands, (2) effectively retooling machinery and modifying operational parameters to address unexpected system failures and maintain product quality and (3) upgrading older systems to integrate new technologies [27]. RMS places a strong emphasis on convertibility (the system can be readily changed or reconfigured), integrability (the modules can function together efficiently), interoperability (the system can work with other systems) and modularity (the system is made up of independent, interchangeable modules) [27]. Modularity, integrability, scalability, convertibility, customisation and diagnosability are the six fundamental characteristics of an RMS [28]. Modularity is a crucial design approach for RMS [11, 27].

36.2.3 Modularity in Manufacturing

Modularity in production encompasses two perspectives: design and product. From a design standpoint, modularity involves segmenting the manufacturing system (factory) into autonomous, interchangeable modules. Each module is a complete entity equipped with distinct functions and interfaces, facilitating the seamless substitution of modules with equivalent components and ensuring minimal disruption to the broader system [11, 27]. On the product side, modularity is applied to create versatile modules that can be integrated across a range of products, enhancing product adaptability and customisation [29]. In manufacturing, modular systems enable businesses to reduce the expenses and investments often involved with managing product diversity by enabling quick and economical responses to market and product changes [30]. Another key benefit is that modular systems eliminate overcapacity and reorganise industrial capacity [31]. By leveraging modular and reconfigurable systems, manufacturers can lower investments, time-to-market and variety costs [32].

36.3 Industrial Cases of Modular Manufacturing

The MODAPTO project aims to advance modular and adaptable production systems, focusing on implementing digital twin technologies and modular approaches in industrial settings. The project's objectives include enhancing flexibility, efficiency and sustainability in manufacturing processes by developing new methods and tools. On that account, it aims to develop an IS that enables modular manufacturing, facilitating interoperability in manufacturing through a framework that connects production modules and systems, employing open interfaces and distributed control. It also aims to include a modular production toolkit for process optimisation and collective intelligence tools for data-driven insights, enhancing manufacturing operations' overall efficiency and adaptability. The developed system will be introduced and evaluated in three distinct manufacturing cases to enable their current manufacturing processes to become modular (at the production and product levels).

Pilot 1 aims to enhance robotic systems by integrating interoperability, sustainability analytics and energy optimisation, utilising industrial digital twins for detailed data on processes like spot welding and roller hemming. The pilot's approach enables achieving sustainability by using precise energy and carbon footprint calculations. It also supports decision-making in selecting appropriate robotic modules and tools for sustainable production, using virtual commissioning for energy-efficient layout design.

Pilot 2 aims to implement modular and reconfigurable production lines that adapt to custom production needs like single-lot production. It will focus on redesigning and optimising production processes, including line balancing, automatic setup and real-time routing, particularly for a specific gear motor production. The system aims to use simulations for redesign decisions and optimised scheduling that integrates predictive maintenance and re-scheduling. It also includes a comparative analysis of identical modules to optimise production based on local parameters and outcomes, enhancing efficiency and responsiveness to market demands.

Pilot 3 focuses on modular logistics and intelligent containers in industrial assembly processes, particularly automotive production. It aims to automate kitting operations using advanced robots and cobots with technologies like vision systems. The project will develop intelligent kits and kit holders to communicate with internal logistics and robots, optimising the delivery of the just-in-time and just-in-sequence of diverse components and enhancing efficiency in assembly lines, from component transportation to assembly and remanufacturing processes.

36.4 Methods

This chapter adopts a multicase study approach, allowing for an in-depth exploration across multiple instances within the reconfigurable manufacturing context [33, 34]. The selected cases represent diverse conditions and industrial environments relevant to the research question to maximise the findings' generalisability and illuminate different aspects of the phenomenon under investigation. The cases were also selected based on the availability and great access to data. The research was carried out following the recommendations and guidelines for qualitative methods as outlined in the literature [35]. Below are the research questions for this study, which have been derived from the literature review and the industrial motivations of companies.

1. How can innovative information systems that enable modular manufacturing enhance companies' value proposition?
2. How do these systems impact companies' digital transformation efforts?

This chapter utilises data from a workshop with stakeholders utilising a structured questionnaire involving technology providers and project pilots (factories aiming to enable modular manufacturing). In order to initially delineate the potential value proposition(s) of modular manufacturing IS, we employed the Value Proposition Canvas (VPC) approach [36]. VPC enables the identification of critical jobs, pains and gains of customers and pinpoint the specific products and services that act as gain creators and pain relievers (Fig. 36.2). A value proposition canvas identifies the main advantage customers get from a product or service and identifies areas for improvement [37]. Through this process, we aim to pinpoint the potential value proposition(s) and examine such a system for market fit. A similar approach to utilising the Value Proposition Canvas (VPC) has also been employed in developing intelligent services for manufacturing firms [38].

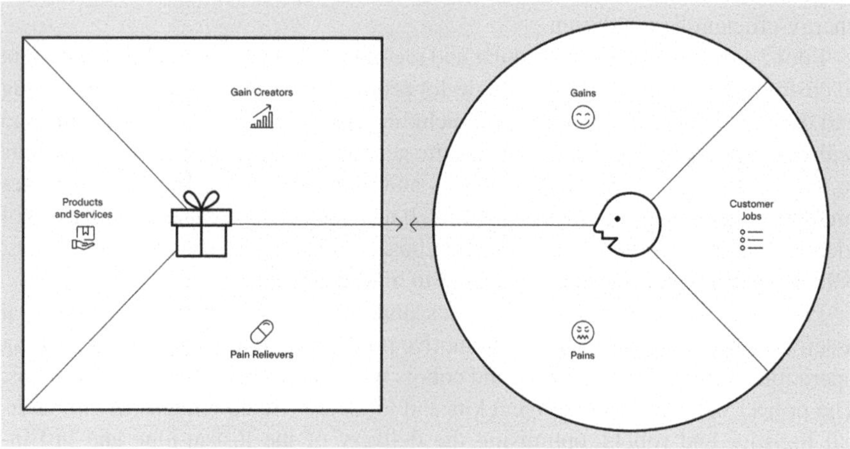

Fig. 36.2 Value proposition canvas (Source: [36])

In this workshop, we engaged with customers (i.e. project pilots) to catalogue and analyse their job functions. In the questionnaire and following the VPC guidelines, we categorised these functions into two primary segments: 'Functional Jobs', where participants detailed their daily tasks (such as creating orders or examining machinery), and 'Supporting Jobs', which encompass the auxiliary tasks that facilitate the main functional jobs (like contacting supervisors or writing reports). Additionally, we delved into identifying 'Pains' associated with these jobs, categorising them as undesired outcomes (like time wastage or report errors), risks (such as failing to meet production deadlines) and obstacles (for instance, lack of specific processes). Furthermore, the questionnaire focused on 'Gains' expected from using a modular manufacturing IS, which we divided into three categories: 'Required Gains' which are essential for the system's functionality, 'Expected Gains' that users anticipate in their functional and supporting jobs and 'Desired Gains' which, though not essential, would enhance the user experience in performing their tasks. Regarding technology partners, we asked them to discuss their products and services as they are envisaged, designed and developed, focusing on general and pilot-specific features. In this line, we engaged them to discuss how the system's features could mitigate indicative customer pains, thus creating customer gains. Additionally, to incorporate the digital transformation perspective, we utilised a practitioner-oriented digital transformation framework by Doukidis et al. [24], employed in previous studies [25, 39]. The data collection and analysis process involved three independent researchers to ensure validity. Additionally, all results were reviewed by four senior managers from four industrial companies to confirm the real-life applicability of the findings.

36.5 Findings

Table 36.1 outlines the specific customer jobs, pains and potential gains for pilots, each grappling with unique challenges in their operational environments. In order to examine for product market fit, each Pain deriving from the customer job(s) and each expected gain is appropriated labelled to be cross-referenced with the respective potential pain alleviators and gain creators of the to-be modular manufacturing system.

Table 36.2 encapsulates the diverse range of products and services (as features of the to-be system) offered by technology partners, along with the associated pain relievers and gain creators.

When examining product-market fit, and according to Osterwalder et al. [36], the pains and expected gains from the market must be met by the value of the products and services brought forth as pain alleviators and gain creators. As such, Table 36.3 presents the observed matching from the workshop output.

Table 36.1 Customer jobs, pains and gains across the pilots

Pilot 1 (P1)	*Customer jobs (CJ)* 1. Conduct robot simulation 2. Choose the fitting robot based on the requirements 3. Transfer of files 4. Communicate with departments *Pains (P)* 1. Loss of potential 2. Requires experience in sustainability parameters 3. Needs experience in calculating carbon footprint/LCA, etc. 4. The system requires high efforts to be adopted in a generic approach *Gains (G)* 1. To have a collection of necessary sensor data for different consumptions 2. To enable decision support in choosing the correct robot based on requirements from the customer 3. Adopting a system that does not require high effort is needed.
Pilot 2 (P2)	*Customer jobs (CJ)* 1. Scheduling/adapting (daily) production orders (under various conditions, e.g. capacity) 2. Analysing maintenance data/scheduling/adapting to preventive maintenance 3. Collecting/reworking data and simulating new processes 4. Use simulation for project investment decisions 5. Simulating global organisation 6. Equipment sustainability 7. Integration human factor *Pains (P)* 1. Scheduling disruptions lead to incorrect delivery times or times when there is problematic supply of products to assemble 2. Suboptimal preventive maintenance schedule 3. Lack of methodology and data to correctly analyse/plan the maintenance 4. Processes cannot be correctly simulated due to lack of models/methodology/data 5. Some solutions do not interface with the company's system 6. Variability treatment *Gains (G)* 1. A system that optimises productivity/delivery time/use of resources 2. A system that adapts to urgent order/process failure 3. A system that optimally plans new processes 4. Methodologies/tools to simulate the new processes/predictive maintenance/scheduling 5. A system that interfaces on a local level with current systems with good UX 6. Signing the organisation 7. Logic to pilot the modules

(continued)

Table 36.1 (continued)

Pilot 3 (P3)	*Customer jobs (CJ)* 1. (Re)design containers adapted to customer requirements 2. Provide containers at the right time for production and monitor deliveries 3. Schedule production takes into account constraints, and optimises flows in the plant 4. Increase safety in the plant for the intralogistics 5. Check notifications regarding the Kitting Robot System (KRS) about quality requirements 6. Monitoring the correct preparation of the Kit Holders 7. Checking about Picking Sequences. Specifically on the correctness of the specific pick-and-place task and of the whole optimisation carried out in the sequence 8. Overviewing changes in the priorities during the preparation of the Kits 9. Be warned about disruptions in the logistics process 10. Reporting about general behaviour and unplanned events of KRS *Pains (P)* 1. Lose of time analysing reports 2. Not understanding reports/intakes 3. Spend resources for integrations with low ROI 4. Few resources for integration (IT staff, knowledge) 5. Issues in the components picking (grasp/release issues) 6. Faults in the correct orchestration of the KRS processes 7. Mistakes/errors in the optimal organisation of the work 8. Issues with the logistics (lack of components in input, output pipeline full, etc.) *Gains (G)* 1. A feasible/understandable, and sensible production plan 2. A clear evaluation of potential solutions: few figures, few criteria 3. A simple view of the fleet of containers, adapted to everyday users 4. A simplified vision of the usage of containers, with few figures 5. Automatisation of the Kitting process 6. A reduction of the number of operators involved in the kitting process 7. A reduction of the non-value-added activities 8. A reduction of the disruptions/issues caused by the KRS during the (automatic) preparation of the Kits 9. Have Easiness and simplification of the kit holder preparation process

Table 36.2 Products/services, pain relievers and gain creators

Products and services (PS)	1. Proactive and reactive support for optimisation of scheduling (offline and real-time) [Pilot 1: Decision support for optimal programming of robotic arms considering energy consumption and carbon emissions/Pilot 3: Decision support tool for optimal picking sequence given the required contents of the intelligent kit holder, and corresponding evaluation mechanising/Pilot 3: Decision Support tool for optimal production re-scheduling] 2. Potential plant design or potential sets of orders/jobs simulation engine (co-simulation) based on workflows provided by (configured by) the pilots on their digital twins (DTs) and schedules./An experimentation environment to support the modelling environment in identifying requirements. 3. A reliability modelling and assessment model: Preventive/predictive maintenance strategies tools (Algorithms) for maintenance planning optimisation 4. Measuring and storage capability for maintainability parament (energy/carbon) on the production module level 5. Tabulation module for energy and carbon usage of robots with the corresponding tool 6. Provision of a modelling bases environment to configure/reconfigure a module that will be used by every pilot to configure their specific DT and schedules (as workflows) 7. UI Building tool: This allows the combination of generic UIs in order to create a web app specific to the needs of each use case in every pilot 8. An advanced visualisations toolkit (AVT) 9. A Production knowledge base (PKB) 10. A DT management platforms 11. Enhanced DT(s) for production modules
Pain relievers (PR)	1. Utilises comprehensive data analysis and a robust knowledge base to improve decision-making processes. 2. Reduces costs and resource consumption through more flexible and easily reconfigurable systems. 3. Offers intuitive modelling tools for users to create their DTs effortlessly, 4. Enables effective simulation of module interactions, facilitating better planning and optimisation. 5. Simplifies the creation of pilot-specific UIs from generic functionalities, improving user experience 6. Supports collaborative exploration of scenarios, enhancing team-based problem-solving 7. Provides a centralised system for storing and retrieving use-case-specific information, ensuring easily accessible data. 8. Facilitates easy analysis of data to uncover valuable insights 9. Allows for easy organisation and leveraging of DTs to capitalise on their capabilities. 10. Focuses on providing high-quality, user-friendly, effective, open-source software for various production roles. 11. Offers tailored training resources to incorporate the human factor, enhancing skill sets across different production roles. 12. Aids robot programmers select precise movements for complex tasks, significantly reducing manual effort and increasing efficiency. 13. Reduces the time needed for warehouse component picking and streamlining inventory management. 14. Enables monitoring of energy usage and accurate prediction of power supply needs for production areas, contributing to sustainability and cost reduction

(continued)

Table 36.2 (continued)

Gain creators (GC)	1. Improves daily production scheduling and allows simulation of different production settings to maximise daily orders. 2. Empower users at pilot sites to configure digital twins, enhancing flexibility 3. Enables simulation of hypothetical modules' impact on production lines and offers reusability of system's functionalities for tailored UI. 4. Facilitates optimal scheduling and pilot experimentation, directly impacting production efficiency. 5. Enhances decision-making with visual data correlation and a repository of generated results for data-driven insights. 6. Focuses on sustainable product production and ROI improvement, contributing to energy savings and carbon footprint reduction. 7. Expand digital models to include emissions and energy values, offering a competitive edge and potential for premium pricing. 8. Simplifies inter-departmental cooperation, enhancing operational synergy and reducing overheads. 9. Provides user-friendly interfaces and generic datasets for ML training, facilitating cross-departmental use and enhancing model accuracy. 10. Suggest optimal maintenance actions and planning, reducing failures and enhancing production performance. 11. Organises important production KPIs in a central place, ensuring easy access and management of performance metrics. 12. Increases customer base through better customisation opportunities and efficient supply chain management, leading to savings and new business models. 13. Reduces picking time and alleviates bottlenecks, making production more responsive to changes.

(continued)

Table 36.3 Pains and expected gains met by the products and services value

	Pains (pain relievers)	Gains (gain creator)
Pilot 1	P2:(PR14) P3:(PR8) P4:(PR10,11)	G1:(GC4,5,9) G2:(GC3,4,5) G3:(GC2,4,9)
Pilot 2	P1:(PR4,6) P2:(PR1,7) P3(PR1,7) P4(PR3,4) P2P5:(PR5) P6:(PR4,8)	G1: (GC1,4,13) G2:(GC3,2) G3:(GC8,9) G4:(GC3,5) G5:(GC9) G6:(GC8) G7:(GC2)
Pilot 3	P1:(PR4,6) P2:(PR1,7) P3:(PR1,7) P4:(PR3,4) P5:(PR12) P6:(PR4,6) P7:(PR1,7) P8:(PR13,8)	G1: (GC1,4) G2:(GC5) G3:(GC9) G4:(GC5) G5:(GC12) G6:(GC13) G7:(GC4,10) G8:(GC3,15) G9:(GC2,9)

36.6 Discussion and Implication

According to Table 36.3, it was observed that the system to be developed supports a modular system that exhibits product-market fit, and its value proposition brought forth by the products and services is equipped to handle the customer jobs and alleviate the current pains by its gain creators. However, to be optimally introduced into an existing rigid (non-modular) production, it must also be examined under the lens of change it requires, thus under a digital transformation lens. To do so, we need to map the products and services that constitute the system's value proposition in its respective expected benefits in the DX context. Based on the practitioner framework of Doukidis et al. [24], companies' DX efforts can be categorised into four pillars: process transformation, customer experience transformation, business model transformation and organisational transformation, which is about the necessary organisational changes.

As it occurs from the above analysis (Table 36.4), the identified value proposition of a modular production system can mainly enable the transformation of business processes at large and, secondly, enable transformation in customer experience and potential transformation in business models. For instance, these technologies can enhance customer experience, as simulation enables testing and developing

Table 36.4 Digital transformation mapping

Product/service	Customer experience (CE)	Business processes (BP)	Business model (BM)
PS1	Indirectly applicable as it leads to better product quality and faster delivery times, enhancing the customer experience.	Directly applicable as it enhances business process efficiency through optimised scheduling and decision support tools.	Directly applicable on Pilot 1 for virtual commissioning of new robots enabling.
PS2	Directly applicable as simulation allows for the testing and developing of more customer-tailored products and services.	Directly applicable as it aims to optimise internal processes through simulation and experimentation.	Directly applicable on Pilot 1 for virtual commissioning of new robots, Pilot 2 for production layout. Pilot 3 for internal logistics, enabling pilots to adapt to customer requirements, leading to mass customisation.
PS3	Indirectly applicable as modelling and assessment tools ensure consistent quality to customers.	Directly applicable for streamlining maintenance planning, optimising asset performance and minimising downtime.	N/A currently

(continued)

Table 36.4 (continued)

Product/service	Customer experience (CE)	Business processes (BP)	Business model (BM)
PS4,5	Directly applicable for Pilot 1, as customers have introduced a new parameter for energy and sustainability.	Directly applicable for Pilot 1, 2 as implementing precise measurement and storage of energy and carbon data at the module level enables businesses to make informed operational efficiency and sustainability decisions.	Indirectly applicable as access to detailed energy and carbon metrics facilitates the transformation of business models to include sustainability as a value proposition for environmentally conscious consumers.
PS6	N/A	Directly applicable as the ability to configure and reconfigure modules streamlines internal processes, making them more adaptable and efficient.	N/A
PS7,8	N/A	Directly applicable as it allows factories to quickly assemble and deploy customised interfaces, improving efficiency and responsiveness to business needs.	N/A
PS9	N/A	Directly applicable as PKB optimises manufacturing processes by centralising essential production knowledge.	N/A
PS10,11	Indirectly applicable as enhanced DTs allow businesses to offer superior products by closely monitoring and adjusting production processes in real-time, ensuring high quality and customisation.	Directly applicable as DTS optimise manufacturing processes through the core services, improving efficiency and enabling agile responses to production challenges.	N/A
ALL PS	Organisational Transformation (OT): Directly applicable as implementing such tools would require changes in organisational structures, processes, digital culture and upskilling or reskilling of staff to utilise the new tools effectively.		

more customer-tailored products and services and modelling and assessment tools to ensure consistent customer quality. Additionally, these technologies can facilitate a mass customisation business model in manufacturing, as demonstrated by various pilots: Pilot 1 for virtual commissioning of new robots, Pilot 2 for production layout optimisation and Pilot 3 for enhancing internal logistics. Each pilot has shown the capability to adapt to specific customer requirements, showcasing the flexible application of these technologies across different manufacturing areas. In particular, by utilising modularity in assembly lines, with the supportive systems of emerging technologies, Pilot 2 can create a variety of gearmotor configurations in a set of standard production line modules, enabling the company to rapidly change its production design to offer a wide range of products without the complexity typically associated with a fixed line manufacturing.

Integrating modular systems requires organisational changes regarding the pillar of organisational transformation. In particular, the digital organisational culture, sponsorship from top management and users' digital skills significantly affect technologies' positive impact and exploitation. This evidence aligns with previous studies in manufacturing [25], which present evidence that emerging manufacturing systems mainly transform business processes. Considering all the above, modularity as an emerging mode of production (and products) can enhance DX efforts by recombining different systems and technologies to support manufacturing firms in the production decision-making process. This transformation also leads to direct and indirect benefits relevant to responsiveness to customer needs by providing critical capabilities that allow manufacturers to shift from mass production to mass customisation (as an example of pilots 1,3), providing new ways to create and capture value and enhancing customer experience.

36.7 Conclusion and Limitation

The contribution of this study lies in identifying the value proposition for modular manufacturing Information Systems, effectively bridging the knowledge gap about their strategic benefits by examining a to-be-developed system with the required features that enable it to fit the market needs [3, 4, 6, 14, 15]. The results of this study contribute to both the technology and innovation management literature and the manufacturing literature, enhancing our understanding of how digital technologies can be effectively implemented and leveraged in industrial settings, motivated by real-life industrial challenges. The research also elucidates how these systems can be integrated into the broader objectives of digital transformation in the manufacturing sector through the mapping of the products and services in a practitioner DX framework, potentially initiating a paradigm shift in how technology is perceived and utilised in industrial operations, what it already meets and what extensions it requires to enable a widespread DX in the firm. This directly responds to the call for further research highlighted in a recent digital transformation literature review by Kraus et al. [16].

The study also provides essential business implications for the four participating industries and companies from similar sectors looking to implement relevant technologies in practice. This study aids in informed decision-making by demonstrating how these systems align with and enhance manufacturing goals. Consequently, this study serves as a guide for manufacturers contemplating the adoption of modular systems, offering insights into their strategic utility. From the analysis, companies can identify industrial companies' potential pains, gains and customer jobs. They can understand how these innovative technologies and information systems can alleviate those pains, create gains and support customer jobs. Additionally, by employing the practitioner-oriented framework proposed by Doukidis et al. [24], organisations can map how these technologies correlate with their existing digital transformation efforts. Companies can also explore the strategic role of technologies to make informed investment decisions and ensure that they achieve their full digital transformation potential [25].

This study also acknowledges certain limitations, primarily due to the ongoing nature of the implementation of such a system. At present, there needs to be more empirical data from the assessment of the system, which restricts our ability to evaluate its long-term impact. Consequently, while the study offers valuable insights, these must be interpreted with an understanding that the actual implications may vary once more empirical data becomes available. Future research will be essential to validate and refine the conclusions drawn in this study, as it will benefit from a richer data set that includes practical results from the system's full-scale operation. In particular, the research will examine the actual impact of these technologies by analysing key performance indicators (KPIs) and conducting interviews with end users to investigate the differences between the potential and actual effects of the technologies. Further research will also involve additional interviews and focus groups with companies and other stakeholders to analyse the business value of these technologies, examine their impact on the business model elements of companies, and explore the possibility of creating new business models.

Finally, this study's methodology, which combines the Value Proposition Canvas with a digital transformation practitioners' framework by Doukidis et al. [24], can be replicated for other technology projects, both EU-funded and industrial, to facilitate the development of digital innovations.

Acknowledgement This research was funded by the H2020 Modapto project, a European Union's Horizon 2020 programme, under grant agreement No. 101091996.

References

1. Wang, B., Tao, F., Fang, X., Liu, C., Liu, Y., Freiheit, T.: Smart manufacturing and intelligent manufacturing: a comparative review. Engineering. **7**(6), 738–757 (2021). https://doi.org/10.1016/j.eng.2020.07.017

2. Jones, M.D., Hutcheson, S., Camba, J.D.: Past, present, and future barriers to digital transformation in manufacturing: a review. J. Manuf. Syst. **60**, 936–948 (2021). https://doi.org/10.1016/j.jmsy.2021.03.006
3. Napoleone, A., Andersen, A.-L., Brunoe, T.D., Nielsen, K.: Towards human-centric reconfigurable manufacturing systems: literature review of reconfigurability enablers for reduced reconfiguration effort and classification frameworks. J. Manuf. Syst. **67**, 23–34 (2023). https://doi.org/10.1016/j.jmsy.2022.12.014
4. Napoleone, A., Negri, E., Macchi, M., Pozzetti, A.: How the technologies underlying cyber-physical systems support the reconfigurability capability in manufacturing: a literature review. Int. J. Prod. Res. **61**(9), 3121–3143 (2023). https://doi.org/10.1080/00207543.2022.2074323
5. Yelles-Chaouche, A.R., Gurevsky, E., Brahimi, N., Dolgui, A.: Reconfigurable manufacturing systems from an optimisation perspective: a focused review of literature. Int. J. Prod. Res. **59**(21), 6400–6418 (2021). https://doi.org/10.1080/00207543.2020.1813913
6. Pansare, R., Yadav, G., Nagare, M.R.: Reconfigurable manufacturing system: a systematic review, meta-analysis and future research directions. J. Eng. Des. Technol. **21**(1), 228–265 (2021). https://doi.org/10.1108/JEDT-05-2021-0231
7. Wang, Y., Ma, H.-S., Yang, J.-H., Wang, K.-S.: Industry 4.0: a way from mass customisation to mass personalisation production. Adv. Manuf. **5**(4), 311–320 (2017). https://doi.org/10.1007/s40436-017-0204-7
8. Kombaya Touckia, J., Hamani, N., Kermad, L.: Digital twin framework for reconfigurable manufacturing systems (RMSs): design and simulation. Int. J. Adv. Manuf. Technol. **120**(7), 5431–5450 (2022). https://doi.org/10.1007/s00170-022-09118-y
9. Sabioni, R.C., Daaboul, J., Le Duigou, J.: An integrated approach to optimise the configuration of mass-customised products and reconfigurable manufacturing systems. Int. J. Adv. Manuf. Technol. **115**(1), 141–163 (2021). https://doi.org/10.1007/s00170-021-06984-w
10. Koren, Y., Gu, X., Guo, W.: Reconfigurable manufacturing systems: principles, design, and future trends. Front. Mech. Eng. **13**(2), 121–136 (2018). https://doi.org/10.1007/s11465-018-0483-0
11. Lameche, K., Najid, N.M., Castagna, P., Kouiss, K.: Modularity in the design of reconfigurable manufacturing systems. IFAC-PapersOnLine. **50**(1), 3511–3516 (2017). https://doi.org/10.1016/j.ifacol.2017.08.939
12. Mollajan, A., Hamedani-KarAzmoudehfar, F., Ashofteh, A., Shahdadi, A., Iranmanesh, S.H.: Design of integrated manufacturing information systems for reconfigurability and adaptability by modularising the system architecture. Int. J. Comput. Integr. Manuf. **37**, 509 (2023). https://doi.org/10.1080/0951192X.2023.2228262
13. Pakkanen, J., Juuti, T., Lehtonen, T.: Identifying and addressing challenges in the engineering design of modular systems–case studies in the manufacturing industry. J. Eng. Des. **30**(1), 32–61 (2019). https://doi.org/10.1080/09544828.2018.1552779
14. Boldt, S., Rösiö, C.: Evaluation of reconfigurability in brownfield manufacturing development. In: Säfsten, K., Elgh, F. (eds.) Advances in Transdisciplinary Engineering. IOS Press (2020). https://doi.org/10.3233/ATDE200190
15. Davendralingam, N., Guariniello, C., Xiao, L.: Assessing benefits of modularity in missions and system of systems. In: Systems Engineering for the Digital Age: Practitioner Perspectives, pp. 775–788. Scopus (2023). https://doi.org/10.1002/9781394203314.ch37
16. Kraus, S., Durst, S., Ferreira, J.J., Veiga, P., Kailer, N., Weinmann, A.: Digital transformation in business and management research: an overview of the current status quo. Int. J. Inf. Manag. **63**, 102466 (2022). https://doi.org/10.1016/j.ijinfomgt.2021.102466
17. Mehrabi, M., Ulsoy, A., Koren, Y.: Reconfigurable manufacturing systems and their enabling technologies. IJMTM. **1**, 114–131 (2000). https://doi.org/10.1504/IJMTM.2000.001330
18. Vial, G.: Understanding digital transformation: a review and a research agenda. J. Strateg. Inf. Syst. **28**(2), 118–144 (2019). https://doi.org/10.1016/j.jsis.2019.01.003

19. Favoretto, C., Mendes, G.H.d.S., Filho, M.G., Gouvea de Oliveira, M., Ganga, G.M.D.: Digital transformation of business model in manufacturing companies: challenges and research agenda. J. Bus. Ind. Mark. **37**(4), 748–767 (2021). https://doi.org/10.1108/JBIM-10-2020-0477
20. Abdallah, Y.O., Shehab, E., Al-Ashaab, A.: Understanding digital transformation in the manufacturing industry: a systematic literature review and future trends. Prod. Manag. Dev. **19**(1), e20200021 (2021). https://doi.org/10.4322/pmd.2021.001
21. Dutta, G., Kumar, R., Sindhwani, R., Singh, R.K.: Digital transformation priorities of India's discrete manufacturing SMEs—A conceptual study in perspective of Industry 4.0. Compet. Rev. **30**(3), 289–314 (2020). https://doi.org/10.1108/CR-03-2019-0031
22. Zhu, Q., Huang, S., Wang, G., Moghaddam, S.K., Lu, Y., Yan, Y.: Dynamic reconfiguration optimisation of intelligent manufacturing system with human-robot collaboration based on digital twin. J. Manuf. Syst. **65**, 330–338 (2022). https://doi.org/10.1016/j.jmsy.2022.09.021
23. Bouchard, S., Gamache, S., Abdulnour, G.: Operationalising mass customisation in manufacturing SMEs—A systematic literature review. Sustainability. **15**(4), Article 4 (2023). https://doi.org/10.3390/su15043028
24. Doukidis, G., Spinellis, D., Ebert, C.: Digital transformation—A primer for practitioners. IEEE Softw. **37**(5), 13–21 (2020). https://doi.org/10.1109/MS.2020.2999969
25. Farmakis, T., Lounis, S., Mourtos, I., Doukidis, G.: Digital twins as an integral part of manufacturing digital transformation. In: Prastacos, G., Pouloudi, N. (eds.) Leading and Managing in the Digital Era. LMDE 2023 Lecture Notes in Information Systems and Organisation, vol. 69. Springer, Cham (2024). https://doi.org/10.1007/978-3-031-65782-5_12
26. Singh, A., Gupta, S., Asjad, M., Gupta, P.: Reconfigurable manufacturing systems: journey and the road ahead. Int. J. Syst. Assur. Eng. Manag. **8**(2), 1849–1857 (2017). https://doi.org/10.1007/s13198-017-0610-z
27. Kim, D.-Y., Park, J.-W., Baek, S., Park, K.-B., Kim, H.-R., Park, J.-I., Kim, H.-S., Kim, B.-B., Oh, H.-Y., Namgung, K., Baek, W.: A modular factory testbed for the rapid reconfiguration of manufacturing systems. J. Intell. Manuf. **31**(3), 661–680 (2020)
28. Prasad, D., Jayswal, S.C.: Reconfigurability consideration and scheduling of products in a manufacturing industry. Int. J. Prod. Res. **56**(19), 6430–6449 (2018). https://doi.org/10.1080/00207543.2017.1334979
29. Shaik, A.M., Rao, V.V.S.K., Rao, C.S.: Development of modular manufacturing systems—A review. Int. J. Adv. Manuf. Technol. **76**(5), 789–802 (2015). https://doi.org/10.1007/s00170-014-6289-2
30. Bejlegaard, M., Brunoe, T.D., Nielsen, K.: Application of module drivers creating modular manufacturing equipment enabling changeability. Procedia CIRP. **52**, 134–138 (2016). https://doi.org/10.1016/j.procir.2016.07.059
31. Lampón, J.F., Cabanelas, P., González-Benito, J.: The impact of modular platforms on automobile manufacturing networks. Prod. Plan. Control. **28**(4), 335–348 (2017). https://doi.org/10.1080/09537287.2017.1287442
32. Brunoe, T.D., Soerensen, D.G.H., Nielsen, K.: Modular design method for reconfigurable manufacturing systems. Procedia CIRP. **104**, 1275–1279 (2021). https://doi.org/10.1016/j.procir.2021.11.214
33. Eisenhardt, K.M.: Building theories from case study research. Acad. Manag. Rev. **14**(4), 532–550 (1989). https://doi.org/10.5465/amr.1989.4308385
34. Yin, R.K.: Discovering the future of the case study method in evaluation research. Eval. Pract. **15**(3), 283–290 (1994). https://doi.org/10.1016/0886-1633(94)90023-X
35. Yin, R.K.: Case Study Research: Design and Methods. SAGE (2009)
36. Osterwalder, A., Pigneur, Y., Bernarda, G., Smith, A.: Value Proposition Design: How to Create Products and Services Customers Want. John Wiley & Sons (2015)
37. Helmold, M.: Creating the value proposition. In: Management for Professionals, Part F440, pp. 89–94. Scopus (2020). https://doi.org/10.1007/978-3-030-46985-6_9

38. Neuhüttler, J., Woyke, I.C., Ganz, W.: Applying value proposition design for developing smart service business models in manufacturing firms. In: Freund, L.E., Cellary, W. (eds.) Advances in the Human Side of Service Engineering, pp. 103–114. Springer International Publishing (2018). https://doi.org/10.1007/978-3-319-60486-2_10
39. Karatzas, S., Farmakis, P., Chassiakos, A., Farmakis, T., Christoforou, Z., Liappi, G.: Development of a multi criteria model for assisting EV user charging decisions. J. Sustain. Dev. Energy Water Environ. Syst. **11**, 1–23 (2022). https://doi.org/10.13044/j.sdewes.d10.0439

Chapter 37
Enhancing User Centricity in the Growth Hacking Process: A Conceptual Framework

Claudia Perillo, Annaluce Mandiello, Anna Bastone, and Federica Zeuli

Abstract Nowadays markets and society in general are rapidly changing leading companies to emphasize innovation management and marketing. In this respect, Growth Hacking emerges as a strategic approach to grow fast through low-cost experiments, enhancing innovation processes. Innovation, indeed, does not guarantee the survival of a company but user-centricity combined with it could do it. For these reasons, this work investigates how a user-centered approach can be applied to growth hacking in order to activate value co-creation. These concepts are central to modern marketing research and innovation management. The authors propose an integration of the Growth Hacking model with the Design Thinking one, given its focus on empathy and humanization of the process. Design Thinking is generally known for its focus on user needs and collaborative innovation. Future steps concern the conduction of a multiple case study of hackathon to investigate how the principles of Design Thinking can help Growth Hacking processes to provide innovation user oriented and effective.

Keywords Growth hacking · User-centricity · Design thinking · Hackathon

C. Perillo (✉)
Department of Communication Science, University of Teramo, Teramo, Italy
e-mail: cperillo@unisa.it

A. Mandiello · A. Bastone · F. Zeuli
Department of Management Studies and Quantitative Methods, University of Naples Parthenope, Naples, Italy

F. Schiavone et al. (eds.), *Advanced Perspectives and Trends in Digital Transformation of Firms, Networks, and Society*, Springer Proceedings in Business and Economics, https://doi.org/10.1007/978-3-031-80692-6_37

37.1 Introduction

37.1.1 Research Context

Companies are paying particular attention and strategic relevance to the management and marketing of innovation, given the fact that the world, and consequently the market, is continually and rapidly changing. In this respect, implementing the Growth Hacking (GH) approach, an established marketing model is increasingly becoming an effective way of innovating as fast as society does. This study is based on this first consideration. Moreover, to gain a competitive advantage, companies are dealing also with market complexity and intense competition across different industries [1]. Companies must grow quickly and adopt a business model that is scalable to achieve success [2]. Once again, the growth challenges of today's businesses could be tackled effectively and efficiently through GH. However, innovation is not the only factor on which firms can build their survival. Moving from the "value co-creation" well-established concept, user centricity [3–5] is becoming more and more relevant in marketing streams of research. It is not possible, hence, thinking about an innovation revolutionizing company's processes without considering its users. Thus, the research question of this study is "How to apply a user-centered approach to GH?." To answer this question, the authors, according to the literature, consider that user-centricity and value co-creation clearly emerge within the Design Thinking (DT) framework.

Thus, the authors identify and merge DT and GH characteristics in a case study to understand how to integrate DT in GH to make it user-centered.

37.2 Objective

37.2.1 Purpose of the Work

Growth Hacking

Nowadays, competitive landscapes strengthened companies' necessity to innovate in order to grow, it has become essential for any business [6]. GH was first defined by Ellis [7] as "a process of rapid experimentation across the funnel to learn the most effective way to scale sustainable customer adoption." Being it a process [7, 8], a marketing strategy, a predefined process [9], and a paradigmatic approach [10, 11]. GH represents a way to pursue innovation aiming at the maximization of user attraction at a low cost [12, 13]. GH concerns the conduction of experiments in business settings [14–16] to enhance business value [17]. Furthermore, scholars recognize GH as a new way of scaling businesses [10, 18, 19], applicable to different types of businesses [11, 20, 21].

Another relevant consideration on the current market is related to the increasing relevance, in terms of driver of competitive advantage, acquired by data collection, storage, analysis, and management, and the subsequent re-shaping of organizational routines [9, 22]. GH makes it possible by integrating digital technologies, such as data, tools, and platforms, into different types of businesses [15, 23]. Indeed, GH leverages advanced digital technologies to achieve fast sustainable growth [10, 11, 15, 18, 24]. In order to align strategy, use big data, and make decisions, GH could be a viable option [25].

Design Thinking

Given the rising relevance of user-centricity in designing and developing solutions, the idea of "one size fits all" has been officially overcome [26]. Thinking out of the box and user-centricity are proper of a well-established problem-solving approach: Design Thinking (DT). DT, indeed, leverages interdisciplinary teams, qualitative user research methods, rapid prototyping, thinking out of the box, and iterative learning cycles to foster innovation and entrepreneurship [27].

Indeed, to face nowadays increasing competition, innovation should meet agility, dynamicity, and organizational capabilities to allow companies to face uncertainty and complexity [28]. In this respect, DT represents an effective solution, being it an approach for creative problem solving [29–31] and a human-centered approach [32–34] for rapid and conceptual prototyping [35] and for business modeling [36].

In 2008, the HPI-Stanford DT Research Program was run to define a DT methodology. It is built on the following steps:

- Empathize—with the users;
- Define—the users' needs, their problems, and the insights collected about it;
- Ideate—by challenging assumptions and creating ideas for innovative solutions;
- Prototype—to start creating solutions;
- Test—solutions.

A possible subsequent phase could be represented by iteration of the process itself. This process allows companies to adjust their offering on customer's needs.

37.3 Methodology

37.3.1 Multiple Case-Study

This research is grounded on a background analysis on GH and DT. The literature overview highlighted strengths, weaknesses, and common roots of the two models. In particular, the derived theoretical hypotheses state that it is necessary to integrate a user-centered approach in GH to ensure customers' satisfaction and thus retention.

In this way, competitive advantage could be achieved. Then, the authors are going to conduct a multiple case-study methodology in the context of hackathons, to investigate the integration of approaches abovementioned. Hackathon could represent an effective approach to generate innovation in which user-centricity and fast growth are pursued simultaneously.

37.4 Results

37.4.1 Expected Findings

The different phases related to both models have been identified and distinguished. GH Pirate Funnel [37] is a GH framework aimed at achieving competitive advantage [1, 38–43], sustainable fast growth [7, 8] and enhanced sales effectiveness. This model shows the different necessary stages [18, 37] to turn a so-called "passenger" into a user. The first phase is called "awareness," meaning that this is the moment in which potential user is in contact with company's product/service or brand. The "acquisition" phase aims at acquiring qualified leads, converting them into paying customers. During the "activation" phase, new customers are activated by inducing them to do "designated actions"; in this way, the user experience and loyalty are activated [10]. To gain user monetization, the "revenue" phase aims to optimize the cash flow strategies increasing customer lifetime value [44]. The last phase is the "retention" one, in which the aim is retaining existing customers. A further phase could be the "referral" one [17]. In this stage, retained users refer to others about the products, acting as ambassadors.

On the other hand, for what concerns DT, the HPI-Standford's DT process (2008) is made of five different phases. The first phase deals with empathizing with the user [45]. Then, it is necessary to collect insights and summarize customers, "defining" the problem [46]. In the third phase, the "ideation" one, it is necessary to ideate the wider possible range of solutions [47]. In the "prototype" phase, the most promising ideas are transformed into prototypes [48]. The last step deals with testing the prototype with real users, collecting feedbacks, and validating assumptions. In conclusion, in order to integrate user-centricity to GH, DT phases and GH ones are merged. In particular, the authors applied:

- Empathy capabilities to the awareness phase.
- Definition techniques to the acquisition phase.
- Ideation phase to the activation phase.
- Prototyping activities to affect revenue monetization.
- Testing procedures to customer retention.

Furthermore, the integrated framework benefits from both the learning phase of DT and the iteration phase of GH and, the next step of this research, see it applied to the hackathon case. Hackathon are digital competitions (hacking + marathon) in which multidisciplinary teams collaborate to solve a problem, innovate, or create a product/service, engineering over a short period of time [49].

GH and DT in the context of a hackathon refers to the development of a solution characterized by both fast growth and user-centricity. More precisely, applying the phases of the integrated GH and DT model allows to guide Hackathon activities:

- Empathy and Awareness: who are the target clients (preferences, behavior, habits, etc.)? How can they come in touch with the product/service or the brand? Which are the potential channels?
- Definition and Acquisition: proper definition of the problem to assure the acquisition of the customer. What is the need to be satisfied?
- Ideation and Activation: what is the design of the user experience and user journey? Which are the necessary "designated action"?
- Prototyping and Revenue: how does the prototype works? What is the monetization architecture?
- Test and Retention: how is it possible to retain customers?

Teams involved into a hackathon can build a functional prototype and also consider critical aspects of user acquisition, engagement, and growth by utilizing the integrated GH and DT model. The chance of a hackathon project becoming a successful and sustainable product or service can be increased by this user-centered approach to fast growth.

37.5 Implications

37.5.1 Expected Implications

This work contributes to literature on growth hacking, innovation management, user-centricity, and design-thinking. The integrated framework, in which DT Standford's phases are integrated into the GH Pirate Funnel could represent an effective checklist for hackathons. The solution engineering, indeed, could be enhanced through the derived model, addressing both fast growth and user-centricity.

References

1. Mariani, M.M., Wamba, S.F.: Exploring how consumer goods companies innovate in the digital age: the role of big data analytics companies. J. Bus. Res. **121**, 338–352 (2020)
2. Hoffman, R., Yeh, C.: Blitzscaling: the Lightning-Fast Path to Building Massively Valuable Companies. Crown Currency (2018)
3. Bhargav-Spantzel, A., Camenisch, J., Gross, T., Sommer, D.: User centricity: a taxonomy and open issues. In: Proceedings of the Second ACM Workshop on Digital Identity Management, pp. 1–10 (2006)

4. Patton, J.: Understanding user centricity. IEEE Softw. **24**(6), 9–11 (2007)
5. Recordon, D., Reed, D.: OpenID 2.0: a platform for user-centric identity management. In: Proceedings of the Second ACM Workshop on Digital Identity Management, pp. 11–16 (2006)
6. Bargoni, A., Smrčka, L., Santoro, G., Ferraris, A.: Highway to hell or paradise city? Exploring the role of growth hacking in learning from innovation failure. Technovation. **131**, 102945 (2024)
7. Ellis, S.: Find a Growth Hacker for Your Startup. Startup Marketing (2010)
8. Ellis, S., Brown, M.: Hacking Growth: How Today's Fastest-Growing Companies Drive Breakout Success. Currency (2017)
9. Bargoni, A., Santoro, G., Petruzzelli, A.M., Ferraris, A.: Growth hacking: a critical review to clarify its meaning and guide its practical application. Technol. Forecast. Soc. Chang. **200**, 123111 (2024)
10. Bohnsack, R., Liesner, M.M.: What the hack? A growth hacking taxonomy and practical applications for firms. Bus. Horiz. **62**(6), 799–818 (2019)
11. Troisi, O., Maione, G., Grimaldi, M., Loia, F.: Growth hacking: insights on data-driven decision-making from three firms. Ind. Mark. Manag. **90**, 538–557 (2020)
12. Jaring, P., Bäck, A., Komssi, M., & Käki, J. (2015). Using Twitter in the acceleration of marketing new products and services. Journal of Innovation Management, **3**(3), 35–56
13. Kemell, K. K., Feshchenko, P., Himmanen, J., Hossain, A., Jameel, F., Puca, R. L., ... & Abrahamsson, P. (2019, August). Software startup education: gamifying growth hacking. In Proceedings of the 2nd ACM SIGSOFT International Workshop on Software-Intensive Business: Start-ups, Platforms, and Ecosystems (pp. 25–30)
14. Brecht, P., Hendriks, D., Stroebele, A., Hahn, C.H., Wolff, I.: Discovery and validation of business models: how B2B startups can use business experiments. Technol. Innov. Manag. Rev. **11**(3) (2021)
15. Conway, T., Hemphill, T.: Growth hacking as an approach to producing growth amongst UK technology start-ups: an evaluation. J. Res. Mark. Entrep. **21**(2), 163–179 (2019)
16. Hu, Y.: Tealeaf CxMobile-replaying real time customer experience. In: 2016 24th International Conference on Geoinformatics, pp. 1–4. IEEE (2016)
17. Feiz, D., Zarei, A., Maleki Minbashrazgah, M., Shaabani, A.: Typology of growth hacking strategies along the growth hacking funnel. Interdiscip. J. Manag. Stud. **14**(2), 331–346 (2021)
18. Cavallo, A., Cosenz, F., Noto, G.: Business model scaling and growth hacking in digital entrepreneurship. J. Small Bus. Manag. **1-28** (2023)
19. Sanasi, S., Ghezzi, A., Cavallo, A.: What happens after market validation? Experimentation for scaling in technology-based startups. Technol. Forecast. Soc. Chang. **196**, 122839 (2023)
20. Coll-Rubio, P., Carbonell, J.M.: Growth communication strategies in the digital age. Am. Behav. Sci. **68**(1), 43–55 (2024)
21. Tumbas, S., Berente, N., Vom Brocke, J.: Born digital: growth trajectories of entrepreneurial organizations spanning institutional fields. In: ICIS (2017)
22. Bertello, A., Ferraris, A., Bresciani, S., & De Bernardi, P. (2021). Big data analytics (BDA) and degree of internationalization: the interplay between governance of BDA infrastructure and BDA capabilities. Journal of Management and Governance, **25**(4), 1035–1055
23. Pascucci, F., Savelli, E., Gistri, G.: How digital technologies reshape marketing: evidence from a qualitative investigation. Ital. J. Mark. **2023**(1), 27–58 (2023)
24. Bargoni, A., Jabeen, F., Santoro, G., Ferraris, A.: Growth hacking and international dynamic marketing capabilities: a conceptual framework and research propositions. Int. Mark. Rev. **41**, 74 (2023)
25. Rosemann, M., & vom Brocke, J. (2015). The six core elements of business process management. In J. vom Brocke & M. Rosemann (Eds.) Handbook on business process management 1 (pp. 105–122). Berlin: Springer
26. Zahid, A., Yadav, U.N., Mistry, S.K.: Digital health services in Bangladesh—The need for a sustainable design framework. World Med. Health Policy. **15**(4), 682–695 (2023)
27. Brown, T.: Design thinking. Harv. Bus. Rev. **1** (2008)

28. Mortati, M., Magistretti, S., Cautela, C., Dell'Era, C.: Data in design: how big data and thick data inform design thinking projects. Technovation. **122**, 102688 (2023)
29. Jaskyte, K., & Liedtka, J. (2022). Design thinking for innovation: Practices and intermediate outcomes. Nonprofit Management and Leadership, **32**(4), 555–575
30. Kamble, S., Rana, N. P., Gupta, S., Belhadi, A., Sharma, R., & Kulkarni, P. (2023). An effectuation and causation perspective on the role of design thinking practices and digital capabilities in platform-based ventures. Technological Forecasting and Social Change, **193**, 122646
31. Magistretti, S., Dell'Era, C., Cautela, C., & Kotlar, J. (2023). Design thinking for organizational innovation at PepsiCo. California Management Review, **65**(3), 5–26
32. Buchanan, R. (2001). Human dignity and human rights: Thoughts on the principles of human-centered design. Design issues, **17**(3), 35–39
33. Chouki, M., de Mozota, B. B., Kallmuenzer, A., Kraus, S., & Dabic, M. (2021). Design thinking and agility in digital production: The key role of user experience design. IEEE transactions on engineering management, **70**(12), 4207–4221
34. Dam, R. F., & Siang, T. Y. (2021, December). What is design thinking and why is it so popular? Interaction Design Foundation
35. Plattner, H.: An Introduction to Design Thinking, Process Guide. Institute of Design at Stanford (2018)
36. Santa-Maria, T., Vermeulen, W.J., Baumgartner, R.J.: The Circular Sprint: circular business model innovation through design thinking. J. Clean. Prod. **362**, 132323 (2022)
37. McClure, S., Shah, S., Shah, S.: Web Hacking: Attacks and Defense. Addison-Wesley Professional (2003)
38. Carneiro, A.: How does knowledge management influence innovation and competitiveness? J. Knowl. Manag. **4**(2), 87–98 (2000)
39. Dereli, D.D.: Innovation management in global competition and competitive advantage. Procedia Soc. Behav. Sci. **195**, 1365–1370 (2015)
40. Hana, U.: Competitive advantage achievement through innovation and knowledge. J. Compet. **5**(1), 82–96 (2013)
41. Kim, W.C., Mauborgne, R.: Strategy, value innovation, and the knowledge economy. MIT Sloan Manag. Rev. (1999)
42. Michalisin, M.D., Smith, R.D., Kline, D.M.: In search of strategic assets. Int. J. Org. Anal. **5**(4), 360–387 (1997)
43. Weerawardena, J., Mavondo, F.T.: Capabilities, innovation and competitive advantage. Ind. Mark. Manag. **40**(8), 1220–1223 (2011)
44. Sineni, S.L.: Growth Hacking: a Deep Look into Online Marketing for Startups. Doctoral dissertation, University of Southern California (2014)
45. Köppen, E., Meinel, C.: Empathy via design thinking: creation of sense and knowledge. In: Design Thinking Research: Building Innovators, pp. 15–28. Springer International Publishing, Cham (2014)
46. Wanasinghe, T. R., Wroblewski, L., Petersen, B. K., Gosine, R. G., James, L. A., De Silva, O., ... & Warrian, P. J. (2020). Digital twin for the oil and gas industry: Overview, research trends, opportunities, and challenges. IEEE access, **8**, 104175–104197
47. Kamrowska-Załuska, D.D., Parteka, T.: Design thinking (DT) for the design and planning education of engineer-architects. World Trans. Eng. Technol. Educ. **18**, 97–101 (2020)
48. Isa, S.S., Liem, A., Steinert, M.: The value of prototypes in the early design and development process. In: DS 80–5 Proceedings of the 20th International Conference on Engineering Design (ICED 15) Vol 5: Design Methods and Tools-Part 1, Milan, Italy, 27–30.07.15, pp. 235–242 (2015)
49. Heller, B., Amir, A., Waxman, R., Maaravi, Y.: Hack your organizational innovation: literature review and integrative model for running hackathons. J. Innov. Entrep. **12**(1), 6 (2023)

Chapter 38
Exploring the Relevance of the Metaverse for Consumers: A Systematic Literature Review

Marcello Risitano, Giuseppe La Ragione, Maria Palazzo, and Francesco Parola

Abstract The rise of digitalization is transforming our world, pushing people to spend considerable time on digital platforms (e.g., social media, branded apps, metaverse). Consumers' desire to participate in value co-creation processes and tools that enable them to feel included, such as the metaverse, are highly encouraging. The metaverse has transitioned from fictional notion to a thriving digital ecosystem, functioning as a collective of virtual spaces where users navigate between digital and physical realities. It fosters social interactions through digital avatars, supports value generation, and experience co-creation by integrating mixed reality, ambient intelligence, and advanced ICTs. This immersive platform not only enhances consumer interactions but also reshapes work, socialization, and experiential dynamics beyond physical constraints. By exploring virtual service dynamics, product interactions, and engaging immersive experiences, the metaverse reveals vast potential for redefining consumer behavior and value creation. It signifies a network of interconnected virtual worlds, continuously evolving as a crucial element in digital interaction and engagement. This chapter aims to decode the metaverse's transformative impact from a consumer perspective, focusing on the primary bibliometric trends, thematic areas, and knowledge structures within this realm.

Keywords Metaverse · Consumer · Systematic literature review

M. Risitano · G. La Ragione (✉)
University of Naples "Parthenope", Naples, Italy
e-mail: giuseppe.laragione001@studenti.uniparthenope.it

M. Palazzo
Mercatorum University, Rome, Italy

F. Parola
Mercatorum University (On Going), Rome, Italy

Transport Regulation Authority, Turin, Italy

F. Schiavone et al. (eds.), *Advanced Perspectives and Trends in Digital Transformation of Firms, Networks, and Society*, Springer Proceedings in Business and Economics, https://doi.org/10.1007/978-3-031-80692-6_38

38.1 Introduction

The rise of digitalization is transforming our world, pushing people to spend considerable time on digital platforms (e.g., social media, branded apps, metaverse). Consumers' desire to participate in value co-creation processes and tools that enable them to feel included, such as the metaverse, are highly encouraging.

The metaverse, a term first coined by Stephenson [1], has evolved from a fictional concept into a burgeoning digital ecosystem. It is seen as a collection of virtual environments where users can seamlessly move between digital and physical realms [2]. This digital space empowers users to interact socially through digital avatars, facilitating the generation of value and co-creation of experiences. The metaverse combines mixed reality (MR) technology, ambient intelligence, and advancements in information and communication technologies (ICTs), thereby providing a bridge between the digital and physical universes [3, 4].

The significance of studying the metaverse lies in its potential to revolutionize how consumers interact, work, socialize, and experience the world. As a parallel virtual reality, the metaverse offers immersive experiences not confined to the physical world's limitations. It allows for the exploration of virtual service businesses, interaction with virtual features of products and services, and engagement in immersive experiences that deepen users' understanding of offerings. The metaverse extends beyond single products or platforms, referring to a host of interconnected virtual worlds that serve as persistent spaces for information interaction and engagement [3].

Research into the metaverse is critical for understanding its impact on various domains, including social interactions, commerce, education, and entertainment. Recently, scholars are focusing on the metaverse facets. For instance, Narin [5] conducted a content analysis on metaverse literature, revealing key themes such as virtual worlds, augmented reality, and the social and economic potential of these digital spaces. Similarly, Kim [6] discussed the metaverse's capacity to enhance interactive advertising, raising important questions about privacy, ethics, and the valuation of goods and services within these virtual environments.

The metaverse's importance is further underscored by its role in facilitating immersive interactions that can mitigate the perceived risks associated with service offerings. By leveraging on virtual reality (VR), augmented reality (AR), and mixed reality (MR) technologies, the metaverse provides multisensory experiences that heighten individuals' sense of engagement and satisfaction [7]. This immersive environment supports social and economic activities and introduces disruptions that redefine consumer behavior and value co-creation [4]. As this digital space continues to evolve, research will play a crucial role in shaping its trajectory, ensuring that the metaverse remains an inclusive, engaging, and valuable component of our digital future.

From these considerations, the purpose of this chapter is to understand the state of the art of the literature regarding the metaverse from a consumer perspective.

Based on previous statements, it is vital to define the following research questions to guide our work:

RQ1: What are the main bibliometric variables, research trends, and recurrent thematic areas in the use of metaverse from a consumer perspective?
RQ2: What are the knowledge clusters in the intellectual structure in this research area?

38.2 Objective

This chapter aims to analyze the literature about the metaverse by understanding its relevance. In particular, by exploring the knowledge clusters, the work aims to explore the relevant theme present in the research area.

38.3 Methods

The chapter adopts a structured approach to examine academic works, utilizing a multi-step procedure known as the systematic literature review (SLR) methodology. Chaudhary et al. [8] highlight the SLR's benefits, including the precise selection of research that matches the inquiry question and contributes theoretical insights and frameworks that spur further investigation. The research employs the PRISMA Protocol, as outlined by Moher et al. [9], to ensure a standardized selection methodology. Conference papers, books, book chapters, and PhD dissertations are excluded due to their minimal impact on academic literature. Following this, a content analysis is conducted on academic articles to extract key bibliographic information from studies focusing on metaverse and consumer. The analysis utilizes Microsoft Excel and VOSviewer Software for further examination. Initially, a bibliometric analysis is undertaken to investigate the research topic, identifying significant journal contributions and categorizing papers by their types and methodologies. Then, trends and thematic areas are explored through a co-occurrence analysis of keywords, providing insights into the evolving research landscape in this field [10]. The PRISMA approach's initial step identifies the dataset and defines the research scope and focus, selecting the Scopus database for its extensive coverage and widespread use in research for its comprehensive selection of scientific journals, using as specific keywords "metaverse and consumer." These keywords have been searched within paper titles, abstracts, or keywords. This process resulted in a preliminary database of 323 records spanning from 2008 to February 2024. After refining through specific exclusion criteria: Ex1, non-article documents; Ex2, documents not in the business, management, or accounting fields; Ex3, non-English documents; Ex4, documents not published in peer-reviewed international journals, the database was narrowed down to 68 articles. The final phases of the PRISMA protocol involve setting eligibility criteria and selecting papers for the final sample after thoroughly reviewing all full-text articles. This led to the exclusion of 30 articles and the compilation of a sample of 38 bibliographic records. All bibliometric information from the final sample is exported to Microsoft Excel, with Scopus facilitating the extraction of essential data such as titles, abstracts, keywords, author names, publication years, citation counts, and journal names. This dataset is then enhanced with additional information relevant to the research objectives, particularly concerning the approach of the analyzed sample.

38.4 Findings

The analysis shows a growing trend of publications exploring the metaverse from a consumer perspective. The distribution across years reveals a noticeable upward trajectory in the number of studies conducted. This trend underlines the escalating interest in this particular field especially in the last 2 years. In relation to research focus, 11 studies adopt a quantitative approach, 16 follow a qualitative methodology, 2 encompass a combination of both quantitative and qualitative methods, and 9 embrace a systematic literature review (Table 38.1).

The preeminent outlet for distributing metaverse research is the *Journal of Consumer Psychology*, which encompasses three publications of 38 of the studies within our sample. Other journals that study this topic can be found in Table 38.2.

Utilizing VOSviewer software [11–13], we conducted a co-occurrence analysis of keywords. This method, grounded in bibliometrics, was developed in the Java programming language. It allows us to chart networks among keywords, resulting in clusters that highlight primary research domains. Through co-occurrence analysis, the chapter pinpoints instances where chosen keywords appear together in articles, signifying their interconnectedness. VOSviewer employs a natural language processing algorithm to gauge keyword associations and form clusters, color-coding them for differentiation [13]. The resultant visual representation showcases "bubbles" their sizes reflective of keyword frequency within selected publications [14]. The findings enable us to distinguish and interconnect four discrete clusters, each comprising a collection of keywords exhibiting co-occurrence associations. A keyword is allocated to a sole cluster exclusively.

Table 38.1 Paper type

Category	n	Percentage
Qualitative	16	42.10
Quantitative	11	28.94
Systematic literature review	9	23.68
Mixed methods	2	5.26

Table 38.2 Top 10 most relevant contributions

Journal	n	Percentage
Journal of Consumer Psychology	3	7.89
International Journal of Consumer Studies	2	5.26
International Journal of Contemporary Hospitality Management	2	5.26
Journal of Business Research	2	5.26
Journal of Retailing and Consumer Services	2	5.26
Psychology and Marketing	2	5.26
Tourism Review	2	5.26
AMS Review	1	2.63
Asia Pacific Journal of Marketing and Logistics	1	2.63
Business Horizons	1	2.63

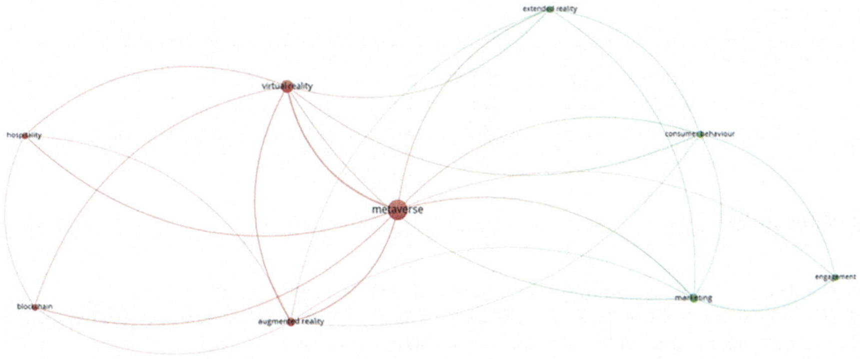

Fig. 38.1 Co-occurrence analysis (VOSviewer)

The bibliographic coupling consists of two clusters (Fig. 38.1), encompassing 9 primary items and interconnected by 25 links. The cumulative strength of these links amounts to 57. In particular:

Cluster 1 (Metaverse and tools) comprises five items: augmented reality, blockchain, hospitality, metaverse, and virtual reality.

Cluster 2 (Metaverse to engage consumers) comprises four items: consumer behavior, engagement, extended reality, and marketing.

38.5 Implications

This work has both theoretical and managerial implications. To the best of the authors' knowledge, this is the first SLR that studies the metaverse and its impact on consumers. The research findings provide theoretical insights to evaluate the relevance of the digital technologies related to the metaverse in creating new consumption models—not confined to the physical world's limitations—to define "hybrid customer experience" through a multi-sensory involvement of the consumer before the purchase phase, during the purchase, and/or after the purchase. Secondly, it allows for an understanding of the current literature's themes on the topic, paving the way for future research, and evaluating the academic challenges in terms of adoption of new research methods (e.g., neuromarketing techniques) and exploration of specific business contexts (e.g., tourism, retail, gaming). For managers, on the other hand, understanding the benefits of metaverse serves as a powerful competitive tool to achieve consumer engagement and involve the customers in the value co-creation of experiences, maximizing consumer–brand relationships in the long term.

Disclaimer The views and opinions expressed in this chapter are those of the author and do not necessarily reflect the official policy or position of the respective institution.

References

1. Stephenson, N.: Snow crash: A novel. New York, NY, Bantam Books. (1992)
2. Zhao, Y., Jiang, J., Chen, Y., Liu, R., Yang, Y., Xue, X., Chen, S.: Metaverse: perspectives from graphics, interactions and visualization. Vis. Inform. **6**(1), 56–67 (2022)
3. Gursoy, D., Malodia, S., Dhir, A.: The metaverse in the hospitality and tourism industry: an overview of current trends and future research directions. J. Hosp. Market. Manag. **31**(5), 527–534 (2022)
4. Buhalis, D., Harwood, T., Bogicevic, V., Viglia, G., Beldona, S., Hofacker, C.: Technological disruptions in services: lessons from tourism and hospitality. J. Serv. Manag. **30**(4), 484–506 (2019)
5. Narin, N.G.: A content analysis of the metaverse articles. J. Metaverse. **1**(1), 17–24 (2021)
6. Kim, J.: Advertising in the metaverse: research agenda. J. Interact. Advert. **21**(3), 141–144 (2021)
7. Koohang, A., Nord, J., Ooi, K., Tan, G., Al-Emran, M., Aw, E., et al.: Shaping the metaverse into reality: multidisciplinary perspectives on opportunities, challenges, and future research. J. Comput. Inf. Syst. **63**(3), 735–765 (2023)
8. Chaudhary, S., Dhir, A., Ferraris, A., Bertoldi, B.: Trust and reputation in family businesses: a systematic literature review of past achievements and future promises. J. Bus. Res. **137**, 143–161 (2021)
9. Moher, D., Shamseer, L., Clarke, M., Ghersi, D., Liberati, A., Petticrew, M., Shekelle, P., Stewart, L.A.: Preferred reporting items for systematic review and meta-analysis protocols (PRISMA-P) 2015 statement. Syst. Rev. **4**, 1 (2015)
10. Tranfield, D., Denyer, D., Smart, P.: Towards a methodology for developing evidence informed management knowledge by means of systematic review. Br. J. Manag. **14**(3), 207–222 (2003)

11. Biggi, G., Giuliani, E.: The noxious consequences of innovation: what do we now? Ind. Innov. **28**(1), 19–41 (2020)
12. Secinaro, S., Calandra, D., Lanzalonga, F., Ferraris, A.: Electric vehicles' consumer behaviours: mapping the field and providing a research agenda. J. Bus. Res. **150**, 399–416 (2022)
13. Van Eck, N.J., Waltman, L., Dekker, R., Van Den Berg, J.: A comparison of two techniques for bibliometric mapping: multidimensional scaling and VOS. J. Am. Soc. Inf. Sci. Technol. **61**(12), 2405–2416 (2010)
14. Van Eck, N.J., Waltman, L.: Software survey: VOSviewer, a computer program for bibliometric mapping. Scientometrics. **84**(2), 523–538 (2010)

Chapter 39
Green Mobility Choices to Build Consumer-Brand Relationships: A Multigroup Analysis

Francesco Calza, Giuseppe La Ragione, Marcello Risitano, and Annarita Sorrentino

Abstract In recent times, the escalating problem of global pollution has demanded urgent attention from various stakeholders, including governments, businesses, and consumers. The digital age has catalyzed a shift in consumer attitudes, increasing awareness about the environmental impacts of their consumption choices and fostering the rise of "green consumers" who prioritize products with minimal environmental footprints. This chapter examines the influence of Green Consumption Values on brand experience, brand trust, and the intention to engage in word of mouth within the mobility sector, including cruise and automotive industries. These sectors have been scrutinized for their environmental impacts, leading to significant sustainability investments to align with consumer expectations and leverage sustainability as a competitive advantage. The multigroup structural equation modeling analysis of responses from 709 participants reveals that green values positively influence brand experience and trust, which in turn, enhance consumer word-of-mouth intentions, underscoring the critical role of environmental commitment in consumer–brand relationships within the mobility sector.

Keywords Green consumer · Consumer–brand relationships · Mobility sectors · Structural equation modeling

39.1 Introduction

In recent times, the problem of global pollution has escalated to a level that demands serious attention. To counteract this issue, diverse stakeholders, including governments, politicians, and business leaders, are actively searching for ways to reduce

F. Calza · G. La Ragione (✉) · M. Risitano · A. Sorrentino
University of Naples "Parthenope", Naples, Italy
e-mail: giuseppe.laragione001@studenti.uniparthenope.it

F. Schiavone et al. (eds.), *Advanced Perspectives and Trends in Digital Transformation of Firms, Networks, and Society*, Springer Proceedings in Business and Economics, https://doi.org/10.1007/978-3-031-80692-6_39

the environmental footprint of the products and services they offer to society. The easy access to information has led to a shift in consumer attitudes thanks to the digital technologies, making them more aware of the environmental consequences of their consumption choices. This shift signifies a change in consumer behavior, with individuals now being identified by researchers as "green consumers." A green consumer is someone who evaluates the environmental impact of a product prior to purchasing it [1, 2]. According to Haws et al. [1], Green Consumption Value (GCV) refers to the efforts made by consumers to decrease pollution by changing their buying habits. These consumers tend to support brands that reflect their ethical standards [3, 4], preferring products that align with their values. In light of this, mobility is one of the most important sectors affected by sustainability [5]. Cruise and automotive companies must showcase their commitment to environmental sustainability to meet consumer expectations supported by digital technologies.

Cruise company and automotive producers have come under scrutiny for their significant environmental footprint, including high levels of fossil fuel emissions, improper wastewater management, and other environmentally unfriendly practices. In response, the mobility industry has made substantial investments in sustainability initiatives in recent years [6, 7], adopting new guidelines, digital technologies for supporting green consumer attitude, and some standards to reduce their carbon footprint and promote sustainable cruising practices. This approach positions sustainability as a key factor for competitive advantage.

The credibility of the companies in the eyes of its consumers is fragile; failing to deliver on environmental promises can severely damage brand trust. Additionally, in the mobility sector, the concept of repeat customers and their influence through word of mouth is vital for business success [8]. It's common for individuals who have previously had a positive brand experience and high brand trust to share their experiences with others, highlighting the importance of maintaining a positive and trustworthy brand image in the industry.

According to previous statements, we hypothesized the following:

H1a: Green Consumption Values impact positively on Brand Experience.
H1b: Green Consumption Values impact positively on Intent to Word of Mouth.
H1c: Green Consumption Values impact positively on Brand Trust.
H2a: Brand Experience impact positively on Brand Trust.
H2b: Brand Experience impact positively on Intent to Word of Mouth.
H3a: Brand Trust impact positively on Intent to Word of Mouth.
H4: Mobility categories moderate the relationships among the construct.

39.2 Objective

This chapter aims to explore the effect of green consumption values on consumer–brand relationships that can lead to an intention to engage in word of mouth. The research was conducted within the mobility sector, offering an innovative comparison between mobility services (e.g., cruises) and mobility products (e.g., automotive).

39.3 Methods

In this chapter, to delve into the observed phenomenon, we employed multigroup structural equation modeling (SEM) as the analytical approach. We distributed a survey comprising four constructs to a participant pool of 709 individuals. The survey was distributed to Cruise passenger and automotive product buyer to evaluate consumer mobility choice. The Green Consumption Values (GCV) were gauged using a 6-item scale [1]. Brand experience (BE) was determined using an 8-item scale (negative item was removed since their factor loading was low) [9]. Brand trust (BT) was determined using an 8-item scale (Delgado-Ballester et al. [10]), and the intention to word of mouth (IWM) was assessed with a 2-item scale [11].

To evaluate the consistency of the measurement model across both mobility choices, we implemented a multigroup confirmatory factor analysis (CFA), following the approach outlined by Steenkamp and Baumgartner [12], using the AMOS 22 software. Initially, to assess measurement invariance, we defined a multigroup model where all factor loadings and variances were equally constrained for the groups from Cruise passenger and automotive buyer. The fit indices for this model demonstrated a good fit (x2/df = 3.691, CFI = 0.924, RMSEA = 0.062), suggesting invariance across the two groups.

Furthermore, in both groups, all standardized factor loadings were above 0.60 and significantly non-zero at the 0.001 significance level. In addition, to confirm the constructs and their distinctiveness, we compared the square root of the average variance extracted (AVE) for each construct against the correlations among constructs. The discriminant validity was affirmed since the square root of AVE for each construct was higher than its correlations with other constructs. Also, AVE values surpassed the generally accepted benchmark of 0.50 [13], reinforcing the constructs' convergent validity.

39.4 Findings

The model exhibits a reasonable fit to the data, evidenced by a CMIN/DF ratio of 3.691, aligning with Schermelleh-Engel et al.'s [14] acceptable criterion of less than 5. The Comparative Fit Index (CFI) of 0.924 and the Tucker-Lewis Index (TLI) of 0.915 both surpass the 0.90 benchmark, suggesting a good fit [15]. The Root Mean Square Error of Approximation (RMSEA) of 0.062 falls within MacCallum et al.'s [16] acceptable range of 0.05–0.08. The significant Chi-square test reflects its sensitivity to large sample sizes, which can lead to significant results even for minor model data discrepancies [17]. Table 39.1 shows that all factor loadings are above

Table 39.1 Factor loading, Cronbach alpha, C.R. and Ave

Construct	Items	Mean	SD	Loadings	Cronbach's alpha(α)	C.R.	AVE
Full model (n = 709)							
GCV	GCV_1	3.94	1.132	0.938	0.979	0.974	0.863
	GCV_2	3.90	1.186	0.953			
	GCV_3	3.80	1.267	0.953			
	GCV_4	3.99	1.134	0.924			
	GCV_5	3.76	1.268	0.947			
	GCV_6	3.84	1.282	0.936			
BT	BT_1	4.17	.826	0.864	0.955	0.960	0.751
	BT_2	4.13	.845	0.872			
	BT_3	4.07	.939	0.865			
	BT_4	4.19	.864	0.887			
	BT_5	4.03	.961	0.859			
	BT_6	3.89	.992	0.773			
	BT_7	3.93	.959	0.873			
	BT_8	3.94	.961	0.86			
BE	BE_1	4.11	.912	0.86	0.931	0.976	0.838
	BE_2	4.09	.911	0.873			
	BE_3	4.07	.912	0.822			
	BE_4	3.90	.948	0.826			
	BE_5	3.88	.991	0.771			
	BE_6	3.64	1.101	0.727			
	BE_7	3.74	1.099	0.746			
	BE_8	4.07	1.086	0.758			
ITW	ITW_1	3.09	1.671	0.994	0.692	0.698	0.537
	ITW_2	2.84	1.690	0.532			
Cruise (n = 404)							
GCV	GCV_1	3.48	1.175	0.927	0.974	0.974	0.863
	GCV_2	3.48	1.247	0.945			
	GCV_3	3.33	1.302	0.94			
	GCV_4	3.63	1.208	0.913			
	GCV_5	3.22	1.288	0.93			
	GCV_6	3.35	1.341	0.919			
BT	BT_1	4.15	.862	0.866	0.960	0.96	0.752
	BT_2	4.12	.894	0.875			
	BT_3	4.04	.993	0.864			
	BT_4	4.25	.920	0.885			
	BT_5	4.03	1.028	0.858			
	BT_6	3.97	.969	0.83			
	BT_7	3.97	.945	0.901			
	BT_8	3.98	.980	0.856			

(continued)

Table 39.1 (continued)

Construct	Items	Mean	SD	Loadings	Cronbach's alpha(α)	C.R.	AVE
BE	BE_1	4.28	.886	0.839	0.732	0.905	0.546
	BE_2	4.31	.853	0.833			
	BE_3	4.29	.842	0.736			
	BE_4	4.03	.915	0.743			
	BE_5	3.95	1.035	0.706			
	BE_6	3.73	1.136	0.676			
	BE_7	3.89	1.126	0.678			
	BE_8	4.49	.867	0.681			
ITW	ITW_1	4.00	1.310	0.782	0.675	0.696	0.535
	ITW_2	3.18	1.733	0.677			
Automotive (n = 305)							
GCV	GCV_1	4.55	.701	0.903	0.974	0.975	0.868
	GCV_2	4.46	.814	0.943			
	GCV_3	4.42	.897	0.949			
	GCV_4	4.48	.803	0.927			
	GCV_5	4.49	.787	0.945			
	GCV_6	4.48	.851	0.921			
BT	BT_1	4.20	.775	0.86	0.974	0.95	0.707
	BT_2	4.14	.777	0.876			
	BT_3	4.10	.862	0.894			
	BT_4	4.11	.778	0.867			
	BT_5	4.02	.866	0.853			
	BT_6	3.78	1.014	0.691			
	BT_7	3.89	.978	0.834			
	BT_8	3.89	.934	0.834			
BE	BE_1	3.88	.897	0.873	0.858	0.959	0.747
	BE_2	3.80	.905	0.9			
	BE_3	3.77	.918	0.881			
	BE_4	3.72	.962	0.909			
	BE_5	3.79	.923	0.888			
	BE_6	3.51	1.042	0.816			
	BE_7	3.54	1.032	0.832			
	BE_8	3.52	1.106	0.808			
ITW	ITW_1	1.88	1.292	0.621	0.630	0.641	0.474
	ITW_2	2.38	1.517	0.75			

Table 39.2 Path model result

Hypothesis	Direct effect	Path coefficient	*p*	Interpretation
Full model				
H1a	GCV - > BE	0.008	0.747	*Not supported*
H1b	GCV - > ITW	−0.356	***	*Supported*
H1c	GCV - > BT	0.09	***	*Supported*
H2a	BE - > BT	0.598	***	*Supported*
H2b	BE- > ITW	0.785	***	*Supported*
H3a	BT - > ITW	0.352	***	*Supported*
Cruise				
H1a	GCV - > BE	0.034	0.280	*Not supported*
H1b	GCV - > ITW	0.120	0.003	*Supported*
H1c	GCV - > BT	0.80	0.002	*Supported*
H2a	BE - > BT	0.590	***	*Supported*
H2b	BE- > ITW	0.486	***	*Supported*
H3a	BT - > ITW	0.621	***	*Supported*
Automotive				
H1a	GCV - > BE	0.362	***	*Supported*
H1b	GCV - > ITW	0.184	0.022	*Supported*
H1c	GCV - > BT	−0.034	0.327	*Not supported*
H2a	BE - > BT	0.693	***	*Supported*
H2b	BE- > ITW	0.207	0.123	*Not supported*
H3a	BT - > ITW	0.131	0.378	*Not supported*

***$p < 0.001$

the 0.70 threshold (except one item of ITW in all three the cases). Furthermore, the values of the construct reliability indices are all higher than 0.70 (with some exceptions that are close to the range of acceptability), while the AVE is always higher than the threshold value of 0.50 [18]. This denotes satisfactory reliability of the individual items and adequate reliability and convergent validity of the respective constructs.

Table 39.2 reports the structural model's main results in terms of path coefficients in detail. All the hypotheses are confirmed with the exception of H1a for the full model. In particular, the study identifies a positive and significant effect of BE on ITW (b = 0.785), confirming the hypotheses H2b. In Table 39.2, also the results of the path coefficient of the two different groups (i.e., Cruise passenger and automotive buyer) are presented. For the first group, all the hypotheses are confirmed with the exception of the H1a. Instead, for the second group H1a, H1b, and H2a are supported.

In Table 39.3, a comparison between the two groups reveals that GCV positively impacts the automotive group. On the other hand, GCV strongly positively impacts the cruise sector compared to the automotive sector. Finally, the effect of BT on ITW was stronger on the Cruise passenger.

Table 39.3 Multigroup analysis

Direct effect	Group 1 Cruise	Group 2 Automotive	|*difference*|	*p*	Decision
GCV - > BE	0.034	0.362	−0.328	***	*Supported*
GCV- > ITW	0.120	0.184	−0.064	0.476	*Not supported*
GCV - > BT	0.80	−0.034	0.834	0.009	*Supported*
BE - > BT	0.590	0.693	0.10875	0.139	*Not supported*
BE- > ITW	0.486	0.207	0.279	0.128	*Not supported*
BT - > ITW	0.621	0.131	0.49	0.009	*Supported*

***$p < 0.001$

39.5 Implications

This chapter carries both theoretical and managerial implications. Firstly, it analyzes consumers' perceptions of green mobility in the context of purchasing products and services supported by digital transformation. In particular, the impact of Green Consumer Value (GCV) on consumer–brand relationships in fostering intention to word of mouth (ITW) has been explored. By constructing a multigroup SEM (structural equation modeling) analysis, the study offers an opportunity to understand the differences between the groups of cruise passengers and automotive buyers. The results are also useful for managers to take on the green and digital challenges. In fact, the distinction between the two groups can inform strategies that are tailored to the different sectors, potentially building a competitive advantage and reducing costs in terms of marketing strategy development.

Acknowledgments This research has been supported by the European Union under the PNRR of NextGenerationEU, National Sustainable Mobility Center (CN00000023, MUR Decree n. 1033—17/06/2022, Spoke 10).

References

1. Haws, K.L., Winterich, K.P., Naylor, R.W.: Seeing the world through GREEN-tinted glasses: green consumption values and responses to environmentally friendly products. J. Consum. Psychol. **24**(3), 336–354 (2014)
2. Nguyen, H.V., Nguyen, C.H., Hoang, T.T.B.: Green consumption: closing the intention-behavior gap. Sustain. Dev. **27**(1), 118–129 (2019)
3. Huang, H.C., Lin, T.H., Lai, M.C., Lin, T.L.: Environmental consciousness and green customer behavior: an examination of motivation crowding effect. Int. J. Hosp. Manag. **40**, 139–149 (2014)
4. Markovic, S., Iglesias, O., Singh, J.J., Sierra, V.: How does the perceived ethicality of corporate services brands influence loyalty and positive word-of-mouth? Analyzing the roles of empathy, affective commitment, and perceived quality. J. Bus. Ethics. **148**(4), 721–740 (2018)
5. Risitano, M., Romano, R., La Ragione, G., Quintano, M.: Analyzing the impact of green consumption values on brand responses and behavioural intention. Business Ethics, the Environment & Responsibility. **32**(3), 1096–1112 (2023)

6. Hoarau-Heemstra, H., Wigger, K., Olsen, J., James, L.: Cruise tourism destinations: practices, consequences and the road to sustainability. J. Destin. Mark. Manag. **30**, 100820 (2023)
7. Klein, R.A.: Responsible cruise tourism: issues of cruise tourism and sustainability. J. Hosp. Tour. Manag. **18**(1), 107–116 (2011)
8. Castillo-Manzano, J.I., Castro-Nuño, M., Pozo-Barajas, R.: Addicted to cruises? Key drivers of cruise ship loyalty behavior through an e-WOM approach. Int. J. Contemp. Hosp. Manag. **34**(1), 361–381 (2022)
9. Brakus, J.J., Schmitt, B.H., Zarantonello, L.: Brand experience: what is it? How is it measured? Does it affect loyalty? J. Mark. **73**(3), 52–68 (2009)
10. Delgado-Ballester, E., Munuera-Aleman, J.L., Yague-Guillen, M.J.: Development and validation of a brand trust scale. Int. J. Mark. Res. **45**(1), 35–54 (2003)
11. Arnold, M.J., Reynolds, K.E.: Affect and retail shopping behavior: understanding the role of mood regulation and regulatory focus. J. Retail. **85**(3), 308–320 (2009)
12. Steenkamp, J.B.E., Baumgartner, H.: Assessing measurement invariance in cross-national consumer research. J. Consum. Res. **25**(1), 78–90 (1998)
13. Fornell, C., Larcker, D.F.: Evaluating structural equation models with unobservable variables and measurement error. J Mark Res. **18**(1), 39–50 (1981)
14. Schermelleh-Engel, K., Moosbrugger, H., Müller, H.: Evaluating the fit of structural equation models: tests of significance and descriptive goodness-of-fit measures. Methods Psychol. Res. Online. **8**(2), 23–74 (2003)
15. Hu, L.T., Bentler, P.M.: Cutoff criteria for fit indexes in covariance structure analysis: conventional criteria versus new alternatives. Struct. Equ. Model. Multidiscip. J. **6**(1), 1–55 (1999)
16. MacCallum, R.C., Browne, M.W., Sugawara, H.M.: Power analysis and determination of sample size for covariance structure modeling. Psychol. Methods. **1**(2), 130 (1996)
17. Bentler, P.M., Bonett, D.G.: Significance tests and goodness of fit in the analysis of covariance structures. Psychol. Bull. **88**(3), 588 (1980)
18. Hair, J.F., Babin, B.J., Krey, N.: Covariance-based structural equation modeling in the journal of advertising: review and recommendations. J. Advert. **46**(1), 163–177 (2017)

Part VII
Organizational Resilience and Performance

Chapter 40
Continuous Improvement and Organizational Resilience: The Role of Digital Maturity in Small and Medium-Sized Enterprises

Raffaella Montera, Leul Girma Haylemariam, and Nicola Cucari

Abstract This chapter aims to explore the relationship between continuous improvement (CI), organizational resilience (OR), and the moderating influence of digital maturity (DM) within SMEs by analyzing if and to what extent SMEs' DM contributes to continuous improvement to recover from disruptions. A moderated hierarchical regression is adopted to capture evidence from 133 Italian SMEs belonging to the textile industry. Findings reveal that people and strategy positively affect the reactive OR with the moderation of the digital intensity by paving the way for operational excellence. Conversely, culture and top management play a crucial role in the proactive resilience of SMEs with the moderation of transformation management intensity by creating the ideal conditions for organizational excellence. The study identifies new critical success factors (CSFs) that directly and indirectly influence the company's ability to face the crises. Moreover, CI and DM are conceptualized as organizational capabilities that not only constitute resilience but also serve as prerequisites for its cultivation. Thus, resilience can be perceived as a meta-capability, herein deconstructed into its constituent elements having a cognitive and behavioral nature.

Keywords Continuous improvement · Organizational resilience · Digital maturity · SMEs

R. Montera (✉)
Pegaso University, Naples, Italy
e-mail: raffaella.montera@unipegaso.it

L. G. Haylemariam · N. Cucari
Sapienza University of Rome, Rome, Italy

F. Schiavone et al. (eds.), *Advanced Perspectives and Trends in Digital Transformation of Firms, Networks, and Society*, Springer Proceedings in Business and Economics, https://doi.org/10.1007/978-3-031-80692-6_40

40.1 Introduction

The current business landscape is characterized by increased turbulence due to increased competition, uncertain environments, and industrial change [1–4]. Small and medium-sized enterprises (SMEs) are more vulnerable than larger firms due to their limited size and resources, as well as their low propensity to anticipate or plan for a crisis [5].

Organizational resilience (OR), as the ability to recover from disruption quickly and effectively [6], creates the need for rapid adaptation and changes management approach and strategy. Increasing resilience capacity requires achieving a competitive advantage based on continuous improvement (CI) [7, 8] as a systematic process of ongoing, incremental, and enterprise-wide change to existing practices [9]. At the same time, digital maturity (DM), as the ability to align an organization's people, culture, structure, and tasks to capture opportunities enabled by technological infrastructure [10], also contributes to building SME resilience [11].

In any case, understanding the impact of DM on SMEs remains limited [12]. Moreover, while the correlations between CI and OR and DM and OR within SMEs remain insufficiently studied [11, 13], the collective study of CI, OR, and DM is notably deficient, highlighting the need to identify practical resilience development strategies, particularly tailored to SMEs [14].

To address these gaps, our study aims to examine the indirect effect of DM on CI and OR by answering the key research questions: whether and to what extent does SMEs' DM contribute to continuous improvement to recover from disruptions?

A moderated hierarchical regression was used to collect evidence from 133 Italian SMEs belonging to the textile industry. We explore the complex relationship between CI, OR, and the moderating influence of DM within SMEs.

The study finds that people and strategy positively influence the reactive OR with the moderation of digital intensity, while culture and top management play a crucial role in the proactive resilience of SMEs with the moderation of transformation management intensity. In doing so, new critical success factors (CSFs), which directly and indirectly influence the company's ability to face the crises, are identified. An implication of our study is the conceptualization of the CI and DM as organizational capabilities that not only constitute but also cultivate resilience, which is perceived as a meta-capability.

The chapter is structured as follows: Section 40.2 introduces the theoretical framework and outlines the hypotheses. Section 40.3 describes the methodology, while Sect. 40.4 presents the results. Section 40.5 provides the discussion and implications, and Sect. 40.6 summarizes the conclusions.

40.2 Theoretical Background and Hypothesis Development

40.2.1 Continuous Improvement and Organizational Resilience

OR is a firm-level capability to adapt and/or dynamically relate to its environment by always finding ways to take chances and advantages of situations [15]. SMEs' resilience is the successful adaptation in maintaining business viability by acquiring, generating, and combining internal and external knowledge resources to perceive, explore, and address rapidly changing environments [16]. It is an umbrella concept [17] that can encompass many dimensions. Jia et al. [18] distinguish between proactive and reactive OR. The former is about anticipating and actively waiting to increase a company's readiness for change; the latter focuses on reacting and recovering from unexpected events.

Herbane [19] urges a rethinking of OR in SMEs by linking resilience to other domains. In this logic, drawing on the field of quality management, CI is one of the key enablers for building resilient SMEs [20]. CI is a cultural ethos that promotes continuous improvement to eliminate waste in business systems and activities [7] through various techniques such as Total Quality Management, Lean Manufacturing, Six Sigma, and Lean Six Sigma. Netland [21] highlights the role of top management in promoting CI through "open door" policies, briefings, and feedback sessions. García et al. [22] highlight the importance of a CI culture in encouraging both operators and management to identify and implement improvement opportunities. Jaca et al. [23] emphasize the importance of identifying critical processes and strategic perspectives, as it is not possible to improve all methods simultaneously. In addition, people orientation in CI includes skills development and training in problem-solving methods [24].

Several investigations have shown that SMEs are more vulnerable to crises than large organizations [25, 26]. Due to crises like natural catastrophes and tough economic shocks, these entities are more fragile and sensitive than big enterprises [27]. However, Six Sigma as a CI method may also be used proactively to enhance SMEs' resilience by reducing risks and being prepared [28]. Zighan et al. [29] suggest that SMEs can maintain competitiveness soon after a crisis by balancing immediate operational measures and long-term strategic planning to enhance their resilience via CI. Moreover, Zighan and Ruel [13] recently found that CI has a fourfold function in developing OR. It is a cyclical process encompassing a crucial cultural factor and may be seen as a corporate philosophy. It also highlights a comprehensive change strategy that makes minor but consistent improvements. However, Núñez-Ríos et al. [30] argue that the effectiveness of CI may be enhanced by focusing on management style and staff development as crucial factors in bolstering SMEs. In consideration of the established relationship between CI and OR, as previously elucidated in the literature, we propose the following hypothesis:

H1: CI positively affects OR.

40.2.2 Moderation Effect of Digital Maturity

DT prioritizes enhancing customer service and market performance by leveraging digital technology to drive efficiency and foster innovation. CI is widely acknowledged as a critical driver of organizational performance improvement, mainly through implementing Total Quality Management in various organizational activities [31]. OR entails complex and collective adaptive capabilities that enable an organization to cope successfully and learn from unexpected events [32]. However, more empirical research needs to be conducted to examine the link between DT and OR in SMEs [11, 33]. Specifically, DT has positively impacted SMEs' organizational performance [34]. It enables operational continuity, which is crucial for SMEs' success [35]. Previous studies have shown that SMEs with advanced digital capabilities exhibit higher levels of operational readiness in terms of situational awareness, vulnerability management, and adaptive capacity [11]. DM enhances the resilience of SMEs during challenging times such as the COVID-19 pandemic [36]. According to Antony et al. [37], it is essential to consider rewarding and recognizing team members for successfully deploying CI through Six Sigma in SME training. In addition, the potential of digitization for enhancing CI has been highlighted in previous research [38]. However, there needs to be more exploration into how quality integration for SMEs functions in a digital environment [39]. Therefore, we hypothesized:

H2: DM positively moderates CI and OR.

In line with the hypotheses of the study, the research model is shown in Fig. 40.1.

40.3 Method

40.3.1 Sample

A random sample of 615 companies was drawn from the AIDA-Bureau Van Dijk database, the most comprehensive database of Italian limited liability companies. We selected Italian firms that have been active from 2002 (commonly regarded as the beginning of the digital age) to 2023, for which complete company profiles were available, belonging to the textile industry (ATECO 2007 code 13). Additionally, we considered the size criterion by selecting Italian firms that meet the European Commission's definition of SMEs, specifically those with up to 250 employees, an annual turnover not exceeding €50 million, or a balance sheet total not exceeding €43 million. This selection was driven by the recognition of SMEs as key contributors to the European economy [40]. The study also identified Italian SMEs as an interesting research context to study OR due to their simultaneous exposure to the pandemic, the Russian invasion of Ukraine, and the resurgence of inflation [41].

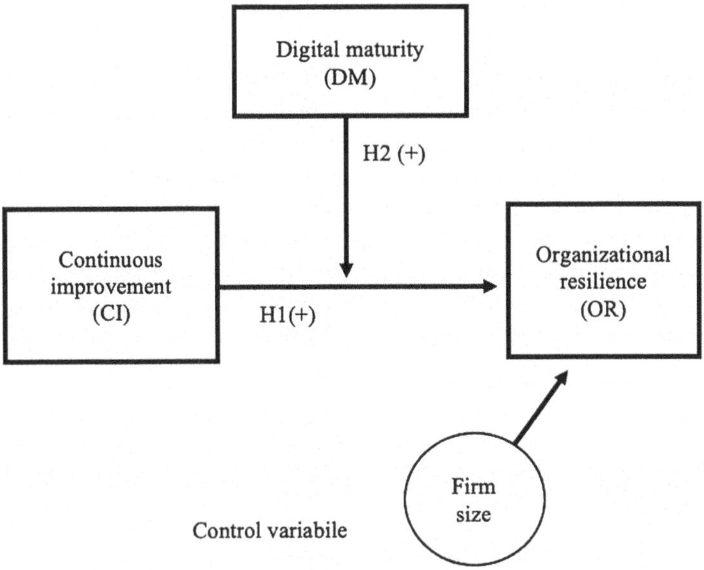

Fig. 40.1 The research model of the study and the relationships hypothesized

Furthermore, the textile and fashion industry emerges as a critical area for examining the resilience of Italian SMEs. This industry stands as a mature, fiercely competitive, and globally interconnected market marked by brief product life cycles and significant fluctuations in demand [42, 43].

40.3.2 Data Collection and Analysis

A web survey was conducted between June and December 2023. Following the key informant approach [44], potential respondents were identified within firms among managers responsible for strategy definition, quality management, and digitalization (e.g., general manager, CEO, quality/improvement manager, chief information officer, chief technology officer, chief digital officer). Following a pre-test on a convenience sample of 30 companies, target respondents were contacted by email, explaining the research objectives and providing a link to the questionnaire. Data protection was assured by promising anonymity and the use of aggregated data. Of the 615 targeted companies, 133 fully completed surveys were collected, resulting in a response rate of 21.6%. The collected data were analyzed by moderated hierarchical regression using SPSS v 22 software.

40.3.3 *Measurement*

The questionnaire consisted of four sections: CI, OR, DM, and respondent details/company information. Responses were measured on a 5-point Likert ordinal scale ranging from (1) strongly disagree to (5) strongly agree. The constructs and variables detailed in Table 40.1 were drawn from existing literature.

CI was the independent variable, OR was the dependent variable, and DM acted as a moderating variable. Firm size, measured by the natural logarithm of full-time employees, served as a control variable to be considered in understanding the adaptive capacity of organizations [18]. Validity and reliability checks for each scale were conducted iteratively, meeting minimum acceptance thresholds (see Table 40.2).

40.4 Findings

Descriptive statistics and correlations for each of the study variables are reported in Table 40.3. The modest correlation coefficients of the variables suggest that multicollinearity is not an issue.

Prior to the regression analyses, the independent and moderator variables were mean centered. A moderated hierarchical regression was used to test the hypotheses. The results followed the four steps for each dimension of the OR: the control variable (model 1), the independent variable (model 2), the moderator variable (model 3), and the interaction effects (model 4) (see Tables 40.4 and 40.5).

Model 1 in Table 40.4 shows a positive and significant effect of firm size on ROR ($\beta = 0.29$, $p < 0.05$), indicating that the number of employees affects the speed and flexibility of an organization's response to a crisis. Model 2 confirms HP1 due to the positive relationships between PEO and ROR ($\beta = 0.28$, $p < 0.10$) and STR and ROR ($\beta = 0.30$, $p < 0.10$) (see Table 40.4), as well as between CUL and POR ($\beta = 0.31$, $p < 0.10$) and TOP and POR ($\beta = 0.26$, $p < 0.05$) (Table 40.5). In model 3, DI does not show a direct significant effect, so the interaction effects are expected to be significant. Model 4 confirms HP2 as the interaction terms for ROR (PEO × DI: $\beta = 0.23$; $p < 0.10$; STR × DI: $\beta = 0.33$; $p < 0.05$) (see Table 40.4) and POR (CUL × TMI: $\beta = 0.27$; $p < 0.05$; TOP × TMI: $\beta = 0.29$; $p < 0.01$) (see Table 40.5) are significant.

40.5 Discussion and Implication

The study confirms a positive and significant relationship between CI and OR within SMEs and verifies the mediating role of DM. The positive relationship between CI and OR revealed in our results is consistent with the literature on SMEs (e.g., [13]).

Table 40.1 Measurements for CI, OR, and DM

Constructs	Variables	Items	Authors
Continuous improvement (CI)	Culture (CUL)	Collaborators are often encouraged to share opinions, experiences, and ideas and to implement them	Lizarelli et al. [45]
		Continuous improvement is stimulated throughout the organization and incorporated into the organization's culture	
		When something goes wrong, the natural reaction of people at all levels is to identify the reasons for going wrong and not blaming individuals	
	People (PEO)	Use of formal method and tools to find and solve problems	
		CI projects are carried out in teams with people from different areas and organizational levels	
		There is training for employees on teamwork and tools and methods for analysis and problem solving	
	Strategy (STR)	Processes are regularly evaluated for implementing changes and improvements	
		There is a formal process for prioritizing improvement projects aligned with the strategy	
		CI projects are measured and evaluated according to their contribution to the company's strategic objectives	
	TOP management (TOP)	There is top management involvement with CI, by eliminating barriers (such as communication), being present, and keeping feedback	
		Provision of resources by top management (time, financial, training) for CI	
		Top management proposes mechanisms for the recognition, communication, and reward of the success of improvement projects	
Organizational resilience (OR)	Proactive organizational resilience (POR)	We created internal awareness for disruptions and made attempts to drive this awareness to our employees	Jia et al. [18]
		We analyzed and assessed both probability and impact of potential disruptions	
		We improved our disruption prevention capabilities	
		We engaged in contingency planning to prepare for potential disruptions	
	Reactive organizational resilience (ROR)	We can quickly recognize that there is a threatening situation	
		We can gather and interpret information of cues to gauge the magnitude, location, and causes of the disruption	
		We can quickly identify, formulate, and evaluate a set of possible responses to disruption	
		We can quickly organize a formal response team of key personnel, both on-site and at corporate level	
		We are very successful at dealing with crises, including addressing public relations issues	

(continued)

Table 40.1 (continued)

Constructs	Variables	Items	Authors
Digital maturity (DM)	Digital intensity (DI)	We are using digital technologies (such as analytics, social media, mobile and embedded devices) to understand our customers better	He et al. [46]
		We use digital channels (such as online, social media, and mobile) to market and distribute products and services	
		We sell our products and services through digital channels	
		We use digital channels to provide customer service	
		Technology is allowing us to support customers and to improve operational processes in new ways	
		Our core processes are automated	
		We have an integrated system to support key operational and customer information	
		We use analytics to make better operational decisions	
		We use digital technologies to increase the performance or added value of our existing products and services	
		We have launched new business models based on digital technologies	
	Transformation management intensity (TMI)	Senior executives have a transformative vision of the digital future of our company	
		Senior executives and middle managers share a common vision of digital transformation	
		There are possibilities for everyone in the company to take part in the conversation around digital transformation	
		The company is promoting the necessary culture changes for digital transformation	
		The company is investing in the necessary digital skills	
		Digital initiatives are coordinated across silos such as functions or regions	
		Roles and responsibilities for governing digital initiatives are clearly defined	
		Digital initiatives are assessed through a common set of key performance indicators	
		IT and business leaders work together as partners	
		The IT unit's performance meets the needs of the company	

Table 40.2 Scale validity and reliability

	Validity		Reliability
Variables	Kaiser-Meyer-Olkin test	Bartlett's test of sphericity	Cronbach's alpha
CUL	0.630	0.000	0.721
PEO	0.594	0.000	0.755
STR	0.607	0.000	0.820
TOP	0.749	0.000	0.770
POR	0.611	0.000	0.743
ROR	0.625	0.000	0.716
DI	0.703	0.001	0.882
TMI	0.646	0.000	0.732

Table 40.3 Descriptive statistics and correlations

Variables	M	SD	1	2	3	4	5	6	7	8
1. CUL	5.68	5.25	1							
2. PEO	4.81	0.97	0.46	1						
3. STR	5.96	0.81	0.48	0.35	1					
4. TOP	5.60	0.76	0.84	0.56	0.43	1				
5. POR	4.00	0.79	0.06	0.38***	0.58	0.61***	1			
6. ROR	5.61	0.82	0.34**	0.48	0.32	0.43	0.50	1		
7. DI	5.38	0.73	0.76	0.48	0.60	0.84	0.59	0.50	1	
8. TMI	2.97	1.31	0.40**	0.38	0.26	0.41	0.46	0.39	0.42	1

Notes: M = mean; SD = standard deviation; $p < 0.05$, $^{**}p < 0.01$, $^{***}p < 0.001$

Table 40.4 Results of regression analysis of ROR

	ROR			
Variables	Model 1	Model 2	Model 3	Model 4
Firm size	0.29**	0.13*	0.18**	0.22*
CUL		0.11	0.06	0.19
PEO		0.28**	0.15*	0.18**
STR		0.30*	0.21*	0.13*
TOP		0.22	0.12	−0.08
DI			0.04	0.11
TMI			−0.26	0.16
CULxDI				0.18
CULxTMI				0.12
PEO x DI				0.3*
PEOxTMI				0.14
STRxDI				0.33**
STRxTMI				0.15
TOPxDI				0.02
TOPxTMI				0.09
F	1.376	19.090	0.058	3.776
R^2	0.069	0.153	0.153	0.146

$^{*}p < 0.10$. $^{**}p < 0.05$

Note. Standardized coefficients are reported

Table 40.5 Results of regression analysis of POR

	POR			
Variables	Model 1	Model 2	Model 3	Model 4
Firm size	0.15	0.11	0.06	0.20
CUL		0.31*	0.24***	0.19*
PEO		0.22	0.17	−0.04
STR		0.06	0.09	0.15
TOP		0.26**	0.12**	0.14*
DI			−0.13	0.21
TMI			0.02	0.10
CULxDI				0.09
CULxTMI				0.27**
PEO x DI				0.06
PEOxTMI				0.15
STRxDI				0.13
STRxTMI				0.11
TOPxDI				0.05
TOPxTMI				0.29***
F	1.601	16.772	0.035	5.897
R^2	0.083	0.251	0.251	0.166

*$p < 0.10$. **$p < 0.05$. ***$p < 0.01$
Note. Standardized coefficients are reported

However, we advance knowledge by decomposing this relationship into its components. Thus, different antecedents emerge for each type of resilience. Therefore, our study falls within the stream of research (e.g., [47]) that considers OR as an outcome and empirically suggests that some organizations are resilient and others are not.

People and strategy contribute to the reactive resilience of SMEs. Responding to a crisis with resilient actions is possible when organizational members are organized according to strategic goals. This is successful when there is a strong digital infrastructure that shortens response times by quickly aligning people and strategy in the face of disruptions. Digital technologies help SMEs to transform internal operations and procedures, improve performance efficiency and effectiveness, redesign business models, mitigate risks, and ensure business survival [48]. The combination of people, strategy, reactive resilience, and digital intensity enables SMEs to achieve operational excellence, which involves optimizing processes, managing resources efficiently, and ensuring the overall effectiveness of business operations to enhance competitiveness without succumbing to the unexpected [49].

Culture and top management play a critical role in building proactive resilience in SMEs. Anticipating, managing, and recovering from a crisis through resilience plans becomes possible when a culture of improvement and top management commitment to CI are deeply rooted in SMEs. These anchorages become stronger as the Transformation Management Intensity (TMI) increases. A high TMI means equipping SMEs with a transformational vision, governance, and culture [46], aligning employee behavior with the vision, and developing resource sharing both inside and outside the firm [50]. The combination of culture, top management, proactive resilience, and TMI enables SMEs to achieve organizational excellence. It refers to the state of achieving superior performance across multiple dimensions within an organization through continuous learning, adaptation to change, and a commitment to exceeding stakeholder expectations [51].

Building upon research findings, we conceptualize CI and DM as organizational capabilities that not only constitute resilience but also serve as prerequisites for its cultivation [17]. In this context, resilience can be perceived as a meta-capability, which we deconstruct into its constituent elements having a cognitive and behavioral nature. Based on our findings, we identify the foundational capabilities for each form of resilience, which collectively shape OR. For reactive OR, these capabilities encompass people, strategy, and digital intensity, while for proactive OR, they involve culture, top management, and transformation management intensity.

The research findings highlight the importance of prioritizing the improvement of DM as a specific focus for practitioners seeking to understand how to effectively foster resilience in SMEs [52]. However, this is a difficult task that requires continuous reassessment of the quality and effectiveness of practices and systems, while SMEs often struggle with the speed of digital transformation and significant constraints such as a lack of digital skills and managerial and financial resources for digital transformation [53]. Moreover, the positive link between CI and OR suggests that textile companies can effectively navigate market fluctuations and maintain agility, competitiveness, and resilience through the implementation of CI practices, especially in times of disruptive events. CI enables the industry to adapt, enhance productivity, minimize waste, and improve worker safety and motivation [54]. As Kazancoglu et al. [55] state, this approach cultivates a corporate culture focused on efficiency and resourcefulness, which is essential for sustained success in the textile sector where innovation and responsiveness to market dynamics are crucial for maintaining competitiveness. Furthermore, the adoption of CI practices sends a powerful message to stakeholders, demonstrating a commitment to excellence, enhancing the organization's reputation, attracting customers, and fostering long-term relationships [56].

40.6 Conclusion and Limitation

Our study analyzes the unexamined interplay between DM, CI, and OR within SMEs by identifying new critical success factors (CSFs) that directly and indirectly affect the firm's ability to cope with crises. Thus, we fill a gap in the SME literature due to the lack of research on the role of DM in coping with exogenous shocks [11, 57] and the need to statistically validate the contribution of CI to SME resilience [13].

However, our study is not without limitations. First, the scope of the research and the research design limit the results to be validated in the specific context of Italian

SMEs, and we acknowledge that the results may differ depending on the sector and the geographical context. This provides an opportunity for future research to validate and compare the results within different sectors, those in other industries, or larger organizations in other countries. In this regard, it could be interesting to further explore the implementation of CI within under-researched emerging countries where the textile industry plays a significant role in economic development and encounters specific opportunities and challenges stemming from unstable business environments. Second, considering the sample size as a limitation, another constraint of this study is its generalizability, which could be addressed by conducting research with larger samples. Third, the regression model includes a control variable (i.e., firm size) to test the robustness of the results. Thus, future studies could consider other control variables, such as contextual dynamics and the quality program implemented (i.e., Lean, Six Sigma, Lean Six Sigma), to improve the explanatory power of the results. Fourth, future developments of the research could explore the integration of specific Industry 4.0 solutions (i.e., data analytics, artificial intelligence, smart manufacturing, and digital supply chain management) with CI practices. Finally, besides the interplay of cognitive and behavioral elements of resilience, contextual factors (e.g., resources, social capital, and power) could be considered due to their importance to the successful accomplishment of resilience as well as the development of resilience capabilities.

References

1. Kraus, S., Kumar, S., Lim, W.M., Kaur, J., Sharma, A., Schiavone, F.: From moon landing to metaverse: tracing the evolution of Technological Forecasting and Social Change. Technol. Forecast. Soc. Chang. **189**, 122381 (2023)
2. Paiola, M., Schiavone, F., Khvatova, T., Grandinetti, R.: Prior knowledge, industry 4.0 and digital servitization. An inductive framework. Technol. Forecast. Soc. Chang. **171**, 120963 (2021)
3. Savastano, M., Zentner, H., Spremić, M., Cucari, N.: Assessing the relationship between digital transformation and sustainable business excellence in a turbulent scenario. Total Qual. Manag. Bus. Excell., 1–22 (2022)
4. Weick, K.E., Sutcliffe, K.M.: Managing the Unexpected: Resilient Performance in an Age of Uncertainty, vol. 8. John Wiley & Sons (2011)
5. Miklian, J., Hoelscher, K.: SMEs and exogenous shocks: a conceptual literature review and forward research agenda. Int. Small Bus. J. **40**(2), 178–204 (2022)
6. Bryce, C., Ring, P., Ashby, S., Wardman, J.K.: Resilience in the face of uncertainty: early lessons from the COVID-19 pandemic. In: Covid-19, pp. 48–55. Routledge (2022)
7. Bhuiyan, N., Baghel, A.: An overview of continuous improvement: from the past to the present. Manag. Decis. **43**(5), 761–771 (2005)
8. Sanchez, L., Blanco, B.: Three decades of continuous improvement. Total Qual. Manag. Bus. Excell. **25**(9-10), 986–1001 (2014)
9. Rijnders, S., Boer, H.: A typology of continuous improvement implementation processes. Knowl. Process Manag. **11**(4), 283–296 (2004)
10. Kane, G. C., Palmer, D., Nguyen-Phillips, A., Kiron, D., Buckley, N.: Achieving digital maturity. Adapting your company to a changing world. Deloitte Insights. https://www2.deloitte.com (2017). Accessed 6 Feb 2023.

11. Robertson, J., Botha, E., Walker, B., Wordsworth, R., Balzarova, M.: Fortune favours the digitally mature: the impact of digital maturity on the organisational resilience of SME retailers during COVID-19. Int. J. Retail Distrib. Manag. **50**(8/9), 1182–1204 (2022)
12. Omrani, N., Rejeb, N., Maalaoui, A., Dabić, M., Kraus, S.: Drivers of digital transformation in SMEs. IEEE Trans. Eng. Manag. **71** (2022)
13. Zighan, S., Ruel, S.: SMEs' resilience from continuous improvement lenses. J. Entrep. Emerg. Econ. **15**(2), 233–253 (2023)
14. Saad, M. H., Hagelaar, G., Van Der Velde, G., Omta, S. W. F.: Conceptualization of SMEs' business resilience: a systematic literature review. Cogent Bus. Manag., 8(1), 1938347 (2021).
15. Aldianto, L., Anggadwita, G., Permatasari, A., Mirzanti, I.R., Williamson, I.O.: Toward a business resilience framework for startups. Sustainability. **13**(6), 1–19 (2021)
16. Ciasullo, M.V., Montera, R., Douglas, A.: Building SMEs' resilience in times of uncertainty: the role of big data analytics capability and co-innovation. Transform. Govern. People Process Policy. **16**(2), 203–217 (2022)
17. Ramezani, J., Camarinha-Matos, L.M.: Approaches for resilience and antifragility in collaborative business ecosystems. Technol. Forecast. Soc. Chang. **151**, 1–26 (2020)
18. Jia, X., Chowdhury, M., Prayag, G., Chowdhury, M.M.H.: The role of social capital on proactive and reactive resilience of organizations post-disaster. Int. J. Disaster Risk Reduct. **48**, 101614 (2020)
19. Herbane, B.: Rethinking organizational resilience and strategic renewal in SMEs. Entrep. Reg. Dev. **31**(5-6), 476–495 (2019)
20. North, K., Varvakis, G.: Competitive strategies for small and medium enterprises. In: Increasing Crisis Resilience, Agility and Innovation in Turbulent Times. Springer, Cham (2016)
21. Netland, T.H.: Critical success factors for implementing lean production: the effect of contingencies. Int. J. Prod. Res. **54**(8), 2433–2448 (2016)
22. García, J.L., Rivera, D.G., Iniesta, A.A.: Critical success factors for Kaizen implementation in manufacturing industries in Mexico. Int. J. Adv. Manuf. Technol. **68**, 537–545 (2013)
23. Jaca, C., Santos, J., Errasti, A., Viles, E.: Lean thinking with improvement teams in retail distribution: a case study. Total Qual. Manag. Bus. Excell. **23**(3-4), 449–465 (2012)
24. Yadav, G., Desai, T.N.: Analyzing lean six sigma enablers: a hybrid ISM-fuzzy MICMAC approach. TQM J. **29**(3), 488–511 (2017)
25. Herbane, B.: Exploring crisis management in UK small-and medium-sized enterprises. J. Conting. Crisis Manag. **21**(2), 82–95 (2013)
26. Kato, M., Charoenrat, T.: Business continuity management of small and medium sized enterprises: evidence from Thailand. Int. J. Disaster Risk Reduct. **27**, 577–587 (2018)
27. Etemad, H.: Managing uncertain consequences of a global crisis: SMEs encountering adversities, losses, and new opportunities. J. Int. Entrep. **18**, 125–144 (2020)
28. Madu, N.C., Kuei, C.H.: Disaster relief supply chain quality management (DRSCQM). Int. J. Qual. Reliab. Manag. **31**(9), 1052–1067 (2014)
29. Zighan, S., Abualqumboz, M., Dwaikat, N., Alkalha, Z.: The role of entrepreneurial orientation in developing SMEs resilience capabilities throughout COVID-19. Int. J. Entrep. Innov. **23**(4), 227–239 (2022)
30. Núñez-Ríos, J.E., Sánchez-García, J.Y., Soto-Pérez, M., Olivares-Benitez, E., Rojas, O.G.: Components to foster organizational resilience in tourism SMEs. Bus. Process. Manag. J. **28**(1), 208–235 (2022)
31. Antunes, M. G., Mucharreira, P. R., Texeira Fernandes Justino, M. D. R., Teixera-Quirós, J.: Total quality management and quality certification on services corporations.International. J. Qual. Res. 14(3), 847–864 (2020).
32. Yao, Y., Fabbe-Costes, N.: Can you measure resilience if you are unable to define it? The analysis of Supply Network Resilience (SNRES). Supply Chain Forum Int. J. 19(4), 255-265 (2018)
33. Viana, C.J.P., Suquillo, E.J., Carmona, C., Arellano, E.J.: Digital maturity as a determinant of business resilience in microenterprises in Ibero-America: a post COVID-19 analysis. Procedia Comput. Sci. **224**, 485–489 (2023)

34. Guo, H., Yang, Z., Huang, R., Guo, A.: The digitalization and public crisis responses of small and medium enterprises: implications from a COVID-19 survey. Front. Bus. Res. China, 14, 1-25 (2020).
35. Papadopoulos, T., Baltas, K.N., Balta, M.E.: The use of digital technologies by small and medium enterprises during COVID-19: implications for theory and practice. Int. J. Inf. Manag. **55**, 102192 (2020)
36. Klein, V.B., Todesco, J.L.: COVID-19 crisis and SMEs responses: the role of digital transformation. Knowl. Process. Manag. **28**(2), 117–133 (2021)
37. Antony, J., Kumar, M., Madu, C. N.: Six sigma in small-and medium-sized UK manufacturing enterprises: some empirical observations. Int. J. Qual. Reliabil. Manag., 22(8), 860-874 (2005)
38. Buer, S.V., Fragapane, G.I., Strandhagen, J.O.: The data-driven process improvement cycle: using digitalization for continuous improvement. IFAC-PapersOnLine. **51**(11), 1035–1040 (2018)
39. Dutta, G., Kumar, R., Sindhwani, R., Singh, R.K.: Digitalization priorities of quality control processes for SMEs: a conceptual study in perspective of Industry 4.0 adoption. J. Intell. Manuf. **32**(6), 1679–1698 (2021)
40. European Commission: Communication from the commission to the European Parliament, the Council, the European Economic and Social Committee and the Committee of the Regions: an SME strategy for a sustainable and digital Europe. Brussels, COM/2020/103 Final (2020). Accessed May 20, 2023. https://eur-lex.europa.eu/legal-content/EN/TXT/?uri=COM:2020:103:FIN
41. Confindustria-Cerved: Rapporto regionale PMI 2023, Rubbettino: Cosenza (2023).
42. Adam, M., Strähle, J., Freise, M.: Dynamic capabilities of early-stage firms: exploring the business of renting fashion. J. Small Bus. Strateg. **28**(2), 49–67 (2018)
43. Galavotti, I., D'Este, C., Fellegara, A.M.: Detecting industry-level patterns of resilience: a financial analysis of Italian SMEs in the fashion industry. Piccola Impresa/Small Bus. **3**, 72–98 (2022)
44. Forliano, C., Orlandi, L.B., Zardini, A., Rossignoli, C.: Technological orientation and organizational resilience to Covid-19: the mediating role of strategy's digital maturity. Technol. Forecast. Soc. Chang. **188**, 122288 (2023)
45. Lizarelli, F. L., de Toledo, J. C., Alliprandini, D. H.: Relationship between continuous improvement and innovation performance: an empirical study in Brazilian manufacturing companies. Total Qual. Manag. Bus. Excell., 32(9-10), 981-1004(2021).
46. He, Z., Huang, H., Choi, H., Bilgihan, A.: Building organizational resilience with digital transformation. J. Serv. Manag. **34**(1), 147–171 (2023)
47. Parker, H., Ameen, K.: The role of resilience capabilities in shaping how firms respond to disruptions. J. Bus. Res. **88**, 535–541 (2018)
48. Razavi Hajiagha, S.H., Alaei, S., Sadraee, A., Nazmi, P.: A perspective of international performance improvement concentrating on innovation and digital resilience of SMEs: the case of an emerging economy. J. Enterp. Inf. Manag. (2023)
49. Muhammad, S., Pan, Y., Agha, M.H., Umar, M., Chen, S.: Industrial structure, energy intensity and environmental efficiency across developed and developing economies: the intermediary role of primary, secondary and tertiary industry. Energy. **247**, 123576 (2022)
50. Westerman, G., Tannou, M., Bonnet, D., Ferraris, P., McAfee, A.: The Digital Advantage: how digital leaders outperform their peers in every industry. MITSloan Manag. **2**, 2–23 (2012)
51. Correia, F.D., Carvalho, A.M., Campbell-Allen, N., Sampaio, P.: A new approach to organisational excellence for small and medium enterprises: the QOE-SME model. Total Qual. Manag. Bus. Excell. **32**(11-12), 1359–1383 (2021)
52. Timans, W., Ahaus, K., van Solingen, R., Kumar, M., Antony, J.: Implementation of continuous improvement based on Lean Six Sigma in small-and medium-sized enterprises. Total Qual. Manag. Bus. Excell. **27**(3-4), 309–324 (2016)
53. Pelletier, C., Cloutier, L.M.: Conceptualising digital transformation in SMEs: an ecosystemic perspective. J. Small Bus. Enterp. Dev. **26**(6/7), 855–876 (2019)

54. Chan, C.O., Tay, H.L.: Combining lean tools application in kaizen: a field study on the printing industry. Int. J. Product. Perform. Manag. **67**(1), 45–65 (2018)
55. Kazancoglu, Y., Lafci, C., Berberoglu, Y., Jagtap, S., Celik, C.C.: The analysis of critical success factors for successful kaizen implementation during the COVID-19 pandemic: a textile industry case study. TQM J., 1754–2731 (2023). https://doi.org/10.1108/TQM-08-2023-0254
56. Bessant, J., Caffyn, S., Gallagher, M.: An evolutionary model of continuous improvement behaviour. Technovation. **21**(2), 67–77 (2001)
57. Çallı, B.A., Çallı, L.: Relationships between digital maturity, organizational agility, and firm performance: an empirical investigation on SMEs. Bus. Manag. Stud. **9**(2), 486–502 (2021)

Chapter 41
Performance and Impact of Digital Academic Spin-Offs

Carmine Passavanti, Simonetta Primario, and Pierluigi Rippa

Abstract This study explores the performance and impact of Digital Academic Spin-offs (DASO) compared to traditional academic spin-offs. Academic entrepreneurship through spin-offs is vital for technological and economic development. While various factors influencing Academic Spin-offs' performance have been examined, the role of digital technologies in academic entrepreneurship remains underexplored. This chapter addresses this gap by analyzing financial and economic data from 294 spin-offs from major Italian Universities over the past decade. Using Generalized Estimating Equations, the study evaluates the impact of digital technologies, intellectual property, gender diversity, governance structures, and company age on spin-off performance. The results show that digital technologies significantly enhance spin-offs' business performance, while the effects of intellectual property are more complex. Gender diversity and governance structures, highlighting areas for further research. The findings suggest that DASO outperforms non-digital peers, emphasizing the need for continued development in digital technology entrepreneurship within academic environments.

Keywords Digital academic spin-offs · Academic entrepreneurship · Digital technologies

41.1 Introduction

The commercialization of university research products, particularly academic entrepreneurship (AE) through spin-off creation, is essential for regional and national technological development and economic growth [1–3].

C. Passavanti (✉) · S. Primario · P. Rippa
Department of Industrial Engineering, University of Naples Federico II, Naples, Italy
e-mail: carmine.passavanti@unibg.it

F. Schiavone et al. (eds.), *Advanced Perspectives and Trends in Digital Transformation of Firms, Networks, and Society*, Springer Proceedings in Business and Economics, https://doi.org/10.1007/978-3-031-80692-6_41

Given its relevance in the scientific community [4], several aspects, factors, and characteristics of academic spin-offs (ASOs) that influence their performance have been studied. Specifically, microlevel analysis has focused on founders' traits and entrepreneurial skills, which enable improved economic performance, growth, and impact [5–7]. At the institutional level, research has been conducted on the effectiveness and added value of technology transfer offices (e.g., [8, 9]) and of the equity owned by the parent university [10]. While Sciarelli et al. [2] analyze the influence of universities, firms, and foundations on performance.

According to Rippa and Secundo [11], the intersection of AE with digital technologies provides a novel conceptualization of AE. Indeed, a recent stream of research on AE analyzes adaptive strategies for meeting market needs created through digital technologies [12]. In this context, several authors argue that the use and creation of digital products have led to a paradigm shift underlying AE (e.g., [13–15]). In particular, creating digital products within the university can contribute significantly to economic growth (e.g., [12, 16]).

In addition, several authors argue that bringing digital products and services to the market is a current necessity (e.g., [17]) and, at the same time, a very challenging task, but with essential benefits for innovation and business performance [18, 19].

Considering the widespread consensus among scholars and policymakers about the positive role of spin-offs, the small body of empirical evidence available to assess the contribution of academic venturing to local development is somewhat surprising [20]. Few studies have compared the performance of ASOs with their nonacademic counterparts [21, 22], showing that academic business ventures do not outperform their nonacademic counterparts. Especially in Europe, several studies have noted that most ASOs remain small and appear to be lifestyle firms rather than high-growth ventures [1]. However, Vincett [3] shows through a longitudinal study that the economic impact generated by ASOs exceeds government funding by substantial margin.

Several studies have shown how ASOs, although belonging to the same university, can differ in performance depending on the department of origin [23, 24], the characteristics of the founding group [25], the type of ownership and governance [2], and the product developed [16]. The few comparative studies conducted on AE have focused on the comparison between ASOs and nonacademic counterparts; however, to the best of our knowledge, no comparative studies have been conducted between different types of ASOs, born and developed within the same context, and therefore, subject to the same institutional logics [26].

Thus, our work seeks to answer the call for a greater understanding of the implications due to the adoption and development of digital technologies in academic entrepreneurship through longitudinal empirical studies [14, 15]. In addition, our comparative study investigates the business performance and economic impact generated by ASOs, comparing them with Digital Academic Spin-offs (DASOs). Specifically, we define a DASO as one that develops one of the three distinct but related elements identified by Nambisan [27]: digital artifacts, digital infrastructure, and digital platforms. In fact, for Deperi et al. [28], digital enterprises are companies that create a significant share of their value by exploiting digital resources and technologies and selling digital products and services.

In light of the increasing commercialization of university research via the creation of business ventures, the pervasive influence of digital technologies in daily life and business processes, and the identified gaps in existing scholarship, our study seeks to address the following research question:

Do DASOs perform better than ASOs, or does digital technology entrepreneurship in universities still need to make progress?

A comprehensive review of the literature has led us to develop the following hypotheses, each aimed at understanding the multifaceted influences on the performance of ASO. These hypotheses are grounded in examining digital technologies, intellectual property, gender impact, governance structure, and the company's age. They reflect the current understanding of what factors might drive the success or failure of these entities in the competitive business landscape.

41.2 Research Hypotheses

41.2.1 Digital Technologies Impact

The exploration into the role of digital technologies highlights their significant effect on ASOs, underscoring an evolving business landscape where digital innovation is key to competitiveness. Studies such as Garcez et al. [12] and Guo et al. [29] support this perspective, illustrating how digital technologies enhance performance through improved innovation and knowledge management.

Hp.1: Using digital technologies positively impacts the business performance of academic spin-offs.

41.2.2 Intellectual Property Impact

Intellectual property (IP) is identified as a pivotal asset, especially within the ASO field. The strategic importance of IP, as demonstrated by Andries and Faems [30], lies in securing a competitive edge and promoting faster growth, reinforcing the value of patents, trademarks, or designs in business success.

Hp.2: Holding and employing intellectual property positively impacts the business performance of academic spin-offs.

41.2.3 Women's Impact

The contribution of women to company success is emphasized through their unique role in fostering inclusive and productive work environments. The research by Del Sarto et al. [31] highlights the positive impact of female participation in ownership on company profitability and performance, showcasing the value of gender diversity in academic enterprises.

Hp.3: The presence of women in the company's ownership positively impacts the business performance of academic spin-offs.

41.2.4 Governance Impact

The governance structure, including the strategic involvement of external companies and academic institutions, is crucial for spin-offs. Recent studies by Ferretti et al. [32] discuss the mixed impacts of university ownership on growth and the positive influence of external companies on financial outcomes, suggesting a nuanced role of governance in spin-off success.

Hp.4.1: The presence of external companies in the ownership structure positively affects the business performance of academic spin-offs.

Hp.4.2: Including the reference university in the ownership structure positively impacts the business performance of academic spin-offs in their early years but may negatively affect growth in later stages.

Hp.4.3: The prestige of the reference university positively influences the business performance of academic spin-offs.

41.2.5 Years of Activity Impact

The age of a company plays a critical role in its adoption of digital technologies and subsequent performance, with younger companies showing a greater propensity for digital innovation. This relationship, explored by Nambisan et al. [33] and supported by the European Digital Transformation Scoreboard [34], suggests that a company's digital maturity is a significant determinant of its financial success.

Hp.7: The years of activity positively impact the business performance of academic spin-offs.

41.3 Methods

41.3.1 Data Collection

In this study, we focused on spin-offs from major universities and technical institutes in Italy to eliminate nationality differences and concentrate on entities significantly involved in technology transfer. This selection limits the scope to a manageable size while ensuring the sample is representative of the most active institutions in this field.

The initial database was created using official lists of ASOs available on their respective websites. This was complemented with descriptions of their activities from their websites and official sources, along with economic and financial indicators extracted using Aida software, all consolidated into a single Excel sheet. ASOs that were dissolved, liquidated, or affiliated with both an Italian and a foreign university without available 2022 financial statements were excluded. A manual review ensured data accuracy, resulting in a final sample of 294 spin-offs.

Scopus was used to identify keywords to filter DASOs due to its comprehensive coverage of scientific, technical, and social sciences literature, including a complete set of metrics. A search string was crafted using Boolean operators to include variations of "Technology" and "Digital," ensuring broad coverage to minimize the exclusion of relevant studies. This initial search yielded over 5.7 million results, which were narrowed down by applying exclusion and inclusion criteria, such as language and relevance to "Business, Management and Accounting" and "Computer Science." Ultimately, 39,829 articles from journals ranked in the AJG 2021 list were included, and the keywords from these articles were analyzed using VOSviewer software.

A database of 121,467 keywords with a minimum threshold of occurrence was created. For usability, keywords with occurrences of 50 or more were considered and then manually filtered to exclude any that could create ambiguity.

The final step involved filtering the initial database for DASOs using a Python script that cross-referenced the keywords with the company activity descriptions in the database. This process identified 171 DASOs and 123 ASOs. A manual review was then conducted to ensure the accuracy of the classifications, especially for those companies with keyword matches below the average (4.77).

This methodology showcases the comprehensive approach to data collection and analysis for understanding the financial and economic characteristics of Italian ASOs over the last 10 years, with a specific focus on distinguishing DASOs from ASOs.

In this study, the financial and economic performance of ASOs in Italy over the past decade is analyzed using ten indicators extracted from the Aida database. These indicators include Revenue, Net Profit, EBITDA, Total Assets, Net Equity, Return on Assets (ROA), Return on Equity (ROE), Debt Ratio, Short-term Debt Ratio, and Long-term Debt Ratio. These metrics were selected for their ability to provide a comprehensive overview of a company's financial health, operational efficiency, and profitability without delving into the specifics of each indicator.

41.3.2 Model

In this chapter, we employ Generalized Estimating Equations (GEE) to analyze the financial and economic performance of Italian ASOs. The GEE method extends generalized linear models (GLM) to accommodate qualitative dependent variables in the presence of correlated observations, such as repeated measures or clustered data. This approach is particularly suitable for longitudinal economic performance studies, offering robust regression coefficient estimates in samples with high correlation among repeated measures.

We specifically apply logistic regression for binary outcomes to model the success or failure of the ASOs. Logistic regression is chosen due to the dichotomous nature of our dependent variable, which cannot satisfy the linear regression assumptions of linearity, homoscedasticity, and normality of errors due to its qualitative nature.

The logistic model transforms the odds of the dependent event (e.g., success of the spin-off) into a logistic function, ensuring the predicted probabilities remain within the 0 and 1 range. This is achieved by applying a logit transformation to the odds ratio, resulting in a model where increases in the independent variable(s) translate to increases or decreases in the probability of the event occurring.

Our model incorporates multiple independent variables to predict the binary outcome Y_{it}, representing the success or failure of the ASOs in year t for event i, as follows:

$$Y_{it} = a + \sum_{j=1}^{J} b_j X_{itj} + \varepsilon_{it}$$

Here, Y is the outcome, i indicates the event, t denotes time, a is the intercept, j represents the number of predictors, b_j is the slope of predictor j, X_{itj} is the predictor, and ε_{it} is the error term.

The variables used in this study are then explicitly included in the model, which can be expressed as:

$$Y = \ln\left(\frac{P(UN_i > 0)}{1 - P(UN_i > 0)}\right) = a + b_1 \text{Digital\_Tech}_{it1} + b_2 \text{Ext\_Part}_{it2}$$
$$+ b_3 \text{Uni\_Comp}_{it3} + b_4 \text{Women}_{it4} + b_5 \text{Int\_Prop}_{it5} + b_6 \text{Uni\_Rank}_{it6}$$
$$+ b_7 \text{Act\_Year}_{it7} + \varepsilon_{it}$$

where t refers to the reference year (2013, …, 2022). This formulation allows for the examination of how various factors, including digital technology adoption, external partnerships, university competition, female participation, intellectual property, university ranking, and years of activity, influence the likelihood of spin-off success.

41.4 Findings

The following results were obtained using SPSS software with a confidence level set at 95%. This implies a significance level of 5%, aligning with the conventional standard that considers a 5% (or 1/20) chance as "quite unlikely" for the observed difference to be due to random chance alone. The analysis includes a total of 1842 cases, all of which were included in the study, as shown in Tables 41.1 and 41.2.

The logistic regression model fitted in this study is expressed as:

$$\ln\left(\frac{P_{Y=1}}{P_{Y=0}}\right) = -0.064 + 0{,}430\text{Digital\_Tech} + 0{,}325\text{Ext\_Part}$$
$$-0{,}067\text{Uni\_Comp} - 0{,}009\text{Women} - 0{,}417\text{Int\_Prop}$$
$$-0.005\text{Uni\_Ranking} - 0.066\text{Activity\_Years}$$

Table 41.1 Parameter estimates

			Hp Test			
Parameter	P	Std. Error	Wald Chi-Square	df	Sign.	Exp (B)
(Intercept)	−0.064	0.6680	0.009	1	0.924	0.938
[Digital_Tech = 1]	0.430	0.1652	6.774	1	0.009	1537
[Int_Prop = 1]	−0.417	0.1334	9.744	1	0.002	0.659
[Ext_Part = 1]	0.325	0.2632	1.521	1	0.218	1383
[Women = 1]	−0.009	0.1717	0.002	1	0.960	0.992
[Uni_Comp_Soc = 1]	−0.067	0.5155	0.017	1	0.897	0.935
Uni_Ranking	−0.005	0.0089	0.323	1	0.570	0.995
Activity_Years	−0.066	0.0146	20.747	1	<0.001	0.936
(scale)	1					

Dep Variable UN
[a]Set to zero as this parameter is redundant

Table 41.2 Parameter estimates for the adjusted logistic regression model

			Hp Test			
Parameter	P	Std. Error	Wald Chi-Square	df	Sign.	Exp(B)
(Intercept)	−0.264	0.1858	2.016	1	0.156	0.768
[Digital_Tech = 1]	0.431	0.1668	6.671	1	0.010	1.539
[Int_Prop = 1]	−0.425	0.1337	10.123	1	0.001	0.654
Activity_Years	−0.070	0.0145	23.300	1	<0.001	0.932
(Scale)	1					

Dep Variable UN
[a]Set to zero as this parameter is redundant

The parameter estimate table reveals which covariates are significant for the model. External partnerships, female presence, university composition, and university ranking do not significantly influence the dependent variable, as their p-values exceed 0.05. This indicates that hypotheses Hp.3, Hp.4.1, Hp.4.2, and Hp.4.3 are not supported. However, the use of digital technologies, possession and use of intellectual property, and years of activity are significant ($p < 0.05$), suggesting a notable effect on the dependent variable.

Removing nonsignificant covariates (see Table 41.2) for a more streamlined model yields:

$$\ln\left(\frac{P_{Y=1}}{P_{Y=0}}\right) = -0.264 + 0,431\text{Digital\_Tech} - 0,425\text{Int\_Prop} - 0.07\text{Activity\_Years}$$

This adjusted logistic regression model maintains consistency with the initial findings. Evaluating the odds ratios (Exp(B)) shows that digital technology use, the only covariate with a value greater than 1, significantly increases the likelihood of achieving a positive net profit, supporting hypothesis Hp.1. Conversely, the possession and use of intellectual property appear to negatively impact net profit, as indicated by an odds ratio significantly less than 1, leading to the rejection of hypothesis Hp.2. The years of activity, with an odds ratio just below 1, present a borderline case that does not fully support rejecting hypothesis Hp.7.

This model demonstrates that, all else being equal, companies utilizing digital technologies have a 54% higher probability of realizing a positive net profit compared to those that do not.

41.5 Conclusion

The regression model analysis reveals significant findings regarding the impact of digital technologies on ASOs. The positive influence of the "Digital_Tech" variable underscores the beneficial effects of adopting digital technologies on economic performance. This aligns with extensive literature that has debated the positive impact of specific digital technologies on business performance, including studies by Tarutė and Gatautis [35], Parida et al. [36], and Eller et al. [37]. These findings are corroborated by more focused research within the Italian context by Cirillo et al. [38], as well as international studies by Rothmann and Koch [39], and insights from Khin and Ho [40] on the competitive advantage conferred by a digital-oriented corporate culture and improved digital capabilities. However, contrasting views are presented by Guo et al. [29], who suggest that while digital technologies positively impact productive factors, they do not universally enhance business performance due to increased operational costs and management expenses.

The results indicate that digital technologies have become integral to daily and corporate life. Even noncore digital enterprises leverage these technologies in numerous secondary processes.

Regarding hypothesis Hp.2, the negative impact of intellectual property ownership and usage on economic performance contradicts many studies within nonacademic enterprises, suggesting intellectual property as a fundamental organizational performance driver and a source of sustainable competitive advantage [41]. This discrepancy may indicate that a dichotomous variable insufficiently captures the value of intellectual property or that ASOs fail to maximize this asset effectively. Moreover, the study did not account for internal university dynamics related to intellectual property usage, existing and forthcoming legislation, university regulations, and the impact of patent ownership on a researcher's career.

For hypotheses Hp.3, Hp.4.1, Hp.4.2, and Hp.4.3, no definitive conclusions can be drawn for the sampled population since the related variables were not significant in the model. This outcome suggests that spin-offs might not fully utilize certain resources, or unique dynamics within the academic environment may not emphasize these variables. The study hints at a complex interplay among the variables, suggesting that even minor external incentives could trigger a chain reaction, similar to investments from Business Angels.

Lastly, the significance of Hp.7 (years of activity) cannot be outright dismissed, but its implications vary case by case. The mixed literature on this topic, with varying focuses on economic and financial performance, complicates the interpretation. The study's results, while not surprising, suggest that the impact of time on business performance is nuanced and that other factors, such as the amount of time entrepreneurs dedicate to their ventures, might be more relevant than merely the years of activity. This highlights the challenges faced by ASOs, including limited resources and the diverse interests and objectives of their founding teams.

References

1. Fini, R., Fu, K., Mathisen, M.T., Rasmussen, E., Wright, M.: Institutional determinants of university spin-off quantity and quality: a longitudinal, multilevel, cross-country study. Small Bus. Econ. **48**(2), 361–391 (2017)
2. Sciarelli, M., Landi, G.C., Turriziani, L., Tani, M.: Academic entrepreneurship: founding and governance determinants in university spin-off ventures. J. Technol. Transf. **46**(4), 1083–1107 (2021)
3. Vincett, P.S.: The economic impacts of academic spin-off companies, and their implications for public policy. Res. Policy. **39**(6), 736–747 (2010)
4. Miller, K., Alexander, A., Cunningham, J.A., Albats, E.: Entrepreneurial academics and academic entrepreneurs: a systematic literature review. Int. J. Technol. Manag. **77**(1–3), 9–37 (2018)
5. El-Awad, Z., Brattström, A., Breugst, N.: Bridging cognitive scripts in multidisciplinary academic spinoff teams: a process perspective on how academics learn to work with non-academic managers. Res. Policy. **51**(10), 104592 (2022)

6. Knockaert, M., Ucbasaran, D., Wright, M., Clarysse, B.: The relationship between knowledge transfer, top management team composition, and performance: the case of science–based entrepreneurial firms. Entrep. Theory Pract. **35**(4), 777–803 (2011)
7. Tagliazucchi, G., Marchi, G., Balboni, B.: A nonlinear relationship between the team composition and performance in university spin-offs. Technol. Forecast. Soc. Chang. **172**, 121061 (2021)
8. Algieri, B., Aquino, A., Succurro, M.: Technology transfer offices and academic spin-off creation: the case of Italy. J. Technol. Transf. **38**(4), 382–400 (2013)
9. Belitski, M., Aginskaja, A., Marozau, R.: Commercializing university research in transition economies: technology transfer offices or direct industrial funding? Res. Policy. **48**(3), 601–615 (2019)
10. Colombo, M.G., D'Adda, D., Piva, E.: The contribution of university research to the growth of academic start-ups: an empirical analysis. J. Technol. Transf. **35**(1), 113–140 (2010)
11. Rippa, P., Secundo, G.: Digital academic entrepreneurship: the potential of digital technologies on academic entrepreneurship. Technol. Forecast. Soc. Chang. **146**, 900–911 (2019)
12. Garcez, A., Silva, R., Franco, M.: Digital transformation shaping structural pillars for academic entrepreneurship: a framework proposal and research agenda. Educ. Inf. Technol. **27**(1), 1159–1182 (2022)
13. Linzalone, R., Schiuma, G., Ammirato, S.: Connecting universities with entrepreneurship through digital learning platform: functional requirements and education-based knowledge exchange activities. Int. J. Entrep. Behav. Res. **26**(7), 1525–1545 (2020)
14. Secundo, G., Rippa, P., Cerchione, R.: Digital academic entrepreneurship: a structured literature review and avenue for a research agenda. Technol. Forecast. Soc. Chang. **157**, 120118 (2020)
15. Toniolo, K., Masiero, E., Massaro, M., Bagnoli, C.: A grounded theory study for digital academic entrepreneurship. Int. J. Entrep. Behav. Res. **26**, 1567 (2020)
16. Ivanović-Đukić, M., Stevanović, T., Rađenović, T.: Does digitalization affect the contribution of entrepreneurship to economic growth. Proc. Rijeka Fac. Econ. J. Econ. Bus. ***36***(2), 653–679 (2019)
17. Gregory, R.W., Kaganer, E., Henfridsson, O., Ruch, T.J.: IT consumerization and the transformation of IT governance. MIS Q. **42**(4), 1225–1253 (2018)
18. Ferreira, J.J., Fernandes, C.I., Ferreira, F.A.: To be or not to be digital, that is the question: firm innovation and performance. J. Bus. Res. **101**, 583–590 (2019)
19. Hanelt, A., Firk, S., Hildebrandt, B., Kolbe, L.M.: Digital M&A, digital innovation, and firm performance: an empirical investigation. Eur. J. Inf. Syst. **30**(1), 3–26 (2021)
20. Iacobucci, D., Micozzi, A.: How to evaluate the impact of academic spin-offs on local development: an empirical analysis of the Italian case. J. Technol. Transf. **40**(3), 434–452 (2015)
21. Ortín-Ángel, P., Vendrell-Herrero, F.: University spin-offs vs. other NTBFs: total factor productivity differences at outset and evolution. Technovation. **34**(2), 101–112 (2014)
22. Wennberg, K., Wiklund, J., Wright, M.: The effectiveness of university knowledge spillovers: performance differences between university spinoffs and corporate spinoffs. Res. Policy. **40**(8), 1128–1143 (2011)
23. Bercovitz, J., Feldman, M.: Academic entrepreneurs: organizational change at the individual level. Organ. Sci. **19**(1), 69–89 (2008)
24. Grimaldi, R., Kenney, M., Siegel, D.S., Wright, M.: 30 years after Bayh–Dole: reassessing academic entrepreneurship. Res. Policy. **40**(8), 1045–1057 (2011)
25. Colombo, M.G., Piva, E.: Firms' genetic characteristics and competence-enlarging strategies: a comparison between academic and non-academic high-tech start-ups. Res. Policy. **41**(1), 79–92 (2012)
26. Urbano, D., Alvarez, C.: Institutional dimensions and entrepreneurial activity: an international study. Small Bus. Econ. **42**(4), 703–716 (2014)
27. Nambisan, S.: Digital entrepreneurship: toward a digital technology perspective of entrepreneurship. Entrep. Theory Pract. **41**(6), 1029–1055 (2017)

28. Deperi, J., Bertrand, O., Meschi, P.X., Nesta, L.: An organizational learning approach to digital and non-digital firm acquisition behavior. Eur. Manag. J. **40**, 873 (2022)
29. Guo, X., et al.: Does digital transformation improve the firm's performance? From the perspective of digitalization paradox and managerial myopia. J. Bus. Res. **163**, 113868 (2023)
30. Andries, P., Faems, D.: Patenting activities and firm performance: does firm size matter? J. Prod. Innov. Manag. **30**(6), 1089–1098 (2013)
31. Del Sarto N, et al. (2021) 'When Women Lead Companies: the Role of Women in Italian Research Spin-off Companies'. ISBN: 978-2-9602195-3-1.
32. Ferretti, M., Ferri, S., Fiorentino, R., Parmentola, A., Sapio, A.: What drives the growth of academic spin-offs? Matching academics, universities, and non-research organizations. Int. Entrep. Manag. J. **16**, 137–163 (2020)
33. Nambisan, S., Wright, M., Feldman, M.: The digital transformation of innovation and entrepreneurship: progress, challenges and key themes. Res. Policy. **48**(8), 103773 (2019)
34. Digital Transformation Scoreboard: EU Businesses Go Digital: Opportunities, Outcomes and Uptake, p. 2018. Publications Office of the European Union (2018)
35. Tarutė, A., Gatautis, R.: ICT impact on SMEs performance. Procedia Soc. Behav. Sci. **110**, 1218–1225 (2014)
36. Parida, V., Sjödin, D., Reim, W.: Reviewing literature on digitalization, business model innovation, and sustainable industry: past achievements and future promises. Sustainability. **11**(2), 391 (2019)
37. Eller, R., Alford, P., Kallmünzer, A., Peters, M.: Antecedents, consequences, and challenges of small and medium-sized enterprise digitalization. J. Bus. Res. **112**, 119–127 (2020)
38. Cirillo, V., et al.: New digital technologies and firm performance in the Italian economy. Ind. Innov. **30**, 159 (2022)
39. Rothmann, W., Koch, J.: Creativity in strategic lock-ins: the newspaper industry and the digital revolution. Technol. Forecast. Soc. Change. **83**, 66–83 (2014)
40. Khin, S., Ho, T.C.F.: Digital technology, digital capability and organizational performance: a mediating role of digital innovation. Int. J. Innov. Sci. **11**(2), 177–195 (2019)
41. Teixeira, A.A., Ferreira, C.: Intellectual property rights and the competitiveness of academic spin-offs. J. Innov. Knowl. **4**(3), 154–161 (2019)

Chapter 42
Cities Organising Smart Innovative Communities

Mauro Romanelli

Abstract Cities of tomorrow are planning smart itineraries for the urban future, adopting a smart city view to support urban innovation by fostering collaborative processes for sustainable urban growth and quality of life improvement. As smart innovative communities, cities are adopting a smart city framework to urban planning for innovation. A smart city helps to shape the city of the future an engine of urban innovation, a space for collaborative and organisational innovation. Cities are organising smart innovative urban communities, planning a smart city framework for social urban growth. The European cities are defining smart urban planning, enabling the smartness an ability to shape organisational and collaborative urban spaces that support social open innovation. Sustainable urban future relies on cities becoming smart communities that support urban innovation for good life, promoting collaborative and multi-actor innovation, following a human-centred approach to smart urban development, and promoting a smart city vision and community to urban growth.

Keywords Smart cities · Smart communities · Urban innovation

42.1 Introduction

Cities of tomorrow will be smart and socially innovative communities that will be able to build collaborative urban ecosystems for innovation [1, 2]. Cities are planning smart urban development, tracking a pathway for sustainable urban wealthy futures, adopting an innovation-led view to smart city strategy [3], dealing with smart environments, techno-centric and human-centred innovations that enable

M. Romanelli (✉)
Department of Business and Economics, University of Naples Parthenope, Naples, Italy
e-mail: mauro.romanelli@uniparthenope.it

F. Schiavone et al. (eds.), *Advanced Perspectives and Trends in Digital Transformation of Firms, Networks, and Society*, Springer Proceedings in Business and Economics, https://doi.org/10.1007/978-3-031-80692-6_42

smart, sustainable, socially innovative society and communities [4]. Sustainable urban development relies on cities that adopt a smart city framework and contribute to public value creation and innovation [5]. As incubators and agents of social innovation, cities adopt a smart city vision in order to shape the processes and address the pathways that contribute to urban innovation, following a human-centric view, combining both technological and social innovation [6], making the city a smart socially innovative community which is able to drive collaborative and multi-actor innovation. Sustainable urban future relies on social innovation for good life through cities that adopt a human-centred smart view to urban growth by promoting multi-actor innovation, and following a collaborative view to smart urban innovation for urban radical changes [7–10].

Intelligent cities foster smart city development and enable people-driven and community-led innovation, and enhance citizen engagement, opening to knowledge-led and technology-driven social innovation, and engaging the urban communities and stakeholders by using the potential offered by technologies. Innovative cities are using the potential of technologies in order to shape human-centred cities that enhance collective intelligence and cooperation [11], leading to net action and collaborative spaces for advancing technological and social innovation, by promoting collaborative and multi-actor processes [8, 12–14]. Smart urban future relies on bottom-up initiatives, citizen engagement, and capabilities offered by evolving and advanced technologies, fostering smart innovative communities [15, 16] in order to improve the quality of life within urban communities.

While there are many studies that focus on smart city visions, few studies investigate the role of smart city planning and vision as a driver of urban innovation. Which is the urban future development the cities do aim to achieve? As intelligent communities, cities are planning the future urban development by selecting a smart view to driving urban growth and innovation. The research question refers to how cities are rethinking the urban future, shaping the city a smart socially innovative urban community. The aim of this study is to elucidate how cities are rethinking the urban future, planning a smart city view to drive the cities smart socially innovative urban communities that enhance urban as well as collaborative and socially open spaces. Three case studies are analysed and presented. Paris, Wien and Florence are rethinking about the urban future by adopting the smart vision for urban planning in order to shape smart-driven, innovation-led and human-centred communities. Future cities develop a smart city and community vision to foster collaboration a driver of sustainable urban innovation. Planning a smart view to urban development helps cities to shape urban communities engines of urban innovation. After introduction and theoretical background, in the third section the methodology used is presented. In the fourth section, findings and issues are elucidated. Finally, discussion and conclusions are set out.

42.2 The Future of Cities as Smart Innovative Communities

As organisational spaces [17], arenas of innovation [13], and special kinds of organisations [18], cities are adopting a smart view, fostering multi-level and multi-actor collaborative innovation [19, 20], shaping sustainable and healthy growth, and socially innovative and inclusive urban communities for good quality of life. Smart cities contribute to driving urban competitiveness, promoting the quality of life, and supporting urban innovation in long-term horizon [16, 21]. Smart city solutions and framework contribute to sustainable urban development and communities [22]. As a *continuum* along several urban stakeholders that contribute to the wellbeing of the city and enable effective and livable urban workplaces [23], smart cities rely on human and organisational assets and facilitate multi-actor collaboration and collective action, and support the efforts of the urban stakeholders in order to make an innovative city and transform the urban community in a significant way [24], driving innovation-led as well as inclusive, sustainable, and people-centred communities [3, 25]. Cities drive smart innovations [26]. Smart cities contribute to shaping urban innovation [3]. Cities support smart social open innovation in order to improve the quality of life and achieve the long-term sustainability [21]. Cities adopt a human-centric view in shaping the smart city a platform for technological and social innovation [6]. A smart city helps to shape the city of the future an engine of social innovation [27], providing solutions for sustainable urban, social and economic development [28] that relies on planning and constructing urban organisational spaces for open innovation-driven cities and urban communities that employ human, technological, social and institutional capabilities [3, 21, 29]. Urban innovation relates to intentional and proactive processes that support creative ideas in order to enhance qualitative urban changes [14]. Smart cities are the symbol of technology-driven urban innovation [30]. Cities are constructing a wealthy future, designing a smart city vision, and leading to smart community development for collaborative and multi-actor innovation [1, 8], employing several capabilities within smart urban ecosystems for innovation [31], and fostering collaborative planning and innovation towards citizen-centred and societal smart city [32]. Cities contribute to driving urban innovation by promoting the smartness as the ability to attract and mobilise human capital in collaboration with urban actors and stakeholders by using information and communication technologies [33]. Innovative cities support collaboration within smart innovative urban communities [15]. Enabling smart city development helps the cities to meet urban sustainability priorities, shaping urban collaborative ecosystems for urban innovation [34]. Cities focus on smart, sustainability-oriented social and human-centred innovation relying on citizen engagement [35], empowering the citizens to proactively support services co-production and innovation [36].

42.3 Methodological Section

The study employed a qualitative, descriptive and exploratory methodology to analyse how cities are rethinking on the future of urban development, shaping smart urban innovation and communities, rediscovering the smart city an opportunity to drive social innovation. The European cities are rethinking on the urban future, adopting a smart city vision, and planning a smart framework to urban innovation in order to revitalise the urban community, and enhance both the impact and the quality of urban growth. The study relies on selecting three European cities that are relevant for cultural, political, and historical urban and societal growth: Paris, Wien and Florence are adopting a smart city strategy to develop the city of the future a socially innovative urban community. In Europe, cities are called to deal with urbanisation in order to enhance sustainable and inclusive growth, by supporting collaborative as well as organisational urban spaces for innovation. They have defined both a smart city road map and elucidated a framework to track the urban future for wealthy urban communities in long-term horizon, shaping the urban spaces engines of social innovation and growth. The study focuses on three illustrative and exploratory case studies concerning the European cities of Paris, Florence and Wien that are rethinking on identifying a smart view to the urban future, by understanding a phenomenon still in progress and emerging in contemporary times, whereas the boundaries between the phenomenon and the context are not clearly evident, and following a multiple case study analysis, whereas the smart city framework and strategy provide a means for urban planning that serves for urban development solutions [30, 37]. The case study methodology refers to an empirical inquiry without the rigidity of a predefined structure for observations and analysis [38]. The research methods involved case selection and data collection from policy documents as smart city strategy frameworks drawn from *Paris Smart and Sustainable Looking ahead to 2020 and beyond*, *Firenze Smart City Plan* and *Smart City Strategy Vienna* [39–41] as information sources that permit to collect and analyse data and information, and to investigate how some European cities are planning the urban future development, leading to urban communities engines of social innovation and growth, coherently with a smart city view. Investigating the smart city strategy design helps to understand the trajectories of development cities are identifying to shape the pathways for advancing towards urban social innovation [28, 29, 42]. The three cities refer to the adoption of a smart city vision as urban development framework and planning in order to revitalise the city a socially innovative community able to proceed towards urban sustainability.

42.4 Findings

Cities are planning the future of urban development by selecting a smart vision to drive urban innovation and social growth within urban spaces. Rethinking smart innovative cities relies on driving smart city urban planning and implementation in order to shape cities smart-driven, innovation-led, and human-centred communities. Planning urban development relies on strengthening the organisational and social spaces for innovation. Three case studies are analysed and presented. The European cities are rethinking about the urban future by adopting a smart vision planning in order to support innovation and drive sustainable urban development, shaping innovative as well as collaborative urban spaces and communities. This analysis elucidates that the European cities are redesigning the urban planning, by rethinking about the future of urban communities, and adopting a smart vision in order to shape the city of the future an innovative community, by strengthening collaborative frameworks among several stakeholders that contribute to engendering social and organisational innovation processes within urban spaces and communities.

The smart city framework *Paris Smart and Sustainable Looking ahead to 2020 and beyond* [39] aims at encouraging all the urban stakeholders to work together to shape a smart and sustainable city by using the potential of information technology in order to support urban innovation, and improve the quality of life within urban communities. Smartness is conceived as a means to drive the city to improve the quality of life and achieve sustainable urban futures. The smart city framework seems to make the city an open platform that supports the experimentation and the development of new innovative services and solutions that improve the quality of life and contribute to the wealthy, safe and healthy urban communities and environments. Developing the city of the future means to bring together social, human, technological and organisational dimensions along three pillars, leading to a smart, innovative and sustainable city: the *connected city* (the tool) for integrating technological advancements; the *open city* (the method) by stimulating citizen participation and collaborative projects that contribute to supporting collaborative urban spaces for innovation, knowledge and information sharing, leading to urban value co-creation among private and public actors (academia, scientists, researchers, business, public administration); the *sustainable city* (the goal) is a place where technology serves the urban communities that contribute to urban innovation design and solutions, and produce the issues of urban planning choices with regards to the co-production of public services and the co-design of policy decision-making processes.

In the *Firenze Smart City Plan* [40] smart urban innovation relies on the city as a living organism in which the urban community of several urban stakeholders and proactive citizens contribute to identifying collaborative spaces for urban development, innovation and quality of life improvement. The smart city plan seems to be an open space for debate and confrontation, information and knowledge exchange and sharing among all urban stakeholders. Smart urban innovation relies on promoting the collaboration that relies on all the involved stakeholders (public administration, research centres, cultural institutions, businesses, associations, groups, citizens) which aim at working together along the urban value chain. A smart city strategy may contribute to social change and innovation. It develops through the four *Is*: the *integration* of competences and background of the municipality jointly with energies and competences of the urban community; the *innovation* as the development of innovative services and solutions through digital and interactive information technology; the *involvement* as a means to enable the stakeholders to identify, plan, share and implement strategic long-terms vision and goals; and the *information* relies on the use of technology as a way to reinforce and consolidate the relationships between municipality and citizens, and monitor the implementation of smart urban innovation processes and strategic issues.

In the *Smart City Strategy Vienna* [41] the smart city vision develops to ensure the quality of living, sustainability, prosperity, and the opportunities for education and workplaces. The role of information technology is to simplify the urban life, enabling urban change and innovation processes. Smart city vision helps the city to drive urban innovation for collaborative and organisational action. Smart socially innovative cities support both urban inclusive communities and civic participation in order to ensure top-level quality of living, safety and security, leisure quality and culture, leading to social openness and cultural diversity. Technology serves for community development and supports the city to behave as innovative and organisation-driven community, opening up to collaborative processes that drive urban value creation and social innovation. Technology supports co-participation, co-determination and co-decision-making as strategic and organisational frameworks in order to shape the city a smart innovative urban community.

42.5 Discussion and Conclusions

As smart innovative urban communities, cities of tomorrow are planning the urban future, making urban collaborative as well as organisational open spaces, shaping urban communities engines of social inclusion and innovation. The Paris sustainable city vision stresses the connected city as a tool, the open city as a method, and the sustainable city as a goal. In Wien, the smart vision relies on the use of technology as a means to enhance co-participation, co-decision-making, and inclusion as strategic and organisational views to urban sustainable development. In Florence, the smart city planning focuses on searching for bringing together integration, innovation, involvement, and information. There are common trends and differences in the approach the cities are following to make smart socially innovative urban communities. Technology is conceived as a means to facilitate and enable urban spaces for dialogue, confrontation, openness, and cooperation. Technologies help connectivity and communication, leading to human-centred urban development. The role of municipal authority is changing in Florence, the municipal authority is in charge of initiating the process that supports the urban community as a meeting place of innovation. In Wien, the municipal authority is in charge of facilitating the involvement of citizens and urban communities, and to support the engagement of several urban stakeholders that contribute to shaping urban spaces as collaborative frameworks that enhance the urban community. In Paris, the municipal authority should act to involve and embed profit and no-profit actors in collaborative and cooperative processes and ways. Technology enables citizen and community participation and engagement for continuous innovation for urban sustainability. Cities of the future will aim at driving sustainable urban innovation, understanding the potential of information technology as a means that enables cities to adopt a human-centred and smart-oriented view and to develop the city an organisational space for action and innovation, a socially innovation-driven, collaborative-led, and knowledge-oriented community. Technology enables citizen and community participation and engagement for continuous innovation for urban sustainability. The study is mainly exploratory. The analysis provides only three illustrative and exploratory case studies to elucidate that smart city-led urban planning helps to make the city a socially innovative urban community, enabling collaborative frameworks that enhance urban spaces for social innovation. Today, cities are facing the challenge of driving social innovation and promoting collaborative approaches for urban future. Urban change relies on cities that evolve as communities, and develop innovation, promoting collaborative frameworks through stakeholder involvement and engagement. The aim of cities is to develop smart city frameworks in order to shape smart innovative urban communities, rethinking the city an engine of social, organisational and collaborative innovation within urban spaces. Further research perspectives aim at investigating how Italian cities are planning the urban future, following a smart and sustainable vision to urban innovation and growth, and dealing with artificial intelligence applications in order to enhance urban, community, and collective intelligence.

References

1. Mora, L., Appio, F.P., Foss, N.J., Arellano-Gault, D., Zhang, Z.: Organizing for smart city development: research at the crossroads. Introduction to the special issues. Org. Stud. **44**(10), 1559–1575 (2023)
2. Dolmans, S.A.M., van Galen, W.P.L., Walrave, B., den Ouden, E., Valkenburg, R., Romme, A., Georges, L.A.: Dynamic perspective on collaborative innovation for smart city development: the role of uncertainty, governance, and institutional logics. Organ. Stud. **44**, 1577–1601 (2023)
3. Nam, T., Pardo, T.A.: Smart city as urban innovation with dimensions of technology, people and institutions. In: Proceedings of the 5th International Conference on Theory and Practice of Electronic Governance, pp. 185–194. ACM (2011)
4. Carayannis, E.G., Dezi, L., Gregori, G., Calo, E.: Smart environments and techno-centric and human-centric innovations for Industry and Society 5.0: a quintuple helix innovation system view towards smart, sustainable, and inclusive solutions. J. Knowl. Econ., 1–30 (2021)
5. Pardo, T.A., Gil-Garcia, J.R., Gascó-Hernández, M., Cook, M.E., Choi, I.: Creating public value in cities: a call for focus on context and capability. In: Estevez, E., Pardo, T.A., Scholl, H.J. (eds.) Smart Cities and Smart Governance. Towards the 22nd Century Sustainable City, pp. 119–140. Springer, Cham (2021)
6. Costales, E.: Identifying sources of innovation: building a conceptual framework of the Smart City through a social innovation perspective. Cities. **120**, 103459 (2022)

7. Yigitcanlar, T.: Smart city beyond efficiency: technology-policy-community at play for sustainable urban futures. Hous. Policy Debate. **31**(1), 88–92 (2021)
8. Sørensen, E., Torfing, J.: Co-initiation of collaborative innovation in urban spaces. Urban Aff. Rev. **54**(2), 388–418 (2018)
9. Andreani, S., Kalchschmidt, M., Pinto, R., Sayegh, A.: Reframing technologically enhanced urban scenarios: a design research model towards human centered smart cities. Technol. Forecast. Soc. Change. **142**, 15–25 (2019)
10. Nilssen, M.: To the smart city and beyond? Developing a typology of smart urban innovation. Technol. Forecast. Soc. Change. **142**, 98–104 (2019)
11. Yigitcanlar, T., Xia, B., Cortese, T.T.P., Sabatini-Marques, J.: Understanding city 4.0: a triple bottom line approach. Sustainability. **16**(1), 326 (2023)
12. Czarniawska, B.: Remembering while forgetting: the role of automorphism in city management in Warsaw. Public Adm. **62**(2), 163–173 (2002)
13. Zhang, S., Wang, X.: Does innovative city construction improve the industry-university-research knowledge flow in urban China? Technol. Forecast. Soc. Change. **174**, 121200 (2022)
14. Meijer, A.J., Thaens, M.: Urban technological innovation: developing and testing a sociotechnical framework for studying smart city projects. Urban Aff. Rev. **54**(2), 363–387 (2018)
15. Caragliu, A., Del Bo, C.F.: Smart innovative cities: the impact of Smart City policies on urban innovation. Technol. Forecast. Soc. Chang. **142**, 373–383 (2019)
16. Appio, F.P., Lima, M., Paroutis, S.: Understanding smart cities: innovation ecosystems, technological advancements, and societal challenges. Technol. Forecast. Soc. Change. **142**, 1–14 (2019)
17. Knox, H.: Cities and organisation: the information city and urban form. Cult. Organ. **16**(3), 185–195 (2010)
18. Rosemann, M., Becker, J., Chasin, F.: City 5.0. Business & information. Syst. Eng. **63**, 71–77 (2021)
19. European Commission: Report from the Commission to the Council on the Urban Agenda for the EU, COM(2017) 657 final, 20.11.2017 (2017)
20. European Commission: Smart cities and communities—European Innovation Partnership, C (2012) 4701 final, Brussels, 10.7.2012. (2012)
21. Paskaleva, K.A.: The smart city: a nexus for open innovation? Intell. Build. Int. **3**(3), 153–171 (2011)
22. Sharifi, A., Allam, Z., Bibri, S.E., Khavarian-Garmsir, A.R.: Smart cities and sustainable development goals (SDGs): a systematic literature review of co-benefits and trade-offs. Cities. **146**, 104659 (2024)
23. Gil-Garcia, J., Pardo, T., Nam, T.: A comprehensive view of the 21st century city: smartness as technologies and innovation in urban contexts. In: Gil-Garcia, J., Pardo, T., Nam, T. (eds.) Smarter as the New Urban Agenda. Public Administration and Information Technology, vol. 11. Springer, Cham (2016)
24. Eger, J.M.: Smart communities, universities, and globalization: educating the workforce for tomorrow's economy. Metrop. Univ. **16**(4), 28–38 (2005)
25. Chang, J.I., Choi, J., An, H., Chung, H.Y.: Gendering the smart city: a case study of Sejong City, Korea. Cities. **120**, 103422 (2022)
26. Joss, S., Sengers, F., Schraven, D., Caprotti, F., Dayot, Y.: The smart city as global discourse: storylines and critical junctures across 27 cities. J. Urban Technol. **26**(1), 3–34 (2019)
27. Anttiroiko, A.-V., Valkam, P., Bailey, S.J.: Smart cities in the new service economy: building platforms for smart services. AI Soc. **29**(3), 323–334 (2014)
28. Mora, L., Bolici, R.: How to become a smart city: learning from Amsterdam. In: Bisello, A., Vettorato, D., Stephens, R., Elisei, P. (eds.) Smart and Sustainable Planning for Cities and Regions, pp. 251–266. Springer, Cham (2017)
29. Angelidou, M.: Smart cities: a conjuncture of four forces. Cities. **47**, 95–106 (2015)

30. Mora, L., Deakin, M., Reid, A., Angelidou, M.: How to overcome the dichotomous nature of smart city research: proposed methodology and results of a pilot study. J. Urban Technol. **26**(2), 89–128 (2019)
31. Gupta, A., Panagiotopoulos, P., Bowen, F.: Developing capabilities in smart city ecosystems: a multi-level approach. Organ. Stud. **44**, 01708406231164114 (2023)
32. Alizadeh, H., Sharifi, A.: Toward a societal smart city: clarifying the social justice dimension of smart cities. Sustain. Cities Soc. **95**, 104612 (2023)
33. Meijer, A., Bolívar, M.P.R.: Governing the smart city: a review of the literature on smart urban governance. Int. Rev. Adm. Sci. **82**(2), 392–408 (2016)
34. Mora, L., Appio, F.P., Foss, N.J., Arellano-Gault, D., Zhang, X.: Boosting urban sustainability through organizing collaborative ecosystems for smart city development. Organ. Stud. (2020)
35. Tura, N., Ojanen, V.: Sustainability-oriented innovations in smart cities: a systematic review and emerging themes. Cities. **126**, 103716 (2022)
36. Martin, C.J., Evans, J., Karvonen, A.: Smart and sustainable? Five tensions in the visions and practices of the smart-sustainable city in Europe and North America. Technol. Forecast. Soc. Chang. **133**, 269–278 (2018)
37. Eisenhardt, K.M.: Building theories from case study research. Acad. Manag. Rev. **14**(4), 532–550 (1989)
38. Yin, R.K.: Case Study Research: Design and Methods. SAGE Publications, Thousand Oaks (2002)
39. Mairie de Paris: Paris Smart and Sustainable Looking Ahead to 2020 and Beyond (2015)
40. Comune di Firenze: Firenze Smart City Plan (2015)
41. City of Vienna: Smart City Strategy Vienna (2022)
42. Sancino, A., Hudson, L.: Leadership in, of, and for smart cities–case studies from Europe, America, and Australia. Public Manag. Rev. **22**(5), 701–725 (2020)

Chapter 43
A Digital and Pluralistic Governance for Transport Companies: The European Way to Climate Mitigation in Maritime Transport Sector

Luca Della Tommasina

Abstract The challenge for a more sustainable maritime transport is above all a matter of corporate governance of transport companies. All the recent developments of Italian and European Business Law highlight that the most effective driver toward sustainability is the corporate structure, including ESG committees, corporate internal controls, and dialogues with institutional investors and other stakeholders. As confirmed by the new Dutch Corporate Governance Code, sustainable long-term value creation also requires awareness and anticipation of developments in new technologies and changes to business models and associated risks, including cybersecurity, supplier and chain dependencies, data protection, and ethically responsible use of new technologies. The chapter aims at focusing these topics from a corporate law perspective, with reference to the European Taxonomy Regulation (and the definition of economic activities which may qualify as "contributing substantially to the sustainable use and protection of water and marine resources"), as well as to the new EU Corporate Sustainability Due Diligence Directive.

Keywords Corporate governance · Adverse impacts · Controls · Corporate arrangements · Transports

---

L. D. Tommasina (✉)
Pisa University, Pisa, Italy
e-mail: luca.dellatommasina@unipi.it

F. Schiavone et al. (eds.), *Advanced Perspectives and Trends in Digital Transformation of Firms, Networks, and Society*, Springer Proceedings in Business and Economics, https://doi.org/10.1007/978-3-031-80692-6_43

43.1 Transport Companies and Protection of Water and Marine Resources from the Financial and Insurance Investments' Point of View

It has been said that key elements to pursue sustainable success for a corporation are investors' perspectives, the role of the board, the reporting model, communication of long-term strategy, outlook, the identification and representation of risks, and, finally, intangibles and alternative measures of performance (C. Busco). Technology is a fundamental driver for all these elements. One need only consider the importance of:

i) Digital platforms for the dialogues with stakeholders, as regards the identification of sustainability risks
ii) Digital platforms for the dialogues with institutional investors, for what concerns the communication of long-term strategy to the market
iii) Digital systems for the circulation of information inside the board of directors, the overall monitoring of business operations, and their benchmarking in terms of impact on the environment

In general, in the transports' field, the matter of sustainability is crucial and strategic. Everybody knows the new financial instruments aimed at promoting sustainability goals in the transport sector: green bonds focusing on blue projects, also known as "blue bonds," above all. Nicholas Pfaff, Deputy CEO and Head of Sustainable Finance at the ICMA, said that blue bonds "have great potential to expand the pool of capital market finance dedicated to the blue economy."

In 2016, United Nations Secretary-General's High-level Advisory Group on Sustainable Transport has defined sustainable transport as follows: "the provision of services and infrastructure for the mobility of people and goods—advancing economic and social development to benefit today's and future generations—in a manner that is safe, affordable, accessible, efficient, and resilient, while minimizing carbon and other emissions and environmental impacts."

This raises many problems for a corporate law researcher, but, above all, a problem of corporate governance from a twofold point of view:

a) The corporate governance of (maritime) transport companies
b) The engagement of banks, insurance companies, and any other financial intermediaries into their investee (transport business) companies

Indeed, for financial and insurance service providers, a matter of sustainability comes into play not only when giving loans to transport companies or covering risks of (maritime) transport business, but also when assuming the status of institutional investor (or asset manager) within a transport corporation (Siri, Zhu; Weber, Baisch; Martina; Genovese; Brozzetti; Calvosa; Driessen).

From the latter point of view, European banks—while placing their financial offer onto the market, including by posting their non-financial reports and engagement reports[1]—have to follow the so-called Taxonomy Regulation. Such a Regulation has fixed harmonized criteria that must be met in order for a business activity to be considered sustainable. The Taxonomy Regulation lists six environmental criteria:

1. Climate change mitigation
2. Climate change adaptation
3. Sustainable use and protection of water and marine resources
4. Transition to a circular economy
5. Pollution prevention and control
6. Protection and restoration of biodiversity and ecosystems

To be deemed as sustainable, an economic activity must substantially contribute to one of the objectives of the Taxonomy Regulation, provided that—at the same time –: (a) it does not create significant harm to any of the other objectives, (b) it is in line with minimum safeguards regarding human right and labor laws, and (c) it fulfills certain performance criteria, the so-called Technical Screening Criteria (TSC), described in the Taxonomy Regulation Delegated Acts.

Therefore, while providing a rating for shares or bonds issued by a maritime transport company, a bank shall take into account such definitions. And it shall consider many other profiles. For example, Article 12 of Taxonomy Regulation states that an economic activity shall qualify as *contributing substantially to the sustainable use and protection of water and marine resources* where that activity either contributes substantially to *achieving the good status* of bodies of water or marine waters or to *preventing the deterioration* of them (if already having good status), by (a) protecting the environment from the *adverse effects of urban and industrial waste water discharges*; (b) protecting human health from the *adverse impact* of any contamination of water intended for human consumption; (c) *improving water management and efficiency, including by protecting and enhancing the status of aquatic ecosystems*, by promoting the sustainable use of water through the *long-term protection* of available water resources, inter alia, through measures such as *water reuse*, by ensuring the progressive *reduction of pollutant emissions* into surface water and groundwater, by contributing to *mitigating the effects of floods and droughts*, or through any other activity that protects or improves the qualitative and quantitative status of water bodies; and (d) ensuring the *sustainable use of marine ecosystem services or contributing to the good environmental status of marine waters*.

In order to meet such requirements a business entity shall assess its own *adverse impacts*. This is a fundamental concept to get to the heart of the new Corporate Sustainability Due Diligence Directive (CSDDD), requiring companies to perform due diligence procedures with the aim of addressing any (relevant) negative effects of their business conducts on human rights and the environment, along their value chains worldwide as well: it is well known that on February 23, 2022, the Commission submitted to the European Parliament and to the Council a proposal

[1] See—for what concerns Italian corporate law—Article 124-*quinquies* d.lgs. No. 58/1998.

for a Directive on Corporate Sustainability Due Diligence and amending Directive (EU) 2019/1937, with the aim of strengthening the sustainability approach of the biggest companies, not only listed ones, at a global European level, and the Council has recently approved the draft of Directive, thereby completing the adoption process. Member States will ensure that the biggest companies (a) take appropriate measures to identify actual and potential adverse impacts arising from their own operations or those of their subsidiaries and, where related to their value chains of activities, those of their business partners; (b) take appropriate measures to prevent, or where prevention is not possible or not immediately possible, adequately mitigate potential adverse impacts; (c) carry out periodic assessments of their own operations and measures, those of their subsidiaries, and, where related to the chains of activities of the company, those of their business partners, to monitor the effectiveness of the identification, prevention, mitigation, bringing to an end, and minimization of the extent of adverse impacts.

Finance, banks, and industry are therefore part of the same challenge for climate transition (and more generally for the transition toward forms of sustainable capitalism): by labeling shares or bonds of a maritime transport company as green products, banks may help to attract investments to green companies (Colaert; Salerno; Rolli; Brozzetti; Cossu; Corvese; Pilaj; Weber, Baisch; Driessen). While benchmarking shares or bonds of a maritime transport corporation, banks (but the same applies to insurance companies) shall consider not only specific actions already carried out by the single corporation (e.g., the replacement of traditional company vans with new electric models), but also—and maybe above all—the global architecture of internal governance and its ability to identify sustainability risks for the future.

Insurance companies play a key role as well. One of the most important tools for these aims is product oversight and governance (POG), provided for by the EU Insurance Distribution Directive [(EU) 2016/97], also called IDD (Martina). In such a framework, insurance companies are *manufacturers*, and they shall arrange a *product approval process* in order to (a) define the customers for which the product is designed and also the customers for whom it is not recommended and (b) implement appropriate measures to ensure that the positive target market will not be changed in the distribution phase. Product oversight and governance have recently been importing ESG factors and sustainability preferences as well. Indeed, Eiopa's Guidelines so provide: "when considering the sustainability factors of products in order to then match them with the customer's sustainability preferences, insurers and insurance intermediaries may rank and group IBIPs included in their product range in terms of: (a) the proportion invested in economic activities that qualify as environmentally sustainable; (b) the proportion of sustainable investments; (c) the consideration of principal adverse impacts."

This implies that financial and insurance intermediaries check whether the governance is in line with the very (European and worldwide) quality standards, including in terms of digitization (Abriani, Schneider; Mosco; Montagnani; Enriques, Zetzsche; Fenwick, Kaal, Vermeulen; Möslein; Fenwick, Mccahery, Vermeulen; Siri, Macchiavello; Luciano; Schneider; Wen, Zhong, Lee; Wu, Li; Zhang, Yang,

Lv; see also *Studies on the relevance and impact of artificial intelligence for company law and corporate governance. Final report*). It is no accident that the new Dutch Corporate Governance Code states that sustainable long-term value creation also requires awareness and anticipation of developments in new technologies and changes to business models and associated risks, including cybersecurity, supplier and chain dependencies, data protection, and ethically responsible use of new technologies (see also Paragraph 1.1.1., *sub viii*, of the same Code). Moreover, in the maritime transport sector, strategic assessments—about the role of new technologies in the management of internal and external information flows (and related risks)—are currently required by the eFTI Regulation, with a view—highlighted by the European Commission—to:

(i) Sharing data with partners through eFTI platforms for more efficient logistics planning, vehicle loading, and routing
(ii) Improving multimodal transport planning and decision-making
(iii) Strengthening the supply chain monitoring processes

43.2 Supply Chain in Maritime Transport Sector, Lieferkettengesetz, and Corporate Governance

Moreover, *governance* is one of the three pillars on which the ESG commitment of any corporation, including transport (and maritime transport) ones, is founded.

The EU directive on corporate sustainability due diligence aims at fostering sustainable corporate behavior by enhancing sustainability assessments in companies' operations and governance. Without a coherent, serious, and structured system of *sustainability governance* within the single corporation (and the inherent group) it is not possible to significantly affect the entire production chain, or better *supply chain*, of the company. Among the first to notice it in Europe there was Germany, with the adoption of the so-called *Lieferkettengesetz*, for the supervision on the supply chain by big industrial corporations. Such a regulation, which was also designed to face climate change and to promote protection of human rights on a global scale, illuminates the collective cultural choices currently being made under the CSDDD. Article 4, about *Sorgfaltspflicht*, is noteworthy: "*Unternehmen sind dazu verpflichtet, in ihren Lieferketten die in diesem Abschnitt festgelegten menschenrechtlichen und umweltbezogenen Sorgfaltspflichten in angemessener Weise zu beachten mit dem Ziel, menschenrechtlichen oder umweltbezogenen Risiken vorzubeugen oder sie zu minimieren oder die Verletzung menschenrechtsbezogener oder umweltbezogener Pflichten zu beenden. Die Sorgfaltspflichten enthalten:*

1. *die Einrichtung eines Risikomanagements,*
2. *die Festlegung einer betriebsinternen Zuständigkeit,*
3. *die Durchführung regelmäßiger Risikoanalysen,*
4. *die Abgabe einer Grundsatzerklärung,*

5. *die Verankerung von Präventionsmaßnahmen im eigenen Geschäftsbereich und gegenüber unmittelbaren Zulieferern,*
6. *das Ergreifen von Abhilfemaßnahmen,*
7. *die Einrichtung eines Beschwerdeverfahrens,*
8. *die Umsetzung von Sorgfaltspflichten in Bezug auf Risiken bei mittelbaren Zulieferern und,*
9. *die Dokumentation und die Berichterstattung*."

In this list, all the points (or almost) affect governance issues: implementation of efficient arrangements (n. 1, n. 5, n. 7, n. 8: suffice it to say that CSDDD itself requires companies to map their own operations and upstream and downstream chains of activities, to conduct periodic in-depth assessments to identify adverse impacts, by using relevant internal and external information sources, and—where needed—to prioritize adverse impacts depending on severity and likelihood); division of powers (n. 2); internal audit and self-assessment (n. 3, n. 7); accountability (n. 9: this latter issue has been regulated by the so-called CSRD, the EU Corporate Sustainability Reporting Directive in 2022. See also Paragraph 1.1.4—about "Reporting by the management board"—of the Dutch Corporate Governance Code: "In the management report, the management board should provide a more detailed explanation of its view on sustainable long-term value creation and the strategy to realise this and describe the contributions made to sustainable long-term value creation in the past financial year. In addition, it describes the formulated objectives, what effects the company's products, services and activities have had on people and the environment, how the interests of stakeholders have been considered, what action has been taken in that context and the extent to which the set objectives have been attained. The management board should report on both short and the long-term developments").

Therefore, it is important to distinguish:

a) The single operations and the overall business strategy, on the one hand
b) The internal corporate governance, on the other hand

As regards operations and the overall business strategy, the engagement of a maritime transport company is extremely vast, potentially endless. It is enough to take into account the many goals outlined by Genco in its non-financial reports (something similar to the Italian "dichiarazione individuale non finanziaria") in 2022.[2]

---

[2]Among the others, we could remember the following goals:

- Continuing to renew fleet by selling older less fuel-efficient ships and redeploying the capital toward modern fuel-efficient tonnage
- Jointly studying the feasibility of ammonia as an alternative marine fuel as part of the objective to decarbonize the global shipping industry
- Outfitting vessels with energy saving devices (ESD) which are meant to reduce fuel consumption (and ESDs include Mewis Ducts, Fins, New efficient Propeller, and Propeller Boss Cap Fins)

However, the inclusion of sustainability profiles in business strategies is a matter of corporate governance as well. Requiring big corporations to consider adverse impacts does not mean setting performance constraints, nor does it lead them to redirect the corporate purpose in a pluralistic way. It means that companies shall assess non-financial profiles and implement tools (therefore, above all governance tools) in order to identify adverse impacts and to take thoughtful and informed decisions.

Corporate sustainability due diligence is not just a legal framework addressed to the CEO and top management, and the inherent obligations are not exhausted in thoroughly making a reconnaissance of all the specific constraints arising from environmental law (e.g., *not to pollute beyond certain thresholds*) or labor law (e.g., *not to employ child labor or underpaid labor*). The goal is to integrate sustainability issues—more than and besides specific legal constraints—into the discussion of the board of directors. It is a legal framework having its interlocutor on the board of directors (as monitoring board) and traceable to the general duty—which is a *duty of the entire board*—to assess the suitability of corporate arrangements (Ringe; Roe, Spamann, Fried, Wang; Genovese; Montalenti; Orciani). This is undoubtedly a matter of high administration: to say it with German literature, a *Letztentscheidungskompetenz des Vorstands über unternehmerische Rahmenbedingungen*. Indeed, there is no doubt that planning governance that allows the inclusion of the interests of many stakeholders in the decision-making processes, so as to take them into account, is a task belonging to the high administration powers of the board as a whole. From this perspective, it is noteworthy that the Chair of the Board—who directly impacts the choices of governance (just consider his or her power to manage internal information flows prior to each meeting of the board of directors)—has recently been considered to be the very lead actor of digital transition in widely held corporations (Calvosa).

43.3 Corporate Internal Controls and Dialogues with Stakeholders

Internal controls are likely to be the real focus of the sustainable transition. It is no accident that the new Dutch Corporate Governance Code states that the supervisory board "should supervise the manner in which the management board implements the strategy for sustainable long-term value creation. The supervisory board should

---

- Installing homogenizers on ships to reduce the sludge formation and enhance the combustion process in the engines onboard
- Continually assessing the feasibility of carbon capture systems, particularly on fully scrubber-fitted Capesize fleet ("We believe this could be an advantageous investment for ships currently on the water as a way to meaningfully reduce carbon emissions as the world awaits further developments regarding future fuels, vessel designs and bunkering infrastructure": this is the engagement by Genco)

regularly discuss the strategy, the implementation of the strategy and the principal risks associated with it. In the report drawn up by the supervisory board, an account is given of its involvement in the establishment of the strategy, and the way in which it monitors its implementation."

The new directive does not directly concern supervisory board's members, but it makes them nevertheless significantly responsible. This is essentially for two reasons.

The first one has already been said: CSDD is a matter of corporate governance, and then supervisory board's members cannot hide behind the excuse that they do not deal with business. It's not a business matter, but a matter of corporate governance. It concerns the global architecture of the internal proceedings, and a supervisory board's member of one of the big companies of the Directive cannot avoid verifying the adequacy of the corporate proceedings and thus their ability to bring into business decisions the consideration of the interests of third parties to the extent that they are related to environment and human rights.

The second reason arises from recent developments regarding the supervisory board.

The latter is increasingly less a silent onlooker of the choices of directors. Over the last 20 years the consideration of the role of the supervisory board in Italy and (more generally) in Europe has changed a lot. A long series of clues validates such an outcome: the approach of judges has become more rigorous, and it is enough to think of the latest rulings of the Italian "Corte di Cassazione" on the liability of supervisory board's members for lack of supervision about the procedures implemented by directors for the so-called related parties transactions (indeed—this has been ruled—the control about these procedures is an assessment of adequacy, suitability, and substantial fairness and not a mere formal legality check). Then the lawmaker has shown the same intent to enhance the role of the supervisory board, and it is enough to think of the latest developments of insolvency and early warning proceedings (see Article 25-*octies* Legislative Decree No. 14/2019). But the overall approach of literature has changed as well.

Here the very issue of sustainability and stakeholder involvement comes into play. Italian literature recently assumed that the supervisory board might even conduct dialogue with institutional investors or other stakeholders, and dialogues—this is well known—represent a strategic tool for identifying interests other than those of shareholders and abstractly relevant, to the extent that they are exposed to adverse impacts by the company's business.

Involving the supervisory board's members in the dialogues postulates that they have the power to monitor and periodically assess the adequacy of corporate arrangements for the purposes of the CSDD Directive. This power becomes wider when Italian corporations adopt the so-called *sistema dualistico di amministrazione e controllo*, a two-tier board system of governance regulated by Articles 2409-*octies* et seq. of the Italian Civil Code. Here the supervisory board appoints directors, approves the annual financial statements, and adopts resolutions about directors' remunerations, but there is even much more: it may decide about strategic plans, both industrial and financial, drafted by the directors; in this regard, however, it is necessary that the by-laws so provide [Article 2409-*terdecies* (f-*bis*) of the Italian Civil Code]. Therefore, in a two-tier board system, the supervisory board may engage in dialogue with institutional investors and any relevant stakeholders about industrial and financial plans or strategic operations as well, if the clause provided for by the Article 2409-*terdecies* (f-*bis*) of the Italian Civil Code occurs.

Another important player in corporate governance, currently making its way into American companies, should not be overlooked to the aims of the CSDDD: it is the so-called Board Observer, who is often engaged in venture capital operations, some sort of spokesperson, *although without voting rights*, who represents the interests of the financing partner inside the board of directors.

The board observer could well lend itself to different uses, even just to convey the interests of certain stakeholders within the board of directors and its meetings. It could be appointed by an environmentalist association or a public body promoting the interests of a specific geographical area to stimulate the debate inside the board about some of the most relevant sustainability topics.

43.4 ESG Committees. High Environmental Impact and Low Environmental Impact Operations

There are other important actors for more sustainable corporate governance that should not be overlooked, which are committees. Moreover, supervisory board and committees are strongly interlinked: supervisory board is empowered to assess the global corporate structure, including committees and their role in the corporate decision proceedings (the corporate report about ownership structures and governance arrangements—which is mandatory for listed companies according to the Article 123-*bis* Legislative Decree No. 58/1998—provides for it). In this regard, many and many are the ways in which committees, both the ones already provided for by the soft law codes and the many atypical committees that could be envisaged in the future, may concretely enhance sustainability in a big maritime transport company. The following remarks concern just two examples of a virtuous use of such a strategic governance tool.

The first one is the role of ESG committees inside the board of directors of a transport company.

This is a crucial role, as a committee may be an effective connecting point between the board and the various relevant stakeholders, also in the view of conducting dialogues (see par. 3). An ESG committee may collect the inputs of any relevant stakeholder, summarize to the board, monitor the non-financial reporting activities of the board, put questions about the consistency of business strategy with ESG factors, CSR parameters and sustainability goals, and oversee the implementation of sustainability policies. Such a strategic role of committees confirms that corporate sustainability due diligence is primarily a matter of procedures and arrangements, of joint assessments and cross-assessments by the most relevant corporate actors.

The constitution of ESG committees paves the way for further governance solutions—and this is the second of the abovementioned topics—thus facilitating compliance with the obligations laid down in the Articles 4 and 5 CSDDD. An interesting regulatory framework from which to take inspiration, to put in place adequate solutions, is the one provided for by the Regulation No. 17221/2010 on related parties' transactions (approved by the Italian Authority supervising listed companies and financial market sector, so-called Consob). And not because operations that impact on the environment (or human rights) are necessarily conflicted; just because such a Regulation is a great role model for the implementation of decision-making processes that enhance the input of committees and a pluralistic overall assessment of the business strategy.

Consob Regulation classifies transactions (or rather, requires listed companies to classify transactions) into three groups:

- Biggest transactions
- Minimum size transactions
- The other ones (neither biggest nor minimum size transactions)

The rules applying to them are different.

Minimum size transactions do not require any specific internal proceeding.

In the middle (neither largest nor minimum size transactions), a committee, appointed inside the board of directors, must express an opinion on the transaction; if the opinion is negative, the board may still decide to perform the operation, but every 3 months must give notice to the market of the transactions concluded despite the negative opinion of the committee. As regards the biggest transactions, the proceeding is the same—setting aside that the committee has to be involved since the very beginning of negotiations, not only to give an opinion on a transaction with already defined contents—but *the negative opinion of the committee prevents the board from implementing the transaction, unless the latter gets approval by the general meeting of shareholders* (with a whitewash procedure, which means that interested shareholders are not entitled to vote: thus, the general meeting of shareholders plays an arbitration role between the board and the committee).

Taking inspiration from this model, the board of directors could distinguish between high environmental impact and low environmental impact operations with regard to applying two different procedural frameworks. For both cases the board could establish a committee (e.g., an ESG committee), which may engage technicians and experts on marine pollution, ecosystems, biodiversity, and other similar issues. On this ground, the board of directors could then provide that:

a) The committee shall be involved in high impact operations since the very beginning of negotiations
b) The negative opinion of the committee shall not prevent the board from approving the low impact operation, it being understood that such a decision should be

noted in the company documents concerning ESG profiles (non-financial statements, above all)

c) The negative opinion of the committee does not prevent the board from approving the high impact transaction, but in this case the board must convene the general meeting of shareholders for a non-binding resolution on the transaction (and this is not a complete novelty: just consider the so-called *say on climate*, an advisory resolution of the general meeting, regarding environmental policies, which is spreading more and more at European level). In this hypothetical regulation about high environmental impact transactions, the board of directors may not provide for a binding resolution of the general meeting of shareholders (but just for an advisory resolution), since there are specific legal constraints—in corporation and especially in listed companies—preventing any blocking power by the shareholders (see Articles 2380-*bis* and 2364, par. 1, No. 5, of the Italian Civil Code).

Maybe there could be scope for further developments within the abovementioned "sistema dualistico." I mean that it is possible to envisage an ESG regulation, adopted by the directors, establishing that, in the event of a negative opinion by the committee, the supervisory board has the power to carry out a second-degree audit, and to refer the draft of the transaction back to the directors, so as to let them to jointly assess and decide again, hopefully by making adjustments (and provided that, if the draft is confirmed by the directors and addressed again to the supervisory board, the latter cannot prevent the final implementation of the transaction).

It is all a work in progress: the search for ESG-oriented governance solutions has just begun.

References

1. *Study on the relevance and impact of artificial intelligence for company law and corporate governance. Final report.* https://op.europa.eu/it/publication-detail/-/publication/13e6a212-6181-11ec-9c6c-01aa75ed71a1/language-en (2021)
2. Abriani, N., Schneider, G.: Il diritto societario incontra il diritto dell'informazione. IT, Corporate Governance e Corporate Social Responsibility. Rivista delle società. **5**(6), 1391 (2020)
3. Abriani, N., Schneider, G.: Diritto delle imprese e intelligenza artificiale (Dalla Fintech alla Corptech). Il Mulino, Bologna (2021)
4. Brozzetti, A.: La transizione verde europea e lo sviluppo sostenibile: rinnovate coordinate di fondo per sistema finanziario e imprese. Diritto della banca e del mercato finanziario. **3**, 411 (2022)
5. Busco, C.: Governare, gestire e comunicare la sostenibilità d'impresa e le metriche ESG. Corp. Gov. **25** (2023)
6. Calvosa, L.: La governance delle società quotate italiane nella transizione verso la sostenibilità e la digitalizzazione. Rivista delle società. **2**, 316 (2022)
7. Calvosa, L.: Say-on-climate e competenze assembleari nella società per azioni quotata. Rivista delle società. **67**, 1073 (2022)
8. Colaert, V.: Integrating sustainable finance into the MiFID II and IDD investor protection frameworks. In: Busch, D., Ferrarini, G., Grünewald, S. (eds.) Sustainable Finance in Europe.

Corporate Governance, Financial Stability and Financial Markets, p. 445. et seq. Palgrave Macmillan, Cham (2021)
9. Corvese, C.G.: La sostenibilità ambientale e sociale delle società nella proposta di Corporate Sustainability Due Diligence Directive (dalla «insostenibile leggerezza» dello scopo sociale alla «obbligatoria sostenibilità» della due diligence). Banca, impresa e società. **41**, 391 (2022)
10. Cossu, M.: Tassonomia finanziaria e normativa dei prodotti finanziari sostenibili e governo societario. Banca, impresa e società. **41**, 433 (2022)
11. Driessen, M.: Sustainable finance: an overview of ESG in the financial markets. In: Busch, D., Ferrarini, G., Grünewald, S. (eds.) Sustainable Finance in Europe. Corporate Governance, Financial Stability and Financial Markets, p. 329. et seq. Palgrave Macmillan, Cham (2021)
12. Enriques, L., Zetzsche, D.A.: Corporate technologies and the tech nirvana fallacy. Hastings Law J. **72**, 90 (2020)
13. Fenwick, M., Kaal, W.A., Vermeulen, E.P.M.: The "unmediated" and "tech-driven" corporate governance of today's winning companies. New York Univ. J. Law Bus. **116**, 75 (2019)
14. Fenwick, M., Mccahery, J.A., Vermeulen, E.P.M.: The end of "corporate" governance: hello "platform" governance. Eur. Bus. Organ. Law Rev. **20**, 171 (2019)
15. Genovese, A.: La "sustainable corporate governance" delle società quotate. Note introduttive. Corp. Gov., 97 (2022)
16. Luciano, A.M.: Riunione assembleare e diritto d'intervento. Collegialità e tecnologia nel diritto azionario. Giuffrè, Milano (2022)
17. Martina, G.: Organo amministrativo delle imprese di assicurazione e Product Oversight and Governance tra codice delle assicurazioni private e regolamenti IVASS. Orizzonti del diritto commerciale. **797** (2021)
18. Montagnani, M.L.: Il ruolo dell'intelligenza artificiale nel funzionamento del consiglio di amministrazione delle società per azioni. Egea, Milano (2021)
19. Montalenti, P.: Società, mercati finanziari e fattori ESG: ultimi sviluppi. Corp. Gov. **9** (2022)
20. Mosco, G.D.: Roboboard. L'intelligenza artificiale nei consigli di amministrazione. AGE. **18**, 254 (2019)
21. Möslein, F.: Robots in the boardroom: artificial intelligence and corporate law. In: Barfield, W., Pagallo, U. (eds.) Research Handbook on the Law of Artificial Intelligence, p. 649. et seq. Elgar (2018)
22. Orciani, L.: Il ruolo dell'assemblea degli azionisti nello sviluppo delle politiche di sostenibilità: *say on pay, say on climate*. Giurisprudenza commerciale. **51**, 1 (2024)
23. Pilaj, H.: The choice architecture of sustainable and responsible investment: nudging investors toward ethical decision-making. J. Bus. Ethics. **140**, 743 (2017)
24. Ringe, W.-G.: Investor-led sustainability in corporate governance. Ann. Corp. Gov. **7**, 99 (2022)
25. Roe, M., Spamann, H., Fried, J., Wang, C.: The sustainable corporate governance initiative in Europe. Yale J. Regul. Bull. **38**, 133 (2021)
26. Rolli, R.: Dalla corporate social resposibility alla sustainability, alla environmental, social and governance (ESG). Corp. Gov. **2022**, 41 (2022)
27. Salerno, M.E.: L'integrazione dei fattori di sostenibilità nelle regole di comportamento dell'intermediario finanziario: un ritorno al modello di distribuzione "orientato al prodotto". Diritto della banca e del mercato finanziario. **36**, 53 (2022)
28. Schneider, G.: Le tecnologie societarie alla prova del governo sostenibile tra ESG, diligenza d'impresa e corporate digital responsibility. Corp. Gov. **1**, 125 (2022)
29. Siri, M., Macchiavello, E.: Sustainable finance and Fintech: can technology contribute to achieving environmental goals? A preliminary assessment of 'green Fintech' and 'sustainable digital finance'. Eur. Co. Financ. Law Rev. **19**, 144 (2022)
30. Siri, M., Zhu, S.: L'integrazione della sostenibilità nel sistema europeo di protezione degli investitori. Banca, impresa e società. **39**, 3 (2020)
31. R.H. Weber, R. Baisch, Finanzdienstleistungsverträge, in: Verbrauchervertragsrecht der Europäischen Union, edited by D. Trüten/T. Baumgartner/A. Brunner, 2017, 251

32. Wen, H., Zhong, Q., Lee, C.C.: Digitalization, competition strategy and corporate innovation: evidence from Chinese manufacturing listed companies. Int. Rev. Financ. Anal. **82**, 102166 (2022)
33. Wu, S., Li, Y.: A study on the impact of digital transformation on corporate ESG performance: the mediating role of green innovation. Sustainability. **15**, 6568 (2023)
34. Zhang, Q., Yang, M., Lv, S.: Corporate digital transformation and green innovation: a quasi-natural experiment from integration of informatization and industrialization in China. Int. J. Environ. Res. Public Health. **19**, 13606 (2022)

Conclusion

Digital transformation has had a pervasive impact on nearly every sector of society and the global economy. Throughout the chapters of this book, we have explored how emerging technologies are redefining business models, innovation, sustainability, and how we interact with the world.

The empirical and conceptual analyses presented offer a comprehensive view of how digitalization is not just a technological process but a social and economic revolution that permeates organizations and industries.

Across the various chapters, we have seen how AI adoption influences business life cycles, from creativity and project management to safeguarding critical sectors like agri-food and healthcare. It is evident that AI, despite its enormous benefits, requires careful consideration of its ethical and managerial aspects, focusing on issues like security and privacy, particularly in highly vulnerable sectors such as agri-food and healthcare.

In the context of sustainability, digital transformation represents a double-edged sword: while it offers new opportunities to improve efficiency and reduce environmental impact, it also presents complex challenges, especially in sectors like maritime industry and international trade. The studies in this volume demonstrate how the integration of digitalization and sustainability can be a powerful engine for change, particularly through the adoption of circular models and innovations in green logistics and mobility.

Another crucial theme that emerged from our journey into digital transformation is organizational resilience. Small and medium-sized enterprises, startups, and academic institutions are learning to leverage digital maturity to improve their performance and face global challenges with greater flexibility. The integration of digital platforms and pluralistic governance models represents a fundamental step forward in improving long-term competitiveness and sustainability.

Lastly, we cannot ignore the impact of digitalization on consumer behavior and marketing strategies. The metaverse, in particular, is redefining concepts of

F. Schiavone et al. (eds.), *Advanced Perspectives and Trends in Digital Transformation of Firms, Networks, and Society*, Springer Proceedings in Business and Economics, https://doi.org/10.1007/978-3-031-80692-6

engagement and customer experience, offering a new mode of interaction between brands and customers.

In conclusion, digital transformation is an ongoing phenomenon with profound, multidimensional implications for industry, society, and the environment. As we move toward an increasingly digital future, the research presented in this book provides valuable tools for understanding and navigating the challenges and opportunities ahead. Only through an integrated and holistic approach can we fully harness the potential of digitalization while promoting sustainable and inclusive development.

International Conference of the Digital Transformation Societydtsdts2024DTS2NaplesItaly20245232024524https://www.disaq.uniparthenope.it/dts2024-conference/

The manufacturer's authorised representative in the EU is Springer
Nature Customer Service Centre GmbH, Europaplatz 3, 69115 Heidelberg,
Germany. If you have any concerns regarding our products, please
contact ProductSafety@springernature.com

Printed and bound by CPI Group (UK) Ltd, Croydon, CR0 4YY
01/10/2026
02225922-0016